Complex Ophthalmic Dosage Forms: Advances in Biomedical Applications and Future Perspectives

Neelesh Kumar Mehra
Editor

Complex Ophthalmic Dosage Forms: Advances in Biomedical Applications and Future Perspectives

Springer

Editor
Neelesh Kumar Mehra
Department of Pharmaceutics
National Institute of Pharmaceutical Education & Research (NIPER)
Hyderabad, Telangana, India

ISBN 978-981-96-6305-7 ISBN 978-981-96-6306-4 (eBook)
https://doi.org/10.1007/978-981-96-6306-4

This Springer imprint is published by the registered company Springer Nature Singapore Pte Ltd.
The registered company address is: 152 Beach Road, #21-01/04 Gateway East, Singapore 189721, Singapore

Neelesh Kumar Mehra
Editor

Complex Ophthalmic Dosage Forms: Advances in Biomedical Applications and Future Perspectives

Springer

Editor
Neelesh Kumar Mehra
Department of Pharmaceutics
National Institute of Pharmaceutical Education & Research (NIPER)
Hyderabad, Telangana, India

ISBN 978-981-96-6305-7   ISBN 978-981-96-6306-4 (eBook)
https://doi.org/10.1007/978-981-96-6306-4

This Springer imprint is published by the registered company Springer Nature Singapore Pte Ltd.
The registered company address is: 152 Beach Road, #21-01/04 Gateway East, Singapore 189721, Singapore

Preface

The field of ophthalmic drug delivery has witnessed remarkable advancements over the years, driven by the need for more effective, patient-friendly, and targeted therapeutic strategies. Traditional ophthalmic dosage forms, such as eye drops and ointments, often suffer from poor bioavailability, rapid drug elimination, and limited penetration to deeper ocular tissues. To address these challenges, researchers and pharmaceutical scientists have continuously explored novel formulations and delivery systems that enhance drug retention, bioavailability, and therapeutic efficacy.

This book, *Complex Ophthalmic Dosage Forms: Advancements, Future Perspectives, and Biomedical Application*, offers a detailed overview of the latest innovations in ophthalmic drug delivery. Beginning with the fundamentals of ocular drug delivery systems and the anatomy of the eye, this book provides an exploration of drug delivery to both the anterior and posterior segments. A dedicated chapter on the fixed-dose combination approach examines its application in ocular diseases, highlighting both its therapeutic potential and associated challenges. Furthermore, the book delves into key aspects of ophthalmic product development, including reverse engineering, the role of preservatives, and the application of Quality by Design (QbD) principles. It includes discussions on the safety of nanoformulations, as well as an in-depth analysis of emerging nano-assisted platforms in ocular drug delivery, highlighting both opportunities and safety concerns. Recognizing the importance of packaging a dedicated chapter explores the critical role of sterility in ocular products, detailing sterility validation processes to ensure product safety and efficacy. This is followed by a thorough discussion on packaging, including the selection of appropriate containers and closure systems. Given that an optimized packaging system is essential for maintaining the stability, sterility, and overall quality of ophthalmic products, this section highlights key considerations in designing effective packaging solutions. Additionally, the book delves into regulatory considerations, challenges in clinical translation, and potential future developments that may redefine ophthalmic therapeutics.

By consolidating cutting-edge knowledge and insights from experts in the field, this work aims to serve as a valuable reference for both academic and industrial applications.

Hope that this book will inspire further innovations in ophthalmic drug delivery and contribute to the development of safer and more effective treatments for various ocular disorders.

Hyderabad, Telangana, India

Neelesh Kumar Mehra

Contents

Editor and Contributors

About the Editor

Neelesh Kumar Mehra is working in the Department of Pharmaceutics, National Institute of Pharmaceutical Education & Research (NIPER), Hyderabad, India. He earned PhD from Dr. H.S. Gour University, Sagar, and PostDoc from Irma Lerma Rangel College of Pharmacy, Texas A & M Health Science Centre, Kingsville, TX, USA. He served as Manager in Product Development, Sentiss Research Centre, Sentiss Pharma Pvt Ltd., Gurgaon, for development, scale-up, and technology transfer of complex ophthalmic, inhalation, and optic pharmaceutical products. He received "TEAM AWARD" for successful commercialization of Ophthalmic Suspension product. He has authored more than 130 peer-reviewed publications in high repute international journals, and >50 book chapter contributions. He has 4 granted and 4 filed patent on manufacturing process and composition to improve therapeutic efficacy for topical delivery. He has guided 5 PhD students and 36 MS students and currently guiding 10 PhD and 5 MS students for their dissertations/research projects. He has received numerous outstanding awards including Young Scientist and Team Awards for their research output. He is a peer reviewer of various international journals and publishers.

He recently published edited book on Multifunctional nanocarriers, Biopharmaceutical and nanotechnology. He has rich research and teaching experience in the formulation and development of complex, innovative ophthalmic and injectable biopharmaceutical products including micro- and nanotechnologies for regulated market and so on.

According to announcement by the Stanford University Scientists based on the database and basis of standardized citation indicators, Dr. Mehra's name is included in top 2% scientists in the world in the area of Pharmacology and Pharmacy, consecutively for 3 years.

Contributors

Naureen Afrose Bengal College of Pharmaceutical Sciences and Research, Durgapur, West Bengal, India

Kuldeep Kumar Bansal Faculty of Science and Engineering, Pharmacy, Solutions for Health, Abo Akademi, Turku, Finland

Wilson Barnabas Department of Pharmaceutics, College of Pharmaceutical Sciences, Dayananda Sagar University, Bengaluru, India

Sonali Batra Department of Pharmaceutical Sciences, Indira Gandhi University, Meerpur, Rewari, Haryana, India

Atreyee Bhattacharyya Department of Pharmaceutics, National Institute of Pharmaceutical Education and Research, Hyderabad, Telangana, India

Rohit Bhawale Pharmaceutical Nanotechnology Research Laboratory, Department of Pharmaceutics, National Institute of Pharmaceutical Education and Research, Hyderabad, Telangana, India

Pratibha Bhowmick Bengal College of Pharmaceutical Sciences and Research, Durgapur, West Bengal, India

Mithun Bhowmick Bengal College of Pharmaceutical Sciences and Research, Durgapur, West Bengal, India

Rideb Chakraborty Bengal College of Pharmaceutical Sciences and Research, Durgapur, West Bengal, India

Ramesh Chandra Department of Chemistry, University of Delhi, Delhi, India

Institute of Nanomedical Sciences, University of Delhi, Delhi, India

Dr. B.R Ambedkar Centre for Biomedical Research, University of Delhi, Delhi, India

Maharaja Surajmal Brij University, Bharatpur, Rajasthan, India

Shilpkala Gade School of Pharmacy, Queen's University Belfast, Medical Biology Centre, Belfast, UK

Arvind Gulbake Department of Pharmaceutics, National Institute of Pharmaceutical Education and Research, Guwahati, Assam, India

Vivek Gupta College of Pharmacy & Health Sciences, St. John's University, Queens, NY, USA

V. T. Iswariya Department of Pharmaceutics, CMR College of Pharmacy, Hyderabad, India

Pratiksha Jadhav Department of Pharmacy, Birla Institute of Technology and Science, Pilani, Hyderabad Campus, Jawahar Nagar, Kapra Mandal, Medchal District, Hyderabad, Telangana, India

Jilu Jaffet Centre for Ocular Regeneration (CORE), Prof. Brien Holden Eye Research Centre (BHERC), LV Prasad Eye Institute, Hyderabad, India

Manipal Academy of Higher Education (MAHE), Manipal, Karnataka, India

Keerti Jain Department of Pharmaceutics, Drug Delivery and Nanomedicine Research Laboratory, National Institute of Pharmaceutical Education and Research (NIPER)-Raebareli, Lucknow, Uttar Pradesh, India

Roshani Jaiswal Department of Pharmaceutics, National Institute of Pharmaceutical Education and Research, Hyderabad, Telangana, India

J. Josephine Leno Jenita Department of Pharmaceutics, College of Pharmaceutical Sciences, Dayananda Sagar University, Bengaluru, India

Ge Jiang School of Pharmacy, Queen's University Belfast, Medical Biology Centre, Belfast, UK

Khushi Sandeep Kachhwaha Department of Pharmacy, Birla Institute of Technology and Science, Pilani, Hyderabad Campus, Jawahar Nagar, Kapra Mandal, Medchal District, Hyderabad, Telangana, India

Sunil Kumar Kadiri College of Pharmaceutical Sciences, Dayananda Sagar University, Bengaluru, Karnataka, India

Ishika Raju Kamdi Department of Pharmaceutics, National Institute of Pharmaceutical Education and Research, Guwahati, Assam, India

Sushil K. Kashaw Department of Pharmaceutical Sciences, Dr. Harisingh Gour University, Sagar, Madhya Pradesh, India

Omar Khan Pharmaceutical Nanotechnology Research Laboratory, Department of Pharmaceutics, National Institute of Pharmaceutical Education and Research, Hyderabad, Telangana, India

Pradeep Kumar Department of Botany, Swami Shraddhanand College, New Delhi, India

Kshitiz Luthra Pharmaceutical Nanotechnology Research Laboratory, Department of Pharmaceutics, National Institute of Pharmaceutical Education and Research, Hyderabad, Telangana, India

Jitender Madan Department of Pharmaceutics, National Institute of Pharmaceutical Education and Research, Hyderabad, Telangana, India

Namish Manchanda Centre for Pharmaceutical Nanotechnology, Department of Pharmaceutics, National Institute of Pharmaceutical Education & Research (NIPER), Mohali, Punjab, India

D. Manjula Department of Pharmaceutics, College of Pharmaceutical Sciences, Dayananda Sagar University, Bengaluru, India

Neelesh Kumar Mehra Pharmaceutical Nanotechnology Research Laboratory, Department of Pharmaceutics, National Institute of Pharmaceutical Education and Research, Hyderabad, Telangana, India

Deepakkumar Mishra School of Pharmacy, Queen's University Belfast, Medical Biology Centre, Belfast, UK

Saurabh Mittal Center of Pharmaceutics, Amity Institute of Pharmacy, Amity University, Noida, Uttar Pradesh, India

Sonali Mohapatra Centre for Ocular Regeneration (CORE), Prof. Brien Holden Eye Research Centre (BHERC), LV Prasad Eye Institute, Hyderabad, India

Atul Mourya Department of Pharmaceutics, National Institute of Pharmaceutical Education and Research, Hyderabad, Telangana, India

Kanika Nawani Department of Medical Laboratory Technology and Sciences, Delhi Pharmaceutical Sciences and Research University, New Delhi, India

Dipak Panchal Department of Pharmaceutics, National Institute of Pharmaceutical Education and Research, Hyderabad, Telangana, India

Chandrakantsing Vijaysing Pardeshi Industrial Pharmacy Laboratory, Department of Pharmaceutics, R. C. Patel Institute of Pharmaceutical Education and Research, Shirpur, India

Shivam Pathak Center of Pharmaceutics, Amity Institute of Pharmacy, Amity University, Noida, Uttar Pradesh, India

Tejaswini Pingali Centre for Ocular Regeneration (CORE), Prof. Brien Holden Eye Research Centre (BHERC), LV Prasad Eye Institute, Hyderabad, India

Manipal Academy of Higher Education (MAHE), Manipal, Karnataka, India

Varnna Prashob Department of Pharmacy, Birla Institute of Technology and Science, Pilani, Hyderabad Campus, Jawahar Nagar, Kapra Mandal, Medchal District, Hyderabad, Telangana, India

Syed Nazrin Ruhina Rahman Department of Pharmaceutics, Royal School of Pharmacy, The Assam Royal Global University, Guwahati, Assam, India

V. V. Rajesham Department of Pharmacology, CMR College of Pharmacy, Hyderabad, India

Vivek Prabhakar Rajkule Industrial Pharmacy Laboratory, Department of Pharmaceutics, R. C. Patel Institute of Pharmaceutical Education and Research, Shirpur, India

Seema S. Rathore Department of Pharmaceutics, College of Pharmaceutical Sciences, Dayananda Sagar University, Bengaluru, India

Arun Kumar Raut Centre for Ocular Regeneration (CORE), Prof. Brien Holden Eye Research Centre (BHERC), LV Prasad Eye Institute, Hyderabad, India

Akhila Reddy Department of Pharmaceutics, National Institute of Pharmaceutical Education and Research, Guwahati, Assam, India

Kritika Sharma Delhi Pharmaceutical Sciences and Research University, New Delhi, India

Sumit Sharma Delhi Pharmaceutical Sciences and Research University, New Delhi, India

Aayush Singh Department of Pharmaceutics, National Institute of Pharmaceutical Education and Research, Hyderabad, Telangana, India

Anita Singh Department of Pharmaceutical Sciences, Sir J.C. Bose Technical Campus, Kumaon University, Nainital, Uttarakhand, India

Snigdha Singh Department of Pharmaceutics, National Institute of Pharmaceutical Education and Research, Hyderabad, Telangana, India

Vivek Singh Centre for Ocular Regeneration (CORE), Prof. Brien Holden Eye Research Centre (BHERC), LV Prasad Eye Institute, Hyderabad, India

Yuvraj Singh Department of Pharmacy, Birla Institute of Technology and Science, Pilani, Hyderabad Campus, Jawahar Nagar, Kapra Mandal, Medchal District, Hyderabad, Telangana, India

Sakshi Soni Department of Pharmaceutical Sciences, Dr. Harisingh Gour University, Sagar, Madhya Pradesh, India

Vandana Soni Department of Pharmaceutical Sciences, Dr. Harisingh Gour University, Sagar, Madhya Pradesh, India

Divyanshu Srivastava Center of Pharmaceutics, Amity Institute of Pharmacy, Amity University, Noida, Uttar Pradesh, India

Vaibhavi Srivastava Pharmaceutical Nanotechnology Research Laboratory, Department of Pharmaceutics, National Institute of Pharmaceutical Education and Research, Hyderabad, Telangana, India

Sushama Talegaonkar Department of Pharmaceutics, School of Pharmaceutical Sciences, Delhi Pharmaceutical Sciences & Research University (DPSRU), Government of NCT of Delhi, New Delhi, Delhi, India

Sofiya Tarannum Department of Pharmaceutics, Drug Delivery and Nanomedicine Research Laboratory, National Institute of Pharmaceutical Education and Research (NIPER)-Raebareli, Lucknow, Uttar Pradesh, India

Raghu Raj Singh Thakur School of Pharmacy, Queen's University Belfast, Medical Biology Centre, Belfast, UK

Prashant Tiwari College of Pharmaceutical Sciences, Dayananda Sagar University, Bengaluru, Karnataka, India

Lalitkumar K. Vora School of Pharmacy, Queen's University Belfast, Medical Biology Centre, Belfast, UK

Ziqi Yu School of Pharmacy, Queen's University Belfast, Medical Biology Centre, Belfast, UK

Ocular Drug Delivery: Introduction and Fundamental Concept

1

Omar Khan, Rohit Bhawale, Kshitiz Luthra, Vaibhavi Srivastava, Kuldeep Kumar Bansal, Vivek Gupta, and Neelesh Kumar Mehra

Abstract

Ocular diseases are one of the most serious concerns worldwide, with more than 45 million cases of blindness, significantly affecting the vision and quality of life of the patients. However, effective drug delivery to the eyes remains challenging, especially to the posterior region of the eye due to the presence of various ocular barriers. This led to various advancements in the development of various novel ocular therapeutics with superior efficacy and safety. This chapter highlights the basics of ocular drug delivery systems (ODDS) including the anatomy of the eyes and their physiological barriers affecting the drug delivery, various invasive and noninvasive routes of ocular drug administration, a brief overview of ocular diseases, their complications, and various conventional and novel ODDS. Regulatory perspectives of ODDS, overview of marketed ophthalmic products, and recently approved patents along with the future perspectives in design and development of safe and effective ODDS have also been addressed in this chapter.

Authors Omar Khan and Rohit Bhawale have been equally contributed to this chapter.

O. Khan · R. Bhawale · K. Luthra · V. Srivastava
Pharmaceutical Nanotechnology Research Laboratory, Department of Pharmaceutics, National Institute of Pharmaceutical Education and Research, Hyderabad, Telangana, India

K. K. Bansal
Faculty of Science and Engineering, Pharmacy, Solutions for Health, Abo Akademi, Turku, Finland

V. Gupta
College of Pharmacy & Health Sciences, St. John's University, Queens, NY, USA

N. K. Mehra (✉)
Pharmaceutical Nanotechnology Research Laboratory, Department of Pharmaceutics, National Institute of Pharmaceutical Education and Research, Hyderabad, Telangana, India

N. K. Mehra (ed.), *Complex Ophthalmic Dosage Forms: Advances in Biomedical Applications and Future Perspectives*, https://doi.org/10.1007/978-981-96-6306-4_1

Keywords

Ocular drug delivery systems · Ocular diseases · Ocular barriers · Intraocular drug delivery · Polymeric nanoparticles · Micelles

Abbreviations

ODDS	Ocular drug delivery system
NPs	Nanoparticles
CNS	Central nervous system
μL	Microliter
RPE	Retinal pigment epithelium
Pgp	P-glycoprotein
BAB	Blood aqueous barrier
BRB	Blood retinal barrier
RCE	Retinal capillary endothelial
GI	Gastro-intestinal
SCT	Subconjunctival
IVT	Intravitreal
PNCs	Polymeric nanocarriers
LNCs	Lipidic nanocarriers
SLNs	Solid lipid nanoparticles
ECN	Econazole
AUC	Area under the curve
DES	Dry eye syndrome
RES	Resveratrol
ROS	Reactive oxygen species
ILs	Interleukins
PLGA	Polylactic-co-glycolic acid
PVP	Polyvinyl pyrrolidone
HET-CAM	Hen's egg test-chorioallantoic membrane
DEX	Dexamethasone
UM	Uveal melanoma
RB	Retinoblastoma
NLCs	Nanostructured lipid carriers
PKPD	Pharmacokinetic and pharmacodynamic
BOB	Blood ocular barrier
NDDS	Novel drug delivery system
CMC	Critical micellar concentration
ASED	Anterior segment eye diseases
PEG	Polyethylene glycol
LAD	Long-acting delivery

1.1 Eye's Anatomy and Barriers for Ocular Drug Delivery

1.1.1 Eye's Anatomy

Ocular diseases significantly affect the vision and quality of life for patients worldwide. Currently, more than 250 million individuals are affected by visually impaired conditions globally, and the prospect for reducing this prevalence remains concerning [1]. Although the eye is one of the most accessible organs in the human body, delivering medication to its target tissues remains an enormous challenge. Just like the brain, the eye is considered "immune privileged" as being the part of the CNS, protected from systemic circulation. Its complex anatomy and numerous ocular barriers make it a highly isolated organ. As a result, treating ocular disorders, especially those affecting the posterior segment, offers numerous difficulties.

The eye is a highly complex organ made up of delicate tissue structures organized into compact multilayered arrangements, as illustrated in Fig 1.1. Anatomically, the ciliary body and lens separate the anterior and posterior segments of the eyeball. The anterior segment comprises the cornea, conjunctiva, iris, ciliary

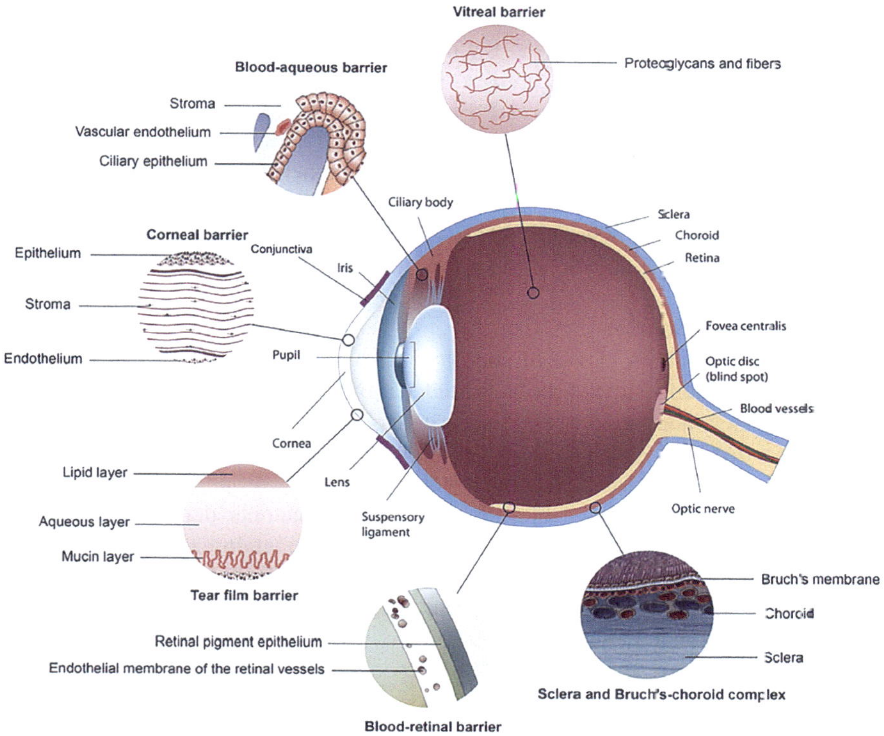

Fig. 1.1 Anatomy of the eye and ocular barriers. (Reproduced with copyright permission from open access Ref. [2])

body, and lens, while the posterior segment comprises the sclera, vitreous humor, retina, choroid, and optic nerve [2, 3].

1.1.2 Barriers for Ocular Drug Delivery

The eye exhibits a complex yet highly organized anatomy. Still, its distinctive characteristics create multiple barriers that can hinder the absorption of drugs, leading to inadequate concentrations at the intended target sites (Fig. 1.1).

1.1.2.1 Tear Film Barrier

A precorneal layer consisting of three distinct layers, an exterior thin lipid layer, a middle aqueous layer, and an interior mucus layer, known as the tear film, serves as an initial permeability barrier, hindering the drug delivery to the anterior segment of the eye [4]. The exterior oil layer prevents evaporation of water and also hinders passage of water-soluble drugs into the sclera and cornea [5]. The middle aqueous layer contains some endogenous water-soluble proteins which can metabolize the drugs, thereby reducing their bioavailability [6]. The interior mucus layer consists of water and negatively charged mucin which traps foreign particulates and acts as a significant barrier for absorption of drugs into the anterior chamber of the eye [7, 8].

In addition to structural barriers, the nasolacrimal duct drainage system and the rapid tear flow display dynamic hurdles. The usual tear volume is around 7 μL but may increase to 30 μL upon topical drug administration, leading to the rapid drainage of excess fluid via the nasolacrimal duct. As a result, more than 85% of the administered drug dose goes away before it even reaches the surface of the cornea. Furthermore, the remaining drug gets further diluted by the rapid tear turnover rate, reducing the concentration gradient. This substantially reduces the bioavailability of intraocular drugs in the aqueous humor, i.e., less than 5% [9, 10].

1.1.2.2 Corneal Barrier

The cornea acts as both chemical and physical barrier for transfer of both water-soluble and -insoluble substances. It is made up of five membranes, namely epithelium, Bowman's membrane, stroma, Descemet's membrane, and endothelium [11, 12]. Uppermost epithelial membrane is highly hydrophobic with tight junctions which act as a barrier for transcorneal permeation of hydrophilic molecules, but these molecules can diffuse through paracellular channels depending on their molecular weight. Furthermore, the presence of drug-metabolizing enzymes and efflux pumps in the epithelium is also responsible for the low bioavailability of drugs administered topically [13, 14]. In contrast, the stroma is hydrophilic in nature and acts as a barrier for hydrophobic molecules which get passed through the epithelium via passive diffusion. Hence, the molecular size, weight, and nature of drug molecules play an important role in their permeability across the cornea. Aqueous humor secreted by the ciliary body (Fig. 1.1) nourishes the anterior eye tissues. This fluid flows toward the cornea and acts as a dynamic barrier which drains off the therapeutic drugs form the ocular tissues. Turnover rate of aqueous humor significantly affects the concentration of the drug in ocular tissues [9, 15].

Instead of the aforementioned corneal pathway, a drug administered topically may transfer into the eye via conjunctiva-sclera-choroid pathway. Transparent membrane covering the ocular surface around the cornea is called the conjunctiva. It is composed of epithelium and stromal layer with rich lymphatic and blood flow. As compared to cornea, conjunctiva is 25–30 times more permeable due to high surface area, larger paracellular transport, and less dense epithelial cells. Nevertheless, due to its high vascularity, a drug may go into systemic circulation which further reduces the drug's ocular bioavailability [16–18].

1.1.2.3 Scleral Barrier

The sclera acts as another significant barrier to ocular drug delivery, due to its limited permeability, complementing the cornea. It is composed of dense, hydrophilic connective tissue made up of collagen, forming the external covering of the eyeball. The scleral membrane consists of an interconnected collagen matrix packed with a negatively charged proteoglycan matrix within the voids between fibrils. The thickness of the sclera plays an important role in assessing the efficiency of trans scleral drug delivery [19, 20].

Furthermore, the highly vascularized choroid layer present between retina and sclera acts as a significant barrier for retinal drug delivery. The innermost layer of the choroid, known as Bruch's membrane, is a thin collagenous membrane, connected to the retinal pigment epithelium (RPE). As the RPE contains many enzymes such as peptidases, esterases, cytochrome P-450, along with Pgp-efflux pump, it can metabolize the administered drugs leading to reduced permeability. Drug delivery via this Bruch's-choroid complex depends on charge (cationic drugs tend to form sustained release depot), molecular weight, and hydrophilic/lipophilic characteristics of the drug [21–23].

1.1.2.4 Blood-Aqueous Barrier

The blood-aqueous barrier (BAB) consists of nonporous vascular endothelial cells present in the anterior region of the eye, lines the blood vessels of the tight junctions and the iris, and connects the upper regions of adjacent epithelial cells in the nonpigmented epithelium of the ciliary body. This barrier restricts the entry of drugs and other substances into the aqueous humor from the plasma. However, the BAB is not entirely impermeable due to the presence of porous vascularization in the stroma of the ciliary body. These capillaries serve as an alternative source of plasma protein leakage into the iris which also helps the absorption of small molecules into the iridial bloodstream [24, 25].

1.1.2.5 Blood-Retinal Barrier

The blood-retinal barrier (BRB) is called the posterior region blood-ocular barrier which restricts the diffusion of drugs from systemic blood circulation into the retina [26]. It can be further classified into outer and inner BRB. Entry of macromolecules and hydrophilic substances is restricted by the inner BRB due to the presence of tightly arranged retinal capillary endothelial (RCE) cells which selectively prevent the transfer of foreign molecules from systemic circulation into the retina [27]. The retinal microvasculature contains strong intercellular junctions which form tight structural barriers that limit the diffusion of molecules into and out of the retina. This is attributed to the distinctive characteristics of the retinal microvascular endothelium, particularly the absence of apertures and an abundance of specialized intercellular junction proteins. The outer BRB, which divides the choroid and Bruch's membrane from the inner retina, is produced by tight junctions within the RPE cells. These tight junctions restrict the penetration of the drug, thereby allowing only 1–2% of the drug being given to reach the eyes posterior region [4, 28].

1.2 Routes for Ocular Drug Administration

Bypassing the aforementioned ocular barriers and achieving high drug concentration at the desired location are essential for the effective management of ocular diseases. These routes include topical, systemic, periocular, and intraocular drug administrations which have their own advantages and challenges.

1.2.1 Topical Drug Delivery

Till date the most acceptable route for the ocular disease management and includes more than 90% today's ophthalmic product market, due to its easy, safe and noninvasive mode of administration [29]. However, the major drawback of this route is insufficient drug concentration at the target site, because of high pre-corneal drug loss due to blinking, tear turnover rate, nasolacrimal drainage, and the presence of ocular barriers. Furthermore, frequent administration of eye drops is required in a single day to keep the effective drug concentration on the eye surface, which leads to more complication and poor compliance. Moreover, despite frequent dosing, the topical administration of conventional eye drops is largely ineffective at delivering drugs to the posterior region of the eye due to the presence of strong ocular barriers. Ocular irritation, preservative damage, and steroidal drug side effects are the prominent issues with the topical ocular drug delivery [30].

1.2.2 Systemic Drug Delivery

It includes oral and intravenous drug administrations. To deliver a drug to the ocular tissues via this route remains a significant challenge due to several reasons, such as if the drug is administered orally, it should not be destroyed in the harsh GI environment or via liver's first-pass metabolism. Moreover, the amount of blood supplied to the eyes is quite less, which leads to a lower concentration of drug accumulation. As previously mentioned, BAB and BRB also restrict the entry of drug into the anterior and posterior regions, respectively. Due to these reasons, systemic administration requires frequent dosing and high doses to maintain effective drug concentrations, which leads to increased chances of systemic side effects and patient incompliance [3, 31].

1.2.3 Periocular Drug Delivery

It includes the administration of drug in the area around the eye, which can accumulate drug in both anterior and posterior region of the eye (Fig. 1.2). This route can provide extended duration of action than the intraocular routes (which mainly includes intravitreal injection) with the larger dose volume administration possibility (up to 1 ml) along with significantly reduced risk of pain, bleeding, and infection [5]. The most common periocular administration strategy includes subconjunctival injection (SCT), which is injected between sclera and bulbar conjunctiva, mainly used for anti-inflammatory and local anesthetic drug administration to the eye's anterior segment. However, SCT medication's ocular bioavailability is usually limited due to higher absorption of drug into the blood streams and lymphatic systems, rather than ocular tissue distribution. To tackle this issue, frequent injections are necessary which may lead to subconjunctival hemorrhage and edema. Other

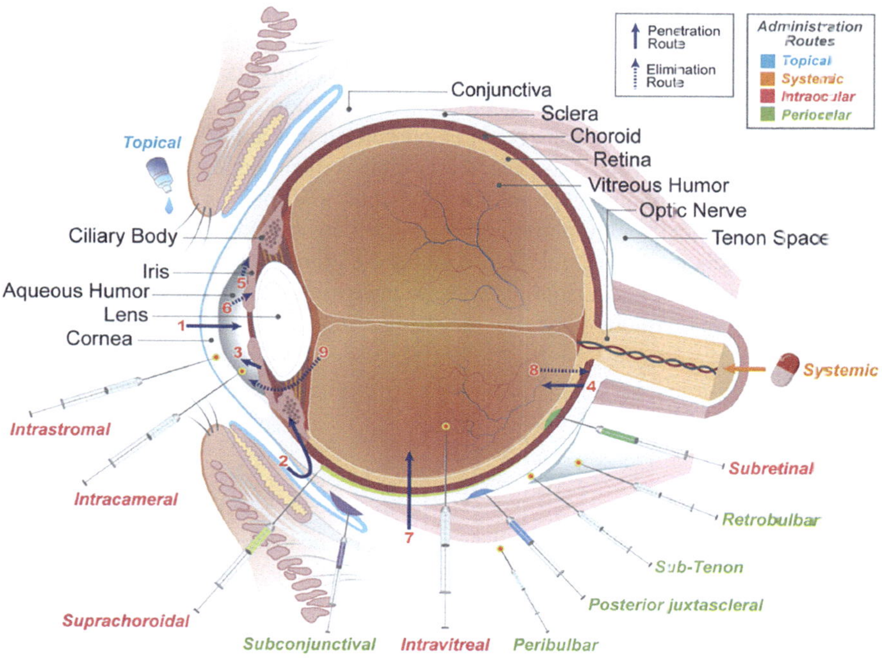

Fig. 1.2 Various routes of ocular drug administration. (Reproduced with copyright permission from open access Ref. [2])

periocular routes include retrobulbar, juxtascleral, sub-tenon, and peribulbar injections which are mainly utilized for local anesthetics administration during eye surgeries. Despite significant advantages of the periocular route as compared to other routes, it still lacks the ability to transfer effective drug concentration into the retina due to drug loss from BRB, periocular space, and the choroidal circulation [32, 33].

1.2.4 Intraocular Drug Delivery

This drug delivery route involves direct injection of drug into the eyeball. It includes intravitreal (IVT), intracameral, subretinal, suprachoroidal, and intrastromal routes (Fig. 1.2). Injection of drug into the anterior chamber is known as intracameral injection, which is used after the eye surgery or for treating keratitis and other anterior chamber disorders. However, this route is not significant in delivering drugs to the posterior region due to the presence of the BRB and other barriers [34]. For this reason, IVT injection is used to penetrate these barriers and deliver the medications into the posterior region of the eye. It includes both IVT implants and injections which have become a prominent strategy since last few decades for treating posterior eye (vitreoretinal) disorders including retinoblastoma. Several potent small and macromolecules, such as steroids, chemotherapeutic agent, growth factors, stem cells, genes, etc., can be delivered using IVT injection into the retina and vitreous humor with high efficiency [35–38]. Nevertheless, damage to ocular tissues such as formation of retinal holes, bleeding, hemorrhage, endophthalmitis, optic nerve damage, and cataract formation are the most serious complications associated with this route owing to deep needle penetration. Furthermore, to deliver drugs to the subretinal space, which is present between photoreceptors and RPE cells, subretinal injections are mostly utilized to cure vision-related disorders caused by mutations in RPE genes and/or photoreceptors. The suprachoroidal space is mainly utilized for microneedles, depots, and implants administration, which leads to reduction in systemic side effects of the formulation. Among the other intraocular injections, the intrastromal injection is considered the least invasive which delivers the drug into the corneal stroma without damaging the surrounding tissues [39, 40].

1.3 Ocular Diseases

1.3.1 Keratitis

It is a type of ocular inflammation affecting the cornea, which may result in conjunctival scar, eyeball atrophy, or even vision loss if not treated properly. Topical antibiotics like antibacterial and antifungal eye drops or lotions are the first-line treatment for keratitis available in the market. However, conventional eye drops have several limitations like low bioavailability, development of antibiotic resistance, and stability issues with the formulation [41]. To tackle these issues, researchers have developed various polymeric and lipidic nanocarriers (PNCs and LNCs) to effectively reduce the pathogen burden by enhancing drug retention and penetration in the anterior region of the eye.

Solid lipid nanoparticles (SLNs) are the type of LNCs having lipidic core and particle size ranging from 10 to 1000 nm. These NPs are proven effective in ocular delivery of various lipophilic drugs. Liang et al. developed both cationic and anionic modified econazole (ECN)-loaded SLNs (E-SLNs) using microemulsification method for the treatment of fungal keratitis. Particle size and zeta potential of these E-SLNs were found to be around 20 nm and ± 25 mV. E-SLNs when compared with conventional ECN suspension showed enhanced corneal permeation, sustained drug release, and superior antifungal properties with no ocular irritation and toxicities. Comparison of the pharmacokinetic profile of cationic and anionic E-SLNs in rabbits showed that cationic E-SLNs possess significantly higher AUC, $t_{1/2}$ values, as well as ocular bioavailability in the cornea and aqueous humor due to their strong mucoadhesive properties [42].

Another strategy to improve ocular retention and bioavailability of drug is to formulate nanogel-based drug delivery systems. Several antifungal drugs, such as itraconazole, voriconazole, fluconazole, etc., have been manufactured as nanogels to improve their therapeutic efficacy for the treatment of fungal keratitis. Lin et al. formulated carbonized nanogels (car-NG) containing lysin (Lys) and quercetin (Qu) (Lys/Qu-CNGs) having enhanced antioxidant as well as antibacterial properties for the management of bacterial keratitis. Lys/Qu-CNGs were prepared using a single-step dry heating process, and they showed superior inherent antioxidant properties and high positive zeta potential. Hemolysis study, MTT assays, and in vivo corneal studies demonstrated good biocompatibility of Lys/Qu-CNGs with the ocular tissues. In vivo efficacy study in *S. aureus*-induced rabbit keratitis model showed these CNGs possess the ability to penetrate tight junctions of the cornea, thereby demonstrating enhanced antibacterial and antioxidant properties [43].

Mucoadhesive and mucopenetrating nanoparticles are proven to be effective in enhancing drug retention and permeability across the cornea. Bhosale et al. prepared mucoadhesive nanoparticles of voriconazole (VZ) using chitosan (VZ-CH) via ionic gelation method, which were further coated with mucopenetrating anionic polymer (sodium alginate) (VZ-CH-SA) for the treatment of fungal keratitis. Hen's egg test-chorioallantoic membrane (HET-CAM) assay was performed to assess the ocular irritabilities of the NPs; it was found that both VZ-CH and VZ-CH-SA NPs showed irritancy score within the acceptable limits. Furthermore, ex vivo corneal permeation study on goat cornea showed 1.18 times higher permeation of VZ-CH-SA NPs compared with VZ-CH NPs due to the escape of NPs from mucin interaction owing to negatively charged sodium alginate coating [44].

1.3.2 Dry Eye Syndrome

Dry eye syndrome (DES) is a complex condition of the ocular surface characterized by the alteration in the homeostasis of the tear film and the occurrence of repeated cycles of ocular inflammation and tear film hyperosmolarity. These cycles result in the damage to the ocular surface, which ultimately leads to disease progression [45, 46]. In recent times, worsening environmental pollution, the increased use of electronic screens, and higher rates of contact lens wear have contributed to a growing prevalence of DES, which affects approximately 350 million people globally [47]. DES significantly impacts patients' quality of life, causing discomfort, visual disturbances, and in severe cases, blindness. It also leads to psychological disturbances like depression, further imposing a substantial economic burden on society. Although the exact mechanisms of DES remain unclear, inflammation is considered as prime factor in its pathogenesis. The inflammatory response in DES involves both acute and chronic immune responses [48].

The primary objective of DES treatment is to restore tear film balance and disrupt the repeating cycle of the disease. Use of artificial tears, topical corticosteroids or secretagogues, and immunosuppressants are the most common treatment available for DES. However, limitations of these treatments include low patient compliance, side effects like cataract formation, elevated intraocular pressure, and glaucoma, as well as ocular discomfort [49]. To address these challenges of ocular drug delivery, advancements in ocular drug delivery systems, particularly PNCs, have shown significant promise. These systems represent innovative and highly effective strategies for the advanced treatment of DES.

As the main characteristic of the DES is lack and instability of tear fluids, nanotherapeutics which imitate the tear are the most suitable options. For this purpose, many researchers have developed liposomes based on phosphatidylcholine as an artificial tear substitute [50]. While in another research, liposomal formulation was developed which restores the outermost layer of the eyes (i.e., lipid layer), which avoids rapid evaporation of the tear film. Furthermore, these nanoformulations were found to be highly biocompatible and biodegradable [51].

Since the core pathogenesis of DES includes inflammation of various ocular regions, treating the inflammation using anti-inflammatory drugs or antibiotics is one of the common strategies. Plant-based polyphenols like resveratrol (RES), curcumin, quercetin, etc. are highly effective in treating inflammation and microbial infection due to their ability to reduce production of ILs, ROS, and TNF-α and prominent antioxidant effect. Luca et al. developed polylactic-co-glycolic acid modified with cationic polymer, polyethyleneimine (PEI-PLGA) NPs loaded with resveratrol (RES-NPs) embedded in the poloxamer 407 hydrogel matrix to enhance corneal residence time of the drug, thereby reducing dosing frequency and increasing bioavailability of the RES. It was found that RES-NPs showed sustained drug release for up to 3 days along with potent anti-inflammatory and antioxidant effects on the epithelial corneal cell lines in vitro. Furthermore, RES-NPs were able to upregulate sirtuin-1, which is essential for upregulation of mitochondrial functions in the cornea [52].

1.3.3 Cataract

According to data, cataract is the most prominent ocular disorder leading to blindness worldwide in the aged population [29]. Etiology of cataract includes congenital disorders, aging, drug's side effects, or trauma to the eye. Surgical intervention is the most effective treatment for cataract till date, which involves complete removal of the opacified lens and replacing it with a biocompatible artificial lens [53]. Many attempts have been done to develop advanced intraocular lens, minimal invasive therapies, targeted drug delivery, etc. to ease its treatment with least complications. Although surgical intervention remains the most effective approach for treating cataracts, the exploration of medical therapies, like the use of nanotechnology in drug delivery, to delay their onset and progression has been a prominent area of research. Numerous synthetic and natural antioxidants, such as resveratrol, curcumin, and lutein, have demonstrated the potential to mitigate oxidative stress and protect the ocular lens from further damage [54]. In one study, chitosan-coated liposomes carrying coenzyme Q10 having anti-oxidative effect have been prepared, which demonstrated to prolong precorneal drug retention by 4.8 times [55]. In another study, Vora and colleagues designed cyclodextrin-based lipid nanoparticles with high polyphenol loading and effective mucoadhesive properties as a promising strategy for cataract management [56].

Diabetic cataract, a severe ocular complication associated with diabetes, occurs due to elevation in oxidative stress and the buildup of advanced glycation end products (AGEs). Chen et al. formulated eye drops containing sodium glucose cotransporter 2 (SGLT2) inhibitor called dapagliflozin (DAP)-encapsulated nanoformulation, made up of polyvinyl pyrrolidone (PVP) and β-cyclodextrin derivative. It was found that the nanoformulation significantly improved aqueous solubility of drug and also showed significantly decreased SGLT2 levels in the cornea of SD rats in vivo [57].

1.3.4 Glaucoma

Glaucoma is characterized by progressive and chronic optic neuropathy marked by the irreversible loss of ganglion cells of the retina and their axons, leading to peripheral vision loss and, in some cases, central vision impairment [58]. It is the leading global cause of irreversible blindness, currently affecting over 76 million people, with prevalence projected to rise to 111.8 million by 2040 [59]. Although pathological mechanisms for this disease are complex, the elevated intraocular pressure remains the primary factor. Increased pressure can impact the biomechanical properties of the posterior globe, exerting stress on axons of ganglion cells and ultimately triggering cell apoptosis. As such, lowering intraocular pressure is the cornerstone of glaucoma treatment. Traditional therapies include the use of muscarinic agents, beta-blockers, prostaglandin analogs, alpha agonists, and carbonic anhydrase inhibitors, which act by either reducing the production of aqueous humor or enhancing its outflow. Recently, a novel therapeutic approach targeting the dysfunction of pressure-regulating Schlemm's canal endothelial cells has emerged. This approach, known as cell-softening therapy, aims to enhance aqueous humor outflow using rho-kinase inhibitors and actin depolymerizers. However, their clinical application is hindered by physiochemical limitations and adverse effects [2, 60].

A prostaglandin analog, bimatoprost (BIM), is widely used to treat glaucoma by reducing elevated IOP in patients, but its marketed formulation suffers from low ocular bioavailability problem. Satyanarayana and coworkers prepared solid lipid nanoparticles (SLNs) of BIM using solvent evaporation/ultrasonication technique. SLNs showed sustained release of drug for up to 12 h. Furthermore, HET-CAM test showed a significant reduction in the irritation score of BIM when encapsulated in SLNs. SLNs, owing to their increased solubility and permeability characteristics, improved the ocular permeability and bioavailability of BIM for the glaucoma management [61].

Ultradeformable vesicular nanocarriers, like transfersomes or transniosomes, are quite promising in drug delivery across the cornea due to their ability to squeeze through the intercellular junctions. Patil et al. developed transniosomal formulation of brinzolamide (BZ) using biocompatible egg lecithin and nonionic surfactants (Span® 60 and Tween® 80) (BZ-TS) via thin-film hydration method. BZ-TS were found to be nonirritant on ocular surface with good ex vivo ocular permeability of 89.17 ± 4.39%, which was 5.5 times higher than the permeability of free drug. Furthermore, in vivo intraocular pressure measurement (IOP) after BZ-TS administration in rabbit glaucoma model showed more than 30% reduction in IOP after 6 h, indicating the effectiveness of transniosomes in the topical ocular delivery [62]. Similarly, Nair et al. prepared ultradeformable bilosomes of BZ (BZ-BL) with uniform particle size and high drug loading. BZ-BL showed trans-corneal permeability of 8.78 ± 0.14 cm/h and 83.09 ± 5.1% corneal permeation in 6 h ex vivo [63]. Moreover, a comparative study was done by Srivastava et al., among the various vesicular nanocarriers including liposomes, transfersomes, niosomes, and transniosomes for the effective management of glaucoma. They found that considering the topical ocular drug delivery, transniosomes are most effective in glaucoma management in terms of ocular permeability and drug release kinetics [64].

1.3.5 Uveitis

Uveitis refers to inflammation of the uveal membrane, which comprises of iris, ciliary body, and choroid [65]. It can be classified as anterior, intermediate, posterior, or panuveitis depending on the region of the eye affected, while infectious or noninfectious (auto-immune) depending of the etiology of the developed disease. It is among the top five leading causes for blindness worldwide and around 10–15% blindness in developed countries [66]. Uveitis tends to affect younger population (around 70% of patients from 20 to 40 years category), unlike other age-related visual disorders like age-related macular degeneration or cataract [67]. This earlier visual impairment can cause a serious impact on the quality of life of an individual. Therefore, prevention and proper management of this condition to prevent blindness and restore vision is essential. As it is an inflammatory disorder, use of systemic or topical steroids like prednisolone, triamcinolone, etc. is the first-line treatment for uveitis. Both of these conventional therapies exhibit some drawbacks like frequent dose administration, low bioavailability for topically administered formulations, and serious systemic side effects of the oral or parenteral steroidal therapy. Due to these reasons, development of novel drug delivery systems with low dosing frequency, high bioavailability, and less systemic side effects is essential for the management of uveitis [68].

Huang et al. developed a dual-drug nanocomposite containing mesoporous silica NPs loaded with berberine (BER) and embedded in PLGA-PEG-PLGA copolymer-based thermogel loaded with steroidal drug dexamethasone (DEX). The main objective of the study was to obtain synergistic anti-inflammatory effect from BER and DEX combination with prolonged duration of action, high bioavailability, and reduced side effects. Nanocomposite showed high drug loading for both the drugs with good biocompatibility suitable for in vivo applications. Nanocomposite showed initial burst release of 10–20% for both the drugs which is good for controlling inflammation at early stage, followed by sustained drug release for 6 weeks (~23% for 8 h). Furthermore, BER and DEX combination therapy showed superior control over the inflammation for up to 28 days in experimental autoimmune uveitis (EAU) rat model with minimum toxicity [69].

1.3.6 Ocular Tumors

Metastases of tumor from other body parts can lead to a variety of tumors in the eye, among which uveal melanoma (UM) and retinoblastoma (RB) are the most prominent intraocular tumors [70].

UM comes under the most aggressive and common primary intraocular tumors, and more than 50% of its cases lead to tumor metastases and ultimately the patient's death [71]. Despite advanced research on the therapies and diagnosis of UM all over the world, there is still a vast gap in effective treatment therapies. Cimino et al. developed both pegylated and uncoated nanostructured lipid carriers (NLCs) of a newly synthesized prodrug, MRJF22, with uniform homogeneity and high drug loading as a potential treatment of UM. In vitro drug release from the NLCs was found to be sustained for more than 24 hrs using Franz diffusion cells. Furthermore, drug-loaded NLCs also showed reduction in the dose-dependent cytotoxicity of prodrug which was confirmed using in vitro cell line studies on human melanoma cells and human corneal epithelium and using HET-CAM assay. Moreover, the drug-loaded NLCs showed significant antiangiogenic and anti-inflammatory effects measured on three-day fertilized eggs after 48 hrs of treatment [72].

Another aggressive intraocular tumor which mostly occurs in infants and children is RB, with the prevalence of 1 in 15,000–20,000 child births worldwide. RB may lead to secondary monocular tumors, loss of vision, and even death. Notable shortcomings of effective RB management include drug resistance, poor pharmacokinetic profile, non-specificity, and low therapeutic efficacy of the drugs which need to the addressed to develop safe and effective chemotherapies for its treatment [73, 74]. Wang et al. prepared inorganic multifunctional nanoparticles with ultrasound-guided immunosynergistic therapy for effective RB management. Multifunctional nanoparticles were made of iron oxide nanoparticles conjugated with hollow gold nanocages (Fe-Au NPs) with inherent magnetic properties to encapsulate perfluoropentane (PF) and muramyl dipeptide (MD). It was found that NPs showed enhanced magnetic resonance and ultrasound imaging in vitro and in vivo, useful for their efficacy evaluations. NPs were found to be releasing MD under the influence of magnetic field upon reaching the target tumor site. It was found that the activation of NPs led to direct tumor cell apoptosis while the release of MD helps in dendritic cell maturation (body's defense mechanism) which helps in recognition and clearance of RB tumor cells in vivo which facilitates the RB treatment with good biosafety and compatibility [75].

1.4 Various Types of Ocular Drug Delivery Systems

Ocular drug delivery systems are meticulously engineered to precisely target specific regions of the eye, thereby enhancing drug bioavailability, therapeutic efficacy, and patient adherence to treatment regimens. They are generally classified into conventional and advanced systems. Various conventional and novel drug delivery systems have been developed to surmount the challenges associated with ocular drug delivery and enhance ocular bioavailability. These include emulsions, ointments, suspensions, aqueous gels, nanomicelles, nanoparticles, liposomes, dendrimers, implants, contact lenses, nanosuspension, microneedles, and in situ thermosensitive gels, all of which are designed to address the previously identified ocular diseases.

1.4.1 Conventional Drug Delivery Systems

The conventional administration methods primarily include topical administration, conjunctival and scleral administration, intracameral administration, intravitreal injection, retrobulbar injection, and systemic approaches. Many conventional ophthalmic dosage forms are quite basic. Typically, drugs that are soluble in water are often administered topically in the form of an aqueous solution, whereas those that are insoluble in water are usually applied as an ointment or aqueous suspension. The main disadvantages of these conventional dose forms include inadequate drug transport to the back of the eye, as well as systemic exposure via nasolacrimal duct draining, which leads to poor ocular absorption. Improving pharmacokinetic and pharmacodynamic (PKPD) results is possible by addressing these issues in novel drug delivery systems. Some physical and structural barriers in the eye system make it hard for medicines to get into the eyes [76]. These include the blood-ocular barrier (BOB), the corneal epithelial membrane, and nasolacrimal drainage. The anterior part of the eye typically receives only 1% or less of a topically applied dosage because of corneal opacification. Pulse entry is a common, yet highly unwelcome, pharmacokinetic characteristic of eye drops. The initial high concentration of the drug in tears, followed by a rapid decline, suggests a potential for toxicity and indicates the necessity for regular dosing. Avoiding the toxicity associated with the high initial concentration is challenging without requiring multiple doses, particularly when working with potent medications. The primary factor contributing to precorneal drug loss, leading to reduced ocular bioavailability, is nasolacrimal drainage. Drugs applied topically can also enter the circulatory system mainly through this pathway, and for potent drugs, systemic exposure through nasolacrimal drainage after topical application can be significant enough to cause systemic toxicity. Timolol serves as a notable example; following topical application, systemic toxicity associated with the ocular solution of timolol has been documented. Solutions, suspensions, and emulsions serve as examples of traditional drug delivery systems, with many being administered as eye drops [77].

1.4.2 Novel Drug Delivery Systems

The treatment of ocular diseases has led to the development of several nanocarriers, including micelles, nanosuspension, nanoemulsion, nanoparticles, liposomes, niosomes, dendrimers, cubosomes, and nanowafers (Fig. 1.3). NDDS is predicted to revolutionize current available ocular drug delivery techniques. Enhancement of bioavailability is one of the many benefits that can be ascribed to NDDS. However, creating NDDS requires stringent development procedures, and selecting the right excipients and materials is crucial to achieving the intended safety and effectiveness.

1.4.2.1 Micelles

Micelles are tiny particles that can carry drugs. They form on their own in a fluid when the polymer/surfactant concentration is higher than the critical micellar concentration (CMC) [78]. Micelles are formed when solutions contain amphiphilic surfactants or diblock polymers that self-assemble at a specific concentration or temperature (Fig. 1.4) [79]. Their sizes vary from 10 to 200 nm, and they can assume various forms, including spherical, cylindrical, or star-shaped. Normal micelles are good carriers for hydrophobic medications in an aqueous solution, whereas reversed micelles may transport hydrophilic pharmaceuticals [80]. Micelles improve the corneal permeability of topically administered medicines, making them ideal candidates for drug delivery to ocular tissues.

Micelles have a lot of problems, including not being able to be scaled up, having trouble adding drugs, and being toxic because they use surfactants, especially ionic-type ones [81].

Cyclosporin A is an immunosuppressant drug that works well and doesn't mix with water. It is used to treat dry eye disease (DED) and keep corneal transplants from being rejected. Extensive research has been conducted to improve the solubility and ocular bioavailability of ocular drug delivery based on cyclosporine A. Polymeric micelles made of the diblock polymer methoxypoly(ethylene

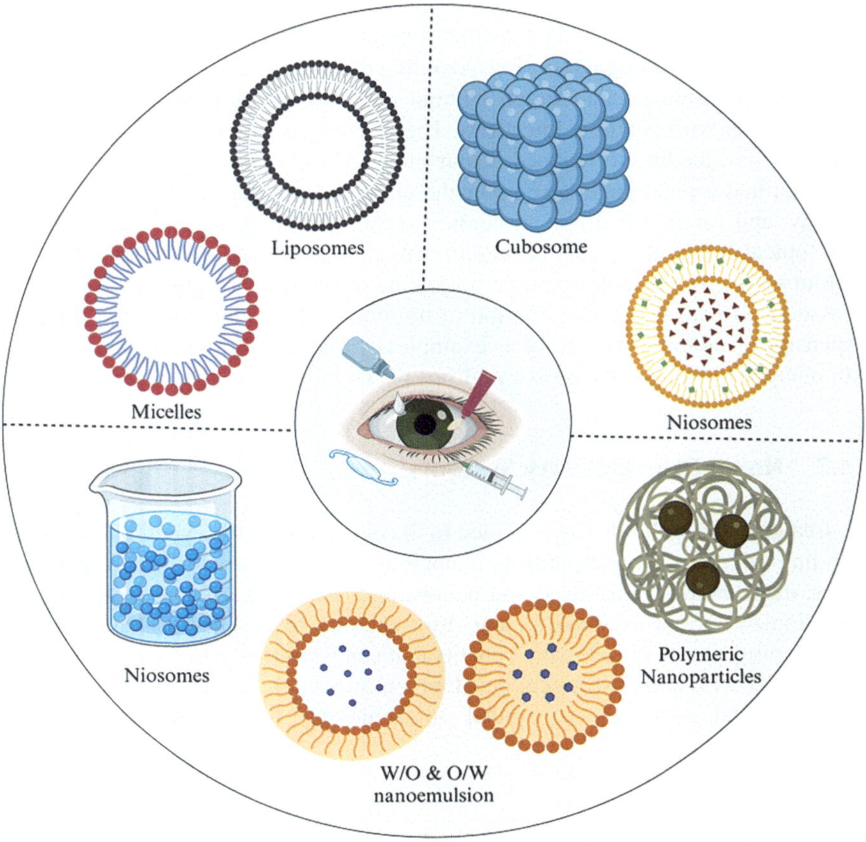

Fig. 1.3 Nanotechnology-based novel drug delivery systems for ocular applications. (Created with biorender.com (licensed version))

glycol)-poly(lactide acid) were created and loaded with cyclosporine A using a thin film dispersion process. The findings indicated that the micellar system increased the retention duration on the precorneal surface by up to 4.5 times and decreased the drug's elimination when contrasted with cyclosporine emulsion [82]. The drug retention and decreased elimination show that cyclosporine micelles may increase the medicine's effectiveness in treating DED. The delivery of cyclosporine A was also investigated and formulated using surfactant micelles that were integrated into in situ gelling systems. When making the micellar delivery systems, two nonionic detergents were used: d-α-tocopherol polyethylene glycol succinate and polyoxyl-40-hydrogenated castor oil. The micelles were subsequently incorporated into the gellan gum dispersion to create a clear and straightforward-to-install aqueous dispersion [83]. When it comes into contact with lacrimal fluid, this dispersion undergoes a gelation process. The combination method prolonged residence duration on

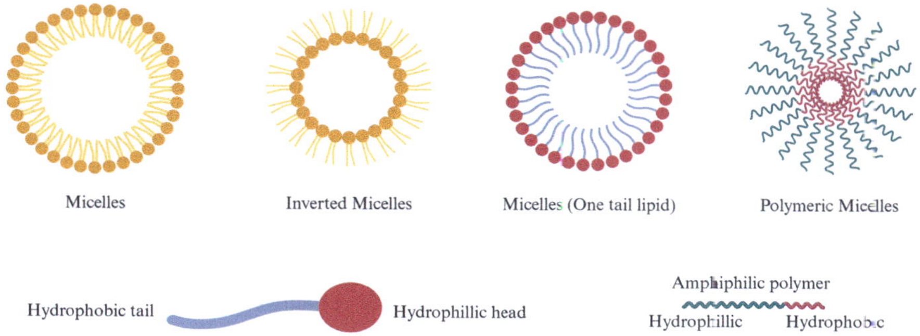

Fig. 1.4 Illustration of various types of micelles and their components: showcasing core-shell structures with hydrophobic cores and hydrophilic shells, reverse micelles, mixed micelles, and polymeric micelles tailored for drug delivery applications. (Created with biorender.com (licensed version))

the ocular surface while causing little harm to corneal cells. When developing eye drops, it is helpful to have the option to use micelles, which can dissolve hydrophobic drugs into a transparent water solution.

A drug that treats glaucoma must stay on the eye's surface for a long time and go deep into the cornea to work effectively and without harm. However, natural physiological defenses, including fast tear clearance and limited cornea permeability, pose considerable hurdles to trans-corneal medication administration. The research by Qiuyu Wei et al. shows that zwitterionic micelles made of OPDEA-PCL amphiphiles could be used as a safe way to deliver drugs through the eye. Because they especially stick to cell walls, these micelles stick better to eye tissues and are harder to tear off. These traits make adsorptive-mediated transcytosis easier, which greatly improves the movement across the cornea and the buildup of the glaucoma drug brinzolamide (BRZ) inside the eye. Consequently, OPDEA-PCL/BRZ formulations are more successful than PEGylated and free BRZ formulations in normalizing intraocular pressure in an open-angle glaucoma rat model. This study highlights OPDEA-PCL micelles as a viable carrier for noninvasive topical trans-corneal medication administration in glaucoma treatment [84].

1.4.2.2 Nanosuspension

Nanosuspensions are colloidal dispersions of recalcitrant drug particles in a dispersion medium that is stabilized by an appropriate surfactant or polymer. To formulate and transport hydrophobic medication aqueous dispersions to intraocular tissues, nanosuspensions are used. They enhance the ocular bioavailability of inadequately water-soluble pharmaceuticals by increasing the solubility, retention time, and sustained drug release [85]. Stabilizers are frequently implemented in the development of ophthalmic nanosuspension, specifically synthetic, natural, or hybrid polymers. Nano solutions aren't perfect for treating anterior segment eye diseases (ASED) because they can become physically unstable (which can lead to sedimentation) and dangerous when detergents are used [81].

Eudragit RL 100-based nanosuspensions were developed for flurbiprofen, an anti-inflammatory agent, through the solvent displacement method [86]. An investigation revealed that the nanosuspension exhibited a prolonged drug release profile and was proven to be stable. To treat ocular irritation, a cationic nanosuspension loaded with diclofenac and composed of chitosan and methoxy poly(ethylene glycol)-poly(ε-caprolactone) diblock copolymer was also studied [87]. In the albino rabbit model, the polymeric nanosuspension had better solubility, as shown by a higher C max in the watery humor and twice as much area under the curve as the commercial diclofenac eye drops. According to the in vivo corneal penetration test, the chitosan-coated nanosuspension demonstrated enhanced corneal penetration and retention. There was no eye irritation. Furthermore, the nanosuspension maintained an exceptionally high level of stability in an aqueous humor solution for 24 hours.

A polymeric nanosuspension made of Eudragit® RL 100 and Eudragit® RS 100 was used to make amikacin more bioavailable in the eye so that it could treat *Staphylococcus aureus* infections of the front of the eye more successfully [88]. The nanoparticles were created using a water-oil-water emulsification solvent evaporation process and then shrunk using high-pressure homogenization. The in vivo pharmacokinetic investigation revealed that the polymeric nanosuspension containing amikacin exhibited a 2.12-fold increase in drug bioavailability compared to the commercially available amikacin ocular drops. However, the nanosuspension required a prolonged time to reach C max. Additionally, the therapeutic efficacy of amikacin was enhanced by the nanosuspension, which demonstrated the ability to induce a symptomatic cure within 4 days. In addition, the treatment of bacterial keratitis was investigated using a combined moxifloxacin and pamoic acid nanosuspension that was composed using the ion coupling technique and possessed mucus-penetrating properties [89]. Instead of using commercial moxifloxacin eye drops three times a day, using moxifloxacin-pamoate nanosuspension once a day was better at stopping and treating *Staphylococcus aureus* infections. The findings of the pharmacokinetic investigation demonstrated that the medication was well distributed throughout the anterior ocular tissues, including the aqueous humor.

Xiaopei Ma et al., choosed cyclosporine A (CsA) as a model medicine for the treatment of DED. Using two methods—PEGylation to get through mucus and cationization for strong cell uptake—together, they made a new CsA nanosuspension (NS@lipid-PEG/CKC) by covering very small drug particles with a mix of lipids (DSPE-PEG2000) and a cationic detergent (CKC). The NS@lipid-PEG/CKC sample, which had a mean size of approximately 173 nm and a positive zeta potential of approximately +40 Mv, exhibited enhanced mucin penetration, excellent cytocompatibility, increased cellular absorption, and protracted precorneal retention without evidence of ocular irritation. Moreover, NS@lipid-PEG/CKC improved tear output and goblet cell density more effectively than a commercial cationic nanoemulsion in a rat model of DED. The results of all the studies suggested that the combination of PEGylation and cationization could be a promising approach to optimize the delivery of drugs to the ocular epithelium in nanomedicine-based eye drops by coordinating mucin penetration and cellular absorption [90].

1.4.2.3 Microemulsion and Nanoemulsion

Microemulsion are colloidal dispersions of oil-in-water or water-in-oil that are stabilized by a surfactant [91]. Nanoemulsion and microemulsion are comparable in terms of their formulation components; however, they differ in their stability. Nanoemulsions are unstable thermodynamically, but microemulsions are stable [92]. Microemulsion and nanoemulsion are gaining popularity for ocular medication delivery owing to their tiny droplet size and large surface area. Some benefits are that the drug dissolves easily, doesn't irritate, improves eye permeability, increases solubility, lasts a long time, is easy to formulate, and can give both hydrophilic and hydrophobic drugs [93]. Nevertheless, the application of nanoemulsion and microemulsion in ocular drug delivery is restricted by the cytotoxicity and physical instability that are a result of the substantial quantity of surfactant [81].

Microemulsions have shown great promise as a method for delivering hydrophilic medications like riboflavin phosphate and hydrophobic pharmaceuticals like triglyceride of docosahexaenoic acid (TG-DHA) to the eyes [94]. A large amount of water was employed to mitigate the cytotoxic effects caused by the surfactant. The resulting microemulsion of the hydrophobic TG-DHA was transparent and dilutable. By administering microemulsion-loaded TG-DHA topically, DED symptoms were alleviated. Furthermore, riboflavin phosphate entered the undamaged corneal tissue and improved the structural stability of the cornea after riboflavin phosphate microemulsion application. The results show that microemulsion may transport hydrophilic and hydrophobic medicines to the eye. To circumvent the problem of several daily administrations of traditional travoprost eye drops for glaucoma treatment, travoprost-loaded contact lenses were studied as a possible solution. Instead of the soaking procedure that is frequently used in loading medications into contact lenses, reliable travoprost microemulsion was employed [95].

When compared to the contact lens that was created using the traditional soaking approach, the drug loading of the contact lens that was loaded with travoprost microemulsion was enhanced, and the in vitro release was extended from 42 to 84 hours. The lens laden with travoprost microemulsion also exhibited enhanced optical properties and edema. In contrast to travoprost eye drops, the contact lens exhibited a longer retention duration on the ocular surface [95]. Thus, microemulsion may improve the drug loading and physical features of drug-loaded contact lenses [96].

1.4.2.4 Polymeric Nanoparticles

Nanoparticles with diameters ranging from 50 to 400 nm are commonly used in delivering treatments to the anterior eye because they may bypass physiological barriers and target medications to particular ocular tissues [97]. They're usually made with safe lipids and polymers, or a mix of the two to make a mixed system [98–100]. To extend the duration of the nanocarrier's residence on the cornea and conjunctiva, it is becoming more prevalent to encapsulate nanoparticles with mucoadhesive polymers, including chitosan, hyaluronic acid, and polyethylene glycol [98, 101]. Nanoparticles provide several benefits, including great scaling potential, increased transcorneal penetration, prolonged drug release, less irritation, targeted drug administration, prevention of premature drug metabolism, and nonspecific uptake [101].

A novel formulation of eye drops, comprising glycol chitosan nanoparticles encapsulated with cerium oxide, has been devised to treat DED symptoms [102]. Some of the symptoms that patients reported after receiving nanoparticle therapy were stabilization of the tear film as well as an improvement in the development and stability of cells in the cornea and conjunctiva. These indications point to a reduction in DED symptoms. The potential of these nanoparticles to reduce the oxidative stress linked to DED was confirmed by the enhancement of antioxidant indicators, such as superoxide dismutase.

To use the advantages offered by both NODS and nanoparticles, hybrid systems consisting of nanoparticles and other nanocarriers may be utilized in drug administration. Triamcinolone acetonide was synthesized in the form of both solid lipid nanoparticles and an in situ gel for the purpose of topical administration [103]. New research examined the mucoadhesion and ocular permeation of three distinct solid lipid nanoparticles with various functional surfaces. Compritol® 888 ATO, poly(2-ethyl-2-oxazoline), and Tween® 80 were utilized to formulate the nanoparticles. The stealth material used to stabilize these nanoparticles was poly(2-ethyl-2-oxazoline), comparable to PEG. After 24 hours, all three versions released about the same amount of ciprofloxacin in a lab setting, ranging from 71% to 75%. Simultaneously, chitosan-coated lipid nanoparticles increased ex vivo permeability and mucoadhesion. Nonetheless, there was a notable enhancement in ocular mucoadhesion observed in the three formulations when compared to the control [98].

1.4.2.5 Liposomes

Liposomes are two-layered vesicles made of phospholipids that have a water-loving center. According to their size and number of bilayers, liposomes are divided into different groups. The classifications of liposomes include small and large unilamellar and multilamellar liposomes [104]. Their composition resembles that of cell membranes, making them very compatible with biological systems [79]. Liposomes are quite popular because of their biodegradability and biocompatibility. Furthermore, ocular medicines with high molecular weight and limited solubility may be integrated into liposomes for efficient delivery [105]. Additionally, the liposomal surface's positive charge promotes interaction with the ocular mucosa, extending corneal penetration and contact duration. Liposomal formulations have been investigated for their efficacy in the treatment of glaucoma, dry eye, and fungal keratitis. The findings indicate significant potential for the clinical application of these delivery systems in the management of ocular diseases [106, 107]. Liposomes for ASED face obstacles such as poor drug-loading capacity and leakage of encapsulated medication, difficulty in scaling up, low stability, and expensive manufacturing costs [81]. Investigations into the potential of a liposomal form of azithromycin for the treatment of dry eye syndrome were initiated when it was shown that these eye drops enhanced the amount, quality, and stability of the tear film [108, 109]. In order to enhance the lipophilicity and loading capacity of azithromycin within the liposomes, the drug was initially combined with cholesteryl hemisuccinate to create an ion pair [107]. In order to increase the medication's stability and solubility in the liposomes and reduce drug leakage, a small amount of medium-chain triglyceride oil was also added. In comparison to azithromycin-loaded liposomes that do not incorporate ion pairing and oil, the average drug loading was observed to rise from 1.5% to 9.2%, while the entrapment efficiency improved from 74% to 96%. The liposomal system exhibited stability for a duration of 3 months, even when maintained at ambient temperature. Furthermore, the liposomes caused twice as much medication to pass through the cornea as the azithromycin solution. Furthermore, the method alleviated DED-related symptoms.

Lactoferrin is a glycoprotein that is naturally found in eye cells. It has been shown to have many medical uses, such as fighting bacteria, preventing scar tissue, reducing inflammation, and fighting oxidative stress [110]. However, its use in ocular administration is limited by low aqueous stability and a strong proclivity for nasolacrimal leakage. To treat DED and ocular inflammation, hyaluronic acid–coated bovine lactoferrin liposomes were studied [111]. The lipid film approach was used in conjunction with high-pressure homogenization. Lactoferrin liposomes were discovered to be nonirritating and stable for more than 60 days. Furthermore, lactoferrin was produced in a controlled and delayed way, and it was able to improve DED symptoms.

Moiseev et al. have recently published findings regarding the design of PEGylated liposomes, which are further modified with maleimide, intended for the ocular delivery of ciprofloxacin. Ex vivo testing of their ability to stick to the eye and conjunctiva of cattle was done using the laser flow-through method. Based on the results, it seems that the capacity of maleimide groups to create covalent connections with thiols found in mucins is responsible for the liposomes' superior retention performance on conjunctival tissue (Fig. 1.5). On the other hand, none of the liposomal versions stuck very well to the cornea. This is because the cornea is less permeable and has fewer mucins [112].

1.4.2.6 Niosomes

Niosomes are stable, vesicular transport systems made up of two layers of nonionic detergents that are amphiphilic. They possess biocompatibility and demonstrate low toxicity as a result of the utilization of nonionic surfactants. Therefore, they are very promising as a means of transporting drugs to the eyes. Additionally, niosomes have been used for tailored transport, long-term drug release, and improved drug absorption [113]. Several factors impede the application of niosomes in drug delivery, including low drug-loading capacity, leakage of encapsulated pharmaceuticals, physical instability, and elevated production costs [114].

Researchers looked at how hyaluronic acid (HA)-modified niosomes could improve the effectiveness of tacrolimus in crossing the cornea and reaching the eye. Tacrolimus is an immunosuppressive medication administered following corneal transplantation surgery to mitigate the likelihood of graft rejection. It has a somewhat large molecular weight (822.5 D) and is extremely hydrophobic, making it difficult to infiltrate the cornea. These characteristics render the formulation and transdermal administration of tacrolimus particularly challenging. This special

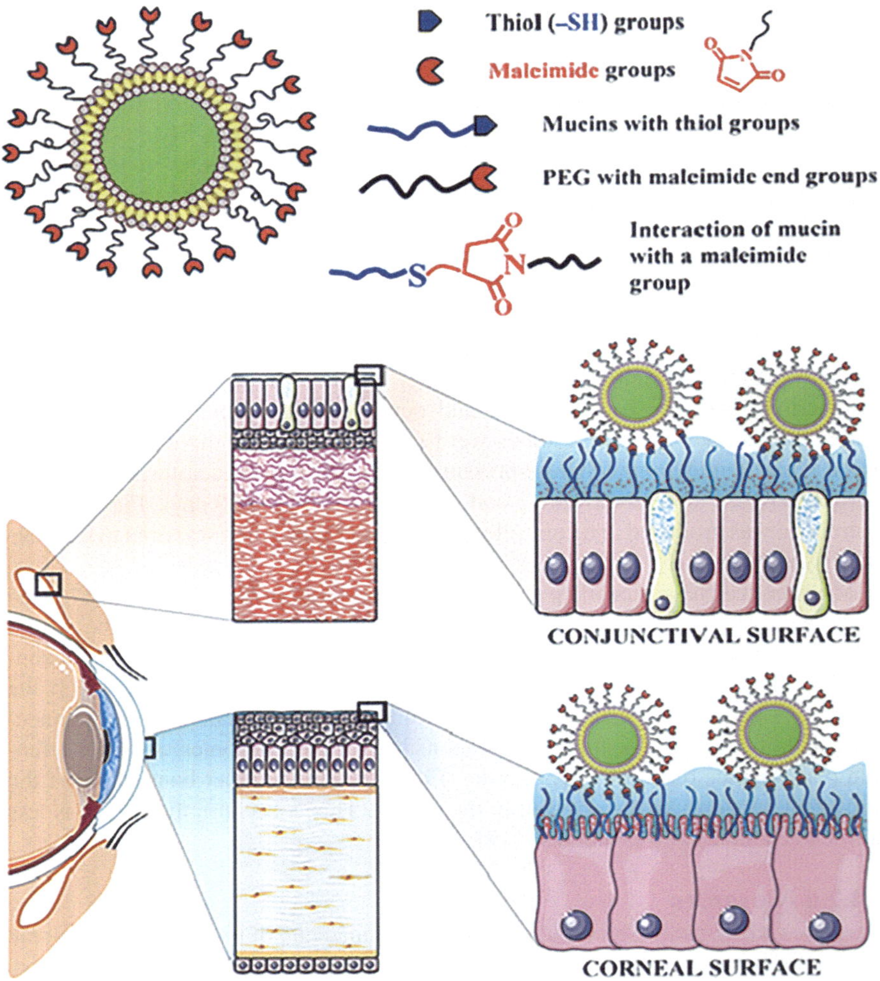

Fig. 1.5 Structure of liposomes functionalized with maleimide groups and their possible reaction with thiol groups present on the cornea and conjunctiva. (Reused with copyright permission from open access Ref. [112])

mixture of niosomes and hyaluronic acid makes the healing effects of tacrolimus better by making it easier for the drug to get into cells. The plasmon resonance investigation verified that the tacrolimus-loaded niosomes coated with hyaluronic acid adhered to the ocular mucins, leading to an increase in transcorneal drug permeability and a prolongation of contact duration. Furthermore, the concentration of the substance in the aqueous humor exhibited a 2.3-fold increase in comparison to the tacrolimus suspension [114].

Similarly, chitosan-coated niosomes containing azithromycin were looked into as a possible way to send the antibiotic drug to the eye in a way that would improve its solubility, corneal permeability, and holding time [115]. They used the thin-film hydration process to create chitosan-coated azithromycin niosomes. In comparison to conventional azithromycin eye drops, the niosomes increased the drug's corneal permeability by up to 2.6 times. Furthermore, the niosomes did not irritate ocular tissues. Many new studies have been done on using niosomes to carry drugs to the eyes, and the results show that niosomal mixtures have a lot of promise to help treat eye diseases [116].

1.4.2.7 Cubosomes

Cubosomes are nanoparticles characterized by a liquid crystalline cubic phase, which are generated through the self-assembly of amphiphilic lipids in an aqueous environment (Fig. 1.6) [113, 118]. The particle diameters range from 100 to 300 nanometers. Cubosomes may be made in two ways: first, by amphiphilic lipids self-assembling in an excess of water; second, by lipids and water emulsifying with a suitable stabilizer using high energy [119, 120]. The amphiphilic lipids commonly utilized in the formulation of cubosomes include monoolein and phytantriol, whereas Pluronics serve as stabilizing agents [121]. Cubosomes are an effective delivery system for hydrophobic medicines. However, they demonstrate minimal effectiveness in entrapping hydrophilic medicines [81].

To improve the therapeutic effectiveness of ciprofloxacin applied directly in the treatment of ocular ulcers and bacterial keratitis, a biodegradable in situ gel was made using ciprofloxacin cubosomes [122]. The parameters of the hydration solution, including the concentrations of ciprofloxacin, phytantriol, and lutrol, as well as its pH, were used to optimize the cubosomal system. The optimized formulation exhibited a significantly improved antibacterial efficacy, being 2.5 times more

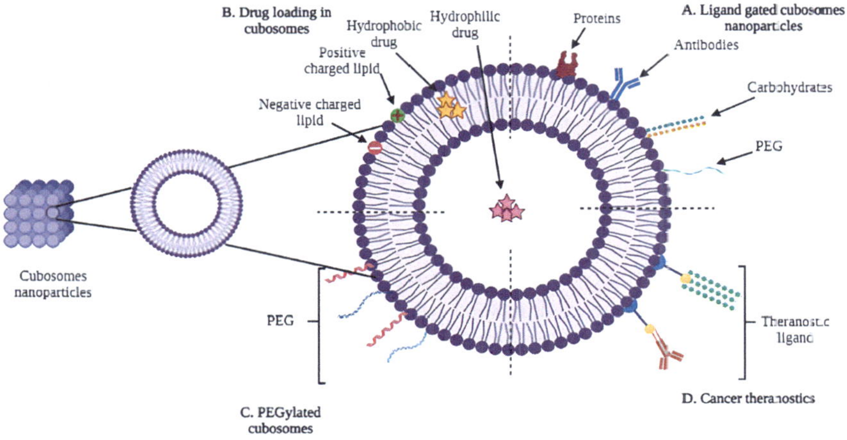

Fig. 1.6 Cubosomes with internal and cubic structures as potential carriers for drug delivery. (Reused with copyright permission from open access Ref. [117])

effective than the commercially available eye drops. Eye drops formulated with ciprofloxacin cubosomes and administered once daily were successful in keeping the medication concentration in the aqueous humor above the minimal inhibitory concentration. According to the researchers, the cubosomes made ciprofloxacin more effective against bacteria by making it stay on the surface of the eye longer and making it easier for bacteria to pass through the cornea. In a separate investigation, the efficacy of natamycin cubosomal nanoparticles was examined as a treatment for fungal keratitis [123]. The probe sonication technique was utilized to obtain a nanosized dispersion of cubosomes. The experts said that the cubosomes made it easier for natamycin to pass through the eye and keep releasing it. When compared to the natamycin solution, the antibacterial test showed that the mixture was more effective at killing *Candida albicans* and *Aspergillus fumigatus*. Furthermore, natamycin cubosomes did not irritate the eye tissues.

1.5 Safety and Regulatory Perspectives

1.5.1 Unique Aspects of Safety Assessment of Long-Acting Delivery

1.5.1.1 Species Selection and Predictivity

Understanding the structure and function of the eye is very difficult for the creation of long-acting delivery (LAD) technologies. Structure-wise, the eye is a very complicated organ, and this makes it very difficult for drugs to work when they are applied locally. The ocular structure is supplied by two separate vascular networks, with certain regions safeguarded by the blood-retinal barrier and the blood-aqueous barrier. The external structure of the eye is proficient in obstructing the infiltration of pharmaceuticals, while the internal environment of the eye is capable of swiftly eliminating drugs through aqueous outflow. It is hard to say if LAD technologies are safe because they can be given in so many different ways and the eye is very complicated structurally. Furthermore, numerous tissues within the eye, particularly the retina and the lens, exhibit a limited capacity for regeneration. Consequently, lesions affecting these tissues are typically irreversible. One of the biggest problems is that there aren't any models to test how safe and harmful LAD technologies are. While in vitro assays are valuable for evaluating local toxicity in specific cell types and uncomplicated tissues, such as the cornea, there currently exist no in vitro or ex vivo models capable of addressing the intricate architecture and fluid dynamics of the eye in its entirety. Consequently, an accurate evaluation of the toxicity associated with LAD technologies necessitates the conduct of in vivo investigations [124, 125].

On the other hand, the implementation of in vivo studies is rendered complex due to the anatomical and physiological disparities of the eye among nonclinical species, as well as their implications for human ocular anatomy. The selection of an appropriate animal model is of paramount importance and may be contingent upon the specific form of left anterior descending artery involvement and the associated toxicities of interest. For example, the nonhuman monkey, which has a vitreous viscosity close to humans, is the only animal that can be used to test particle movement, which is a big problem with intravitreal therapy (IVT) solution. In contrast, the enlarged ocular structure of the pig proves to be more advantageous for the safety evaluation of devices designed for human dimensions. While the larger ocular structure of a canine was employed in the context of the fluocinolone acetonide implant (Retisert), it is noteworthy that the canine eye is susceptible to sterile inflammation after surgical intervention [126]. The rabbit is frequently employed as a lower-order species in research; conversely, rodents are typically not advantageous for evaluating intraocular LAD due to the diminutive size of their eyes and the anatomical and pharmacokinetic disparities between their ocular structures and those of humans and larger animals. Typically, only low dose or exposure multiples can be attained, even in large animal models, due to the diminutive size of the eye. This anatomical limitation restricts the volume that can be administered and presents manufacturing and safety challenges associated with high-concentration formulations, such as stability, pH, osmolality, and viscosity. Table 1.1 shows the different animals used in the preclinical study of ocular drug delivery systems and the advantages and limitations of respective animals from different perspectives.

1.5.1.2 Chronic Exposure

The purpose of continuous release is to expose a healing molecule for a long time, usually several weeks to months. To achieve this objective, a depot of pharmaceuticals that facilitates a gradual and sustained release is necessary. The pharmaceutical agent may be extricated from the target tissues, as exemplified by a slow-release device or polymer implant, or it may persist for an extended duration within or in proximity to the target tissue, as observed with sustained-release formulations of triamcinolone. In either scenario, the persistent presence of the drug, along with any associated polymers or excipients, may result in distinctive toxicities, including particle migration or immune responses. Based on our observations, the migration of particles to the anterior segment occurs within minutes following the administration of IVT, whereas immune responses typically develop over a period ranging from 2 to 6 weeks. To make sure these technologies are safe, long-term hazard tests are needed, which take a lot of time and cost a lot of money.

1.5.1.3 Complex Systems

The intricacy of certain sustained-delivery technologies exacerbates the challenges associated with their safety evaluation. The objective is to forecast the implications for human safety throughout all phases of clinical application. Consequently, it is imperative to delineate the safety of intricate systems, encompassing the processes of implantation and/or injection, drug release, polymer clearance, and subsequent administrations. In the case of polymers physically degrading inside the eye, evaluation is required throughout the degradation process, not only during the first phase of drug release. We need to check if any polymer builds up on repeat doses of non-biodegradable devices to see if it fits with their intended clinical use. There is further cause for worry over the drug's effect in these nonclinical models. For example, humanized protein therapeutics, particularly monoclonal antibodies (mAbs) and

Table 1.1 A comparative overview of animal models' advantages and limitations in preclinical ophthalmic research with applications

Sr. no.	Animal model	Purpose	Advantages	Limitations	Suitable for use	References
1.	Rabbit	Uveitis, corneal studies, drug delivery	Large eyes with identical tear film dynamics, easy handling	Restricted resemblance to the anatomical anatomy of the human retina and choroid	Uveitis models, assessment of ocular bioavailability	[127]
2.	Mouse	Angiogenesis, genetic studies, retinal disease	Small size, extensive genetic tools	Minuscule ocular dimensions, challenging for surgical interventions	AMD studies, retinal degeneration	[128]
3.	Rat	Glaucoma, retinal degradation, uveitis	Genetic modification can achieve, low cost	Small eyes, less anatomically translatable	Glaucoma studies, precorneal residence time	[129]
4.	Monkey	Pharmacokinetics, retinal diseases	Most analogous to the anatomy and physiology of the human eye	Elevated expenses and ethical concerns	Macular degeneration, retinal drug delivery	[130]
5.	Dog	Pharmacokinetics, glaucoma	Human-like aqueous humor dynamics	Ethical issues and excessive expenses	IOP studies and glaucoma research	[131]
6.	Pig	Vitreous and retinal studies	Comparable ocular dimensions and anatomy to humans	Maintenance expenses, restricted genetic models	Vitreoretinal drug delivery and retinal surgery models	[132]
7.	Cat	Retinal physiology and pharmacology	Retinal structures that are analogous to those found in humans	Ethical dilemmas, restricted accessibility	Retinal pharmacokinetics, visual pathway studies	[133]
8.	Guinea pig	Dry eye, allergic conjunctivitis	The eye's immune system reacts similarly to humans'	Small eyes have little significance to retinal research	Dry eye, allergic response studies	[134]

fragment antigen-binding proteins (Fabs), frequently exhibit immunogenicity in nonclinical models, a reaction that is further intensified by extended drug release. Evaluating the interplay between intricate formulations and humanized proteins in nonclinical species frequently presents significant challenges [135–138].

The nonclinical toxicologist who is checking the safety of extended eye drug delivery systems will have a hard time because of these problems, which will be discussed in more detail below. Nonetheless, a robust study design, judicious selection of animal models, and a comprehensive comprehension of the predominant toxicities associated with each route of exposure and type of technology constitute an exemplary foundation in the pursuit of a safe sustained-delivery technology for the treatment of ocular diseases.

1.5.2 Regulatory Aspects

A significant regulatory consideration pertains to the classification of the ophthalmic medication delivery product as either a drug or a combination product. If the transportation method is just a way to get the medicine to the right place, it will probably be regulated as a drug. Nevertheless, if the delivery system constitutes a device with its specific indication, it would probably be classified as a combination product. Consequently, Vitrasert (New Drug Application (NDA) 20,569) and Retisert (NDA 21737), both of which are non-erodible implants, underwent a comprehensive evaluation and received approval as pharmaceutical agents. Visudyne (verteporfin injection, NDA 21–119) necessitates the utilization of lasers operating at a designated light frequency. Each laser was assessed and authorized as a device (Premarket Approval nos. P990048 and P990049 for Zeiss VISULAS 690 s and Coherent Opal Photoactivator, respectively), while Visudyne is classified as a combined product. This is compliant with the US rules (21 CFR 3.2(e), fda.gov/oc/combination) [139]. To date, based on knowledge, very few ophthalmic combo products like Simbrinza®, DuoTrav®, Combigan®, and Cosopt® are presently authorized.

While contact lenses are not presently an authorized medium for the administration of pharmaceutical agents, their potential application for this purpose warrants consideration. By all accounts, the contact lens would be classified as a drug if its only purpose was to encapsulate the substance. But if the product did both deliver drugs and correct vision, it would probably be called a combo product [140–142].

One strategy for drug delivery involves utilizing existing molecules that have received regulatory approval and are no longer under patent protection to create an ophthalmic drug delivery system. This methodology, which does not introduce a novel molecular entity, may utilize the "505(b)(ii)" provision, enacted as a component of the "Drug Price Competition and Patent Term Restoration Act" of 1984 (commonly referred to as Waxman-Hatch) [143].

These applications are predicated on the sponsor's reliance on the prior determinations of safety and efficacy established by the United States Food and Substance Administration for a reference-listed substance, as delineated in 21 CFR 314.54.

Generally, these represent reformulations characterized by distinct dosing regimens while maintaining comparable indications. For instance, Timoptic-XE (timolol gel-forming suspension, NDA 20–330) and Istalol (timolol maleate ophthalmic solution, NDA 21–516) are examples that are based on Timoptic timolol maleate ophthalmic solution (NDA 18–086). Assuming that the product is no longer under patent protection at the time of New Drug Application (NDA) approval, the sponsor may generally reference the preclinical and clinical research pertaining to the innovator product. More preliminary work needs to be done on the new product's safety, effectiveness, and effects in the area where it will be used. It is also important to show through a clinical study that the new product is safe and effective in the same ways as the innovator's product. This "505(b)(ii)" strategy might reduce development risks while potentially saving years of labor and millions of dollars. Certainly, the comprehensive development of the pharmaceuticals and the production of the new product must be executed in accordance with established good manufacturing practices [144].

For both methodologies, it is imperative to evaluate the ocular safety of the excipients employed, such as those utilized in the composition of the ophthalmic product or involved in the drug delivery system. In the event of significant systemic absorption, it may be necessary to consider the implications for systemic safety. The variations in the developmental trajectories may also result in differing levels of exclusivity granted by the Food and Drug Administration.

1.6 Patents and Marketed Products

Ocular drug delivery has gained significant attention in recent years due to the growing prevalence of eye-related disorders and the need for effective, targeted therapies. The field has witnessed a surge in innovative patented technologies and commercial products designed to overcome the eye's unique anatomical and physiological barriers. Patents in ocular drug delivery often focus on advanced formulations such as sustained-release systems, biodegradable implants, nanocarrier-based systems, and novel routes of administration that enhance bioavailability and therapeutic efficacy. Marketed products such as Ozurdex® (dexamethasone implant) and Iluvien® (fluocinolone acetonide implant) exemplify the successful translation of patented innovations into commercial success (Tables 1.2 and 1.3) [166]. These products provide sustained drug release, reducing the frequency of administration and improving patient compliance. Emerging nanotechnology-based products and gene delivery systems further enrich the competitive landscape, demonstrating the dynamic interplay between patent filings and market-driven solutions in ocular therapeutics.

Table 1.2 An overview of marketed formulations for ocular diseases

Marketed name	Formulation type	Drug	Therapeutic indication	US FDA approval year	References
Pred forte	Suspension (eye drops)	Prednisolone acetate (corticosteroid)	Steroid-responsive inflammatory conditions	2024	
Miebo	Eye drop	Perfluorohexyloctane	Dry eye disease	2023	[145]
Omlonti	Topical eye drops	Omidenepag isopropyl	Open-angle glaucoma	2022	[146]
Durysta	Intracameral implant	Bimatoprost	Open-angle glaucoma	2020	[147]
Eysuvis	Ophthalmic suspension	Loteprednol etabonate	Dry eye disease	2020	[148]
Rocklatan	Ophthalmic solution	Netarsudil and latanoprost	Open-angle glaucoma	2019	[149]
Xelpros	Preservative-free ophthalmic emulsion	Latanoprost	Open-angle glaucoma	2018	[150]
Cequa®	Eye drop (Nanomicellar ophthalmic solution)	Cyclosporine A	Dry eye disease	2018	[151]
Dextenza	Intracanalicular insert	Dexamethasone	Ocular inflammation and pain following ophthalmic surgery	2018	[152]
Simbrinza	Suspension (eye Drops)	Brinzolamide (carbonic anhydrase inhibitor) + brimonidine (alpha-2 agonist)	Open-angle glaucoma	2013	[153]
Ozurdex	Intravitreal implant (insert)	Dexamethasone (corticosteroid)	Treatment of macular edema following retinal vein occlusion (RVO), diabetic macular edema (DME), and noninfectious uveitis	2009	[154]

(continued)

Table 1.2 (continued)

Marketed name	Formulation type	Drug	Therapeutic indication	US FDA approval year	References
Durezol	Emulsion (eye drops)	Difluprednate (corticosteroid)	Inflammation and pain associated with ocular surgery	2008	[155]
Brinzolamide and Timolol	Suspension (eye drops)	Brinzolamide (carbonic anhydrase inhibitor) + timolol (beta-blocker)	Open-angle glaucoma	2006	[156]
Restasis	Emulsion (eye Drops)	Cyclosporine A (immunomodulator)	Keratoconjunctivitis sicca (dry eye)	2002	[157]
Xalatan	Solution (eye drops)	Latanoprost (prostaglandin analog)	Open-angle glaucoma	1998	[158]

Table 1.3 An overview of patents in drug formulation: Applications, innovations, and current status

Sr. no.	Application no.	Title of invention	Formulation	Disease	Conclusion	Publication year	Current status (year)	References
1	CN202411364311A	Preparation method of griseofulvin eye drops	Eye drop	Fungal infections	The invention outlines a method for preparing griseofulvin eye drops, involving dissolving hemoglobin in water, griseofulvin in an organic solvent, dopamine hydrochloride, Tris buffer solution, and cell-penetrating peptide. This process creates a nanostructure that can deliver medicine to the posterior eye segment in a local eye drop mode	2024	Pending	[159]
2	CN202411311408A	Corneal contact lens and drug slow-release method	Contact lens	Ocular inflammation	Arc-shaped curved surface multilayer corneal contact lens includes a micro-control circuit layer, drug release layer, and active oxygen detection layer. The microprocessor controls the heating circuit, while the active oxygen detection layer releases medicine molecules. The lens's multilayer structure allows for accurate treatment of ocular inflammation, with the drug molecules acting on the cornea through gaps	2024	Pending	[160]

(continued)

Table 1.3 (continued)

Sr. no.	Application no.	Title of invention	Formulation	Disease	Conclusion	Publication year	Current status (year)	References
3	CN202411111197A	Nanometer emulsion eye cream containing traditional Chinese medicine compound extract and preparation method thereof	Eye cream	NA	The invention pertains to skin care products, specifically nanoemulsion eye cream containing a traditional Chinese medicine compound extract. The extract is made from dogwood, weeping forsythia leaves, pseudo-ginseng, schisandra berry, cassia seed, and dendrobium nobile. The preparation method involves mixing the extract, baking, crushing, and sieving to obtain a powder. The product has stable performance, mild effects, good safety, and no side effects, making it safe and non-harmful to the skin	2024	Pending	[161]
4	CN202410988984A	Sustained release drug delivery system for eyes	Implant		The invention outlines an ophthalmic sustained-release drug delivery system, consisting of a push injector and a loader. The injector implants sustained-release drugs into the eyes, while the loader loads the drugs into the injector. The injector has an ejection button and a pushing mechanism, making it convenient and quick. The loader is quick and accurate and includes a carrier component and a medicine delivery assembly. The system allows for easy placement and loading of slow-release drugs	2024	Active	[162]

5	CN202410539349A	S-(+)-flurbiprofen tromethamine salt eye drops and preparation method and application thereof	Eye drop	Ocular inflammation	The invention relates to pharmaceutical preparations, specifically S-(((+)-flurbiprofen tromethamine salt eye drops. The preparation method involves 100 mL of salt, tackifier, pH regulator, osmotic pressure regulator, and water. These eye drops offer excellent drug effects, no irritation, good compliance, and easy industrial production	2024	Pending	[163]
6	CN202410494091A	Drug-loaded nanomicelle based on hyaluronic acid and preparation method and application thereof	Eye drop	Fundus vascular proliferative diseases	The invention outlines a hyaluronic acid–based drug-loaded nanomicelle preparation method for eye drop drug administration. The process involves activating hyaluronic acid, reacting with hexylamine, maleimide, and peptides, and adding PAMAM/HIF-1 alpha siRNA and bevacizumab. The nanomicelle has good penetrability, safety, targeting capability, and antiangiogenesis effect, making it suitable for treating fundus vascular proliferative diseases	2024	Active	[164]

(continued)

Table 1.3 (continued)

Sr. no.	Application no.	Title of invention	Formulation	Disease	Conclusion	Publication year	Current status (year)	References
7	CN202410115769A	Nanometer eye drop capable of delivering medicine to posterior segment of eye, preparation method, and application thereof	Eye drop	NA	The invention relates to biological medicines and introduces a nano eye drop for delivering medicines to the posterior eye segments. The drop, made of a liposome inner core and chitosan shell layer, effectively loads hydrophilic and hydrophobic drug molecules, prolonging drug retention time and enhancing drug penetration. This high-efficiency delivery strategy promotes clinical application of new ophthalmic drugs	2024	Active	[165]

1.7 Future Perspective and Concluding Remarks

The necessity for creative ways to overcome the physiological and anatomical obstacles of the eye is driving the ongoing, difficult, and quickly changing area of ocular medication administration. This field's future is predicated on the integration of precision medicine, advanced biomaterials, and nanotechnology. Novel strategies with great potential include gene therapy, stimuli-responsive drug delivery systems, and the use of artificial intelligence in medication research and delivery planning [167, 168]. These improvements seek to improve medication targeting accuracy, reduce systemic adverse effects, and increase patient compliance by using sustained and less invasive delivery techniques.

Furthermore, creating biodegradable and biocompatible polymers specifically for ocular applications will be critical. Smart systems that can react to physiological signals like pH or temperature changes can revolutionize the treatment of chronic visual disorders such as glaucoma, uveitis, and age-related macular degeneration. Furthermore, personalized medicine, made possible by advances in genomics and proteomics, will most certainly play an important role in adapting medicines to particular patient demands, assuring both safety and effectiveness [169].

Finally, ocular medication delivery continues to be a field with enormous promise and room for innovation. Future developments may considerably improve therapy results and patient quality of life by addressing current limits and harnessing transdisciplinary research. The combination of cutting-edge technology and a thorough knowledge of ocular physiology and pathology will be critical in overcoming present obstacles and propelling the field ahead. This chapter presents a core knowledge of the concepts that underpin ocular medication delivery, acting as a springboard for future research and development in this interesting and important topic.

References

1. Vedadghavami A, Zhang C, Bajpayee AG. Overcoming negatively charged tissue barriers: drug delivery using cationic peptides and proteins. Nano Today. 2020;34:100898.
2. Han H, Li S, Xu M, Zhong Y, Fan W, Xu J, et al. Polymer- and lipid-based nanocarriers for ocular drug delivery: current status and future perspectives. Adv Drug Deliv Rev. 2023;196:114770.
3. Janagam DR, Wu L, Lowe TL. Nanoparticles for drug delivery to the anterior segment of the eye. Adv Drug Deliv Rev [Internet]. 2017 Dec 1 [cited 2025 Jan 15]:122:31–64. Available from: https://pubmed.ncbi.nlm.nih.gov/28392306/
4. Cunha-Vaz J, Bernardes R, Lobo C. Blood-Retinal Barrier. https://doi.org/10.5301/EJO20106049 [Internet]. 2010 Nov 11 [cited 2025 Jan 16];21(SUPPL.6):3–9. Available from: https://journals.sagepub.com/doi/abs/10.5301/EJO.2010.6049
5. Mandal A, Bisht R, Rupenthal ID, Mitra AK. Polymeric micelles for ocular drug delivery: from structural frameworks to recent preclinical studies. J Control Release. 2017;248:96–116.
6. Gachon AM, Richard J, Dastugue B. Human tears: normal protein pattern and individual protein determinations in adults. Curr Eye Res [Internet] 1982 [cited 2025 Jan 16];2(5):301–308. Available from: https://www.tandfonline.com/doi/abs/10.3109/02713688209000774
7. Khan O, Bhawale R, Vasave R, Mehra NK. Ionic liquid-based formulation approaches for enhanced transmucosal drug delivery. Drug Discov Today. 2024;29(9):104109.

8. Khan O, Yadav R, Mehra NK. Mucoadhesive forms. In: Polymers for oral drug delivery technologies; 2025. p. 703–57.
9. Alshaikh RA, Waeber C, Ryan KB. Polymer based sustained drug delivery to the ocular posterior segment: barriers and future opportunities for the treatment of neovascular pathologies. Adv Drug Deliv Rev. 2022;187:114342.
10. Subrizi A, del Amo EM, Korzhikov-Vlakh V, Tennikova T, Ruponen M, Urtti A. Design principles of ocular drug delivery systems: importance of drug payload, release rate, and material properties. Drug Discov Today. 2019;24(8):1446–57.
11. Micali A, Pisani A, Puzzolo D, Nowińska A, Wylegala E, Teper S, et al. Macular corneal dystrophy: in vivo confocal and structural data. Ophthalmology. 2014;121(6):1164–73.
12. Ruberti JW, Sinha Roy A, Roberts CJ. Corneal biomechanics and biomaterials. Annu Rev Biomed Eng [Internet] 2011 Aug 15 [cited 2025 Jan 16];13(Volume 13, 2011):269–295. Available from: https://www.annualreviews.org/content/journals/10.1146/annurev-bioeng-070909-105243
13. Mofidfar M, Abdi B, Ahadian S, Mostafavi E, Desai TA, Abbasi F, et al. Drug delivery to the anterior segment of the eye: a review of current and future treatment strategies. Int J Pharm. 2021;607:120924.
14. Jumelle C, Gholizadeh S, Annabi N, Dana R. Advances and limitations of drug delivery systems formulated as eye drops. J Control Release. 2020;321:1–22.
15. Hughes PM, Olejnik O, Chang-Lin JE, Wilson CG. Topical and systemic drug delivery to the posterior segments. Adv Drug Deliv Rev. 2005;57(14):2010–32.
16. Nayak K, Misra M. A review on recent drug delivery systems for posterior segment of eye. Biomed Pharmacother. 2018;107:1564–82.
17. Kim SH, Csaky KG, Wang NS, Lutz RJ. Drug elimination kinetics following subconjunctival injection using dynamic contrast-enhanced magnetic resonance imaging. Pharm Res [Internet] 2008 Aug 3 [cited 2025 Jan 16];25(3):512–520. Available from: https://link.springer.com/article/10.1007/s11095-007-9408-z
18. Bock F, Maruyama K, Regenfuss B, Hos D, Steven P, Heindl LM, et al. Novel anti(lymph) angiogenic treatment strategies for corneal and ocular surface diseases. Prog Retin Eye Res. 2013;34:89–124.
19. Summers Rada JA, Shelton S, Norton TT. The sclera and myopia. Exp Eye Res. 2006;82(2):185–200.
20. Sun S, Li J, Li X, Lan B, Zhou S, Meng Y, et al. Episcleral drug film for better-targeted ocular drug delivery and controlled release using multilayered poly-ε-caprolactone (PCL). ActaBiomater. 2016;37:143–54.
21. Duvvuri S, Majumdar S, Mitra AK. Role of metabolism in ocular drug delivery. Curr Drug Metab [Internet]. 2005 Mar 18 [cited 2025 Jan 16];5(6):507–15. Available from: https://www.eurekaselect.com/article/7360
22. Hillenkamp J, Hussain AA, Jackson TL, Cunningham JR, Marshall J. Taurine uptake by human retinal pigment epithelium: implications for the transport of small solutes between the choroid and the outer retina. Invest Ophthalmol Vis Sci. 2004;45(12):4529–34.
23. Kalia YN, Naik A, Garrison J, Guy RH. Iontophoretic drug delivery. Adv Drug Deliv Rev. 2004;56(5):619–58.
24. Coca-Prados M. The blood-aqueous barrier in health and disease. J Glaucoma [Internet] 2014 Dec 10 [cited 2025 Jan 16];23(8 Suppl 1):S36–S38. Available from: https://pubmed.ncbi.nlm.nih.gov/25275903/
25. Lee J, Pelis RM. Drug transport by the blood–aqueous humor barrier of the eye. Drug Metab Dispos. 2016;44(10):1675–81.
26. Huang D, Chen YS, Rupenthal ID. Overcoming ocular drug delivery barriers through the use of physical forces. Adv Drug Deliv Rev. 2018;126:96–112.
27. Motiejunaite R, Kazlauskas A. Pericytes and ocular diseases. Exp Eye Res. 2008;86(2):171–7.
28. Cholkar K, Dasari SR, Pal D, Mitra AK. Eye: anatomy, physiology and barriers to drug delivery. In: Ocular transporters and receptors: their role in drug delivery. Woodhead Publishing; 2013. p. 1–36.

29. Bourne RRA, Steinmetz JD, Flaxman S, Briant PS, Taylor HR, Resnikoff S, et al. Trends in prevalence of blindness and distance and near vision impairment over 30 years: an analysis for the global burden of disease study. Lancet Glob Health [Internet] 2021 Feb 1 [cited 2025 Jan 16];9(2):e130–e143. Available from: http://www.thelancet.com/article/S2214109X20304253/fulltext
30. Baudouin C, Labbé A, Liang H, Pauly A, Brignole-Baudouin F. Preservatives in eyedrops: the good, the bad and the ugly. Prog Retin Eye Res. 2010;29(4):312–34.
31. Shah Shalin S, Denham LV, Elison JR, Bhattacharjee PS, Huq T, Clement C, et al. Drug delivery to the posterior segment of the eye for pharmacologic therapy. Expert Rev Ophthalmol [Internet] 2010 Feb 1 [cited 2025 Jan 16];5(1):75–93. Available from: https://pubmed.ncbi.nlm.nih.gov/20305803/
32. Kang-Mieler JJ, Osswald CR, Mieler WF. Advances in ocular drug delivery: emphasis on the posterior segment. Expert Opin Drug Deliv [Internet] 2014 Oct 1 [cited 2025 Jan 16];11(10):1647–1660. Available from: https://www.tandfonline.com/doi/abs/10.1517/17425247.2014.935338
33. Wai-Yip Lee T, Robinson JR. Drug delivery to the posterior segment of the eye IV: theoretical formulation of a drug delivery system for subconjunctival injection. https://home.liebertpub.com/jop [Internet]. 2009 Mar 2 [cited 2025 Jan 16];25(1):29–37. Available from: https://www.liebertpub.com/doi/10.1089/jop.2008.0010
34. Raghava S, Hammond M, Kompella UB. Periocular routes for retinal drug delivery. Expert Opin Drug Deliv [Internet]. 2004 Nov [cited 2025 Jan 16];1(1):99–114. Available from: https://www.tandfonline.com/doi/abs/10.1517/17425247.1.1.99
35. Emre E, Yüksel N, Duruksu G, Pirhan D, Subaşi C, Erman G, et al. Neuroprotective effects of intravitreally transplanted adipose tissue and bone marrow–derived mesenchymal stem cells in an experimental ocular hypertension model. Cytotherapy. 2015;17(5):543–59.
36. Tan G, Liu D, Zhu R, Pan H, Li J, Pan W. A core-shell nanoplatform as a nonviral vector for targeted delivery of genes to the retina. ActaBiomater. 2021;134:605–20.
37. Ryu M, Nakazawa T, Akagi T, Tanaka T, Watanabe R, Yasuda M, et al. Suppression of phagocytic cells in retinal disorders using amphiphilic poly(γ-glutamic acid) nanoparticles containing dexamethasone. J Control Release. 2011;151(1):65–73.
38. Jiang P, Chaparro FJ, Cuddington CT, Palmer AF, Ohr MP, Lannutti JJ, et al. Injectable biodegradable bi-layered capsule for sustained delivery of bevacizumab in treating wet age-related macular degeneration. J Control Release. 2020;320:442–56.
39. Rai UDJP, Young SA, Thrimawithana TR, Abdelkader H, Alani AWG, Pierscionek B, et al. The suprachoroidal pathway: a new drug delivery route to the back of the eye. Drug Discov Today. 2015;20(4):491–5.
40. Luo LJ, Lin TY, Yao CH, Kuo PY, Matsusaki M, Harroun SG, et al. Dual-functional gelatin-capped silver nanoparticles for antibacterial and antiangiogenic treatment of bacterial keratitis. J Colloid Interface Sci. 2019;536:112–26.
41. Date AA, Kates M, Yoshida T, Babu T, Afzal U, Kanvinde P, et al. Preclinical evaluation of a hypotonic docetaxel nanosuspension formulation for intravesical treatment of non-muscle-invasive bladder cancer. Drug DelivTransl Res [Internet] 2021 Oct 1 [cited 2025 Jan 16];11(5):2085–2095. Available from: https://link.springer.com/article/10.1007/s13346-020-00870-w
42. Liang Z, Zhang Z, Lu P, Yang J, Han L, Liu S, et al. The effect of charges on the corneal penetration of solid lipid nanoparticles loaded Econazole after topical administration in rabbits. Eur J Pharm Sci. 2023;187:106494.
43. Lin HY, Wang SW, Mao JY, Chang HT, Harroun SG, Lin HJ, et al. Carbonized nanogels for simultaneous antibacterial and antioxidant treatment of bacterial keratitis. Chem Eng J. 2021;411:128469.
44. Bhosale VA, Srivastava V, Valamla B, Yadav R, Singh SB, Mehra NK. Preparation and evaluation of modified chitosan nanoparticles using anionic sodium alginate polymer for treatment of ocular disease. Pharmaceutics 2022, Vol 14, Page 2802 [Internet]. 2022 Dec 14 [cited 2025 Jan 21];14(12):2802. Available from: https://www.mdpi.com/1999-4923/14/12/2802/htm

45. Bron AJ, de Paiva CS, Chauhan SK, Bonini S, Gabison EE, Jain S, et al. TFOS DEWS II pathophysiology report. Ocul Surf. 2017;15(3):438–510.
46. Pflugfelder SC, de Paiva CS. The pathophysiology of dry eye disease: what we know and future directions for research. Ophthalmology. 2017;124(11):S4–13.
47. Craig JP, Nichols KK, Akpek EK, Caffery B, Dua HS, Joo CK, et al. TFOS DEWS II Definition and Classification Report. Ocul Surf [Internet]. 2017 July 1 [cited 2025 Jan 16];15(3):276–83. Available from: https://pubmed.ncbi.nlm.nih.gov/28736335/
48. Perez VL, Stern ME, Pflugfelder SC. Inflammatory basis for dry eye disease flares. Exp Eye Res. 2020;201:108294.
49. Jones L, Downie LE, Korb D, Benitez-del-Castillo JM, Dana R, Deng SX, et al. TFOS DEWS II management and therapy report. Ocul Surf. 2017;15(3):575–628.
50. Vicario-de-la-Torre M, Caballo-González M, Vico E, Morales-Fernández L, Arriola-Villalobos P, de las Heras B, et al. Novel Nano-Liposome Formulation for Dry Eyes with Components Similar to the Preocular Tear Film. Polymers (Basel) [Internet]. 2018 Apr 11 [cited 2025 Jan 16];10(4):425. Available from: https://pmc.ncbi.nlm.nih.gov/articles/PMC6415276/
51. Ben BF, Margarint E, Obeid H. Regulation, supervision and European banking performance. Int J Entrep Small Bus. 2022;47(1):1–19.
52. De Luca I, Di Cristo F, Conte R, Peluso G, Cerruti P, Calarco A. In-situ thermoresponsive hydrogel containing resveratrol-loaded nanoparticles as a localized drug delivery platform for dry eye disease. Antioxidants [Internet]. 2023 May 1 [cited 2025 Jan 13];12(5):993. Available from: https://www.mdpi.com/2076-3921/12/5/993/htm
53. Chen X, Xu J, Chen X, Yao K. Cataract: advances in surgery and whether surgery remains the only treatment in future. Adv Ophthalmol Pract Res. 2021;1(1):100008.
54. Thrimawithana TR, Rupenthal ID, Räsch SS, Lim JC, Morton JD, Bunt CR. Drug delivery to the lens for the management of cataracts. Adv Drug Deliv Rev. 2018;126:185–94.
55. Zhang J, Wang S. Topical use of coenzyme Q10-loaded liposomes coated with trimethyl chitosan: tolerance, precorneal retention and anti-cataract effect. Int J Pharm. 2009;372(1–2):66–75.
56. Vora D, Heruye S, Kumari D, Opere C, Chauhan H. Preparation, characterization and antioxidant evaluation of poorly soluble polyphenol-loaded nanoparticles for cataract Treatment. AAPS PharmSciTech [Internet] 2019 Jul 1 [cited 2025 Jan 16];20(5). Available from: https://pubmed.ncbi.nlm.nih.gov/30993475/
57. Chen YY, Chen CK, Wu TT, Ho CY, Yeh TC, Sun GC, et al. Attenuation of epithelial-mesenchymal transition via SGLT2 inhibition and diabetic cataract suppression by dapagliflozin nanoparticles treatment. Life Sci. 2023;330:122005.
58. Stein JD, Khawaja AP, Weizer JS. Glaucoma in adults—screening, diagnosis, and management: a review. JAMA [Internet] 2021 Jan 12 [cited 2025 Jan 16];325(2):164–174. Available from: https://jamanetwork.com/journals/jama/fullarticle/2774838
59. Tham YC, Li X, Wong TY, Quigley HA, Aung T, Cheng CY. Global prevalence of glaucoma and projections of glaucoma burden through 2040: a systematic review and meta-analysis. Ophthalmology. 2014;121(11):2081–90.
60. Pardue MT, Allen RS. Neuroprotective strategies for retinal disease. Prog Retin Eye Res. 2018;65:50–76.
61. Satyanarayana SD, Abu Lila AS, Moin A, Moglad EH, Khafagy ES, Alotaibi HF, et al. Ocular delivery of bimatoprost-loaded solid lipid nanoparticles for effective management of glaucoma. Pharmaceuticals [Internet] 2023 Jul 1 [cited 2025 Jan 13];16(7):1001. Available from: https://www.mdpi.com/1424-8247/16/7/1001/htm
62. Patil RK, Srivastava V, Bhawale R, Tryphena KP, Khatri DK, Doijad N, et al. Revamping the corneal permeability and antiglaucoma therapeutic potential of brinzolamide using transniosomes: optimization, in vitro and preclinical evaluation. Nanomedicine [Internet] 2024 [cited 2025 Jan 21];19(11):947–964. Available from: https://www.tandfonline.com/doi/abs/10.2217/nnm-2023-0280

63. Nair VS, Srivastava V, Bhavana V, Yadav R, Rajana N, Singh SB, et al. Exploring penetration ability of carbonic anhydrase inhibitor–loaded ultradeformable bilosome for effective ocular application. AAPS PharmSciTech [Internet] 2023 Aug 1 [cited 2025 Jan 21];24(6):1–15. Available from: https://link.springer.com/article/10.1208/s12249-023-02617-5
64. Srivastava V, Patil RK, Mehra NK. A one-platform comparison study of brinzolamide-loaded liposomes, niosomes, transfersomes, and transniosomes for better management of glaucoma. Int J Pharm. 2024;666:124741.
65. Leclercq M, Desbois AC, Domont F, Maalouf G, Touhami S, Cacoub P, et al. Biotherapies in Uveitis. Journal of Clinical Medicine 2020, Vol 9, Page 3599 [Internet]. 2020 Nov 8 [cited 2025 Jan 16];9(11):3599. Available from: https://www.mdpi.com/2077-0383/9/11/3599/htm
66. Durrani OM, Tehrani NN, Marr JE, Moradi P, Stavrou P, Murray PI. Degree, duration, and causes of visual loss in uveitis. Br J Ophthalmol [Internet] 2004 Sept [cited 2025 Jan 16];88(9):1159–1162. Available from: https://pubmed.ncbi.nlm.nih.gov/15317708/
67. Tsirouki T, Dastiridou A, Symeonidis C, Tounakaki O, Brazitikou I, Kalogeropoulos C, et al. A focus on the epidemiology of uveitis. OculImmunolInflamm [Internet] 2018 Jan 2 [cited 2025 Jan 16];26(1):2–16. Available from: https://www.tandfonline.com/doi/abs/10.1080/09273948.2016.1196713
68. Ratay ML, Bellotti E, Gottardi R, Little SR. Modern Therapeutic Approaches for Noninfectious Ocular Diseases Involving Inflammation. Adv Healthc Mater [Internet]. 2017 Dec 6 [cited 2025 Jan 16];6(23). Available from: https://pubmed.ncbi.nlm.nih.gov/29034584/
69. Huang C, Zhang Z, Gu J, Li D, Gao S, Zhang R, et al. Combined therapy of experimental autoimmune uveitis by a dual-drug nanocomposite formulation with berberine and dexamethasone. Int J Nanomedicine [Internet]. 2023 [cited 2025 Jan 15];18:4347–63. Available from: https://www.tandfonline.com/action/journalInformation?journalCode=dijn20
70. Chai P, Jia R, Li Y, Zhou C, Gu X, Yang L, et al. Regulation of epigenetic homeostasis in uveal melanoma and retinoblastoma. Prog Retin Eye Res. 2022;89:101030.
71. Chattopadhyay C, Kim DW, Gombos DS, Oba J, Qin Y, Williams MD, et al. Uveal melanoma: from diagnosis to treatment and the science in between. Cancer [Internet]. 2016 Aug 1 [cited 2025 Jan 17];122(15):2299–312. Available from: https://onlinelibrary.wiley.com/doi/full/10.1002/cncr.29727
72. Cimino C, Sánchez López E, Bonaccorso A, Bonilla L, Musumeci T, Badia J, et al. In vitro and in vivo studies of ocular topically administered NLC for the treatment of uveal melanoma. Int J Pharm. 2024;660:124300.
73. Farhat W, Yeung V, Ross A, Kahale F, Boychev N, Kuang L, et al. Advances in biomaterials for the treatment of retinoblastoma. BiomaterSci [Internet] 2022 Sept 27 [cited 2025 Jan 17];10(19):5391–5429. Available from: https://pubs.rsc.org/en/content/articlehtml/2022/bm/d2bm01005d
74. Fabian ID, Onadim Z, Karaa E, Duncan C, Chowdhury T, Scheimberg I, et al. The management of retinoblastoma. Oncogene 2018 37:12 [Internet]. 2018 Jan 11 [cited 2025 Jan 17];37(12):1551–60. Available from: https://www.nature.com/articles/s41388-017-0050-x
75. Wang M, Yang Q, Li M, Zou H, Wang Z, Ran H, et al. Multifunctional nanoparticles for multimodal imaging-guided low-intensity focused ultrasound/immunosynergistic retinoblastoma therapy. ACS Appl Mater Interfaces [Internet]. 2020 Feb 5 [cited 2025 Jan 16];12(5):5642–57. Available from: https://pubs.acs.org/doi/full/10.1021/acsami.9b22072
76. del Amo EM, Rimpelä AK, Heikkinen E, Kari OK, Ramsay E, Lajunen T, et al. Pharmacokinetic aspects of retinal drug delivery. Prog Retin Eye Res. 2017;57:134–85.
77. Munroe WP, Rindone JP, Kershner RM, D'Arcy PF. Systemic side effects associated with the ophthalmic administration of timolol. Drug IntellClin Pharm. 1985;19(2):85–9.
78. Gote V, Sikder S, Sicotte J, Pal D. Ocular drug delivery: present innovations and future challenges. J PharmacolExpTher. 2019;370(3):602–24.
79. Cholkar K, Vadlapudi AD, Dasari S, & Mitra A. Ocular drug delivery. 2014;219–63.
80. Trivedi R, Kompella UB. Nanomicellar formulations for sustained drug delivery: strategies and underlying principles. Nanomedicine. 2010;5(3):485–505.

81. Gorantla S, Rapalli VK, Waghule T, Singh PP, Dubey SK, Saha RN, et al. Nanocarriers for ocular drug delivery: current status and translational opportunity. RSC Adv. 2020;10(46):27835–55.
82. Yu Y, Chen D, Li Y, Yang W, Tu J, Shen Y. Improving the topical ocular pharmacokinetics of lyophilized cyclosporine A-loaded micelles: formulation, *in vitro* and *in vivo* studies. Drug Deliv. 2018;25(1):888–99.
83. Terreni E, Zucchetti E, Tampucci S, Burgalassi S, Monti D, Chetoni P. Combination of Nanomicellar technology and in situ gelling polymer as ocular drug delivery system (ODDS) for cyclosporine-A. Pharmaceutics. 2021;13(2):192.
84. Wei Q, Zhu C, Yuan G, Jin J, Zhang J, Fan W, et al. Active trans-corneal drug delivery with ocular adhesive micelles for efficient glaucoma therapy. J Control Release. 2025;377:578–90.
85. Jacob S, Nair AB, Shah J. Emerging role of nanosuspensions in drug delivery systems. Biomater Res. 2020;24(1):3.
86. Boddeda B, Boddu P, Avasarala H, Jayanti V. Design and ocular tolerance of flurbiprofen loaded nanosuspension. Pharm Nanotechnol. 2015;3(1):56–67.
87. Shi S, Zhang Z, Luo Z, Yu J, Liang R, Li X, et al. Chitosan grafted methoxy poly(ethylene glycol)-poly(ε-caprolactone) nanosuspension for ocular delivery of hydrophobic diclofenac. Sci Rep. 2015;5(1):11337.
88. Sharma UK, Verma A, Prajapati SK, Pandey H, Pandey AC. In vitro, in vivo and pharmacokinetic assessment of amikacin sulphate laden polymeric nanoparticles meant for controlled ocular drug delivery. Appl Nanosci. 2015;5(2):143–55.
89. Josyula A, Omiadze R, Parikh K, Kanvinde P, Appell MB, Patel P, et al. An ion-paired moxifloxacin nanosuspension eye drop provides improved prevention and treatment of ocular infection. BioengTransl Med. 2021;6(3)
90. Ma X, Liu Y, Wang J, Liu H, Wei G, Lu W, et al. Combination of PEGylation and cationization on phospholipid-coated cyclosporine nanosuspensions for enhanced ocular drug delivery. ACS Appl Mater Interfaces. 2024;16(21):27040–54.
91. Ghosh P, Murthy R. Microemulsions: a potential drug delivery system. Curr Drug Deliv. 2006;3(2):167–80.
92. Kai Seng K, Voon LW. Introductory chapter: from microemulsions to nanoemulsions. In: Nanoemulsions – properties, fabrications and applications. IntechOpen; 2019.
93. Gawin-Mikołajewicz A, Nartowski KP, Dyba AJ, Gołkowska AM, Malec K, Karolewicz B. Ophthalmic nanoemulsions: from composition to technological processes and quality control. Mol Pharm. 2021;18(10):3719–40.
94. Lidich N, Garti-Levy S, Aserin A, Garti N. Potentiality of microemulsion systems in treatment of ophthalmic disorders: keratoconus and dry eye syndrome—in vivo study. Colloids Surf B Biointerfaces. 2019;173:226–32.
95. Xu B, Liu T. Travoprost loaded microemulsion soaked contact lenses: improved drug uptake, release kinetics and physical properties. J Drug Deliv Sci Technol. 2020;57:101792.
96. Srivastava V, Pardhi ER, Yadav R, Singh V, Khatri DK, Mehra NK. QbD-driven thymoquinone laden nanoemulsion for glaucoma management: in vitro, ex vivo, and pre-clinical evaluation. J Drug Deliv Sci Technol. 2024;94:105493.
97. Almeida H, Amaral MH, Lobão P, Silva AC, Loboa JMS. Applications of polymeric and lipid nanoparticles in ophthalmic pharmaceutical formulations: present and future considerations. J Pharm Pharm Sci. 2014;17(3):278.
98. Onugwu AL, Attama AA, Nnamani PO, Onugwu SO, Onuigbo EB, Khutoryanskiy VV. Development and optimization of solid lipid nanoparticles coated with chitosan and poly(2-ethyl-2-oxazoline) for ocular drug delivery of ciprofloxacin. J Drug Deliv Sci Technol. 2022;74:103527.
99. Liu D, Lian Y, Fang Q, Liu L, Zhang J, Li J. Hyaluronic-acid-modified lipid-polymer hybrid nanoparticles as an efficient ocular delivery platform for moxifloxacin hydrochloride. Int J BiolMacromol. 2018;116:1026–36.
100. Dave V, Tak K, Sohgaura A, Gupta A, Sadhu V, Reddy KR. Lipid-polymer hybrid nanoparticles: synthesis strategies and biomedical applications. J Microbiol Methods. 2019;160:130–42.

101. Sánchez-López E, Espina M, Doktorovova S, Souto EB, García ML. Lipid nanoparticles (SLN, NLC): overcoming the anatomical and physiological barriers of the eye—Part I—Barriers and determining factors in ocular delivery. Eur J Pharm Biopharm. 2017 Jan;110:70–5.
102. Yu F, Zheng M, Zhang AY, Han Z. A cerium oxide loaded glycol chitosan nano-system for the treatment of dry eye disease. J Control Release. 2019;315:40–54.
103. Tatke A, Dudhipala N, Janga KY, Balguri SP, Avula B, Jablonski MM, et al. In situ gel of triamcinolone acetonide-loaded solid lipid nanoparticles for improved topical ocular delivery: tear kinetics and ocular disposition studies. Nano. 2018;9(1):33.
104. Akbarzadeh A, Rezaei-Sadabady R, Davaran S, Joo SW, Zarghami N, Hanifehpour Y, et al. Liposome: classification, preparation, and applications. Nanoscale Res Lett. 2013;8(1):102.
105. Kaur IP, Garg A, Singla AK, Aggarwal D. Vesicular systems in ocular drug delivery: an overview. Int J Pharm. 2004;269(1):1–14.
106. Hathout RM, Gad HA, Abdel-Hafez SM, Nasser N, Khalil N, Ateyya T, et al. Gelatinized core liposomes: a new Trojan horse for the development of a novel timolol maleate glaucoma medication. Int J Pharm. 2019;556:192–9.
107. Ren T, Lin X, Zhang Q, You D, Liu X, Tao X, et al. Encapsulation of azithromycin ion pair in liposome for enhancing ocular delivery and therapeutic efficacy on dry eye. Mol Pharm. 2018;15(11):4862–71.
108. Liu Y, Kam WR, Ding J, Sullivan DA. One man's poison is another man's meat: using azithromycin-induced phospholipidosis to promote ocular surface health. Toxicology. 2014;320:1–5.
109. Arita R, Fukuoka S. Efficacy of azithromycin eyedrops for individuals with meibomian gland dysfunction–associated posterior blepharitis. Eye Contact Lens. 2021;47(1):54–9.
110. Ahmed K, Saikat A, Moni A, Kakon S, Islam M, Uddin M. Lactoferrin: potential functions, pharmacological insights, and therapeutic promises. J Adv Biotechnol Exp Ther. 2021;4(2):223.
111. López-Machado A, Díaz-Garrido N, Cano A, Espina M, Badia J, Baldomà L, et al. Development of lactoferrin-loaded liposomes for the management of dry eye disease and ocular inflammation. Pharmaceutics. 2021;13(10):1698.
112. Moiseev RV, Kaldybekov DB, Filippov SK, Radulescu A, Khutoryanskiy VV. Maleimide-decorated PEGylated mucoadhesive liposomes for ocular drug delivery. Langmuir. 2022;38(45):13870–9.
113. Bachu R, Chowdhury P, Al-Saedi Z, Karla P, Boddu S. Ocular drug delivery barriers—role of nanocarriers in the treatment of anterior segment ocular diseases. Pharmaceutics. 2018;10(1):28.
114. Zeng W, Li Q, Wan T, Liu C, Pan W, Wu Z, et al. Hyaluronic acid-coated niosomes facilitate tacrolimus ocular delivery: mucoadhesion, precorneal retention, aqueous humor pharmacokinetics, and transcorneal permeability. Colloids Surf B Biointerfaces. 2016;141:28–35.
115. Eid HM, Naguib IA, Alsantali RI, Alsalahat I, Hegazy AM. Novel chitosan-coated niosomal formulation for improved management of bacterial conjunctivitis: a highly permeable and efficient ocular nanocarrier for azithromycin. J Pharm Sci. 2021;110(8):3027–36.
116. Gugleva V, Titeva S, Rangelov S, Momekova D. Design and in vitro evaluation of doxycycline hyclate niosomes as a potential ocular delivery system. Int J Pharm. 2019;567:118431.
117. Umar H, Wahab HA, Gazzali AM, Tahir H, Ahmad W. Cubosomes: design, development, and tumor-targeted drug delivery applications. Polymers (Basel). 2022;14(15):3118.
118. Gan L, Han S, Shen J, Zhu J, Zhu C, Zhang X, et al. Self-assembled liquid crystalline nanoparticles as a novel ophthalmic delivery system for dexamethasone: improving preocular retention and ocular bioavailability. Int J Pharm. 2010;396(1–2):179–87.
119. Han S, Shen JQ, Gan Y, Geng HM, Zhang XX, Zhu CL, et al. Novel vehicle based on cubosomes for ophthalmic delivery of flurbiprofen with low irritancy and high bioavailability. Acta Pharmacol Sin. 2010;31(8):990–8.
120. El-Gendy MA, Mansour M, El-Assal MIA, Ishak RAH, Mortada ND. Delineating penetration enhancer-enriched liquid crystalline nanostructures as novel platforms for improved ophthalmic delivery. Int J Pharm. 2020;582:119313.

121. Gaballa S, El Garhy O, Abdelkader H. Cubosomes: composition, preparation, and drug delivery applications. J Adv Biomed Pharm Sci. 2019;0(0):0–0.
122. Alharbi WS, Hosny KM. Development and optimization of ocular in situ gels loaded with ciprofloxacin cubic liquid crystalline nanoparticles. J Drug Deliv Sci Technol. 2020;57:101710.
123. Dhakne R, Dehghan MH, Kazi M. Ocular delivery of natamycin based on monoolein/span 80/poloxamer 407 nanocarriers for the effectual treatment of fungal keratitis. J Res Pharm. 2020;24(2):251–63.
124. Wilson SL, Ahearne M, Hopkinson A. An overview of current techniques for ocular toxicity testing. Toxicology. 2015;327:32–46.
125. Rönkkö S, Vellonen KS, Järvinen K, Toropainen E, Urtti A. Human corneal cell culture models for drug toxicity studies. Drug DelivTransl Res. 2016;6(6):660–75.
126. Taylor MM, Kern TJ, Riis RC, McDonough PL, Erb HN. Intraocular bacterial contamination during cataract surgery in dogs. J Am Vet Med Assoc. 1995;206(11):1716–20.
127. Barot M, Bagui MR, Gokulgandhi MK, Mitra A. Prodrug strategies in ocular drug delivery. Med Chem (Los Angeles). 2012;8(4):753–68.
128. Fletcher EL, Jobling AI, Vessey KA, Luu C, Guymer RH, Baird PN. Animal models of retinal disease; 2011. p. 211–86.
129. Lopez NN, Clark AF, Tovar-Vidales T. Isolation and characterization of human optic nerve head astrocytes and lamina cribrosa cells. Exp Eye Res. 2020;197:108103.
130. Wu WC, Kao YH, Hu PS, Chen JH. Geldanamycin, a HSP90 inhibitor, attenuates the hypoxia-induced vascular endothelial growth factor expression in retinal pigment epithelium cells in vitro. Exp Eye Res. 2007;85(5):721–31.
131. Park SA, Komáromy AM. Biomechanics of the optic nerve head and sclera in canine glaucoma: a brief review. Vet Ophthalmol. 2021;24(4):316–25.
132. McCall MA. Pig models in retinal research and retinal disease. Cold Spring Harb Perspect Med. 2024;14(4):a041296.
133. Loiseau A, Raîche-Marcoux G, Maranda C, Bertrand N, Boisselier E. Animal models in eye research: focus on corneal pathologies. Int J Mol Sci. 2023;24(23):16661.
134. Zhang R, Pandzic E, Park M, Wakefield D, Di Girolamo N. Inducing dry eye disease using a custom engineered desiccation system: impact on the ocular surface including keratin-14-positive limbal epithelial stem cells. Ocul Surf. 2021;21:145–59.
135. Adamson P, Wilde T, Dobrzynski E, Sychterz C, Polsky R, Kurali E, et al. Single ocular injection of a sustained-release anti -VEGF delivers 6 months pharmacokinetics and efficacy in a primate laser CNV model. J Control Release. 2016;244:1–13.
136. de Zafra CLZ, Sasseville VG, Matsumoto S, Freichel C, Milton M, MacLachlan TK, et al. Inflammation and immunogenicity limit the utility of the rabbit as a nonclinical species for ocular biologic therapeutics. Regul Toxicol Pharmacol. 2017;86:221–30.
137. Wakshull E, Quarmby V, Mahler HC, Rivers H, Jere D, Ramos M, et al. Advancements in understanding immunogenicity of biotherapeutics in the intraocular space. AAPS J. 2017;19(6):1656–68.
138. Bantseev V, Erickson R, Leipold D, Amaya C, Miller PE, Booler H, et al. Nonclinical safety assessment of anti-factor D: key strategies and challenges for the nonclinical development of intravitreal biologics. J Ocul Pharmacol Ther. 2018;34(1–2):204–13.
139. Lauritsen K, Nguyen T. Combination products regulation at the FDA. Clin Pharmacol Ther. 2009;85(5):468–70.
140. Kaufman HE, Uotila MH, Gasset AR, Wood TO, Ellison ED. The medical uses of soft contact lenses. Trans Am Acad Ophthalmol Otolaryngol. 1971;75(2):361–73.
141. Gebhardt BM, Kaufman HE. Collagen as a delivery system for hydrophobic drugs: studies with cyclosporine. J Ocul Pharmacol Ther. 1995;11(3):319–27.
142. Kaufman HE, Steinemann TL, Lehman E, Thompson HW, Varnell ED, Jacob-Labarre JT, et al. Collagen-based drug delivery and artificial tears. J Ocul Pharmacol Ther. 1994;10(1):17–27.
143. Johnston G, Williams RL. 505(b)(2) applications: history, science, and experience. Drug Inf J. 2002;36(2):319–23.

144. Kaufman B, Novack GD. Compliance issues in manufacturing of drugs. Ocul Surf. 2003;1(2):80–5.
145. MIEBO® (perfluorohexyloctane ophthalmic solution) Dry eye drops [Internet]. [cited 2025 Jan 13]. Available from: https://www.miebo.com/
146. Santen and UBE Received FDA approval for OMLONTI ® (Omidenepag Isopropyl Ophthalmic Solution) 0.002% for the Reduction of Elevated Intraocular Pressure in Patients with Primary Open-Angle Glaucoma or Ocular Hypertension. 2022 [cited 2025 Jan 13]; Available from: https://www.glaucoma.org/glaucoma/glaucoma-facts-and-stats.php
147. DURYSTA® (bimatoprostintracameral implant) | Patient website [Internet]. [cited 2025 Jan 13]. Available from: https://www.durysta.com/
148. EYSUVIS® Eye drops for short-term dry eye treatment [Internet]. [cited 2025 Jan 13]. Available from: https://eysuvis.myalcon.com/
149. ROCKLATAN® Eye Drops | Patient Safety, Savings, Use Information [Internet]. [cited 2025 Jan 13]. Available from: https://rocklatan.myalcon.com/
150. Glaucoma treatment and care | XELPROS® [Internet]. [cited 2025 Jan 13]. Available from: https://www.xelpros.com/
151. CEQUA® for Chronic dry eye | Patient Information [Internet]. [cited 2025 Jan 13]. Available from: https://www.cequa.com/
152. Home – DEXTENZA [Internet]. [cited 2025 Jan 13]. Available from: https://www.dextenza.com/
153. CHMP. Simbrinza, INN-brinzolamide+brimonidine tartrate.
154. Home Page [Internet]. [cited 2025 Jan 22]. Available from: https://www.ozurdex.com/
155. Pharmaceuticals Corporation N. Highlights of prescribing information. [cited 2025 Jan 22.]; Available from: www.fda.gov/medwatch
156. Dry eye drops, gel, and treatment devices | Alcon [Internet]. [cited 2025 Jan 22]. Available from: https://www.myalcon.com/eye-care-products/dry-eye/
157. Highlights of prescribing information full prescribing information: contents* 1 indications and usage 2 dosage and administration 3 dosage forms and strengths 4 contraindications 5 warnings and precautions 5.1 potential for eye injury and contamination 5.2 use with contact lenses. [cited 2025 Jan 22.]; Available from: www.fda.gov/medwatch
158. High Eye Pressure/Intraocular Pressure (IOP) | XALATAN® (latanoprost ophthalmic solution) | Safety [Internet]. [cited 2025 Jan 22]. Available from: https://www.xalatan.com/
159. CN118846131A – Preparation method of griseofulvin eye drops – Google Patents [Internet]. [cited 2025 Jan 13]. Available from: https://patents.google.com/patent/CN118846131A/en?oq=CN202411364311A
160. CN118859556A – Corneal contact lens and drug slow release method – Google Patents [Internet]. [cited 2025 Jan 13]. Available from: https://patents.google.com/patent/CN118859556A/en?oq=CN202411311408A
161. CN118831136A – Nanometer emulsion eye cream containing traditional Chinese medicine compound extract and preparation method thereof – Google Patents [Internet]. [cited 2025 Jan 13]. Available from: https://patents.google.com/patent/CN118831136A/en?oq=CN202411111197A
162. CN118512302A – Sustained release drug delivery system for eyes – Google Patents [Internet]. [cited 2025 Jan 13]. Available from: https://patents.google.com/patent/CN118512302A/en?oq=CN202410988984A
163. CN118436596A – S- (+) -flurbiprofen tromethamine salt eye drops and preparation method and application thereof – Google Patents [Internet]. [cited 2025 Jan 13]. Available from: https://patents.google.com/patent/CN118436596A/en?oq=CN202410539349A
164. CN118059044A – Drug-loaded nano micelle based on hyaluronic acid and preparation method and application thereof – Google Patents [Internet]. [cited 2025 Jan 13]. Available from: https://patents.google.com/patent/CN118059044A/en?oq=CN202410494091A
165. CN117643573A – Nanometer eye drop capable of delivering medicine to posterior segment of eye, preparation method and application thereof – Google Patents [Internet]. [cited 2025 Jan 13]. Available from: https://patents.google.com/patent/CN117643573A/en?oq=CN202410115769A

166. Pharmacoeconomic Review Report: Dexamethasone (Ozurdex). Pharmacoeconomic Review Report: Dexamethasone (Ozurdex): (Allergan Inc): Indication: For the treatment of adult patients with diabetic macular edema who are pseudophakic [Internet]. 2018 [cited 2025 Jan 13.]; Available from: https://www.ncbi.nlm.nih.gov/books/NBK539204/
167. Vora LK, Gholap AD, Jetha K, Thakur RRS, Solanki HK, Chavda VP. Artificial intelligence in pharmaceutical technology and drug delivery design. Pharmaceutics. 2023;15(7):1916.
168. Puccetti M, Pariano M, Schoubben A, Giovagnoli S, Ricci M. Biologics, theranostics, and personalized medicine in drug delivery systems. Pharmacol Res. 2024;201:107086.
169. Singh AV, Chandrasekar V, Paudel N, Laux P, Luch A, Gemmati D, et al. Integrative toxicogenomics: advancing precision medicine and toxicology through artificial intelligence and OMICs technology. Biomed Pharmacother. 2023;163:114784.

Eye: Anatomy, Physiology, and Disease

2

Jilu Jaffet, Tejaswini Pingali, Arun Kumar Raut,
Sonali Mohapatra, and Vivek Singh

Abstract

The human eye is a complex organ responsible for the sense of vision, allowing the perception of light, color, and depth. It is composed of multiple structures that work in concert to capture light, convert it into neural signals, and transmit these signals to the brain for interpretation. The eye being a specialized organ, the homeostasis is maintained by integration of the anatomical features with various physiological processes within each part of the eye. In this chapter, the eye is explored in terms of its key components, including the cornea, lens, retina, optic nerve, lacrimal functional unit, and supporting muscles, its structural and functional importance, and how they work together to enable vision. Understanding ocular anatomy and physiology is critical for the effective management and prevention of ocular diseases, underscoring the importance of eye health in overall well-being.

Authors Jilu Jaffet and Tejaswini Pingali have equally contributed to this chapter.

J. Jaffet · T. Pingali
Centre for Ocular Regeneration (CORE), Prof. Brien Holden Eye Research Centre (BHERC), LV Prasad Eye Institute, Hyderabad, India

Manipal Academy of Higher Education (MAHE), Manipal, Karnataka, India

A. K. Raut · S. Mohapatra · V. Singh (✉)
Centre for Ocular Regeneration (CORE), Prof. Brien Holden Eye Research Centre (BHERC), LV Prasad Eye Institute, Hyderabad, India
e-mail: viveksingh@lvpei.org

N. K. Mehra (ed.), *Complex Ophthalmic Dosage Forms: Advances in Biomedical Applications and Future Perspectives*, https://doi.org/10.1007/978-981-96-6306-4_2

Keywords

Tear film · Cornea · Limbus · Conjunctiva · Sclera · Trabecular meshwork · Aqueous humor · Uvea · Choroid · Lens · Vitreous humor · Retina · Optic nerve · Lacrimal gland · Meibomian gland · Ocular muscles

2.1 Introduction: Eye

Eye is a complex and highly specialized organ to capture and process visual information, thus enabling to interact with the surrounding environment. It is composed of several anatomical structures and functional layers, each contributing to vision and ocular health. The human eye consists of three coats or layers namely the fibrous, vascular, and neural coat. The eye is structurally divided into anterior and posterior segments primarily consisting of cornea, lens, aqueous and vitreous humors, retina, and optic nerve. These structures work together to refract, focus, and detect light before transmitting signals to the brain. Supporting systems, including the lacrimal functional unit, extraocular muscles, and ocular appendages like the eyelids ensure protection, lubrication, and precise movement. The optical power of the relaxed human eye is approximately 60 Diopters (D). The cornea and lens contribute about 70% and 30% to the refractive power of the eye respectively, focusing light onto the retina. The retina, a ten-layered neural tissue, translates light into electrical signals through photoreceptor cells, which are transmitted via the optic nerve to the visual cortex. Additionally, vascularized uvea and protective sclera maintain structural integrity and physiological balance in the eye. A delicate interplay between these structures maintains homeostasis, ensuring visual clarity while protecting against environmental and pathological challenges. This synergy between components preserves visual clarity and ocular health, emphasizing the ocular complexity and clinical importance. Disorders affecting any component of the eye can compromise vision, highlighting the importance of understanding ocular anatomy and physiology for clinical advancements. This chapter divides the anatomy of the eye into anterior and posterior segments which will be discussed in detail.

2.2 Anterior Segment

2.2.1 Cornea

The human cornea is an avascular, biologically well-engineered structure characterized by transparency and resilience. The anterior part of the eye plays an important role in light transmission and refraction accounting for 65–75% of the total refractive power of the eye [14]. Architecturally, the main structural protein of the cornea is collagen which forms a complex fibril network that maintains the transparency and mechanical strength of the cornea. Proteoglycans, such as decorin and lumican, play a crucial role in regulating fibril spacing and hydration, ensuring corneal transparency and biomechanical stability. The cornea is of approximately 550–640 μm thickness which decreases with age. It is made up of both acellular and cellular components, where the acellular components include collagen and glycosaminoglycans, whereas the cellular components include epithelial cells, keratocytes, and endothelial cells. The epithelial cells are derived from epidermal ectoderm whereas the keratocytes and endothelial cells are derived from neural crest. These components actively play an important role in maintaining corneal homeostasis and repair [8] (see Figs. 2.1 and 2.2d). Structurally, the cornea comprises of five distinct layers, the outermost epithelium, Bowman's layer, stroma, Descemet's membrane, and the innermost endothelium [17].

2.2.2 Epithelium

The corneal epithelium is about 50 μm in thickness and composed of five to seven layers with three different cell types namely the basal cells (single layer), wing cells (—two to three layers), and superficial cells (—two to three layers). It serves as a protective barrier preventing the entry of dust, pathogens, and other harmful particles. It also interacts with the external environment and underlying stroma to maintain corneal transparency. Epithelium is self-renewing, with superficial cells being shed into the tear film and replaced by basal cells undergoing mitosis which are

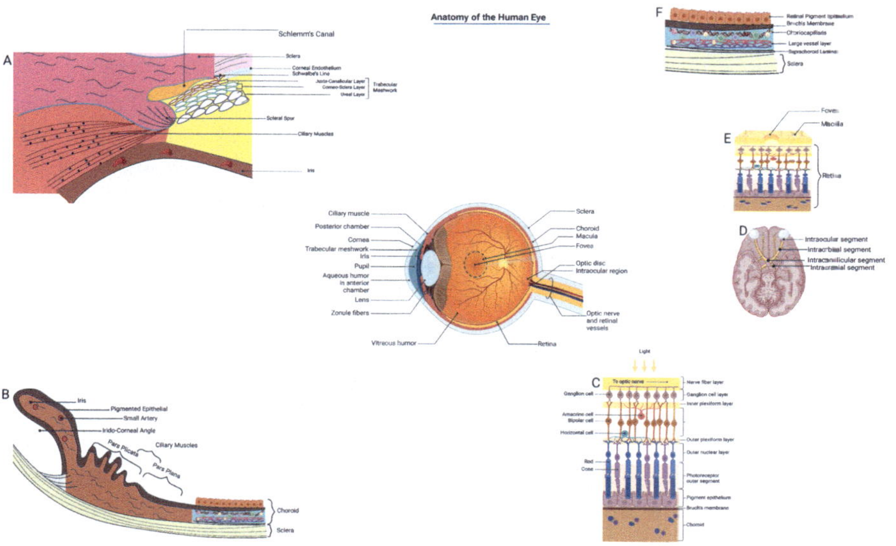

Fig. 2.1 Schematic illustrating the **whole human eye anatomy** highlighting the structural details of the (**a**) trabecular meshwork, (**b**) uveal tract of the anterior segment, (**c**) retinal layers, (**d**) optic nerve, (**e**) fovea and macula, and (**f**) choroid from the posterior segment of the eye. (Created with biorender.com licensed version)

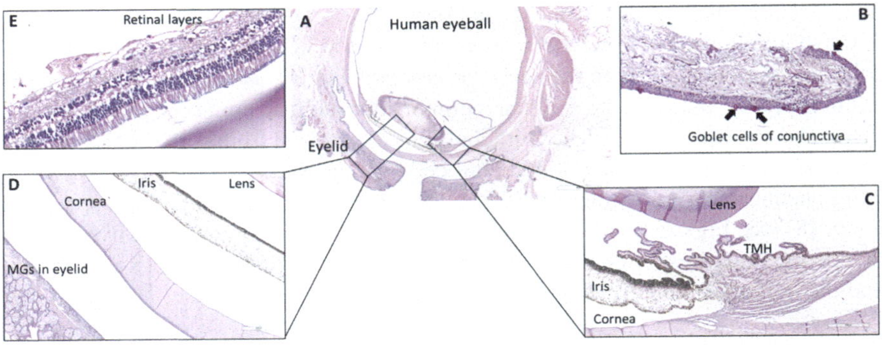

Fig. 2.2 Hematoxylin and eosin staining of (**a**) lateral view of the **whole human eye** highlighting the structural details of the (**b**) goblet cells of conjunctiva, (**c**) trabecular meshwork (TMH), iris, and cornea, (**d**) meibomian glands (MGs) in the eyelid, and (**e**) retinal layers

continuously regenerated from limbal stem cells. The corneal epithelium acts as a major barrier to drug delivery, necessitating permeation-enhancing strategies for effective drug treatment. Lipophilic drugs penetrate efficiently due to the lipid-rich outermost layer. Prodrug modifications, iontophoresis, and microemulsions are often employed to enhance the absorption significantly.

2.2.3 Bowman's Layer

Bowman's layer is a thin, acellular, and collagenous zone of about 8–12 μm located beneath the epithelial basement membrane. This layer is composed of densely packed collagen fibrils, which provide structural support to the cornea but do not regenerate if damaged. The exact function of Bowman's layer is still a topic of research, but it is believed to contribute to the biomechanical stability of the cornea. Damage to this layer, as seen in various corneal diseases, can lead to scarring and vision impairment.

2.2.4 Stroma

Corneal stroma constitutes approximately 90% of the corneal thickness (450–500 μm) and is composed of collagen fibrils arranged in a highly organized, lamellar structure. The stroma is rich in collagen type I and V, which are essential for maintaining its transparency and structural integrity. The stroma also contains keratocytes, which are specialized cells responsible for maintaining the extracellular matrix and facilitating wound healing. These primary cells of the stroma produce and organize these collagen fibrils. The average diameter of the collagen fibril is around 30 nm which is smaller than the wavelength of the light, hence making the cornea more transparent. The precise "orthogonal" arrangement of collagen fibrils is critical for maintaining corneal transparency, as it minimizes light scattering. Any disruption to the orderly arrangement of the collagen fibrils can result in reduced visual acuity. The stromal layer has a limited ability to regenerate, which pose challenge in wound healing and repair after injury or surgery.

2.2.5 Descemet's Membrane

Descemet's membrane is a thin but resilient layer that lies between the stroma and the endothelium. It measures about 3–4 μm in thickness during birth and thickens with age to about 10–12 μm during adulthood. It serves as the basement membrane for the endothelial cells and provides additional structural support to the cornea. Descemet's membrane is unique in its ability to regenerate, although significant damage can still compromise corneal function. This structure plays a key role in protecting the cornea from injury and infection.

2.2.6 Endothelium

The endothelium is the innermost layer of the cornea and about 5 μm in thickness. It consists of a single layer of hexagonal cells which are critical for maintaining corneal transparency by regulating the fluid balance within the stroma. They actively pump excess water out of the corneal stroma, which is achieved by sodium-potassium pumps (Na^+/K^+-ATPase), which actively transport sodium ions out and potassium ions inside the cells thus preventing swelling and clouding. Unlike the epithelium, endothelial cells have very limited regenerative capacity, and damage to this layer can lead to corneal edema and loss of vision. Fuchs' endothelial dystrophy is a well-known condition that affects this layer, often necessitating corneal transplantation.

A relatively recent discovery in corneal anatomy is the Dua's layer, which was first described by Harminder Dua in 2013. It is a thin, acellular, collagenous layer located between the stroma and Descemet's membrane with a thickness of about 10–15 μm. Despite its thinness, Dua's layer is believed to play a critical role in maintaining corneal integrity, particularly in conditions involving the posterior part of the cornea, such as corneal hydrops and Descemet's membrane detachments. Identification of this layer has enhanced the understanding of the corneal structure, influencing surgical approaches in procedures like deep anterior lamellar keratoplasty (DALK) by providing an additional landmark and layer of support during dissection [6].

2.3 Limbus

The limbus is the narrow transitional zone between the cornea and the conjunctiva, typically measuring around 1–2 mm in width. It plays a critical role in ocular health by housing limbal stem cells (LSCs), which are responsible for the continuous regeneration of the corneal epithelium. This region serves as a structural and functional boundary that maintains the integrity of the corneal surface while preventing the conjunctival epithelium from growing into the corneal surface. The limbus is composed of both epithelial and stromal components. Its structure is characterized by an undulating, ridged appearance known as the "Palisades of Vogt," which provides a unique niche for limbal epithelial stem cells. These palisades are rich in blood vessels, nerves, and extracellular matrix (ECM) components, forming a microenvironment that supports the survival, proliferation, and migration of LSCs. LSCs reside in specialized niches in the basal layer of the limbal epithelium. These cells are undifferentiated, highly mitotic, and play a vital role in maintaining corneal homeostasis. Upon receiving appropriate signals from the surrounding stromal niche, LSCs proliferate, differentiate, and migrate centripetally toward the central cornea, where they replace aged or damaged epithelial cells. The integrity of these stem cells is critical for the long-term health and transparency of the cornea. The stroma of the limbus contains a dense network of collagen fibers, fibroblasts, and immune cells, which provides structural support and protects the stem cell niche from environmental stress, such as UV radiation and mechanical injury. The blood vessels in limbal stroma also provide nutrients and oxygen to the LSCs, facilitating their ability to regenerate the corneal epithelium.

Limbal stem cell deficiency (LSCD) occurs when the LSC population is depleted or dysfunctional, compromising the regenerative capacity of the corneal epithelium. Without adequate LSCs, the corneal epithelium cannot renew itself, leading to a process called "conjunctivalization" of the cornea. This results in vascularization, scarring, and a loss of corneal transparency, which severely impairs vision. LSCD can result from a variety of causes, including chemical burns, ocular trauma, and various inflammatory diseases. Drug delivery systems such as contact lenses and polymer-based films embedded with drugs *(such as Cyclosporin A)* are being utilized for the treatment of LSCD.

2.4 Conjunctiva

The conjunctiva is a thin, transparent mucous membrane that covers the sclera (bulbar conjunctiva) and lines the inside of the eyelids (palpebral conjunctiva), playing a critical role in ocular surface homeostasis. This tissue is highly vascularized and contains numerous immune cells, including lymphocytes and macrophages, contributing to its immune surveillance function. Structurally, the conjunctiva consists of an epithelial layer and an underlying stromal layer, which contains blood vessels, lymphatics, and immune cells. The conjunctival epithelium includes goblet cells, which secrete mucins-glycoproteins that are essential for the formation of the tear film thereby maintaining tear film stability and ocular surface lubrication. Disruptions in this mucin production, as seen in conditions such as dry eye disease, can lead to ocular surface damage and discomfort. The conjunctiva is crucial in providing a smooth ocular surface during blinking and eye movements and in responding to foreign antigens through immune responses (see Figs. 2.1 and 2.2b).

The conjunctiva is also a site for several pathological conditions, including conjunctivitis, pterygium, and neoplasms. Conjunctivitis, or inflammation of the conjunctiva, can be caused by infections (bacterial, viral, or fungal), allergens, or irritants. Pterygium is a benign, fibrovascular growth that extends onto the cornea, potentially leading to visual impairment if left untreated. Surgical excision is typically required in advanced cases, and new research is focused on using anti-angiogenic therapies to prevent recurrence [9]. The conjunctiva facilitates the systemic absorption of topically administered drugs such as eye drops, gels, and inserts to deliver anti-inflammatory or immunomodulatory drugs. The mucous membrane also supports liposomal and nanoparticle formulations for sustained drug release.

2.5 Sclera

The sclera is the opaque, dense, and fibrous outer layer of the eye that provides structural support and protection to the inner ocular components. It covers almost 80% of the eye's surface extending from the cornea to the optic nerve posteriorly and is composed predominantly of irregularly arranged collagen fibers, which give it strength and rigidity. Unlike the cornea, the collagen fibrils of the sclera are randomly oriented, leading to its opaque appearance. The primary role of the sclera is to maintain the shape of the eye ball and to serve as an attachment point for the extraocular muscles, which facilitate eye movement. The thickness of the sclera varies across different regions, being thickest at the posterior region and thinnest at the central region, which correlates with its mechanical functions. It also protects the eye from external injury and contributes to maintaining intraocular pressure. The sclera is less cellular and vascularized compared to other ocular tissues, making it more resistant to infection but also slower to heal after injury [26].

Scleral diseases include scleritis, an inflammatory condition that is often associated with systemic autoimmune disorders such as rheumatoid arthritis. Scleritis can cause severe pain and visual impairment if left untreated. Another condition, staphyloma, occurs when the sclera becomes thinner and bulges outward, often as a result of high intraocular pressure in diseases such as glaucoma. The fibrous composition of the sclera serves as a gateway for posterior segment drug delivery via transscleral routes, which includes sub-tenon's injection, scleral plugs, and microneedle arrays delivering directly to vitreous or retina. Recent research in scleral tissue engineering has explored the development of synthetic materials that mimic the biomechanical properties of the sclera. These materials could be used to repair scleral defects or reinforce thinning scleral tissue in conditions like high myopia [2] (see Fig. 2.1).

2.6 Trabecular Meshwork

Sclera has a wedge-shaped, fibrous structure composed of collagen I and III termed scleral spur which serves as the attachment site for trabecular meshwork (TMH). TMH can be anatomically divided into three regions: (inner to outer) the uveal meshwork, the corneoscleral meshwork, and the cribriform meshwork (see Figs. 2.1a and 2.2c) [7, 11].

2.6.1 The Uveal Meshwork

The uveal meshwork is the innermost layer of TMH adjacent to the anterior chamber, made up of —one to three layers of lamellar sheets. The lamellae of uveal and corneoscleral meshwork ultra-structurally have densely packed collagen type I, III, and elastin fibers, providing resistance to tension and elasticity. This is followed by a layer composed of long-spacing collagens (III, IV, V, VI, and VII), heparin sulfate, laminin, and fibronectin which are surrounded by a continuous endothelial sheath maintaining the structural integrity [11].

2.6.2 The Corneoscleral Meshwork

The corneoscleral meshwork is the thickest of all TMH layers, with each lamellar sheet originating from the scleral spur and attached to juxtacanalicular tissue through desmosomes. The spaces between the cytoplasmic processes are bridged by phagocytic trabecular cells. The lamellar sheets possess elliptical trans trabecular perforations, the size of which progressively reduces from the uveal meshwork layer to corneoscleral meshwork, and serve as the passage for the aqueous humor [11].

2.6.3 The Cribriform Meshwork

The cribriform meshwork is the smallest TMH layer located adjacent to inner endothelial lining, also known as juxtacanalicular tissue. It differs from other regions, in lacking the lamellar arrangement and having star-shaped trabecular cells embedded in a loose connective extracellular matrix. The composition of the extracellular matrix is defined by collagens (I, III, and IV), fibronectin, elastic fibers, and a variety of glycosaminoglycans including heparin sulfate, hyaluronic acid, dermatan sulfate, chondroitin sulfate, and keratan sulfate. The elastic fibers form the characteristic cribriform plexus and the loose fibrillar extracellular matrix gives open spaces, crucial to define the degree of resistance to the aqueous humor [11].

Trabecular meshwork forms the blood-aqueous barrier and critically resists the entry of various drugs into the eye; thus, the majority of the lipophilic drugs are administered topically for ease of delivery into the TMH [4]. In case of intravitreal and intracameral injections, the drugs get cleared from the anterior chamber through the conventional outflow pathways of the aqueous humor, while preventing reflux of blood and other potential toxic substances back into the anterior chamber [4]. Irrespective of the flow rate of aqueous humor and the contraction frequency of ciliary muscle, maintenance of intraocular pressure (IOP) is one of the most critical roles of TMH.

Any obstruction to the aqueous humor flow leads to increase in IOP and causes glaucoma, more commonly primary open-angle glaucoma (POAG). The increase in outflow resistance in POAG could be a result of certain morphological changes in TMH such as increase in thickness of extracellular matrix or trabecular sheets, formation of fibrous deposits in outflow route, and decrease in trabecular cell density. Steroid-induced glaucoma is commonly reported secondary open-angle glaucoma in individuals exposed to steroids (glucocorticoids, adrenocorticoids, or cortisone). Steroids administered through topical, intraocular, and periocular routes have shown prominent elevation of IOP compared to systemic routes. The presence of steroid receptors on the surface of TMH cells increases the susceptibility to alterations in their structure, migration, phagocytic nature, and composition of extracellular matrix, thereby leading to an increase in resistance of aqueous humor outflow in this condition [25].

2.7 Aqueous Humor

The aqueous humor is the fluid that fills the anterior chamber of the eye, the space between the cornea and the front surface of the crystalline lens (see Fig. 2.1). It is produced by the ciliary epithelium, and it reaches the anterior chamber through the crystalline lens and the iris, passing through the pupil. Aqueous humor is essentially a modified blood plasma, containing water, electrolytes, and nutrients that support the avascular cornea and lens. Its refractive index is approximately 1.333, slightly lower than that of the cornea (1.376) and the crystalline lens (1.406 to 1.386, depending on the region of the lens). This difference in refractive indices, along with the curvature of optical surfaces like the cornea and lens, contributes to the bending of light, facilitating proper focus on the retina [24].

The flow of aqueous humor serves two essential functions:

- **Nutrient Delivery and Waste Removal**: The aqueous humor serves as a surrogate vascular system for the cornea and lens, delivering essential nutrients like oxygen and glucose while removing metabolic waste. The endothelial layer of the cornea facilitates the diffusion of these nutrients from the aqueous humor. Since the cornea and lens lack blood vessels, this fluid is vital for maintaining their transparency and functionality.
- **Intraocular Pressure Regulation**: Aqueous humor also plays a key role in maintaining the IOP, which is essential for preserving the shape and optical integrity of the globe. As aqueous humor circulates, it exerts pressure within the eye, typically around 15 mmHg. This pressure is essential for supporting the

overall structure of the eye and ensuring the proper function of other ocular components. Excessive resistance or overproduction of aqueous humor can cause elevated IOP, which may result in optic nerve damage, as seen in glaucoma.

Glaucoma is a progressive eye disease that results from elevated intraocular pressure, typically caused by impaired drainage of aqueous humor [5]. When IOP exceeds the tolerance of the optic nerve, it can lead to optic nerve damage and subsequent vision loss. There are several types of glaucoma, each with distinct characteristics:

- ***Primary Open-Angle Glaucoma (POAG):*** The most common form, where the drainage angle between the iris and cornea remains open, but the trabecular meshwork gradually becomes less efficient at draining aqueous humor. Vision loss is typically slow and initially affects peripheral vision.
- ***Angle-Closure Glaucoma (ACG):*** A less common but more acute condition where the angle between the iris and cornea becomes blocked, leading to a sudden rise in IOP. This can be accompanied by severe pain and requires immediate medical intervention.
- ***Low-Tension Glaucoma (LTG):*** Even with normal IOP, optic nerve damage may still occur, suggesting that some individuals have a lower tolerance for pressure.

2.8 Lens

It is a transparent, avascular structure of the eye with a diameter of 9–10 mm and thickness of 4–5 mm. It is made up of four parts: lens capsule, epithelial cells, lens fibers, and lens zonules. Lens capsule is a collagenous barrier enclosing and shaping the lens during accommodation. It is majorly composed of collagen type IV, laminin, perlecan, and entactin and shows variable thickness with a maximum around the equator and thinnest near the poles. Lens epithelium is simple cuboidal in shape and restricted to the anterior lens surface. The anterior epithelial cells are connected by gap junctions to allow exchange of small molecules and ions. Lens epithelial cells in the pre-equatorial and equatorial regions exhibit maximum mitotic activity marking this zone as the germinative zone. The cells become elongated columnar in shape at the equator with their apical portions lying deep within each other. These elongated cells are termed as lens fibers. Lens fibers appear as ordered concentric shells, tightly packed with little or no intercellular spaces in the nucleus of the lens. Lens fibers and epithelial cells are abundant in crystalline protein along with other cytoskeletal components (actin, myosin, microtubules, and vimentin). The lens zonule are delicate fibers arising from the pars plana region of the ciliary body and holding the lens through its capsule. They are synthesized by the nonpigmented cells of ciliary epithelium [18] (see Figs. 2.1 and 2.2a).

The organized arrangement of lens components contributes to the proper transmittance and focusing of light on the retina. Although its refractive power (15 diopters) is lower than the cornea (43 diopters), the contracting ability of the lens allows flexibility in its refractive power. Lens allows an efficient transmission of 1200 nm wavelength of light but very little transmission below 390 nm. The colorless lens gradually gains yellow pigmentation with age, suspected due to the deposition of tryptophan derivatives [18].

The most common degenerative condition affecting the lens is the cataract. The degeneration of the lens proteins causes slow opacification of the lens which starts affecting bilaterally in the fourth and fifth decade of life in varying severity. Based on factors involved in cataract development, it is categorized as (i) congenital cataract, having association with placental nutrition deficiency or low oxygen; (ii) senile cataract, most common age-related form; (iii) subcapsular cataract when there is fibrous or plaque-like appearance in the anterior and posterior lens capsule, which gives clinical complaints of halos, glares, and photophobia; (iv) cortical cataract involves opacity of the cortex region which is seen as wedge-shaped opacification; (v) endocrine factors include some underlying metabolic or systemic condition like diabetes, myotonic dystrophy, atopic dermatitis, neurofibromatosis type 2, and hypoparathyroidism giving rise to lens deformities and cataract development; (vi) secondary cataract is associated with other ocular conditions like myopia, hereditary fundus dystrophies, chronic anterior uveitis, acute congestive angle closure, etc. At the molecular level, the amino acids like tyrosine released from the breakdown of lens protein crystallin get further converted to melanin or adrenaline causing the shift in lens pigmentation from yellow to brown to black.

Additionally, rare developmental disorders associated with lens include aphakia (lens being absent), microspherophakia (abnormally shaped lens, usually round), ectopialentis (disposition of the lens), and lens coloboma (absence of the parts of lens) [15].

2.9 Uveal Tract

Uvea is the middle vascular layer of the eye, divided into two sections: Anterior uvea consisting of the iris and the ciliary body and the posterior uvea composed of the choroid. The continuous layer of the iris, ciliary body, and choroid terminates anteriorly at the pupil and posteriorly at the optic nerve (see Figs. 2.1b and 2.2c, d).

2.9.1 Iris

Iris is a thin circular, pigmented diaphragm, present in the coronal plane, anterior to the lens. It separates the aqueous part of the eye into anterior and posterior chambers. From anterior to posterior, it can be divided from anterior to posterior into anterior border layers, stroma, dilator muscle, and pigmented posterior epithelium. Anterior border layer is composed of fibroblasts and melanocytes with few collagen fibers. Stroma consists of loose connective tissue made up of fibroblasts, melanocytes, and collagen (types I and III) fibers along with resident immune cells like mast cells and macrophages forming bulk of the iris. The loose nature of connective tissues in this layer allows free flow of fluids through the anterior border layer during contractions and dilations. The dilator muscles have partially pigmented myoepithelial cells while the pigmented posterior epithelium has heavily pigmented cuboidal epithelial cells in radially arranged furrows [27, 28].

Additionally, from pupil to ciliary body, iris can be subdivided into three zones: pupil, mid, and root. Pupil, the opening, has a varying diameter from 1 to 8 mm and regulates the amount of light entering the eye. The size of the pupil is controlled by the dilator and sphincter pupillae muscles. During low light intensities or in a state of fear and excitement, the pupils undergo dilation or mydriasis through dilator pupillae muscles, while in an illuminated scenario or while sleeping, the pupils undergo contraction or miosis through sphincter pupillae muscles. Iris root aids in attachment at the iridocorneal angle of the anterior chamber while also being the site of entry for ciliary arteries (temporal and nasal) to form an incomplete major arterial circle. The capillaries of the iris are characterized by non-fenestrated endothelial cells joined by tight junctions, reducing the permeability of different solutes, and forming a critical component of blood-ocular barrier [7, 27, 28].

2.9.2 Ciliary Body

Ciliary body is a 5–6 mm wide tissue located adjacent to the iris and divided into two zones: anterior pars plicata and posterior pars plana. The pars plicata is composed of numerous (70–75) meridional folds termed as ciliary processes. Ciliary processes are made up of an inner capillary core, covered by loose stroma embedded with fenestrated blood capillaries, and a final double-layered epithelium. These processes have a vital role in establishing blood-aqueous barrier and immune-privilege site in the eye. Pars plana, on the other hand, has a flat surface extending from the ciliary processes to the orra serrata, where the ciliary epithelium transitions to neural retina. Pars plana along with the region between the ciliary processes is known to be the site for origin of zonular fibers that support the lens [7, 27, 28].

Histologically, ciliary body can be divided into three regions: epithelium, stroma, and muscle. The non-pigmented ciliary epithelium is associated with the secretion of zonular fibers, extracellular components of vitreous like collagen and hyaluronic acid. The stromal region of the ciliary body is embedded with melanocytes, fibroblasts, and a few immune cells (mast cells, macrophages, and lymphocytes). Ciliary muscles occupy two-third of the anterior portion of ciliary body, embedded in the vascular connective tissue of the ciliary stroma. The contraction of ciliary muscles causes relaxation and change in the shape of the lens zonules [7, 27, 28].

Anterior uvea or iris-ciliary body forms the site of action for the topically administered drugs. The melanin deposition in these regions binds to the drugs and significantly reduces their availability. The blood-ocular barrier prevents entry of the majority of the drugs into the anterior uvea [4].

Uveitis is the most common pathological condition associated with the uvea. Based on the affected region, it is termed either anterior uveitis affecting the iris and ciliary body or posterior uveitis affecting the choroid. Anterior uveitis majorly caused by herpes simplex virus, leprosy, or Lyme disease is characterized by a decrease in IOP, inflammation of iris and ciliary body, and increase in iris pigmentation. Posterior uveitis is caused by syphilis, toxocariasis, or cytomegalovirus characterized by retinal detachment and infiltration of immune cells into the choroid. The pan-uveitis is inflammation of all the parts of uvea, as sequelae of tuberculosis, syphilis, or candida infection [27, 28].

2.10 Posterior Segment

2.10.1 Choroid

Choroid is a thin, vascular, highly pigmented, loose connective tissue situated between the sclera and the retina. It extends from the optic nerve to the orra serrata and varies in thickness based on the flow of blood. The blood flow washes away a large portion of the low permeable or hydrophilic drugs administered through the non-corneal route and only a small fraction crosses the blood-retinal barrier. The choroid can be divided anatomically into the Bruch's membrane, the suprachoroid, and the choriocapillaris. The Bruch's membrane is a connective tissue layer composed of collagen, elastic fibers, and endothelial cells. Choriocapillaris is the network of fenestrated capillaries providing nutritional support to the retina. Suprachoroid marks the transition zone between the sclera and the choroid. It has lamellae embedded with melanocytes, fibrocytes, collagen, few ganglion cells, blood vessels, and nerves along with a thin, avascular suprachoroidal space (See Fig. 2.1f) [7, 27, 28].

The primary function of the choroid is to nourish the outer part of the retina through its choriocapillaris. The blood vessels of the choroid have thermoregulatory role, influence the intraocular pressure, and absorb the drugs in the systemic circulation. The numerous melanocytes and choroidal pigments aid in light absorption, prevent reflection of light from the retina, while providing high resistance to the drug flow. The systemic circulation of the choroid aids in the transport of periocular injected drugs to the posterior segment of the eye [4].

2.10.2 Vitreous Humor

The vitreous humor is a transparent, gel-like substance that fills the posterior portion of the eye, occupying the space between the lens and the retina. It plays a critical role in maintaining the structure of the eye and providing support to the retina. Composed of a delicate network of collagen fibrils suspended in hyaluronic acid, the vitreous humor allows light to pass through the eye while keeping the posterior segment stable (see Fig. 2.1).

The vitreous humor, or vitreous body, fills the posterior segment of the eye, making up about two-thirds of its internal volume. It is primarily composed of collagen fibrils, which create a fine network for structural support, and hyaluronic acid, a molecule that binds water and gives the vitreous its gel-like consistency. This combination forms a transparent, elastic gel that is essential for maintaining the eye's shape and ensuring that light can travel unobstructed to the retina.

The vitreous humor is loosely attached to the retina at key points: it has a gentle connection at the optic nerve head and macula, while it is strongly attached at the oraserrata, located at the peripheral retina just behind the ciliary body. This strong attachment helps maintain the stability of the vitreous body and separates the anterior and posterior segments of the eye.

These attachment points are essential for maintaining the structural integrity of the eye and keeping the vitreous positioned against the retina [21]. This separation ensures that the aqueous humor in the anterior chamber does not mix with the fluids in the posterior chamber, preserving the functional environment of both regions.

The primary function of vitreous humor is to provide structural support to the eye. Its gel-like consistency helps the eye maintain its spherical shape, while its attachment points keep the retina in close contact with the underlying layers of the globe. This support is crucial for the retina's function and health, ensuring that the light-sensitive tissue remains properly aligned to capture images. Like the aqueous humor, the vitreous humor plays a critical role in light transmission. It is completely transparent, allowing light to pass from the lens to the retina without scattering, thus ensuring clear vision [29].

As part of the natural aging process, the vitreous humor undergoes syneresis, leading to its gradual liquefaction and contraction. This breakdown of collagen fibrils results in small pockets of fluid forming within the vitreous, causing it to lose structural integrity and shrink away from the retina at its attachment points. This common occurrence, known as posterior vitreous detachment (PVD), is usually benign but can lead to the appearance of floaters—small shapes that drift across the visual field due to particles or cells breaking free and casting shadows on the retina. However, excessive traction from the shrinking vitreous can create complications, such as retinal tears or detachment, which may manifest as flashes of light, a sudden increase in floaters, or even vision loss. Overall, while PVD is typically harmless, it can increase the risk of serious retinal issues that threaten vision [16].

Diagnosis of vitreous conditions typically involves ophthalmoscopy for visual inspection and ultrasound imaging to assess cloudiness from issues like hemorrhage. While most age-related changes, such as floaters or posterior vitreous detachment, require no treatment, severe cases may need intervention. Vitrectomy can remove the vitreous humor to restore vision or treat retinal detachment, while laser photocoagulation or cryotherapy can address retinal tears and prevent detachment [23].

2.10.3 Retina

2.10.3.1 Retinal Layers

The retina is the innermost layer located at the posterior segment of the eye, surrounded by the vascular choroid and fibrous sclera. It extends from the optic disc to the oraserrata, covering the inner wall of the eye behind the lens and vitreous body. The retina is a ten-layered structure consisting of various specialized cell types in the eye that capture incoming light and transmit the signals along neuronal pathways to the brain, allowing it to perceive vision. The retina is made up of two parts, the macula and the peripheral retina. Specifically, the peripheral retina is subdivided into ten distinct layers as follows (see Figs. 2.1c and 2.2e):

I. **Inner Limiting Membrane:** It is the innermost layer of the retina, essential for retinal function, supporting both structural integrity and the regulation of the retinal microenvironment.

II. **Retinal Nerve Fiber Layer:** It is a critical layer of the retina that contains the unmyelinated axons of the ganglion cells, which converge at the optic disc to form the optic nerve. These axons carry visual information from the retina to the brain.

III. **Ganglion Cell Layer:** Ganglion cells receive processed signals from bipolar cells and their axons form the optic nerve. They encode visual information such as brightness, contrast, and motion and transmit this information to the brain for further processing.

IV. **Inner Plexiform Layer:** Contains synaptic connections between bipolar cells, ganglion cells, and amacrine cells. This layer facilitates complex processing of visual information, including temporal and spatial integration, and contributes to the formation of visual receptive fields.

V. **Inner Nuclear Layer:** Composed of cell bodies of bipolar cells, horizontal cells, and amacrine cells. Horizontal cells provide lateral inhibition, which is a visual process to enhance contrast and sharpness, color discrimination, and light adaptation of the visual signals, while amacrine cells modulate the transmission of signals, influencing ganglion cell activity and contributing to temporal processing, which is a process to encode spatial information through luminescence modulations caused by eye movements.

VI. **Outer Plexiform Layer:** Contains synapses between photoreceptors and bipolar cells. This layer facilitates the transmission of signals from photoreceptors and plays a role in integrating visual signals through horizontal cells, which provide lateral connections that enhance visual contrast.

VII. **Outer Nuclear Layer:** Composed of the cell bodies of photoreceptors. This layer is essential for capturing light, as the photoreceptors contain visual pigments that undergo conformational changes in response to light, initiating the phototransduction cascade.

VIII. **Outer Limiting Membrane:** It is a thin layer of the retina situated between the outer nuclear layer and the outer plexiform layer that separates the cell bodies of the photoreceptors from their inner segments, providing structural support.

IX. **Photoreceptor Layer:** This is the layer containing the light-sensitive cells, the rods and cones responsible for detecting the visual stimuli. Rods detect low light levels and are responsible for night vision (scotopic vision), while cones enable color vision (photopic vision) and high-acuity vision under bright light. They contain photopigments that absorb photons and initiate the phototransduction cascade, converting light into electrical signals.

X. **Retinal Pigment Epithelium (RPE):** RPE is the outermost layer of the retina, adjacent to the choroid. It consists of a single layer of pigmented cells that support and nourish the photoreceptors. It supports photoreceptors by absorbing scattered light, preventing light scattering that could distort the image. The RPE also recycles visual pigments such as retinol and is involved in the maintenance of the blood-retinal barrier, nutrient transport, and waste removal.

The area located between the RPE and the photoreceptor layer is known as the subretina. Administration of the drugs via subretinal route provides direct contact with the photoreceptor's membrane, RPE cells, rendering it to be the preferred site for drug delivery for the treatment of retinal diseases.

2.10.3.2 Macula Lutea and Fovea Centralis

The macula, also known as the macula lutea due to its yellowish pigment, is the most sensitive area of the retina and provides the clearest vision. It is located to the side of the optic disc when viewed during a fundoscopic exam. The macula is rich in cone photoreceptors, which are responsible for high visual acuity and color vision. The yellow color comes from carotenoids called lutein and zeaxanthin, which have anti-inflammatory properties and can filter blue light. In the center of the macula is the fovea centralis, a 1.5 mm diameter area without blood vessels that contains a high number of cone cells. It is characterized by a very high density of cones and a lack of rod cells, along with a reduced number of retinal layers, allowing light to strike the photoreceptors directly. Together, the macula lutea and fovea centralis play crucial roles in central vision, providing the ability to perceive fine details and color (see Fig. 2.1e).

2.10.4 The Optic Nerve: Structure, Function, and Clinical Relevance

The cranial nerve II (CN II), also known as the optic nerve, is responsible for transmitting afferent impulses of vision to the brain, serving as a link between the retinal layer and visual cortex of the brain [22]. In addition to its primary role in vision, the optic nerve also plays a part in various ocular reflexes, helping the eye respond to changes in light.

The optic nerve is a bundle composed of over one million nerve fibers, glial cells, and connective tissue. These fibers, which are extensions of the ganglion cells in the retina, known as "retinal ganglion cell" (RGC) axons, are responsible for carrying visual information from the eye to the brain. The lamina cribrosa provides passage and protection for these fibers as they exit the eye. The optic nerve also controls the pupillary light reflex by signaling the brainstem to constrict the pupils in response to light. The optic nerve extends from the back of the eyeball to the brain's visual cortex, which can be divided into four distinct segments (see Fig. 2.1d).

2.10.4.1 Intraocular Segment

The intraocular segment of the optic nerve, originating at the optic disc (or nerve head), marks the convergence of all retinal nerve fibers. This region also serves as the entry and exit point for the central retinal vessels. The absence of photoreceptors at the optic disc creates a physiological blind spot in the visual field. Typically, the disc is slightly vertically oval (approximately 2.7 mm^2) with a horizontal diameter of about 1.5 mm. Its coloration is yellowish-orange, with the temporal half generally paler than the nasal half. The optic disc margin is sharply defined, though the nasal edge may appear less distinct due to higher nerve fiber density. The intraocular segment encompasses the neuroretinal rim, composed of existing optic nerve fiber bundles, exhibiting characteristic variations in width (narrowest temporally, broadest superiorly and inferiorly). Within the disc lies the optic cup, a small, flattened depression lacking nerve fibers. The cup-to-disc ratio, reflecting the relative size of the cup, is clinically significant as an indicator of potential nerve fiber loss. While variability in disc size precludes absolute cup/disc ratio values, enlargement of the cup generally signifies pathological changes [1].

2.10.4.2 Intraorbital Segment

The intraorbital segment of the optic nerve begins as it exits the eye through the lamina cribrosa, a sieve-like structure composed of scleral connective tissue. This segment measures approximately 25–30 mm in length and is surrounded by protective layers of orbital fat and the bony structure of the orbit. Within the orbit, the optic nerve follows an S-shaped course, allowing flexibility to accommodate extreme eye movements without causing strain or injury. The nerve fibers in this segment are myelinated, which significantly increases the diameter of the nerve and enhances the efficiency of signal transmission. The intraorbital optic nerve passes within the optic muscle cone and through the bony optic canal, ultimately accessing the cranial cavity. This structural arrangement not only protects the nerve but also ensures a stable path for signal propagation from the eye to the brain.

2.10.4.3 Intracanalicular Segment

The optic nerve then passes through the optic canal, a narrow bony channel in the skull, it enters the intracanalicular segment. Measuring approximately 4–10 mm in length, this portion of the nerve is enclosed by the meningeal layers—dura mater, arachnoid mater, and pia mater—which provide stability and protection during eye movements. The intracanalicular segment ensures a seamless transition for the nerve as it progresses toward the cranial cavity [19].

2.10.4.4 Intracranial Segment

As the optic nerve exits the optic canal, it enters the intracranial segment. This portion is about 10 mm long and extends to the optic chiasm. The intracranial segment, like the other parts of the optic nerve, is covered by the meningeal layer which protects it and maintains its structure.

Damage to the optic nerve can lead to various vision problems. Anterior ischemic optic neuropathy (AION) is one of the examples that occurs when the blood supply to the optic nerve is reduced, resulting in sudden vision loss. Another issue, papilledema, is swelling of the optic disc due to increased pressure in the brain. The common treatment approaches are via intravitreal, retro orbital, and suprachoroidal routes that are known to deliver the drug within 2 mm of the optic nerve [27, 28].

2.10.5 Ocular Appendages

The ocular appendages are the orbital structures situated adjacent to the eye to ensure protection and support the eye. These mainly include: eyelids, lacrimal functional unit, orbital bone, and extraocular muscles.

2.10.5.1 Eyelid

Eyelids form the most periocular structure protecting the eye from external damage and pathogen entries. The eyelid epidermis has keratinized squamous epithelial cells, while dermis has papillary and reticular layers. The dermis forms the base for hair follicles, sebaceous glands (Meibomian and Zeis glands), and sweat glands (Moll glands). Medial and lateral canthi are the points of interaction for the upper and lower eyelids; with palpebral fissure, an elliptical opening between both the eyelids and lacrimal puncta located at the medial ends help in the drainage of tears (see Figs. 2.2a and 2.3c).

The movement of the eyelid is regulated by set of muscles: orbicularis oculi palpebral muscle, levator palpebrae muscle, and retractor muscle. The striated skeletal muscle of orbicularis is involved in the eyelid closure while the pull of levator-palpebrae on the upper lid and retractor muscle on the lower lid ensures eyelid opening. The upper lid covers the corneoscleral junction in the open state and undergoes maximum displacement in the eyelid blinking.

Eyelids act as a wiper to evenly spread the tears and remove any debris from the ocular surface. The punctal opening maintains balance between tear retention and drainage. Eyelashes further enhance protection from dust and allergens in the environment. The modified sebaceous glands aid in the secretion of the outermost lipid layer of the tear film and thus retain tears on the surface.

Being the most exposed structure of the orbit, eyelids are infected by various pathogens, animal bites, trauma, or insect bites. The most common clinical presentations include cellulitis, canaliculitis, necrotizing fasciitis, and hordeolum [3, 7].

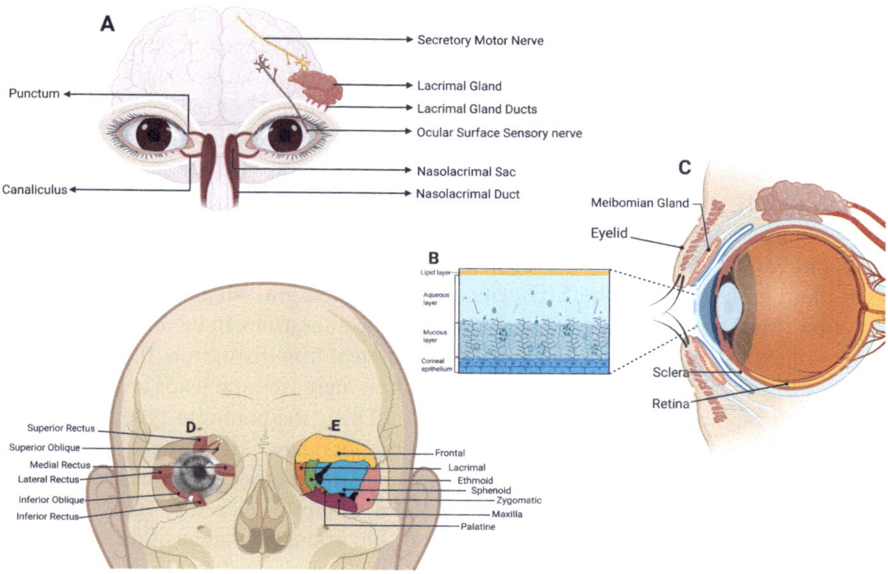

Fig. 2.3 Schematic summarizing the **human ocular appendages**: (**a**) the entire anatomy of lacrimal functional unit, (**b**) three distinct layers of tear film, (**c**) anatomy of eyelid and meibomian gland, and the location and alignment of (**d**) the six extraocular muscles and (**e**) seven orbital bones in the human orbit. (Self-created figures from authors. Created with biorender.com licensed version)

2.10.5.2 Lacrimal Functional Unit

2.10.5.2.1 Tear Film

The lacrimal functional unit (LFU) is a complex integrated unit of the ocular surface and the tear glands interconnected by the sensory and the motor nerves. LFU maintains a healthy ocular surface by replenishing the tears and lubricating the ocular surface. Abnormality in any part of the LFU can lead to altered tear composition, affect the tear film stability, and cause dry eye. Precorneal tear film is a 2–5.5 µm thick layer spread over the outer mucosal surface of the eye in order to maintain ocular surface homeostasis. The tear film is a thin fluid layer that is about 3–10 µL in volume and is secreted at a rate of 1–2 µL/min. The tear film contributes to maintaining the transparency of the cornea, providing nutrition by acting as an interface medium for the environmental oxygen and protecting the ocular surface. The tear film is primarily composed of two layers: the mucoaqueous layer and the lipid layer (see Fig. 2.3b). The mucoaqueous layer is a combination of secretions from conjunctival mucins and a larger part being the aqueous secretions from the lacrimal gland (LG). The mucins within the mucoaqueous layer have a thickness range of 0.5–1 µm, whereas the larger aqueous component extends up to 2–5 µm. The meibum is the major component of the lipid layer and is secreted by the meibomian glands (MG) embedded on the eyelid rim inside the tarsal plate. Meibum and the aqueous tear components mix while blinking to form the precorneal tear film. The lipid layer thickness ranges around 15–157 nm with an average thickness of 42 nm.

The lipid layer is the thinnest layer present on the outer surface and protects the aqueous tears from rapid evaporation and prevents spillage of tears at lid margin. This is achieved by creating a water-tight seal along with the lid closure [12]. The lipid layer is predominantly composed of cholesterol, wax esters, fatty acids, diesters, triacylglycerol, and phospholipids. The fully differentiated meibocytes in the meibomian glands produce a lipid-rich secretion known as meibum. The meibum from individual MGs reaches the ocular surface through the ductules and the central duct that finally opens at the MG orifice in the lid margin. Meibomian gland dysfunction (MGD) is one of the major causes for the alterations in the basic composition or thickness of the lipid layer that can lead to tear film disruption. In the absence of a healthy lipid layer, the rate of tear evaporation increases and leads to a condition known as evaporative dry eye disease (EDE). EDE is the common type of dry eye among the population, and it is about 60% of the total cases of dry eye.

The aqueous layer of tear film is 98% water and the other 2% contains various proteins such as lysozyme, lactoferrin, lipocalin, albumin, secretory IgA, IgG, and electrolytes like calcium, sodium, potassium, bicarbonate, magnesium, and chloride. The high number of antibacterial proteins in the aqueous layer protects the eye from infections. Aqueous component is produced by the secretions from the lacrimal gland as well as the accessory glands; glands of Krause and Wolfring. Lacrimal gland is an exocrine gland located in the upper outer part of the orbit, i.e., the bony socket of the eye. It has a roughly almond-shaped structure and is responsible for producing the aqueous layer of the tear film.

The mucin layer is produced by the goblet cells of conjunctiva and the glands of Manz, located in the crypts of Henle and in the bulbar conjunctiva. It is composed primarily of mucins, which are glycoproteins that provide a gel-like consistency, which is crucial for maintaining a healthy ocular surface and preventing dry eye conditions. MUC4, MUC5AC, and MUC1 are the major mucins present in the mucin layer. This layer helps to stabilize the tear film, ensuring that tears adhere to the surface of the eye, particularly the cornea and conjunctiva. The mucin layer is involved in maintaining the ocular surface hydrated and reduces friction during blinking. Also, it serves as a protective barrier, preventing pathogens and irritants from adhering to the ocular surface. By enhancing the adhesion of the aqueous and lipid layers, it contributes to the overall stability of the tear film.

2.10.5.3 Lacrimal Gland

The human lacrimal apparatus consists of the main lacrimal gland, the accessory glands; glands of Wolfring and Krause and the lacrimal passages. The main lacrimal gland is majorly divided into two lobes: the larger orbital lobe located in the upper part of the orbit and the smaller palpebral lobe found within the upper eyelid. The orbital lobe is almost double the size of the palpebral lobe. The main LG is approximately 20 x 25 mm and has a thickness of 3 mm in the palpebral lobe and 5 mm in the orbital lobe (see Fig. 2.3a). The primary function of the LG is to produce tears, which contain water, electrolytes, proteins, and enzymes. This aqueous layer helps to keep the eye moist, nourishes the cornea, and provides a protective barrier against pathogens. The LG is innervated by the autonomic nervous system, with parasympathetic stimulation primarily from the facial nerve being responsible for tear production. Tears produced by the lacrimal gland are transported through a series of small ducts that lead to the conjunctival sac, where they spread across the surface of the eye during blinking.

The LG is mainly composed of acinar epithelial cells (80% of the total gland), ductal epithelial cells, myoepithelial cells, mesenchymal stem cells [10], and endothelial cells. The orbital and palpebral lobes of LG are composed of lobules that are separated from each other through loose connective tissue. Each lobule consists of round LG acini that has pyramidal acinar cells and the long intralobular ducts which have columnar ductal epithelial cells arranged within. The loose connective tissue between the acini and the ducts contains interlobular ducts, blood vessels, nerve fibers, fibroblasts, plasma cells, and a few lymphocytes. Interlobular ducts collect the secretions from the acini and the intralobular ducts and transfer them to approximately 12 excretory ducts, which open into the conjunctival fornix. The presence of a healthy ocular surface epithelium also plays an important role in transport of lacrimal secretions to the ocular surface.

A damage or dysfunction of LG acini and/or ducts can affect the production and transport of lacrimal secretions to the ocular surface and hence disrupt the tear film homeostasis. An abnormal tear film is unstable and prevents the lubrication of the ocular surface, leading to a condition known as aqueous deficient dry eye disease (ADDE).

2.10.5.4 Meibomian Gland

Meibomian glands are specialized sebaceous glands located between the eyelashes and the bulbar conjunctiva that secrete a mixture of fat and oil onto the ocular surface. They play a crucial role in maintaining the health and stability of the tear film by producing meibum, an oily secretion that forms the outermost layer of the tear film. Each eye contains approximately 20–30 meibomian glands on the lower eyelid with a rough length of 5.5 mm and 40–50 on the upper eyelid with a length of 2 mm, which are embedded in the tarsal plates (the cartilage-like structure of the eyelids). The meibomian glands are arranged vertically through the tarsal plates in both the superior and inferior eyelids (see Figs. 2.2d and 2.3c). Each gland consists of approximately 10–15 acini filled with secretory meibocytes that produce meibum, which contributes to the lipid layer of the tear film. The round acini cluster around the central duct, which channels the meibum to the ocular surface. Surrounding the end of each gland is Riolan's muscle that contracts during each blink, along with the orbicularis oculi, to help push the meibum out of the duct and onto the lid margin, where it exits through the orifice.

The contents of the meibum include wax and sterol esters (comprising approximately 77%), including fatty acids, fatty alcohols and cholesterol, phospholipids (8%), and diglycerides, triglycerides, and hydrocarbons (9%). These glands are long, tubular structures that appear as grape-like clusters attached to a central stalk that open onto the mucocutaneous junction of eyelid margin through tiny orifices.

Meibomian gland constantly synthesizes and secretes the meibum as part of the basal secretion, and it is also influenced by blinking, which pushes the meibum onto the ocular surface. This process also helps to spread the lipid layer evenly across the tear film. The essential lipids in the meibum help prevent the evaporation of the tear film and maintain a stable tear film.

Dysfunction of the meibomian glands can lead to conditions such as meibomian gland dysfunction, which is a common cause of dry eye syndrome. MGD has a widespread prevalence of 39–50% of the total population with the increase in age. MGD can result in insufficient lipid production, leading to increased tear evaporation and discomfort. Also, if the MG orifice is blocked due to lid margin keratinization or any other ocular conditions, the meibum builds up within the glands and it leads to gland atrophy. The gland dropout and the duct dilatation are two key structural features observed in patients with MGD. Meibomian glands are essential for proper tear film function, ocular surface lubrication, and overall ocular health.

2.10.5.5 Goblet Cells of Conjunctiva

Goblet cells are specialized epithelial cells located in the conjunctiva, the mucous membrane that covers the sclera of the eye and lines the inner eyelids. These cells play a vital role in maintaining the health and function of the ocular surface. The mucin layer secreted by the goblet cells is the innermost layer of the tear film, which helps to spread the aqueous layer evenly across the eye and prevents evaporation of tears. Goblet cells are primarily responsible for producing and secreting soluble mucins, which are glycoproteins that form a gel-like layer. The two major classes of mucins present at the ocular surface are the secreted mucins like Muc2 and Muc5AC and the membrane-associated mucins such as Muc1, Muc4, and Muc16. Mucin layer is crucial for stabilizing the tear film and ensuring that tears adhere properly to the ocular surface. Additionally, these mucins help clear pathogenic bacteria and particulate contaminants resulting in a smooth refractive corneal surface, which minimizes light scatter and facilitates clear vision.

Goblet cells are predominantly found in the conjunctival epithelium, especially in the forniceal conjunctiva and less in the bulbar conjunctiva (see Fig. 2.2b). In response to irritation or inflammation, the number and activity of goblet cells can increase, producing more mucin to help protect and heal the ocular surface. The maintenance of an adequate number of conjunctival goblet cells and its function is important in preserving a healthy tear film and ocular surface. Overall, goblet cells are essential for maintaining the health and comfort of the eye by ensuring proper tear film formation and providing a protective barrier against irritants.

2.10.6 Extraocular Muscles and Ocular Movements

The eye is supported by seven extraocular muscles, six of them are called true muscles as they are involved in the movement of the eye and additional levator palpebrae superioris is a standalone muscle associated with the eyelid movement. The true muscles of the globe are four rectus muscles, namely medial, lateral, superior,

Table 2.1 Different types of extraocular muscles and associated ocular movements

Ocular movements	Description	Muscles associated
Abduction	Movement of eye away from the nose, around the vertical axis	Lateral rectus, superior oblique, inferior oblique
Adduction	Movement of eye toward the nose, around the vertical axis	Medial rectus, superior rectus, inferior rectus
Supraduction	Elevation of the eye, around the horizontal axis	Superior rectus, inferior oblique
Infraduction	Depression of the eye, around the horizontal axis	Inferior rectus, superior oblique
In-cycloduction	Inward torsional rotation of the eye, around the median axis	Superior rectus, superior oblique
Ex-cycloduction	Outward torsional rotation of the eye, around the median axis	Inferior rectus, inferior oblique

and inferior rectus, and two oblique muscles namely superior and inferior obliques (see Fig. 2.3d) [20]. The extraocular muscles are highly vascularized and embed large amounts of loose connective tissue rich in elastic fibers and reticulin. The ophthalmic artery, the lacrimal artery, and the infraorbital artery supply blood to the extraocular muscles. There is a similar venous drainage system with superior and inferior ophthalmic veins draining into the cavernous sinuses. The communications have a clinical significance as they mark the routes for the entry of pathogens or infections [7].

The main function of the ocular muscles is the ocular movements along three axes: horizontal, vertical, and torsional. Details about each of the ocular movements and the muscles associated are listed in Table 2.1.

2.10.7 Orbital Bones

Orbits are the symmetrical pear-shaped cavities housing the eye, extraocular muscles, lacrimal apparatus, adipose tissue, tenons, and vascular and nervous connections. The fat present in the orbit acts as a cushion aiding in smooth movement and protection of ocular adnexa. Optic nerve is situated at the stem of the orbit entering through the middle cranial fossa along with the ophthalmic artery (see Fig. 2.3e) [13]. The orbital margin is formed by seven bones, and the location details of these are mentioned in Table 2.2.

2.11 Summary and Conclusion

Anatomically, the human eye is divided into anterior and posterior segments, with a collaborative association of various parts in each segment being integral to the maintenance of the visual axis.

Table 2.2 Types and location of different orbital bones

Orbital region	Orbital bones
Roof	Frontal bone (orbital plate), sphenoid (lesser wing)
Floor	Maxillary bone, zygomatic bone (orbital plate), palatine bone
Medial wall	Maxillary bone, lacrimal bone, ethmoid bone, sphenoid (lesser wing)
Lateral wall	Sphenoid (greater wing), zygomatic bone

- The cornea and lens of the anterior segment, through their structural complexity, maintain transparency and refractive power; alterations to these structures can cause scarring or edema impairing vision.
- The gradient of intercellular pore size in trabecular meshwork layers regulates intraocular pressure; obstruction of this flow can lead to glaucoma, potentially damaging the optic nerve and causing blindness.
- The retinal layers, with their inverted orientation in the posterior segment, not only act as blood-ocular barrier but also convert the light signals to electrical signals.
- After refractive errors, cataract, glaucoma, corneal opacity, diabetic retinopathy, and conjunctivitis remain the most common eye diseases, often treated through topical, intravitreal, or intracameral drug administration.

References

1. Albert DM, Miller JW, Azar DT. Albert & Jakobiec's principles & practice of ophthalmology: 4-volume Set (expert consult—online and print), 3e. 5502. 2008. https://books.google.com/books/about/Albert_Jakobiec_s_Principles_and_Practic.html?id=iTw_SQAACAAJ
2. Bulloch G, Seth I, Sukumar S, Chen Y, Zhu Z. Scleral thinning causes, diagnosis, and management: a narrative review. Cont Lens Anterior Eye. 2023;46(3):101825. https://doi.org/10.1016/j.clae.2023.101825.
3. Cochran ML, Lopez MJ, Czyz CN. Anatomy, head and neck: eyelid. StatPearls; 2023. https://www.ncbi.nlm.nih.gov/books/NBK482304/
4. del Amo EM. Topical ophthalmic administration: can a drug instilled onto the ocular surface exert an effect at the back of the eye? Front Drug Deliv. 2022;2:954771. https://doi.org/10.3389/FDDEV.2022.954771.
5. Dietze J, Blair K, Zeppieri M, Havens SJ. Glaucoma. StatPearls; 2024. https://www.ncbi.nlm.nih.gov/books/NBK538217/
6. Dua HS, Faraj LA, Said DG, Gray T, Lowe J. Human corneal anatomy redefined. Ophthalmology. 2013;120(9):1778–85. https://doi.org/10.1016/j.ophtha.2013.01.018.
7. Forrester JV, Dick AD, McMenamin PG, Lee WR. The eye: basic sciences in practice. 4th ed. London: Elsevier; 2016.
8. Gonzalez-Andrades M, Argüeso P, Gipson I. Corneal anatomy. In: Essentials in ophthalmology/essentials ophthalmology; 2019. p. 3–12. https://doi.org/10.1007/978-3-030-01304-2_1.
9. Gupta N, Chhibber-Goel J, Gupta Y, Mukherjee S, Maitra A, Sharma A, Tandon R. Ocular conjunctival microbiome profiling in dry eye disease: a case control pilot study. Indian J Ophthalmol. 2023;71(4):1574–81. https://doi.org/10.4103/ijo.ijo_1756_22.

10. Jaffet J, Mohanty A, Veernala I, Singh S, Ali MJ, Basu S, Vemuganti GK, Singh V. Human lacrimal gland derived mesenchymal stem cells – isolation, propagation, and characterization. Invest Ophthalmol Vis Sci. 2023;64(10):12. https://doi.org/10.1167/iovs.64.10.12.
11. Johnstone MA. Aqueous humor outflow system overview. In: Becker-Shaffer's diagnosis and therapy of the Glaucomas: eighth edition; 2009. p. 25–46. https://doi.org/10.1016/B978-0-323-02394-8.00003-6.
12. Kayal A. The physiology of tear film. In: Dry eye syndrome-modern diagnostic techniques and advanced treatments. In Tech; 2022.
13. Luibil N, Lopez MJ, Patel BC. Anatomy, head and neck, orbit. StatPearls; 2023. https://www.ncbi.nlm.nih.gov/books/NBK539843/
14. Meek KM, Knupp C. Corneal structure and transparency. Prog Retin Eye Res. 2015;49:1–16. https://doi.org/10.1016/j.preteyeres.2015.07.001.
15. Nizami AA, Gurnani B, Gulani AC. Cataract. In: Principles and practice in ophthalmic assisting: a comprehensive textbook; 2024. p. 511–22. https://doi.org/10.1201/9781003525899-32.
16. Popescu S-I, Munteanu M, Patoni C, Musat AMA, Dragoescu V, Cernat C-C, Popescu M-N, Musat O. Role of the vitreous in retinal pathology: a narrative review Cureus. 2023;15(8) https://doi.org/10.7759/CUREUS.43990.
17. Raut AK, Mohapatra S, Siddiqui G, Rajak SK, Sonar R, Joshi V, Basu S, Singh V. The human cornea: unravelling its structural, chemical and biochemical complexities. Chem Biodivers. 2024; https://doi.org/10.1002/cbdv.202402224.
18. Ruan X, Liu Z, Luo L, Liu Y. The structure of the lens and its associations with the visual quality. BMJ Open Ophthalmol. 2020;5(1) https://doi.org/10.1136/BMJOPHTH-2020-000459.
19. Salazar JJ, Ramírez AI, De Hoz R, Salobrar-Garcia E, Rojas P, Fernández-Albarral JA, López-Cuenca I, Rojas B, Triviño A, Ramírez JM, Salazar JJ, Ramírez AI, De Hoz R, Salobrar-Garcia E, Rojas P, Fernández-Albarral JA, López-Cuenca I, Rojas B, Triviño A, Ramírez JM. Anatomy of the human optic nerve: structure and function. In: Optic nerve; 2018. https://doi.org/10.5772/INTECHOPEN.79827.
20. Shumway CL, Motlagh M, Wade M. Anatomy, head and neck, eye extraocular muscles. StatPearls; 2022. https://www.ncbi.nlm.nih.gov/books/NBK519565/
21. Silva AF, Alves MA, Oliveira MSN. Rheological behaviour of vitreous humour. Rheol Acta. 2017;56(4):377–86. https://doi.org/10.1007/S00397-017-0997-0/FIGURES/10.
22. Smith AM, Czyz CN. Neuroanatomy, cranial nerve 2 (optic). StatPearls; 2022. https://www.ncbi.nlm.nih.gov/books/NBK507907/
23. Sohn EH, Strohbehn A, Stryjewski T, Brodowska K, Flamme-Wiese MJ, Mullins RF, Eliott D. Posteriorly inserted vitreous base: preoperative characteristics, intraoperative findings, and outcomes after vitrectomy. Retina. 2020;40(5):943. https://doi.org/10.1097/IAE.0000000000002482.
24. Sunderland DK, Sapra A. Physiology, aqueous humor circulation. StatPearls; 2023. https://www.ncbi.nlm.nih.gov/books/NBK553209/
25. Varshney T, Dudani P, Gupta V. Steroid-induced glaucoma. In: Childhood glaucoma: a case based color and video Atlas; 2023. p. 235–43. https://doi.org/10.1007/978-981-19-7466-3_22.
26. Watson PG, Young RD. Scleral structure, organisation and disease. A review. Exp Eye Res. 2004;78(3):609–23. https://doi.org/10.1016/s0014-4835(03)00212-4.
27. Yanoff M, Sassani JW. Optic nerve. Ocular Pathol. 2020a;494–518.e5. https://doi.org/10.1016/B978-0-323-54755-0.00013-9.
28. Yanoff M, Sassani JW. Uvea. Ocular Pathol 2020b;357–79.e5. https://doi.org/10.1016/B978-0-323-54755-0.00009-7.
29. Zong Y, Gao QY, Hui YN. Vitreous function and intervention of it with vitrectomy and other modalities. Int J Ophthalmol. 2022;15(6):857–67. https://doi.org/10.18240/IJO.2022.06.02.

Drug Delivery to the Anterior Segment of Eye

3

Shilpkala Gade, Deepakkumar Mishra, Ziqi Yu, Ge Jiang,
Lalitkumar K. Vora, and Raghu Raj Singh Thakur

Abstract

The human eye's complex anatomy and protective barriers pose significant challenges for effective drug delivery, with conventional methods like eye drops exhibiting low bioavailability (<5%) and transient therapeutic effects. This chapter examines recent advancements in ocular drug delivery systems designed to overcome these limitations, focusing on minimally invasive strategies for anterior segment disorders. Innovations such as long-acting biodegradable and non-biodegradable implants (e.g., Ozurdex®, Durysta®) enable sustained drug release, reducing administration frequency and enhancing compliance. Electrospun nanofiber patches and particulate systems (liposomes, nanoparticles) improve corneal penetration and retention, while microneedle technology offers precise, painless delivery of biologics and small molecules. Drug-eluting contact lenses integrate therapeutic release with vision correction, leveraging stimuli-responsive materials for controlled delivery. Despite their promise, challenges persist in scalability, regulatory compliance, and clinical translation. Future directions emphasize smart materials, combination therapies, and personalized approaches to optimize efficacy and safety. Collectively, these technologies represent a paradigm shift in ocular therapeutics, addressing chronic conditions like glaucoma, uveitis, and dry eye syndrome with enhanced bioavailability and reduced patient burden. By bridging gaps between innovation and clinical application, these systems hold transformative potential for improving patient outcomes and quality of life in ocular healthcare.

S. Gade · D. Mishra · Z. Yu · G. Jiang · L. K. Vora (✉) · R. R. S. Thakur (✉)
School of Pharmacy, Queen's University Belfast, Medical Biology Centre, Belfast, UK
e-mail: l.vora@qub.ac.uk; r.thakur@qub.ac.uk

N. K. Mehra (ed.), *Complex Ophthalmic Dosage Forms: Advances in Biomedical Applications and Future Perspectives*, https://doi.org/10.1007/978-981-96-5306-4_3

Keywords

Anterior segment · Ocular disorders · Advanced drug delivery systems · Long-acting drug delivery platforms · Eye-drops

3.1 Introduction

The human eye has a complex structure and protective mechanisms, which pose significant challenges for effective drug delivery. Traditional topical formulations like eye drops, despite their widespread use and patient acceptance, suffer from low bioavailability (<5%) and brief retention times, limiting their efficacy in treating chronic ocular conditions [1].

Over the past decade, these challenges have led to remarkable innovations in ocular drug delivery technologies [2]. The field has experienced a fundamental shift towards developing long-acting systems that can provide sustained therapeutic effects while reducing treatment frequency and enhancing patient compliance. This chapter explores the cutting-edge advancements in ocular drug delivery, focusing on minimally invasive approaches for the anterior segment of the eye disorders [3].

Recent years have witnessed the emergence of several promising technologies aimed at revolutionizing ocular drug delivery. Long-acting preformed implants, in situ forming eye drops, microneedle technology, and electrospun patches represent some of the key innovations in this field [4–9]. These advanced delivery systems strive to overcome the limitations of conventional therapies by offering sustained drug release, improved penetration through ocular barriers, and reduced administration frequency. Each of these technologies presents unique advantages and shows potential for improving the treatment of various ocular disorders, ranging from chronic to acute conditions of the anterior segment.

This chapter explores the cutting-edge advancements in ocular drug delivery, focusing on minimally invasive approaches, underlying principles, advantages, limitations, and potential clinical applications for the anterior part of eye disorders. Furthermore, the challenges in translating these technologies from bench to bedside and the future directions that may further revolutionize ocular drug delivery have been discussed in detail. The ongoing evolution of these technologies holds great promise for improving patient outcomes and quality of life for those suffering from chronic ocular disorders. By providing a comprehensive overview of the current state and future potential of this rapidly evolving field, this chapter aims to contribute to the continued advancement of ocular drug delivery systems for the anterior segment of eye diseases and their impact on patient care.

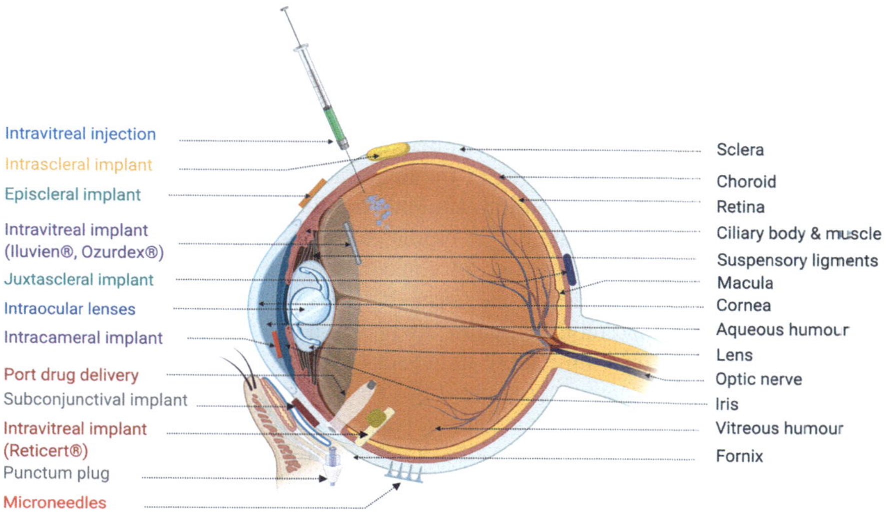

Fig. 3.1 Advanced drug delivery systems for intraocular applications. (Reproduced with copyright permission from Ref. [10])

3.2 Advanced Drug Delivery Systems

Ocular drug delivery has seen remarkable advancements in recent years, driven by the limitations of conventional methods. Traditional approaches often struggle with low bioavailability and require frequent dosing, placing a significant burden on patients. To address these challenges, researchers have leveraged progress in material science and biotechnology to develop innovative drug delivery systems based on biodegradable biomaterials. These cutting-edge platforms include preformed implants, in situ forming eye drops, electrospun patches, nanoparticulate systems, microneedle-based delivery, and drug-eluting contact lenses (Fig. 3.1). Each of these approaches aims to provide sustained, controlled release of therapeutic agents over extended periods, thereby enhancing drug efficacy while reducing administration frequency.

A key objective of these advanced delivery systems is to alleviate the socioeconomic burden on patients with chronic ocular conditions. By improving bioavailability, targeting specific ocular tissues, and minimizing side effects, these technologies hold great promise for revolutionizing the treatment of eye diseases. Their development represents a significant leap forward in ocular therapeutics, offering the potential for improved patient outcomes and quality of life.

3.3 Preformed Ocular Implants

Preformed implant devices are typically designed to directly deliver drugs to the site of action to achieve therapeutic effects with lower drug concentration [11]. These devices allow the drug released in a controlled manner over a longer period, ranging from several days to years [12]. One of the major advantages is that they assist in avoiding the need for frequent administrations and thus achieving extended drug release with no reliance on patients' compliance. In addition, the associated side effects due to the invasive penetration of a needle into the eye may be minimized [13], including the risk of infection and tissue damage [14]. The insertion of devices itself is quite invasive and requires healthcare professionals; however, the strengths mentioned above can overcome these drawbacks [14]. In recent years, preformed ocular implants with sustained release profiles have been widely explored. A summary of implants for eye diseases at various phases of clinical trials in recent years is shown in Table 3.1. The timeline of FDA-approved preformed implants is concluded in Fig. 3.2.

3.3.1 Biodegradable Implants

Biodegradable implants are made from natural or synthetic biodegradable polymers, such as polylactic acid (PLA), polycaprolactones (PCL), polylactic-co-glycolic acid (PLGA), and polyglycolic acid (PGA) [12]. The primary degradation mechanism of biodegradable implants is non- or enzymatic hydrolysis, where two main processes happen: denominated degradation and erosion [13]. The former refers to the breakdown of main chains into nontoxic smaller fragments, which can be safely absorbed or excreted by the body [13], while the latter describes the transformation of hydrophobic polymers into water-soluble materials and solubilized

Table 3.1 Ocular implants at various clinical research phases in recent years

Name	Company	Trial number	Status	Indication	Study type
PA5108	PolyActiva Pty Ltd.	NCT04060758	Active, not recruiting	Glaucoma	Phase I
IBE-814	Ripple Therapeutics Pty Ltd.	NCT04576689	Completed	DMO, RVO	Phase II
OTX-TIC	Ocular Therapeutix, Inc.	NCT05335122	Active, not recruiting	OAG, OHT	Phase II
ENV515	Envisia Therapeutics	NCT02371746	Completed	Glaucoma, OHT	Phase II
NT-501	Neurotech Pharmaceuticals, Inc	NCT06397131	Not yet recruiting	MacTel	Phase III

DMO diabetic macular oedema, *RVO* retinal vein occlusion, *OAG* open-angle glaucoma, *OHT* ocular hypertension, *nAMD* neovascular age-related macular degeneration, MacTel macular telangiectasia type 2

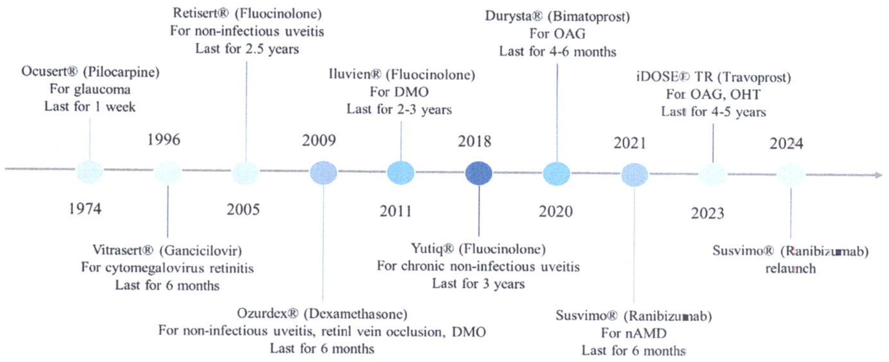

Fig. 3.2 Ocular implants development timeline

in vivo [13, 14]. The major advantage of these devices is that they do not require removal after use, as the body can naturally degrade them [14, 15]. The rate of degradation is influenced by various factors, such as molecular weight, surface properties, and in vivo pH and temperature, which in turn affect drug release rates [14]. In addition, the drug release is also controlled by diffusion, or a combination of degradation and diffusion, including the drug solubility and permeability, and the drug load [12, 14].

However, these systems often face some drawbacks. For example, the "final burst" stage of the device occurs when polymer hydrolysis reaches a certain point and results in uncontrolled release of the remaining drug load [13]. Another challenge is that biodegradable implants are complex and expensive to develop due to a limited range of materials with optimal degradation kinetics and mechanical properties, and strict regulatory requirements, since the material remains in the body [13, 16].

Ozurdex®, a PLGA-based intravitreal implant used for treating noninfectious uveitis and DMO, was approved by the FDA in 2009 as the first approved biodegradable implant [17, 18]. The implant loaded 0.7 mg dexamethasone and can provide the drug release for 6 months [19]. Ultimately, PLGA can hydrolyze into lactic acid and glycolic acid, which can then metabolize into water and carbon dioxide and be eliminated from the body [17]. Durysta®, produced by Allergan plc (Dublin, Ireland), was approved by the FDA for the use of intraocular pressure (IOP) reduction in OAG in March 2020 [20, 21]. The FDA approval was based on the findings of two phase III ARTEMIS studies, which evaluated the implant over 20 months [20]. 10 μg of bimatoprost was held in a PLGA polymer matrix based on Allergan's NOVADUR® platform, allowing for long-term and consistent drug release [21, 22]. It was designed to release bimatoprost for a period of 3–4 months; however, the actual release duration was found to be longer [20, 23].

3.3.2 Nonbiodegradable Implants

Nonbiodegradable implants are commonly fabricated by polymers, such as ethylene-vinyl acetate (EVA), polyvinyl alcohol (PVA), poly(siloxanes), and polyurethanes (PU), due to their exceptional mechanical, physical, and biocompatible properties [13, 14]. This type of implant is structurally durable and maintains integrity over time; however, they need to be removed once the drug content is depleted and sometimes may lead to infections or cosmetic issues [13, 24]. The main driving force of nonbiodegradable implants is passive diffusion and thus can be classified into monolithic and reservoir systems [14]. Drugs are dispersed homogeneously within the nonbiodegradable polymer matrix in monolithic systems, while reservoir systems consist of drugs surrounded by a nonbiodegradable polymer membrane [14]. The drug release performance is controlled by the composed material and drug formulation [24].

Yutiq® (Eyepoint Pharmaceuticals, Watertown, MA, USA) is an injectable device approved by the FDA in 2018 for patients with uveitis [13, 25]. It contains 0.18 mg fluocinolone acetonide and can continuously release drugs for a period of 36 months at a speed of 0.2 mg/day [25].

iDOSE® TR (Glaukos, San Clemente, CA, USA) sustained-release 75 μg travoprost titanium intracameral implant was approved by the FDA in December 2023 [26, 27]. The device can deliver travoprost directly to the anterior chamber to reduce IOP in OAG and OHT [26]. Remarkably, iDOSE® TR release can last for 4–5 years and can be exchanged after it depletes the drug load [27]. The approval was based on the results of one phase II and two phase III clinical trials [18].

3.4 Electrospun Patches

Electrohydrodynamic atomization (EHDA), commonly referred to as the electrospinning method, facilitates the creation of functional polymeric nanofibers with unique morphological characteristics for pharmaceutical uses. This straightforward technique involves applying a high voltage to a polymer solution that is injected at a steady rate, allowing the resulting fibers to be collected on a grounded plate [28]. The diameter and characteristics of the droplets can be modified by adjusting formulation parameters such as drug concentration, polymer concentration, and viscosity or by altering process parameters like voltage, nozzle type, or the distance to the collector plate. In contrast to electrospinning, which propels a continuous liquid jet to form fibers, electrospraying breaks the solvent jet into smaller droplets and requires a higher voltage to disrupt the Taylor cone into elongated particles [29, 30]. A key distinction between these methods lies in the concentration of the polymer solution used; electrospraying necessitates a more concentrated solution. Various nozzle types—such as single, coaxial, printing, multiple, or concentric—can be utilized in the electrospinning setup (Fig. 3.3). When subjected to an electric field, the flowing polymer solution forms a stable Taylor cone that ultimately deposits randomly arranged nanofibers, creating a patch. Electrospun nanofibers and

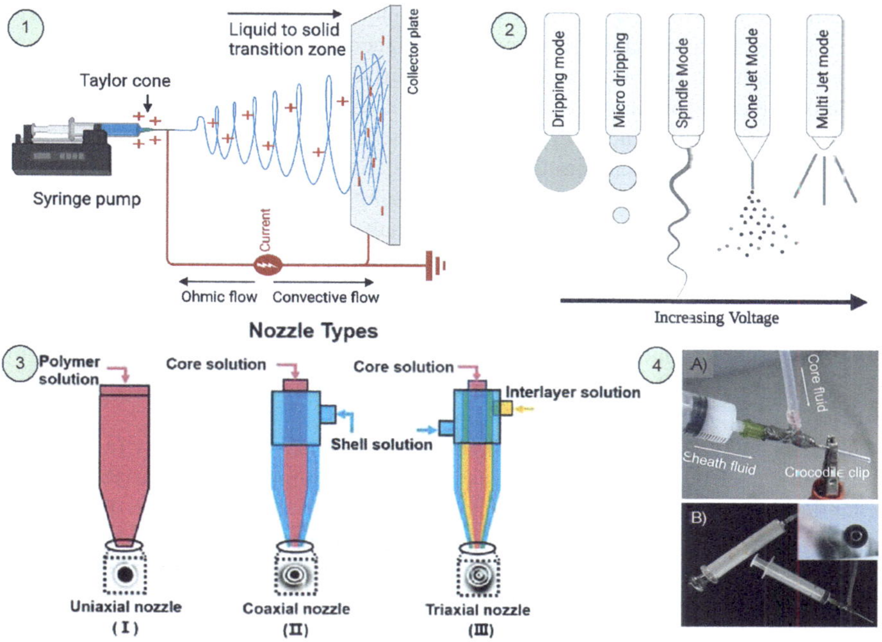

Fig. 3.3 (1) Basic setup of electrospraying and electrospinning techniques to fabricate nano-/microparticles or fibers using polymeric solutions. (2) Schematic diagram depicting the changes in electrospraying mode with increase in voltage. (3) Schematic representation of coaxial setup. (Taken from [39]). (4) (Modified coaxial electrospinning setup—syringe connections to the spinneret taken from [40])

nanoparticles can deliver a range of therapies along with small molecule drugs, genes, and proteins. The scalability of these electrospun patches highlights their potential for commercialization across numerous biomedical applications.

Andreadi et al. developed in situ gelling electrospun films of timolol maleate made up of poly(vinyl alcohol) (PVA) and poloxamer 407 for the management of intraocular pressure (IOP). The drug-loaded electrospun fibers were capable of in situ gelling (due to the presence of poloxamer) and prolonged drug release for 24 h [31]. Various modifications have been done in the electrospinning to control the release of active ingredients. For example, the release of an anti-inflammatory agent ofloxacin was prolonged by PVA/chitosan single nanofibers and PVA/chitosan/Eudragit RL 100 multilayer nanofibers [32]. The cylindrical collector was used to collect the nanofibers after electrospinning, and the second layer was coated onto the first layer for slower drug release. The PK analysis of single-layer and multiple-layer electrospun patches shows that the T_{max} is prolonged from 2 h to 5 h for single-layer and multiple-layer ofloxacin-loaded patches with AUC_{0-96} values of 3191 ± 117 and 2320 ± 60 mg*h/l, respectively, with an enhanced residence time of nearly 18%. Hasbiyani et al. developed collagen with varying amounts of hyaluronic acid and polyethylene oxide-based scaffolds via single-nozzle electrospinning. The uniform nature of electrospun nanofibers and resemblance to corneal surface morphology exhibited high compatibility with the ocular surface [28].

Coaxial electrospinning has emerged as an effective technique for encapsulating sensitive biomolecules such as proteins, vaccines, and other macromolecules, offering protection against denaturation and aggregation often associated with conventional methods [33]. The electrospun fibers using poly-e-caprolactone sheath as a core layer prolonged the release of bevacizumab. The prepared 500 nm fibers displayed sustained drug release upon subconjunctival administration. Three-layer nanofibers with an inner and outer layer of gelatin and a middle layer of PCL were successfully prepared with a 25 μm fiber diameter using a multiple nozzle system [34]. Yu prepared nanofibers made up of ethyl cellulose with varying amounts of ketoprofen in three layers of nanofibers. The prepared nanofibers demonstrated zero-order drug release over a period of 20 h [35].

Electrospun nanofiber patches show great promise for drug delivery to the anterior segment of the eye, offering sustained release, improved bioavailability, and potential for reduced dosing frequency compared to conventional eye drops [36]. Future directions for these patches include developing stimuli-responsive nanofibers for targeted drug delivery, incorporating multiple drugs for combination therapies, and optimizing the materials for enhanced mucoadhesion and corneal permeation [37]. Additionally, research efforts should focus on scaling up production methods and conducting more extensive in vivo studies to validate the efficacy and safety of these innovative ocular drug delivery systems [38].

3.5 Particulate Systems in the Management of Ocular Diseases

Particulate systems have emerged as a promising approach in the management of ocular diseases. They offer enhanced drug delivery and improved therapeutic outcomes. Nanocarriers have been categorized based on their fabrication methods and compositions, each with unique properties that can be optimized for specific applications as shown in Fig. 3.4. These systems, including liposomes, nanoparticles, and dendrimers, have shown significant potential in overcoming the challenges associated with conventional ocular drug delivery methods [41].

Liposomes have demonstrated remarkable efficacy in ocular drug delivery. For instance, Tan et al. developed a novel liposomal formulation for the delivery of cyclosporine A in the treatment of dry eye disease. Their study revealed that the liposomal formulation increased the drug's corneal residence time by 3.5-fold compared to conventional eye drops, resulting in improved therapeutic efficacy. Similarly, Wang et al. utilized mucin-1 aptamer-functionalized liposomes for the delivery of cyclosporin A in dry eye diseases. This approach led to a sevenfold increase in drug retention time in human corneal epithelial cells, demonstrating the potential of targeted liposomal systems in ocular therapeutics [42].

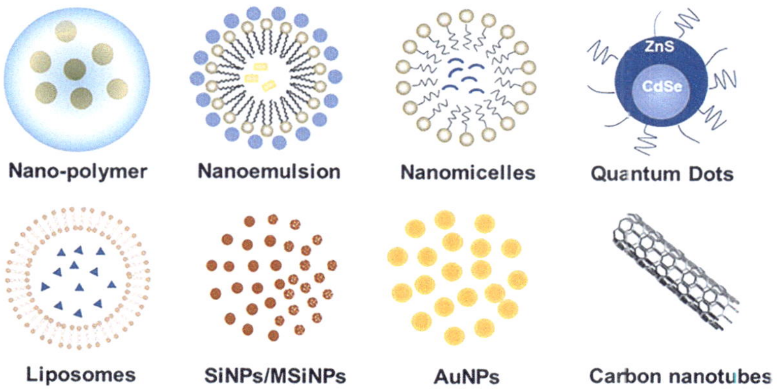

Fig. 3.4 Different types of nanoparticles used in ocular drug delivery systems. (Reproduced with copyright permission from Ref. [41])

Nanoparticles have also shown promise in ocular drug delivery. Elmowafy et al. developed metazolamide-loaded mPEG-PCL micelles for glaucoma treatment. The prepared nanoparticles demonstrated nearly 93% encapsulation efficiency and significantly improved the drug's bioavailability in the eye [43]. In another study, Karumanchi et al. prepared bevacizumab-loaded liposomes that provided prolonged drug release and increased patient compliance in the treatment of ocular angiogenesis [44].

Dendrimers, with their unique structural properties, have gained attention in ocular drug delivery. Wang et al. utilized PAMAM dendrimer-based drug delivery systems for glaucoma treatment. The study showed that dendrimer formulations could prolong precorneal retention and enhance corneal permeation, allowing for multidrug loading and hierarchical structure construction [45]. Additionally, Souza et al. investigated the effect of cationic and anionic dendrimers on corneal penetration efficiency. They found that anionic PAMAM G4 dendrimers increased the solubility of dexamethasone by 3.9-fold and led to a 6.6-fold increase in drug concentration in the aqueous humor compared to conventional formulations [46].

Table 3.2 presents a comprehensive overview of commercial particulate systems developed for the treatment of various ocular disorders, along with their pharmacokinetic (PK) and pharmacodynamic (PD) findings. These formulations represent significant advancements in ocular drug delivery, offering improved therapeutic outcomes compared to conventional methods. The table highlights the advancements in particulate systems for ocular drug delivery, showcasing their potential to improve drug bioavailability and therapeutic outcomes in treating ocular diseases. Particulate systems have shown significant advances in recent years for ocular drug delivery, demonstrating their potential to revolutionize the management of various eye disorders by improving drug bioavailability, increasing retention time, and enhancing therapeutic efficacy.

Table 3.2 Commercial particulate systems intended for the treatment of ocular disorders and their PK/PD findings

Product name	Drug and drug loading	Indication	Type of formulation	Polymer(s)/lipids	PK/PD findings	Route of administration	Phase of approval
Restasis®	Cyclosporin A	Dry eye syndrome	Nanoemulsion	Castor oil	$AUC_{0-72\,h}$ 14,333.2 ng/g.h, corneal clearance of 1.4 g/h [47]	Topical	Approved
Cyclokat®	Cyclosporin A	Dry eye syndrome	Cationic nanoemulsion	Castor oil	AUC_{0-72h} 26,477 ng/g.h Systematic absorption below LOD (0.1 ng/ml) [48]	Topical	Approved
Cequa®	Cyclosporin A	Keratoconjunctivitis sicca	Nano micellar solution	Octoxynol-40, polyoxyl 40 hydrogenated castor oil	$AUC_{0-1\,h}$ 828.25 ± 53.2 ng/g.h [49]	Topical	Approved
Visudyne®	Verteporfin	Choroidal neovascularization	Liposomes	EPG and DMPC (3:5 molar ratio)	AUC_{0-t} (μg.h/ml) 1.62, clearance 99.6 [50]	Injection	Approved
Lacrisek®	Vitamin A and E	Dry eye syndrome	Liposomal spray	Hydrogenated phospholipids	The tear blink intervals were more with Lacrisek® compared to Artelac rebalance® [51]	Topical	Approved
Artelac rebalance®	Vitamin B12	Dry eye syndrome	Liposomal eye drops	Hyaluronic acid, polyethylene glycol 8000	The aqueous-based Artelac Rebalance® was found to improve TBUT [51]	Topical	Approved

Ikervis®	Cyclosporin A	Keratitis in dry eye disease	Cationic nanoemulsion	Medium chain triglycerides, glycerol, cetalkonium chloride, poloxamer, tyloxapol	$AUC_{0-72\,h}$ 26,703.0 ng/g.h, corneal clearance of 0.8 h/h [47]	Topical	Approved
OCS-01	Dexamethasone and cyclodextrin	Postoperative corneal inflammation	Nanoparticles	Cyclodextrin	51% of patients (post cataract surgery) experienced absence of anterior inflammation vs. 19.6% with vehicle control and 72.5% vs. 54.9% with no pain with OCS-01 and vehicle control, respectively [52, 53]	Topical	Phase II
SeeQ	CdSe nanoparticle	Retinitis pigmentosa (RP)	Nanoparticles	Cadmium selenium	BCVA was decreased in patients with RP at least 6 lines in 1 h after IVT [54]	Intravitreal injections	NA
LE-MPP	Loteprednol etabonate	Postoperative inflammation and pain	MPP (mucus penetrating particles)	Pluronic F127	$AUC_{0-12\,h}$ of LE-MPP in cornea and conjunctiva was 1.5-fold higher than lotemax 0.5% eye drops [55]	Topical	Phase III

3.6 Drug-Eluting Contact Lenses

Drug-eluting contact lenses (DECLs) represent a significant advancement in ocular drug delivery systems, addressing the fundamental limitations of traditional methods such as eye drops. These innovative devices are designed to release therapeutic agents directly to the ocular surface, thereby enhancing bioavailability and providing sustained drug delivery over extended periods [56]. The integration of drug delivery capabilities with vision correction has opened new possibilities in the treatment of various ocular conditions, leading to improved patient compliance and therapeutic outcomes. Traditional eye drops, while widely used, suffer from poor bioavailability with typically less than 5% of the applied drug reaching the target tissue, making the development of more efficient delivery systems crucial for optimal therapeutic outcomes [57].

3.6.1 Types of Drug-Eluting Contact Lens and Their Manufacturing Process

The development of drug-eluting contact lenses has evolved through various approaches, each employing distinct materials and fabrication techniques to optimize drug release profiles and therapeutic efficacy. Silicone hydrogel-based systems have emerged as a prominent platform due to their high oxygen permeability and comfort during extended wear [58].

The most prominent method of loading drug molecules in contact lens is often seen as soaking method where the contact lens is soaked in drug solution and releases drug due to corneal concentration gradient. These CLs often show limited improvement in sustaining the release of drug; hence, various other manufacturing methods have been explored to sustainably improve the drug release kinetics of drug-eluting contact lens [59]. Various studies have shown that contact lenses containing varied proportions of crosslinking co-monomers can significantly enhance drug loading capacity as well as drug release by improving the interaction of drug within the contact lens matrix [60, 61]. Research has demonstrated that contact lenses containing 2.5 mol % exhibited enhanced loading of ofloxacin and neomycin, showing 18- and 53-fold increases, respectively, compared to conventional pHEMA contact lenses. These improvements are primarily attributed to charge interactions between the co-monomers and the drug molecules [61].

The integration of nanotechnology has introduced novel possibilities in DECL fabrication. Researchers have developed contact lenses embedded with drug-loaded zwitterionic nanogels using cast molding methods. These nanogels, typically ranging from 220 to 300 nm in diameter, provide superior drug loading capacity while maintaining lens transparency [62]. Studies have demonstrated sustained drug release for up to 6 days under physiological conditions, representing a significant improvement over conventional delivery methods. Lai and Shiau reported significant improvements in ophthalmic drug delivery through pH-triggered release mechanisms, with their system achieving sustained release for up to 120 h under physiological conditions. The large-pore mesoporous structure, combined with surface modification techniques, enabled precise control over drug release kinetics while responding to environmental pH changes [63].

While the traditional soaking method remains widely used in drug incorporation, its limitations of initial burst release and restricted drug loading capacity have prompted exploration of alternative approaches. Molecular imprinting and supercritical carbon dioxide impregnation have emerged as promising techniques [58]. Recent work by Harsolekar et al. demonstrated the importance of hydrogel technology in enhancing the sustained release of ganciclovir from therapeutic contact lenses, achieving drug loading efficiencies of up to 85% with controlled release profiles extending beyond 72 hours [64]. The incorporation of ionic monomers into the polymer matrix has shown considerable improvements in drug loading and release kinetics, as evidenced by the work of Minami et al., who reported a threefold increase in drug retention compared to conventional methods [65].

The development of stimuli-responsive systems represents another significant advancement in DECL technology. These smart materials can respond to various environmental triggers such as temperature, pH, or specific biomolecules, enabling on-demand drug release [62]. Thermo-sensitive nanogel-laden systems have shown promise, with research demonstrating controlled release of therapeutic agents in response to temperature changes at the ocular surface. These systems utilize the phase transition behavior of specific polymers to modulate drug release rates, providing enhanced control over therapeutic delivery [66].

3.6.2 Applications and Therapeutic Potential

Drug-eluting contact lenses have demonstrated significant promise across various therapeutic applications. In glaucoma management, these devices have shown remarkable efficacy in delivering pressure-lowering medications. Groundbreaking research by Ciolino et al. demonstrated that latanoprost-eluting contact lenses could maintain therapeutic levels for up to a month in animal models, representing a substantial improvement over traditional eye drop administration [67].

The application of DECLs in myopia control has also shown promising results, with studies demonstrating sustained drug release for up to 240 hours while maintaining optimal refractive correction. This extended release profile offers significant advantages for patient compliance and therapeutic outcomes. Their system achieved steady-state drug release for up to 2 weeks while maintaining optimal refractive correction. This dual-functionality approach has particular significance in pediatric populations, where compliance with traditional eye drop regimens can be challenging.

The treatment of ocular infections represents another significant application area. Studies conducted by Kang et al. have demonstrated that antibiotic-releasing contact lenses achieve superior corneal drug distribution compared to topical solutions, with sustained therapeutic concentrations maintained for extended periods. Their work with polydopamine-coated composites showed enhanced antibacterial efficacy against common ocular pathogens while maintaining good biocompatibility [68]. In the management of inflammatory conditions, research has shown considerable success in utilizing DECLs for delivering anti-inflammatory agents. Behl et al. successfully demonstrated the delivery of dexamethasone sodium phosphate using chitosan nanoparticle-loaded lenses, achieving sustained therapeutic levels with reduced systemic exposure [69].

Recent developments have also explored the potential of DECLs in treating anterior segment diseases. The incorporation of cyclosporine A, a challenging drug to deliver effectively to the ocular surface, has shown promise in treating dry eye syndrome and other inflammatory conditions [70]. Studies have demonstrated successful loading and release of this hydrophobic drug through various formulation strategies, including the use of vitamin E as a diffusion barrier and cyclodextrin complexation to enhance drug solubility [62].

3.6.3 Current Limitations and Challenges

Despite their promising potential, drug-eluting contact lenses face several significant challenges that must be addressed for widespread clinical adoption. Technical constraints remain a primary concern, particularly regarding the initial burst release of drugs observed with conventional soaking methods (Hsu et al.). This burst release phenomenon can result in suboptimal therapeutic levels during the intended treatment period, necessitating careful formulation design and drug-loading strategies to achieve controlled release profiles [71].

The potential alteration of lens optical properties during drug incorporation requires careful consideration, as noted by Ioniță. Drug loading can affect critical lens parameters such as light transmission, water content, and oxygen permeability. The maintenance of lens transparency and refractive properties while achieving therapeutic drug loads remains a delicate balance that requires optimization through careful material selection and processing techniques [72].

Manufacturing challenges present substantial hurdles in the commercialization of DECLs. The complexity of fabrication processes affects scalability, with current methods often requiring multiple steps and precise control of environmental conditions. Quality control requirements are particularly stringent, as variations in drug distribution within the lens matrix can affect both therapeutic efficacy and optical performance. Regulatory compliance presents additional challenges, with authorities requiring extensive stability data and demonstration of consistent drug release profiles across different manufacturing batches [70].

Clinical implementation faces its own set of challenges. Contact lens discomfort during extended wear periods remains a significant concern, particularly for drug-loaded lenses compared to conventional contact lenses. The risk of corneal toxicity with prolonged use necessitates careful consideration of drug loading levels and release patterns. Patient compliance issues, as highlighted by Rykowska et al., are particularly relevant in chronic conditions requiring long-term therapy. While

DECLs can improve drug delivery efficiency, the need for proper lens handling and maintenance may present barriers to adoption in certain patient populations [73].

3.6.4 Future Directions and Prospects

The future of drug-eluting contact lenses appears increasingly promising, driven by continuous advances in materials science and drug delivery technologies. Smart materials capable of responding to environmental stimuli represent a particularly exciting frontier. Research by Kim et al. has demonstrated the integration of wireless sensors with drug delivery systems, enabling real-time monitoring of intraocular pressure while maintaining controlled drug release patterns for glaucoma therapy [74].

Advanced drug delivery systems incorporating novel polymer compositions show significant promise. Recent developments in multilayer lens designs have achieved sequential drug release profiles, with different therapeutic agents released at predetermined intervals. Work by Desai et al. demonstrated successful incorporation of multiple drugs with minimal cross-interference, maintaining consistent release profiles throughout the therapeutic duration [75].

The integration of bio-responsive elements into DECL designs offers potential for precise therapeutic control. Researchers have developed enzyme-responsive systems that modulate drug release based on specific biological triggers present in tear fluid. These smart systems can potentially adjust therapeutic delivery based on disease progression or healing status, offering personalized treatment approaches. Additionally, the incorporation of biomarkers for real-time monitoring of drug levels could enable feedback-controlled release systems, optimizing therapeutic outcomes while minimizing side effects.

Emerging fabrication technologies, such as 3D printing and microfluidic approaches, are opening new possibilities in DECL manufacturing. These advanced techniques offer improved control over lens architecture and drug distribution, potentially addressing current limitations in drug loading uniformity and release kinetics. Torun et al. (2022) demonstrated the feasibility of microfluidic-based lens fabrication, achieving precise control over drug distribution patterns while maintaining optimal optical properties [76]. Drug-eluting contact lenses represent a revolutionary approach to ocular drug delivery, offering unique advantages over traditional methods. The integration of advanced materials science with innovative fabrication techniques has enabled the development of sophisticated delivery systems capable of addressing various ocular conditions. While significant challenges remain in terms of manufacturing scalability, drug stability, and clinical implementation, ongoing research continues to advance solutions to these limitations [58].

The future of DECLs appears particularly promising with the emergence of smart materials, personalized medicine approaches, and advanced manufacturing technologies. As research progresses and regulatory pathways become better defined, these devices are likely to play an increasingly important role in ophthalmology. The potential for combining therapeutic delivery with diagnostic capabilities suggests a future where ocular treatment becomes more precise, personalized, and patient-friendly [62].

3.7 Microneedles

Microneedles (MNs) are micron-scale arrays characterized by needle lengths of 50 µm to 2 mm and outer diameters below 300 µm [77]. Pioneered by Hashmi et al. in 1995 via silicon-based wet etching [78], MN fabrication has expanded to materials like stainless steel, dextrin, ceramics, and biodegradable polymers. These devices, attached to patches or syringes, penetrate tissues without stimulating nerve endings, enabling pain-free delivery. Conventional hypodermic needles are plagued by pain, infection risks, and biohazard waste. MNs enhance patient compliance through self-administration and targeted delivery [79].

Different types of microneedles include solid, dissolving, hollow, coated, hydrogel-forming, and particle-loaded designs (Fig. 3.5). Solid MNs, often metal or silicon-based, create microchannels for subsequent drug application. Coated MNs employ dip-coated tips for one-step delivery of proteins, DNA, or nanoparticles. Dissolving MNs, crafted from biodegradable polymers like polyvinyl alcohol (PVA), release therapeutics as they degrade, eliminating device removal. Hollow MNs, such as Clearside Biomedical's SCS Microinjector®, enable direct suprachoroidal delivery, exemplified by Xipere® (triamcinolone acetonide) for uveitis and ongoing trials for diabetic macular edema (PEACHTREE) [80, 81].

MNs overcome ocular barriers to deliver biologics like anti-VEGF agents. Kim et al. developed stainless steel MNs (400 µm length) coated with bevacizumab-loaded carboxymethylcellulose, achieving targeted stromal delivery in corneal neovascularization models [82]. Dissolving MNs encapsulate drugs within biocompatible polymers, though limited loading capacity challenges high-dose requirements. Hydrogel-forming MNs, such as PVA-based curved arrays, sustain immunoglobulin G1 release for 4–6 weeks, outperforming intravitreal injections. Amer et al. engineered interlocking hydrogel MNs with 80% enhanced adhesion, enabling zero-order IgG1 release over 6 weeks [80, 83].

Advanced MN systems include bilayer, 4D-printed, and particle-loaded designs. Lee et al. created detachable MNs with porous PVA/PVP sacrificial layers, achieving rapid separation post-application. Particle-loaded MNs, fabricated via water-in-oil-in-water emulsions, localized ovalbumin-PLGA nanoparticles in scleral tissues for 24 hours and sustained protein release over 2 months [83]. Song et al.'s spring-loaded corneal applicator delivered sunitinib malate effectively in neovascularization models, outperforming traditional needle-dipped methods [83].

MNs are revolutionizing ocular therapy by enhancing precision and reducing invasiveness. Their versatility spans vascular, gastrointestinal, and cardiac applications, with cost-effective production promising scalable use. Continued innovation in materials and designs such as adjustable drug loading and hybrid systems will address current limitations, strengthening MNs as a foundation of next-generation drug delivery.

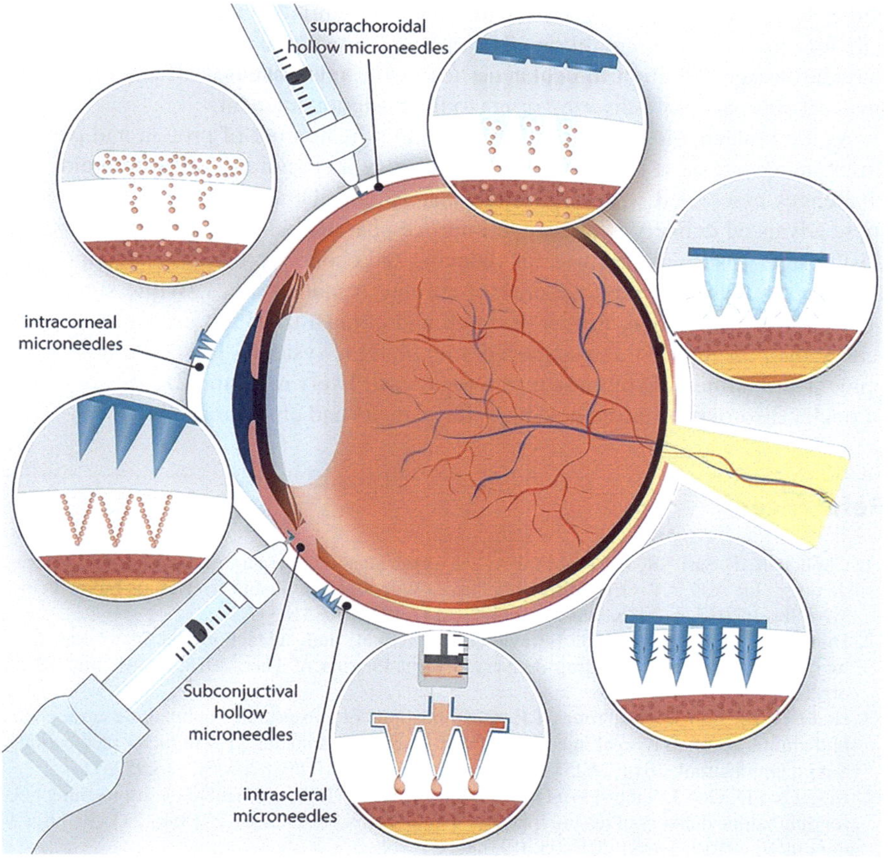

Fig. 3.5 Schematic representation of different types of microneedles for ocular application. Reproduced with copyright permission from Ref. [10]

3.8 Conclusions and Future Perspectives

The field of ocular drug delivery has made significant strides in recent years, driven by the need to overcome the limitations of conventional delivery methods. While topical administration remains a preferred route due to its simplicity and patient acceptance, its low bioavailability (<5%) and inability to effectively treat posterior segment diseases have spurred the development of innovative delivery technologies.

This chapter has explored a range of advanced ocular drug delivery systems, including preformed and in situ forming implants, nanoparticulate systems, and microneedle technology. These emerging technologies offer promising solutions for sustained drug release, enhanced tissue penetration, and improved pharmacokinetic profiles. Preformed implants and in situ forming systems have demonstrated the ability to provide long-term controlled release, potentially extending dosing intervals for chronic ocular conditions. Nanoparticulate systems have shown enhanced permeation and retention in ocular tissues, while microneedles offer a minimally invasive approach for delivering drugs to the posterior segment.

As the biotechnology industry advances and the number of protein and peptide drugs increases, these delivery systems will play a crucial role in overcoming the challenges associated with large molecular weight therapeutics. The potential of these advanced delivery strategies to improve pharmacokinetic profiles and extend dosing intervals holds promise for reducing the economic burden and improving patient compliance in the treatment of chronic ocular diseases. While significant progress has been made, further research and clinical trials are necessary to fully realize the potential of these advanced drug delivery systems. The continued development and refinement of these technologies will likely revolutionize the treatment of ocular disorders, offering new hope for patients and clinicians alike.

References

1. Cholkar K, Dasari SR, Pal D, Mitra AK. 1 - eye: anatomy, physiology and barriers to drug delivery. In: A.K.B.T.-O.T, Mitra R, editors. Woodhead publishing series in biomedicine. Woodhead; 2013. p. 1–36. https://doi.org/10.1533/9781908818317.1.
2. Hartman RR, Kompella UB. Intravitreal, subretinal, and suprachoroidal injections: evolution of microneedles for drug delivery. J Ocul Pharmacol Ther. 2018;34:141. https://doi.org/10.1089/JOP.2017.0121.
3. Haddock LJ, Ramsey DJ, Young LH. Complications of subspecialty ophthalmic care: endophthalmitis after intravitreal injections of anti-vascular endothelial growth factor medications. Semin Ophthalmol. 2014;29:257–62. https://doi.org/10.3109/08820538.2014.959616.
4. Li L, Guo D, Guo J, Song J, Wu Q, Liu D, Bi H, Xie X. Thermosensitive in-situ forming gels for ophthalmic delivery of tea polyphenols. J Drug Deliv Sci Technol. 2018;46:243–50. https://doi.org/10.1016/J.JDDST.2018.05.019.
5. Rohde F, Walther M, Wächter J, Knetzger N, Lotz C, Windbergs M. In-situ tear fluid dissolving nanofibers enable prolonged viscosity-enhanced dual drug delivery to the eye. Int J Pharm. 2022;616:121513. https://doi.org/10.1016/J.IJPHARM.2022.121513.
6. Islam MS, Ang BC, Andriyana A, Afifi AM. A review on fabrication of nanofibers via electrospinning and their applications. SN Appl Sci. 2019;1:1–16. https://doi.org/10.1007/S42452-019-1288-4/TABLES/1.
7. Vora LK, Sabri AH, Naser Y, Himawan A, Hutton ARJ, Anjani QK, Volpe-Zanutto F, Mishra D, Li M, Rodgers AM, Paredes AJ, Larrañeta E, Thakur RRS, Donnelly RF. Long-acting microneedle formulations. Adv Drug Deliv Rev. 2023;201:115055. https://doi.org/10.1016/J.ADDR.2023.115055.
8. Celik N, Khoramnia R, Auffarth GU, Sel S, Mayer CS. Complications of dexamethasone implants: risk factors, prevention, and clinical management. Int J Ophthalmol. 2020;13:1612. https://doi.org/10.18240/IJO.2020.10.16.
9. Lee J-H, Pidaparti RM, Atkinson GM, Moorthy RS. Design of an implantable device for ocular drug delivery. J Drug Deliv. 2012;2012:1–8. https://doi.org/10.1155/2012/527516.
10. Glover K, Mishra D, Gade S, Vora LK, Wu Y, Paredes AJ, Donnelly RF, Singh TRR. Microneedles for advanced ocular drug delivery. Adv Drug Deliv Rev. 2023;201:115082. https://doi.org/10.1016/j.addr.2023.115082.
11. Kang-Mieler JJ, Rudeen KM, Liu W, Mieler WF. Advances in ocular drug delivery systems. Eye. 2020;34:1371–9. https://doi.org/10.1038/s41433-020-0809-0.

12. Kumar A, Pillai J. Implantable drug delivery systems. In: Nanostructures for the engineering of cells, tissues and organs. Elsevier; 2018. p. 473–511. https://doi.org/10.1016/B978-0-12-813665-2.00013-2.
13. García-Estrada P, García-Bon MA, López-Naranjo EJ, Basaldúa-Pérez DN, Santos A, Navarro-Partida J. Polymeric implants for the treatment of intraocular eye diseases: trends in biodegradable and non-biodegradable materials. Pharmaceutics. 2021;13:701. https://doi.org/10.3390/pharmaceutics13050701.
14. Stewart SA, Domínguez-Robles J, Donnelly RF, Larrañeta E. Implantable polymeric drug delivery devices: classification, manufacture, materials, and clinical applications. Polymers (Basel). 2018;10:1379. https://doi.org/10.3390/polym10121379.
15. Ng XW, Liu KL, Veluchamy AB, Lwin NC, Wong TT, Venkatraman SS. A biodegradable ocular implant for long-term suppression of intraocular pressure. Drug Deliv Transl Res. 2015;5:469–79. https://doi.org/10.1007/s13346-015-0240-4.
16. Rafeef M, Rana O, Mo'tasem A, Bashar KM, Rawda A, Mohammed Q. Technical insight into biodegradable polymers used in implants, Jordan. J Pharm Sci. 2018;11
17. Emilia U, Sarah A, Camila P, Dominguez-Robles J, Eneko L. Classification, material types, and design approaches of long-acting and implantable drug delivery systems. In: Larrañeta E, Thakur RRS, Donnelly RF, editors. Long-acting drug delivery systems: pharmaceutical, clinical, and regulatory aspects. 1st ed. Matthew Deans; 2021. p. 17–59.
18. Singh IP, Berdahl JP, Sarkisian SR, Voskanyan LA, Ang RE, Doan LV, Applegate D, Shen Y, Katz LJ, Kothe AC, Navratil T. Long-term safety and efficacy evaluation of travoprost intracameral implant based on pooled analyses from two phase III trials. Drugs. 2024;84:1299–311. https://doi.org/10.1007/s40265-024-02074-9.
19. Iyer S, Radwan AE, Hafezi-Moghadam A, Malyala P, Amiji M. Long-acting intraocular delivery strategies for biological therapy of age-related macular degeneration. J Control Release. 2019;296:140–9. https://doi.org/10.1016/j.jconrel.2019.01.007.
20. Fea AM, Vallino V, Cossu M, Marica V, Novarese C, Reibaldi M, Petrillo F. Drug delivery systems for glaucoma: a narrative review. Pharmaceuticals. 2024;17:1163. https://doi.org/10.3390/ph17091163.
21. Tsung T-H, Chen Y-H, Lu D-W. Updates on biodegradable formulations for ocular drug delivery. Pharmaceutics. 2023;15:734. https://doi.org/10.3390/pharmaceutics15030734.
22. Belamkar A, Harris A, Zukerman R, Siesky B, Oddone F, VerticchioVercellin A, Ciulla TA. Sustained release glaucoma therapies: novel modalities for overcoming key treatment barriers associated with topical medications. Ann Med. 2022;54:343–58. https://doi.org/10.1080/07853890.2021.1955146.
23. Medeiros FA, Walters TR, Kolko M, Coote M, Bejanian M, Goodkin ML, Guo Q, Zhang J, Robinson MR, Weinreb RN, Agar A, Coote M, Bathijia R, Liu L, Roberts T, Faschinger C, Vass C, Collignon N, Alves Pereira AC, Belfort de Mattos R, Dantas FJ, Lopes da Silva MJ, Kanadani F, dos Santos Silva LM, Prata T, Bach-Holm D, Kolko M, Lai J, Tham C, Bátor G, Szalczer L, Varsányi B, Blumenthal E, Geyer O, Lavartovsky S, Pedut-Kloizman T, Shoham-Hazon N, Lujan S, Abela B, Ang RE, Leuenberger EU, Uy H, Yap-Veloso MI, Fryczkowski P, Jurowski P, Kalużny B, Kalużny J, Misiuk-Hojlo M, Raczynska K, Tomczyk-Dorozynska W, Wasyluk J, Zalewski S, Zarnowski T, Feijoó JG, Giménez-Gómez R, Griño EM, López AA, Miralles MG, Moreno JM, Polo V, Taulet EC, Zúñiga BP, Chen Y-Y, Lee Y-C, Alpern L, Berlin MS, Brubaker J, Caldwell D, Camp A, Cantor LB, Caronia R, Crane CJ, Day D, Duzman E, Elfervig J, El-Harazi S, Evans R, Fisher AC, Flynn WJ, Foster CS, Frenkel R, Goyal R, Gross R, Hartman PJ, Haynes WL, Jerkins G, Kim J, Kim M, Kwapiszeski B, Lambright B, Larsen C, Lehmann J, Levenson JH, Logan D, McMillan B, Martel JR, Mayer H, Medeiros F, Moroi S, Moyes A, Myers J, Nairn J, Nielsen S, Ortiz DP, Paauw J, Pai V, Panarelli J, Park A, Qazi MA, Ragusa N, Rhee DJ, Rothman R, Sampson R, Seltzer S, Shrivastava A, Simmons ST, Sims A, Slabaugh MA, Smetana S, Smith O, So SC, Stalmans I, Swarup J, Wallshein J, Walters TR, Zaman F, Zhang R. Phase 3, randomized, 20-month study of bimatoprost implant in open-angle glaucoma and ocular hypertension (ARTEMIS 1). Ophthalmology. 2020;127:1627–41. https://doi.org/10.1016/j.ophtha.2020.06.018.

24. Santos A, Sinn Aw M, Bariana M, Kumeria T, Wang Y, Losic D. Drug-releasing implants: current progress, challenges and perspectives. J Mater Chem B. 2014;2:6157–82. https://doi.org/10.1039/C4TB00548A.
25. Mahmud H, Ahmad TR, Gonzales JA, Stewart JM. Efficacy of the Fluocinolone Acetonide (Yutiq) Intravitreal implant as monotherapy for uveitis. Ocul Immunol Inflamm. 2023;31:1603–7. https://doi.org/10.1080/09273948.2022.2076131.
26. Teymoorian S, Kaur J. Travoprost Intracameral implant in eyes with glaucoma or ocular hypertension: early short-term real-world outcomes. Clin Ophthalmol. 2025;19:157–66. https://doi.org/10.2147/OPTH.S498431.
27. Ichhpujani P, Thakur S. iDose TR sustained-release travoprost implant for the treatment of glaucoma. US Ophthalmic Rev. 2023;17:4. https://doi.org/10.17925/USOR.2023.17.1.4.
28. Hasbiyani NAF, Hikmawati D, Siswanto S. Electrospun collagen-based scaffold as therapeutic agent for ocular chemical injury. J Phys Conf Ser. 2020;1445:12022. https://doi.org/10.1088/1742-6596/1445/1/012022.
29. Guo Q, Mather JP, Yang P, Boden M, Mather PT. Fabrication of polymeric coatings with controlled microtopographies using an Electrospraying technique. PLoS One. 2015;10:e0129960.
30. Pawar A, Thakkar S, Misra M. A bird's eye view of nanoparticles prepared by electrospraying: advancements in drug delivery field. J Control Release. 2018;286:179–200. https://doi.org/10.1016/j.jconrel.2018.07.036.
31. Andreadis II, Karavasili C, Thomas A, Komnenou A, Tzimtzimis M, Tzetzis D, Andreadis D, Bouropoulos N, Fatouros DG. In situ gelling electrospun ocular films sustain the intraocular pressure-lowering effect of Timolol maleate: in vitro, ex vivo, and pharmacodynamic assessment. Mol Pharm. 2022;19:274–86. https://doi.org/10.1021/acs.molpharmaceut.1c00766.
32. Mirzaeei S, Taghe S, Asare-Addo K, Nokhodchi A. Polyvinyl alcohol/chitosan single-layered and polyvinyl alcohol/chitosan/Eudragit RL100 multi-layered electrospun nanofibers as an ocular matrix for the controlled release of Ofloxacin: an in vitro and in vivo evaluation. AAPS Pharm Sci Tech. 2021;22:170. https://doi.org/10.1208/s12249-021-02051-5.
33. Xie J, Ng WJ, Lee LY, Wang C-H. Encapsulation of protein drugs in biodegradable microparticles by co-axial electrospray. J Colloid Interface Sci. 2008;317:469–76. https://doi.org/10.1016/j.jcis.2007.09.082.
34. Liu W, Ni C, Chase DB, Rabolt JF. Preparation of multilayer biodegradable nanofibers by Triaxial electrospinning. ACS Macro Lett. 2013;2:466–8. https://doi.org/10.1021/mz4000688.
35. Yu D-G, Li X-Y, Wang X, Yang J-H, Bligh SWA, Williams GR. Nanofibers fabricated using Triaxial electrospinning as zero order drug delivery systems. ACS Appl Mater Interfaces. 2015;7:18891–7. https://doi.org/10.1021/acsami.5b06007.
36. Mishra D, Gade S, Raghu T, Singh R. Ocular disorders and associated drug development challenges: the current scenario. EC Ophthalmol. 2020;11:92–101. https://pure.qub.ac.uk/en/publications/ocular-disorders-and-associated-drug-development-challenges-the-c. Accessed 31 May 2024.
37. Das A, Shetty S, Chethan KN, Shetty R, SuranjanSalins S. Electrospun nanofibers: transformative innovations in biomedical applications and future prospects in healthcare advancement. Cogent Eng. 2024;11 https://doi.org/10.1080/23311916.2024.2433147.
38. Nadaf A, Gupta A, Hasan N, Fauziya N, Ahmad S, Kesharwani P, Ahmad FJ. Recent update on electrospinning and electrospun nanofibers: current trends and their applications. RSC Adv. 2022;12:23808–28. https://doi.org/10.1039/D2RA02864F.
39. Feng K, Huangfu L, Liu C, Bonfili L, Xiang Q, Wu H, Bai Y. Electrospinning and electrospraying: emerging techniques for probiotic stabilization and application. Polymers. 2023;15:2402. https://doi.org/10.3390/POLYM15102402.
40. Qian C, Liu Y, Chen S, Zhang C, Chen X, Liu Y, Liu P. Electrospun core–sheath PCL nanofibers loaded with nHA and simvastatin and their potential bone regeneration applications. Front Bioeng Biotechnol. 2023;11:1205252. https://doi.org/10.3389/FBIOE.2023.1205252/BIBTEX.

41. Yang C, Yang J, Lu A, Gong J, Yang Y, Lin X, Li M, Xu H. Nanoparticles in ocular applications and their potential toxicity. Front Mol Biosci. 2022;9:931759. https://doi.org/10.3389/FMOLB.2022.931759/BIBTEX.
42. Wong KY, Liu Y, Zhou L, Wong MS, Liu J. Mucin-targeting-aptamer functionalized liposomes for delivery of cyclosporin a for dry eye diseases. J Mater Chem B. 2023;11:4684–94. https://doi.org/10.1039/D3TB00598D.
43. Elmowafy E, Gad H, Biondo F, Casettari L, Soliman ME. Exploring optimized methoxy poly(ethylene glycol)-block-poly(ε-caprolactone) crystalline cored micelles in anti-glaucoma pharmacotherapy. Int J Pharm. 2019;566:573–84. https://doi.org/10.1016/J.IJPHARM.2019.06.011.
44. Karumanchi DK, Skrypai Y, Thomas A, Gaillard ER. Rational design of liposomes for sustained release drug delivery of bevacizumab to treat ocular angiogenesis. J Drug Deliv Sci Technol. 2018;47:275–82. https://doi.org/10.1016/J.JDDST.2018.07.003.
45. Wang J, Li B, Kompella UB, Yang H. Dendrimer and dendrimer gel-derived drug delivery systems: breaking bottlenecks of topical administration of glaucoma medications. MedComm Biomater Appl. 2023;2:e30. https://doi.org/10.1002/MBA2.30.
46. Souza JG, Dias K, Silva SAM, De Rezende LCD, Rocha EM, Emery FS, Lopez RFV. Transcorneal iontophoresis of dendrimers: PAMAM corneal penetration and dexamethasone delivery. J Control Release. 2015;200:115–24. https://doi.org/10.1016/J.JCONREL.2014.12.037.
47. Lallemand F, Schmitt M, Bourges J-L, Gurny R, Benita S, Garrigue J-S. Cyclosporine a delivery to the eye: a comprehensive review of academic and industrial efforts. Eur J Pharm Biopharm. 2017;117:14–28. https://doi.org/10.1016/j.ejpb.2017.03.006.
48. Lallemand F, Daull P, Benita S, Buggage R, Garrigue J-S. Successfully improving ocular drug delivery using the cationic nanoemulsion, novasorb. J Drug Deliv. 2012;2012:604204. https://doi.org/10.1155/2012/604204.
49. Mandal A, Gote V, Pal D, Mitra AK, Ogundele A, Mitra AK. Ocular pharmacokinetics of a topical ophthalmic nanomicellar solution of cyclosporine (Cequa®) for dry eye disease. Pharm Res. 2019;36:36.
50. Houle J-M, Strong A. Clinical pharmacokinetics of Verteporfin. J Clin Pharmacol. 2002;42:547–57. https://doi.org/10.1177/00912700222011607.
51. Garrigue J-S, Amrane M, Faure M-O, Holopainen JM, Tong L. Relevance of lipid-based products in the management of dry eye disease. J Ocul Pharmacol Ther. 2017;33:647–61. https://doi.org/10.1089/jop.2017.0052.
52. Wong CW, Metselaar JM, Storm G, Wong TT. A review of the clinical applications of drug delivery systems for the treatment of ocular anterior segment inflammation. Br J Ophthalmol. 2021;105:1617–22.
53. Papangkorn K, Truett KR, Vitale AT, Jhaveri C, Scales DK, Foster CS, Montieth A, Higuchi JW, Brar B, Higuchi WI. Novel dexamethasone sodium phosphate treatment (DSP-Visulex) for noninfectious anterior uveitis: a randomized phase I/II clinical trial. Curr Eye Res. 2019;44:185–93.
54. Jackson TL, Mandava N, Quiroz-Mercado H, Benage M, Garcia-Aguirre G, Morales-Canton V, Wilbur L, Olson J. Intravitreal quantum dots for retinitis pigmentosa: a first-in-human safety study. Nanomedicine. 2021;16:617–26. https://doi.org/10.2217/nnm-2020-0471.
55. Schopf L, Enlow E, Popov A, Bourassa J, Chen H. Ocular pharmacokinetics of a novel LoteprednolEtabonate 0.4% ophthalmic formulation. Ophthalmol Ther. 2014;3:63–72. https://doi.org/10.1007/s40123-014-0021-z.
56. Patel KD, Silva LVB, Park Y, Shakouri T, Keskin-Erdogan Z, Sawadkar P, Cho KJ, Knowles JC, Chau D, Kim H. Recent advances in drug delivery Systems for Glaucoma Treatment. Mater Today Nano. 2022;18:100178. https://doi.org/10.1016/j.mtnano.2022.100178.
57. Zhang S, Zhu J, Zhao Y, Zhang W, Miao Z, Guo J, Bu H. Recent advances in ocular drug delivery systems. ZhongguoLinchuangYaolixueZazhi. 2015;31:580–3. https://doi.org/10.13599/j.cnki.1001-6821.2015.07.028.

58. Yang H, Zhao M, Xing D, Zhang J, Fang T, Zhang F, Nie Z, Liu Y, Yang L, Li J, Wang D. Contact lens as an emerging platform for ophthalmic drug delivery: a systematic review. Asian J Pharm Sci. 2023;18:100847. https://doi.org/10.1016/j.ajps.2023.100847.
59. Sarmout M, Xiao Y, Hu X, Rakhmetova A, Koole LH. A novel approach to achieve semi-sustained drug delivery to the eye through asymmetric loading of soft contact lenses. Heliyon. 2023;9:e16916. https://doi.org/10.1016/J.HELIYON.2023.E16916.
60. Hiratani H, Alvarez-Lorenzo C. The nature of backbone monomers determines the performance of imprinted soft contact lenses as timolol drug delivery systems. Biomaterials. 2004;25:1105–13. https://doi.org/10.1016/S0142-9612(03)00622-7.
61. Lee D, Lee N, Kwon I. Efficient loading of ophthalmic drugs with poor loadability into contact lenses using functional comonomers. Biomater Sci. 2018;6:2639–46. https://doi.org/10.1039/C8BM00586A.
62. Wang Z, Li X, Zhang X, Sheng R, Lin Q, Song W, Hao L. Novel contact lenses embedded with drug-loaded Zwitterionic Nanogels for extended ophthalmic drug delivery. Nano. 2021;11:2328. https://doi.org/10.3390/NANO11092328.
63. Lai C, Shiau F-J. Enhanced and extended ophthalmic drug delivery by pH-triggered drug-eluting contact lenses with large-pore mesoporous silica nanoparticles. ACS Appl Mater Interfaces. 2023;15:18630–8. https://doi.org/10.1021/acsami.2c22860.
64. Harsolekar M, Ansari M, Supe S, Singh K. Formulation development and evaluation of therapeutic contact lens loaded with ganciclovir. Int Ophthalmol. 2023; https://doi.org/10.1007/s10792-022-02618-6.
65. Minami T, Ishida W, Kishimoto T, Nakajima I, Hino S, Arai R, Matsunaga T, Fukushima A, Yamagami S. In vitro and in vivo performance of Epinastine hydrochloride-releasing contact lenses. PLoS One. 2019;14:e0210362. https://doi.org/10.1371/journal.pone.0210362.
66. Lee S-H, Kim H, Kim D-H, Chang W, Vales TP, Kim J-W, Kim KH, Kim J. Thermo-sensitive Nanogel-laden Bicontinuous microemulsion drug-eluting contact lenses. J Biomed Mater Res B Appl Biomater. 2018;107:1159–69. https://doi.org/10.1002/jbm.b.34209.
67. Ciolino JB, Ştefănescu C, Ross AE, Salvador-Culla B, Cortez P, Ford EM, Wymbs KA, Sprague SL, Mascoop DR, Rudina SS, Trauger SA, Cade F, Kohane DS. In vivo performance of a drug-eluting contact lens to treat glaucoma for a month. Biomaterials. 2014; https://doi.org/10.1016/j.biomaterials.2013.09.032.
68. Kang M, Moon K-J, Lee J, Jeong Y. Antibacterial activity of antibiotic-releasing Polydopamine-coated nephrite composites for application in drug-eluting contact lens. Materials. 2022;15:4823. https://doi.org/10.3390/ma15144823.
69. Behl G, Iqbal J, O'Reilly NJ, McLoughlin P, Fitzhenry L. Synthesis and characterization of poly(2-Hydroxyethylmethacrylate) contact lenses containing chitosan nanoparticles as an ocular delivery system for dexamethasone sodium phosphate. Pharm Res. 2016;33:1638–48. https://doi.org/10.1007/s11095-016-1903-7.
70. Rykowska I. Drug-modified contact lenses – properties, release kinetics, and stability of active substances with particular emphasis on Cyclosporine A: a review. Molecules. 2024;29:2609. https://doi.org/10.3390/molecules29112609.
71. Hsu K-H, Carbia BE, Plummer CE, Chauhan A. Dual drug delivery from Vitamin E loaded contact lenses for glaucoma therapy. Eur J Pharm Biopharm. 2015;94:312–21. https://doi.org/10.1016/j.ejpb.2015.06.001.
72. Ioniță M. Advances in therapeutic contact lenses for the management of different ocular conditions. J Pers Med. 2023;13:1571. https://doi.org/10.3390/jpm13111571.
73. Rykowska I, Nowak I, Nowak R. Soft contact lenses as drug delivery systems: a review. Molecules. 2021;26:5577. https://doi.org/10.3390/molecules26185577.
74. Kim TY, Mok J, Hong SH, Jeong SH, Choi H, Shin S, Joo C, Hahn SK. Wireless Theranostic smart contact lens for monitoring and control of intraocular pressure in glaucoma. Nat Commun. 2022;13 https://doi.org/10.1038/s41467-022-34597-8.
75. Desai AR, Maulvi FA, Desai DM, Shukla M, Ranch KM, Vyas B, Shah S, Sandeman S, Shah DO. Multiple drug delivery from the drug-implants-laden silicone contact lens: addressing the issue of burst drug release. Mater Sci Eng C. 2020;112:110885. https://doi.org/10.1016/j.msec.2020.110885.

76. Torun H, Fazla B, Arman S, Özdalgiç B, Yetisen AK. Microfluidic contact lenses for ocular diagnostics and drug delivery. Nano Select. 2022;4:79–89. https://doi.org/10.1002/nano.202200202.
77. Bariya SH, Gohel MC, Mehta TA, Sharma OP. Microneedles: an emerging transdermal drug delivery system. J Pharm Pharmacol. 2012;64:11–29. https://doi.org/10.1111/j.2042-7158.2011.01369.x.
78. Hashmi S, Ling P, Hashmi G, Reed M, Gaugler R, Trimmer W. Genetic transformation of nematodes using arrays of micromechanical piercing structures. BioTechniques. 1995;19:765–70.
79. Henry S, McAllister DV, Allen MG, Prausnitz MR. Microfabricated microneedles: a novel approach to transdermal drug delivery. J Pharm Sci. 1998;87:922–5. https://doi.org/10.1021/js980042+.
80. Thakur RRS, Tekko IA, Al-Shammari F, Ali AA, McCarthy H, Donnelly RF. Rapidly dissolving polymeric microneedles for minimally invasive intraocular drug delivery. Drug Deliv Transl Res. 2016;6 https://doi.org/10.1007/s13346-016-0332-9.
81. Kim YC, Grossniklaus HE, Edelhauser HF, Prausnitz MR. Intrastromal delivery of bevacizumab using microneedles to treat corneal neovascularization. Invest Ophthalmol Vis Sci. 2014;55:7376–86. https://doi.org/10.1167/iovs.14-15257.
82. Owens CD, Gasper WJ, Walker JP, Alley HF, Conte MS, Grenon SM. Safety and feasibility of adjunctive dexamethasone infusion into the adventitia of the femoropopliteal artery following endovascular revascularization. J Vasc Surg. 2014;59:1016–24. https://doi.org/10.1016/j.jvs.2013.10.051.
83. Wu Y, Vora LK, Wang Y, Adrianto MF, Tekko IA, Waite D, Donnelly RF, Thakur RRS. Long-acting nanoparticle-loaded bilayer microneedles for protein delivery to the posterior segment of the eye. Eur J Pharm Biopharm. 2021;165:306–18. https://doi.org/10.1016/j.ejpb.2021.05.022.

Drug Delivery to the Posterior Segment of Eye

4

Sofiya Tarannum and Keerti Jain

Abstract

Diseases associated with posterior segment of the eye like age-related maculopathy and diabetic retinopathy are prime reasons that extremely menace normal vision as well as healthy well-being, and their therapy is highly difficult for ophthalmologists, researchers, and specialists. Delivery of drugs for the treatment of posterior segment oculopathy using noninvasive procedures is the need of time; however, it is very challenging. Recently, several new topical systems have been formulated which are promising due to the improved patient compliance as the intervention of surgery or needles is negligible. This chapter primarily focuses on the different means of drug delivery for the therapy of conditions associated with the posterior segment of the eye. In the first part of this chapter, the anatomy of ocular posterior segment and the associated diseases are crisply discussed. Further, the biological ocular barriers that prevent the drug delivery are mentioned. Afterward, both invasive and noninvasive routes explored by the researchers for the therapy of posterior segment oculopathy are discussed. As a follow-up, new strategies in the delivery of therapeutics via the noninvasive route using different nanovehicles and contact lenses are extensively elaborated which also draws the conclusion that the potential of nanoformulations for the therapy of posterior oculopathy must be explored at the full extent with systematic and comprehensive efforts by researchers.

S. Tarannum · K. Jain (✉)
Department of Pharmaceutics, Drug Delivery and Nanomedicine Research Laboratory, National Institute of Pharmaceutical Education and Research (NIPER)-Raebareli, Lucknow, Uttar Pradesh, India
e-mail: keertijain.02@niperraebareli.edu.in

N. K. Mehra (ed.), *Complex Ophthalmic Dosage Forms: Advances in Biomedical Applications and Future Perspectives*, https://doi.org/10.1007/978-981-96-6306-4_4

Keywords

Drug delivery · Posterior segment of eye · Noninvasive route · Nanoparticles · Contact lens

Abbreviations

BRB:	Blood retinal barrier
CNV:	Choroidal neovascularization
NLCs:	Nanostructured lipid carriers
NPs:	Nanoparticles
PCL:	Poly(ε-caprolactone)
PLA:	Poly (L-lactide)
PLGA:	Poly(lactic-co-glycolic acid)
PNPs:	Polymeric nanoparticles
PVA:	Polyvinyl alcohol
RPE:	Retinal pigment epithelium
SLNs:	Solid lipid nanoparticles

4.1 Introduction

The healthy eyes as a sensory organ are a precious gift from nature which can collect images and turn them into electrical signals which are then sent to the brain via the second cranial nerve [1]. Several conditions like inflammatory reactions or infections impact the normal functioning of the eye. Anterior oculopathies are treatable when exposed to therapeutics at a high dose. But conditions or diseases associated with ocular posterior segment are challenging as they are unreachable [2].

The posterior section of the eye primarily includes **(a) sclera** (white portion) which is strong, elastic fibrous tissue rich in collagen originating from the pachymeninx of the central nervous system and nonvascular in nature. **(b)** The **choroid** is sandwiched between sclera and retinal pigment epithelium (RPE) that comprises three layers: suprachoroid, different-sized blood vessels, and lamina vitrea. **(c) RPE** cells approximately 3.5 million in number are there in a single eye that form the occluding junctions and are responsible for maintaining the eye's defense mechanism as well as neutralizing the harm caused by the redox imbalance by producing cytokine enzymes like superoxide dismutase, hydroperoxidase, glutathione, and melanin. **(d)** The **retina** is made of parts like macula, optic nerve head, fovea centralis, and side vision area of the retina. This internal layer of the eye globe is comprised of photoreceptors called rods (around 7.7 million) and cones (5 million). The ocular posterior part is mainly covered by **(e) vitreous humor**, which is a gel-like fluid that aids in maintaining the structure of the eyeball (Fig. 4.1) [3].

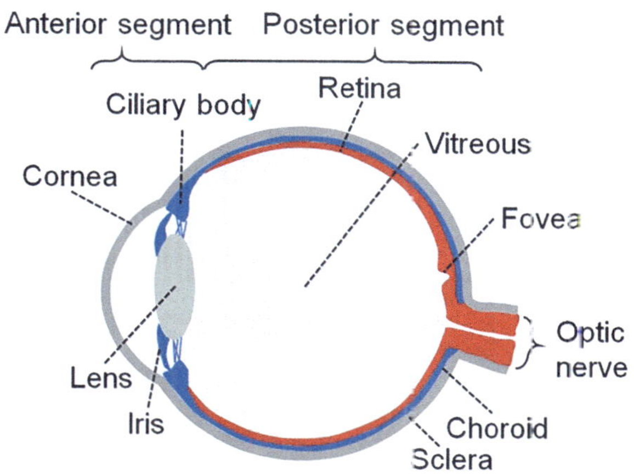

Fig. 4.1 Structure of the eye showing the anterior and posterior segments. (Figure reproduced from Chuang, Fields, Del Priore. 2017 [4] under a Creative Commons license)

Several posterior oculopathies that could become the reason for blindness if remained untreated are enlisted here. These are age-associated maculopathy, swelling in macula, glaucoma, diabetic macular edema; proliferative vitreoretinopathy; cytomegalovirus-induced retinal inflammation; fungal, bacterial, or viral infections in the inner coat of the eye; diabetic vitreoretinopathies; retinal dislocation; hole or slit in retina; uveitis; etc. [5]. Glaucoma, age-related macular degeneration, and diabetic retinopathy are the highly common oculopathies causing total vision loss in 39 million individuals, approximately as per the reports given by the International Federation on Ageing in 2012 [6, 7]. Therapeutics for ocular diseases are commonly available as solutions, suspensions, and ointments that are primarily applied topically. Apart from topical, routes like oral, intravenous, intraocular, subconjunctival, and implants are also explored for delivering therapeutics to the eye [8]. Targeting the ocular tissues is somewhat difficult because of the presence of ocular barriers that prevent the efficient delivery of drugs. Currently, intravitreal route is more common as it has the potential to maximize the interaction between the drugs and the posterior ocular diseased site. Anyhow, repetitive use of intravitreal route led to side effects like retinal dislocation, hemorrhage, inner eye infections, and elevated fluid pressure [9]. Although intravitreal implants can prevent frequent ruptures and release the drug in a prolonged manner, the insertion procedure is problematic due to incisional stress, inflammation, bleeding, as well as patient unacceptance [10, 11]. Therefore, therapies which can be administered without the aid of injections are mostly preferred, so that patient incompliance and side effects can be minimized.

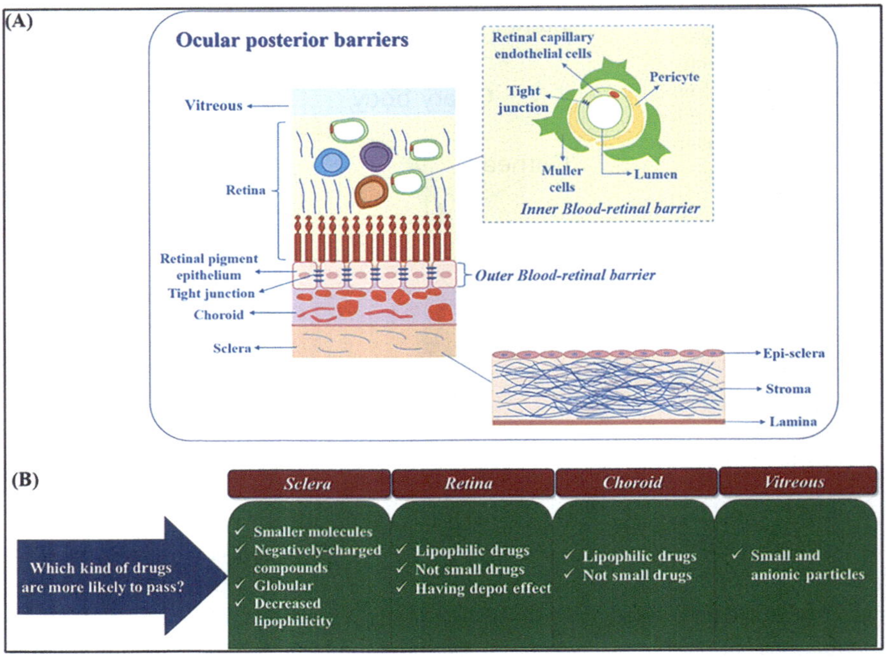

Fig. 4.2 (**a**) The diagrammatic representation of barriers for drug delivery to the posterior segment of the eye. (**b**) Listing of structural properties of therapeutics or nanovehicles influencing penetrability in different tissues which act as the barrier for drug delivery to the posterior segment of the eye. (Reproduced with permission from Qi et al. [3])

4.2 Reasons for Difficulty in Treating Diseases Associated with the Posterior Segment of the Eye

Static barriers (sclera, Bruch's membrane, blood-retinal barrier, efflux pumps) and dynamic barriers (choroidal blood and lymph circulation) present in the ocular posterior segment prevent the drugs' penetration into the targeted diseased site of ocular tissues as shown in Fig. 4.2.

4.2.1 Sclera

Sclera acts as the protective shield and stops the influx of foreign entities into the posterior segment of the eye. Primarily, properties of exogenous agents like size, physicochemical parameters, and surface charge decide the extent of permeability across sclera. As the hydrophobicity and radius of the molecules increase, their ability to permeate across sclera progressively decreases. Therefore, large-sized and highly hydrophobic drugs are not good candidates for the therapy of posterior oculopathies as they cannot cross the hydrophilic scleral minute openings. The positively charged drugs are not able to permeate because of their interaction and entrapment within the negatively charged pores and proteoglycan matrix of sclera. Additionally, the thickening of sclera at the posterior region makes the drug molecules' permeation nearly impossible [12].

4.2.2 Bruch's Membrane

Bruch's membrane/lamina vitrea plays a significant role in modulating the transport of nutrients, fluids, excrements, oxygen, and other biological molecules between the choroid (vascular layer) and retina [13]. With aging, the lamina vitrea gets thickened due to the deposition of calcium and mesh formation by the overly produced collagen fibers further making the permeability to retinal tissues more challenging [14]. Furthermore, the majority of therapeutics could not make it to Bruch's membrane and get lost in the choroid. Oculopathy like age-associated maculopathy is characterized by the lodging of excess amount of drusen bodies which can lead to damage due to the presence of free radicals as well as long-term inflammation to lamina vitrea. This drusen bodies accumulation lowers the exchange of nutrients and therapeutics, eventually resulting in the induction of secondary metabolic stress [15].

4.2.3 Blood Retinal Barrier

Blood retinal barrier (BRB) is formed by the zonulae occludent of RPE cells and retinal capillary endothelial cells. Fenestration-free BRB ensures that apart from O_2, CO_2, and tiny hydrophobic molecules, no drug molecule can passively diffuse to inner retinal tissues from the choroid. Thus, receptors or ATP-mediated transport mechanisms like fluid phase pinocytosis can be explored [16].

4.2.4 Efflux Pumps

Efflux pumps like P-glycoprotein (P-gp) or multidrug resistance protein (MRP) are up-regulated on both apical and basal parts of RPE. At the cellular level, these efflux pumps acting as the main barrier make sure that the efflux of exogenous moieties like drugs is properly done and very few molecules of drug reach within the cells [17].

4.2.5 Choroidal Blood and Lymph Circulation

The choroid is richly supplied with blood (around 80% of the entire blood supply to the eye) and lymph vessels [18]. Hydrophobic therapeutic agents are mostly drained into the choroidal and systemic bloodstream, restricting their further entry into the retinal tissues. This can eventually cause the availability of drug at subtherapeutic concentration in inner ocular tissues, thereby no intended medical effect is anticipated [19]. Sodium fluorescein like polar drugs is reported to have the least chance of being drained into the choroid ensuring high permeation.

4.3 Different Routes for Ocular Posterior Drug Delivery

A suitable route through which drugs can be delivered to maintain a therapeutic level at the diseased site is desirable for the attainment of therapeutic efficacy. The routes explored for ocular disease targeting are broadly categorized as invasive and noninvasive. Invasive administration comprises intravitreal, periorbital, and parenteral injections, whereas noninvasive routes include topical and oral routes [20]. Clinically, the intravitreal route is the most preferred route for the therapy of conditions associated with posterior ocular segment as the topical and other systemic routes have several limitations. Drug administration via the periorbital route has been explored, but its invasive nature is still a matter of concern [12]. The development of topical formulations which can effectively cure the posterior oculopathies has encouraged the researchers to do further research in this area. A schematic representation of different routes for ocular drug delivery is shown in Fig. 4.3.

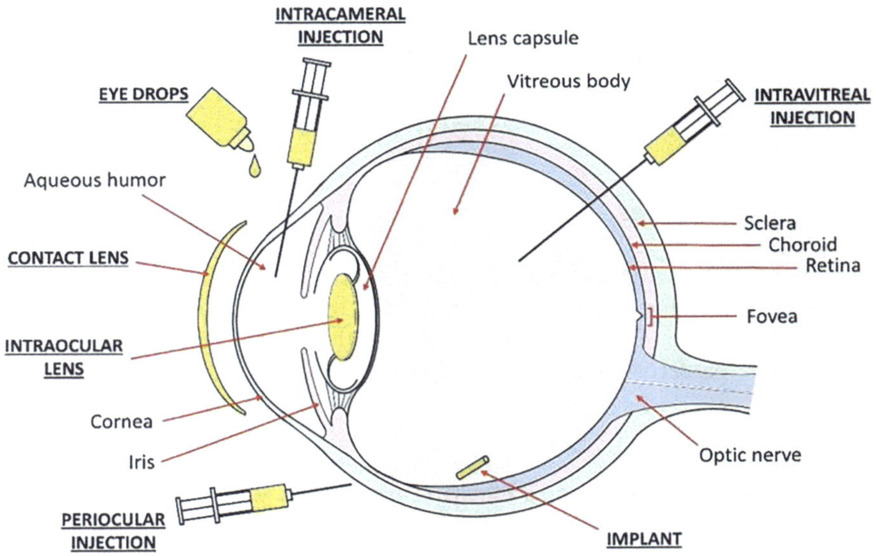

Fig. 4.3 Schematic representation of different routes for ocular drug delivery. (Figure reproduced from Toffoletto, Saramago, Serro. 2020 [21] under a Creative Commons license)

4.3.1 Invasive Administration Routes

4.3.1.1 Intravitreal Route

As the vitreous is in the vicinity of the retina, the intravitreal route is commonly used for administering drugs or drug-loaded nanocarriers for the therapy of posterior oculopathies involving both intravitreal injections and implants. Nowadays, the majority of available medications are administered via these routes due to advantages like attainment of high drug concentration in the vitreous, negligible systemic absorption, and maximum injection volume (0.1 mL), etc. [22]. The treatment using antivascular endothelial growth factor drugs or steroids is usually managed via this route [23]. Formulations administered using this route can be of solution type, hydrogels, microparticles, nanoparticles, and implants [24]. Out of these, implants and solution injections are commonly used for the therapy of posterior oculopathies [25]. The requirement of frequent dosing of drugs administered via this route and their untoward effects are the main limitations, which lead to patient incompliance. Additionally, frequent corticosteroid injections can accelerate the rate of fluid build-up in the eye [26]. Implants act as a promising delivery system as they can release the drugs in a sustained manner in the vitreous humor, prolong the residence time of the drug, and most importantly reduce the frequency of injections significantly [27].

Iluvien® is an intravitreal rod-shaped implant in which the tube is made up of nonbioerodable polyimine and one end is made impermeable due to the coating of silicone, whereas its interior is filled with polyvinyl alcohol (PVA) matrix and drug, fluocinolone acetonide. The other end is plugged with PVA which is biodegradable in nature and releases the drug in a zero-order manner for up to 3 years for the therapy of diabetic macular edema in patients with insignificant increase in intraocular pressure (Fig. 4.4) [28]. Ozurdex® is another FDA-approved intravitreal

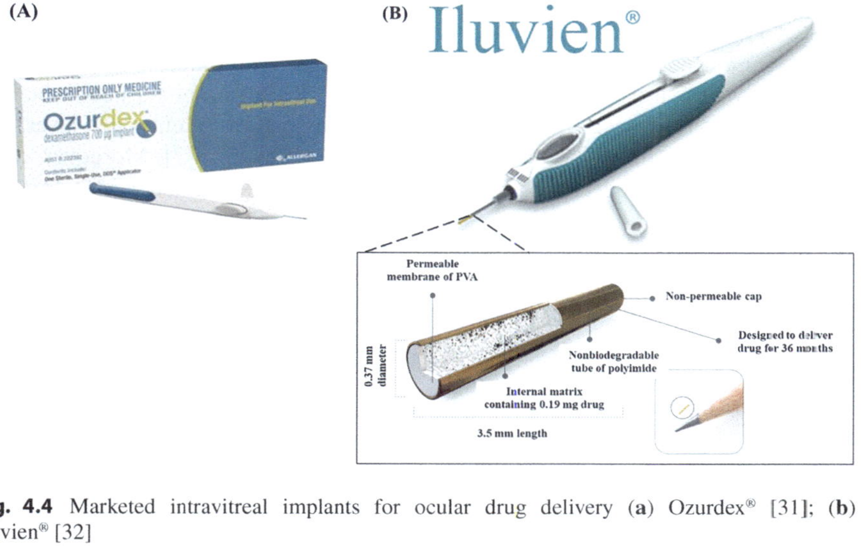

Fig. 4.4 Marketed intravitreal implants for ocular drug delivery (**a**) Ozurdex® [31]; (**b**) Iluvien® [32]

biodegradable implant, made up of poly(lactic-co-glycolic acid) (PLGA) loaded with dexamethasone, which is used for the treatment of diabetes-associated swelling of macula (Fig. 4.4) [29]. Nowadays, implants are becoming popular for the therapy of posterior ocular ailments, but the need for minor surgical interventions for insertion, removal, and replacement of some of the implants results in patient incompliance. Apart from this, the rate of reporting of numerous side effects like fundus disease, retinal separation, as well as damage is increasing with the use of implants. Surgical procedures can result in the blockage of blood vessels particularly in patients with comorbidities like heart and circulatory diseases [11]. However, despite these challenges, this route is commonly preferred as it ensures direct and local therapy [30].

4.3.1.2 Periocular Route

Periocular administration of drugs includes drug delivery through the spaces surrounding the eye, like subconjunctival, episcleral, intraconal, and extraconal. This route minimizes the pain during the administration as it is less invasive than the intravitreal route and increases the drug efficacy for the therapy of posterior oculopathies [33, 34]. This route can target the retina without damaging the sclera, thereby, significantly increasing patient compliance. However, side effects like inflammation and pain in the inner coat of the eye cannot be neglected and the need for medical specialists makes the use of this route less common [35]. As the drug is delivered in the periocular region, it must cross numerous barriers to reach the targeted diseased site in the posterior region of the eye.

4.3.1.3 Parenteral Route

The parenteral route is very uncommon in ocular disorders [20]. Liposomes of photosensitizer, verteporfin (Visudyne®), are available in market and administered via intravenous route for photodynamic therapy of exudative macular degeneration (Table 4.1) [36]. After delivery, the drug eventually reaches the vitreous body after circumventing the conjunctiva, sclera, and choroid via the bloodstream; however, the majority of active agents are unable to bypass the prime blood-ocular barrier [12]. As the choroid is rich in blood vessels, the drug can easily enter this layer, but its further penetration in the retinal blood vessel is difficult [37]. Moreover, drug degradation in the bloodstream and systemic side effects are also the limitations of this route [38].

4.3.2 Noninvasive Routes of Administration

4.3.2.1 Topical Route

Topical route for the delivery of drugs to the posterior segment of the eye is insufficient as less than 5% of the dose administered topically reaches the eye, of which only 0.001% is estimated to reach the posterior diseased site [39, 40]. Reasons like limited dosing (30 μL), very less residence time on the ocular surface, presence of numerous barriers, metabolic enzymes in tears, etc. make this route inefficient [41].

Table 4.1 Commercialized formulations for the therapy of posterior oculopathy

Administration routes	Dosage forms	Name of the drugs	Marketed product name	Indications	Efficacy duration	Approved year	Company
Intravitreal injection	Solution	Pegaptanib sodium	Macugen®	AMD	6 weeks	2004	NexStar Pharmaceuticals
	Suspension	Ranibizumab	Lucentis®	wAMD, RVO, DME, CNV	1 month	2006	Genemtech (US Biotech unit of Roche)
	Implant	Dexamethasone	Ozurdex®	RVO, DME, noninfectious uveitis	6 months	2009	Allergan
	Solution	Aflibercept	Eylea®	AMD, DME	2 months	2011	Bayer、Regeneron
		Conbercept	Langmu®	wAMD, DME, CNV	3 months	2013	Chengdu Kanghong Biotechnologies Co. Ltd.
	Implant	Fluocinolone Acetonide	Iluvien®	DME, uveitis	2 ~ 3 years	2014	Alimera Sciences
		Fluocinolone Acetonide	Yutiq®	Chronic noninfectious uveitis	3 years	2018	EyePoint
	Solution	Brolucizummab	Beovo®	wAMD	3 months	2019	Novartis
		Bevacizumab	Avastin®	wAMD	Off label	–	Genemtech (US biotech unit of Roche)
Surgical intravitreal implantation	Implant	Ganciclovir	Vitrasert®	Cytomegalovirus retinitis	6 months	1996	Bausch & Lomb
		Fluocinolone Acetonide	Retisert®	Chronic noninfectious uveitis	2.5 years	2004	
Intravenous injection	Liposome	Verteporfin	Visudyne®	wAMD	3 months	2002	Novartis

Reproduced with permission from Qi et al. [3]

Abbreviations: *AMD* age-related macular degeneration, *wAMD* wet age-related macular degeneration, *RVO* retinal vein occlusion, *DME* diabetic macular edema, *CNV* choroidal neovascularization

However, strategies are being explored by the researchers to overcome such limitations, so that noninvasive topical administration can be used and the systemic side effects of frequent injections can be avoided.

4.3.2.2 Oral Route

The main requirement for the therapy of posterior oculopathies with oral route is that the drug should be highly bioavailable [20, 42]. A combination of oral and topical delivery known as adjuvant therapy is also very common [37]. Lutein is a dietary supplement that is classically used for ocular disorders like cataracts and macular degeneration and is taken orally. It is an antioxidant as well as a crucial component of macular pigment, reported to be effective in lowering the probability and reducing the advancement of age-related macular degeneration [43, 44]. As the oral absorption proceeds, the therapeutic agents present in the circulatory system should be able to bypass the gut mucosal barrier, blood-aqueous barrier (BAB), and BRB. Moreover, numerous side effects and delayed onset of action are also the main disadvantages that need to be tackled for effective therapy via oral route [3].

4.4 Recent Advancements in Noninvasive Topical Drug Delivery Systems for Ocular Posterior Disease Therapy

The presence of distinctive and sensitive ocular barriers made drug delivery to the posterior segment of the eye extremely difficult [45]. The major advantages of the noninvasive route of administration are that it does not need the assistance of medical professionals, thereby convenient for the patients and safer than the injections. Although, limitations like low drug concentration at the diseased site due to poor bypassing of several ocular barriers need to be addressed [46]. Hence, advanced noninvasive nanoconstructs for the therapy of posterior oculopathies are the need of the time, and several have been studied preclinically already [47]. The next section of this chapter focuses on the topical nanoformulations based on micelles, nanoparticles (NPs), liposomes, dendrimers, and so on and the contact lens for the effective therapy of posterior eye diseases via noninvasive routes [48–51]. The diagrammatic representation of different drug-loaded nanocarriers is shown in Fig. 4.5.

4.4.1 Nanoparticles (NPs)

NPs are useful drug delivery systems with assured application safety, which can be categorized into lipidic NPs and polymeric nanoparticles (PNPs) as per their constituting components [52, 53]. Physiological and bioerodible solid lipids or combinations of solid/liquid lipids making solid lipid nanoparticles (SLNs) and nanostructured lipid carriers (NLCs), respectively, are very common [54, 55]. Biodegradable PNPs are composed of numerous polymers that incorporate drug molecules by complexation, entrapment, or conjugation. Properly prepared and characterized NPs for ocular diseases are easy to apply and convenient for patients because of the higher

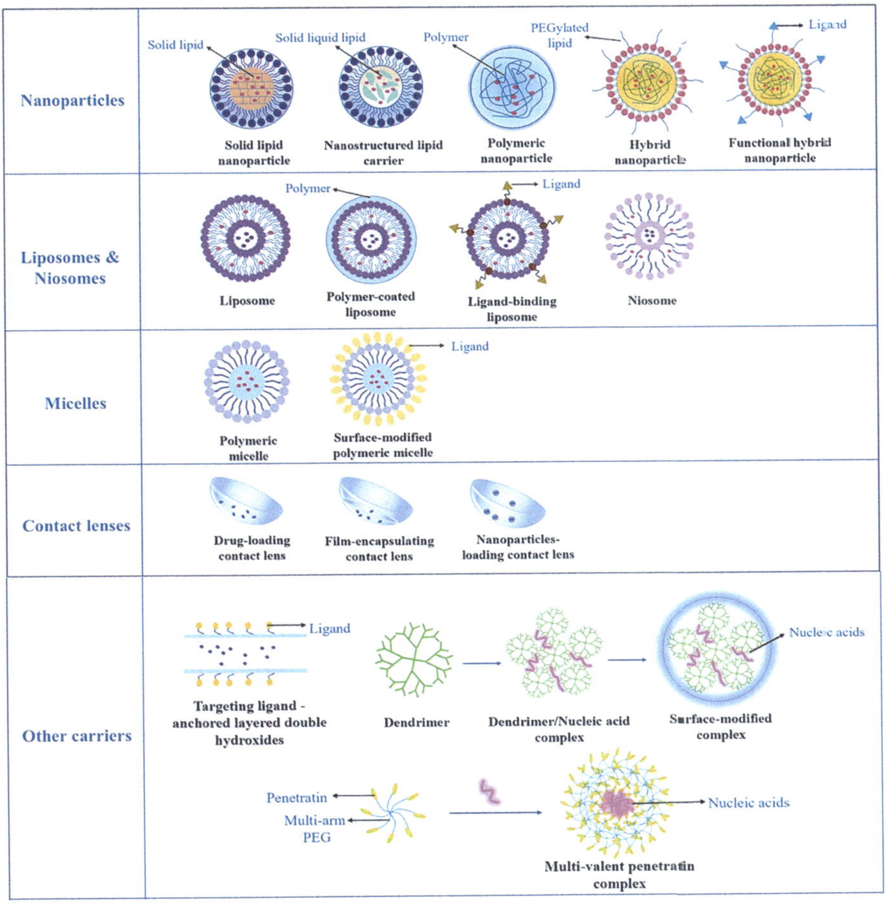

Fig. 4.5 Diagrammatic illustrations of drug-laden nanoplatforms formulated for noninvasive topical delivery for the therapy of posterior oculopathy. (Reproduced with permission from Qi et al. [3])

retention time at diseased sites with reduced dosing frequency [56]. NPs can efficiently deliver the drugs at the posterior eye segment and can also enhance their bioavailability; therefore, a lot of research is being focused on this area.

4.4.1.1 Lipidic Nanoparticles

Lipidic NPs are generally non-immunogenic as they are made up of physiological and biodegradable lipids [57]. These NPs are promising nanovehicles of drugs and are majorly categorized into two types, i.e., SLNs and NLCs. Lipidic NPs are extensively being explored for the therapy of eye disorders to deliver drugs, genes, and diagnostics [58]. The lipid constituents form the matrix of lipidic NPs and are primarily responsible for the properties exhibited by these colloidal nanocarriers. Free fatty acids, long-chain alcohols, acylglycerols, waxes, phospholipids, glycolipids, and sphingolipids are some of the commonly used lipids in the formulation of lipidic NPs. Some of these lipids can act as surfactants also [59, 60]. Octadecanoic acid, benzoic acid glycerol, palmitin, cetyl palmitate, glycerol monostearate, and triglyceride are the majorly explored solid lipids for SLNs, whereas liquid lipids like octadec-9-enoic acid and triacylglycerols are preferred in NLC formulations [61].

SLNs can be the ideal nanocarriers for ophthalmic drug delivery in the form of drops that can be easily administered by the patients as they are safe. SLNs can be sterilized, provide controlled release of drugs, and give the scope of active targeting with high entrapment efficiency. Their production and commercialization are easy compared to other nanostructures [62]. SLNs can permeate the corneal layer and enhance the bioavailability of the loaded drugs as well as their residence time and can also control the release pattern [63]. However, SLN instability is a major concern that needs to be addressed by the researchers. NLCs are considered a better alternative to SLNs as it is comprised of a combination of solid and liquid lipids. Currently, numerous topical SLNs and NLCs are under investigation for the safe and effective therapy of posterior oculopathies [64].

Triamcinolone acetonide-loaded NLCs were developed using high-pressure homogenizer and tested in mice via eye drop instillation [65]. The NLCs successfully delivered the drug to the posterior ocular site for a prolonged period without any invasion or puncture. Other topical NLCs loaded with fluorescence agents were formulated by conjugating fluorescein amine with stearic acid (solid lipid) [66]. The NLCs were tracked using the microscope, and their journey starting from their diffusion at the corneal level to their spread at the posterior sites (sclera and retina) was reported. These studies supported the potential of NLCs as a suitable carrier system for the management of posterior oculopathies.

4.4.1.2 Polymeric Nanoparticles (PNPs)

PNPs are usually solid colloidal nanocarriers of 10–1000 nm made up of either natural or synthetic polymers like chitosan, hyaluronic acid, poly(L-lactide) (PLA), PLGA, poly(ε-caprolactone) (PCL), and many more [67, 68]. Polymers with mucoadhesive properties are good option for development of ocular targeted PNPs, as they can increase the residence time within the cornea as well as at conjunctival epithelium [69–71]. Some of the advantages offered by the PNPs as ophthalmic drug delivery agents include that they can control the release and enhance the stability as well as bioavailability of loaded therapeutics [56, 72]. Recently, the approach of combining conventional ocular preparations with in situ gels and nanosystems is attracting the focus of researchers as they can ensure both enhanced residence time and sustained release. A topical formulation of riluzole-loaded PLGA NPs in situ gel was developed for providing neuroprotection in glaucoma patients. The prepared drug-loaded PLGA NPs in situ gel reached the posterior ocular site after 24 h of topical administration [73].

Furthermore, several polymers are modified using targeting agents for actively targeting the receptors expressed at the posterior ocular site and delivering the drugs. Axitinib-loaded polyethylene glycol (PEG)-distearoyl phosphatidylethanolamine

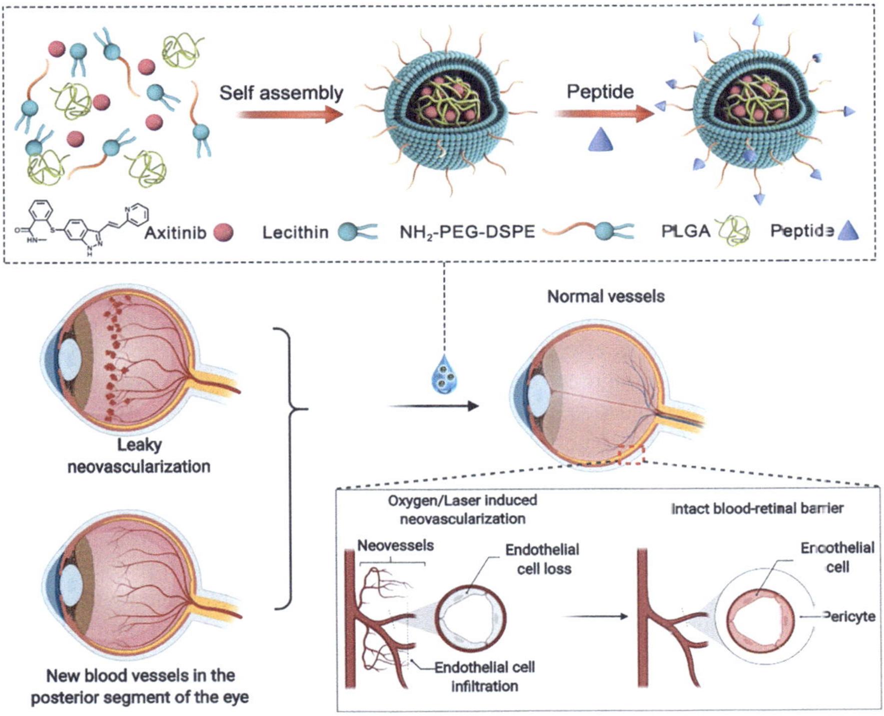

Fig. 4.6 Penetrating peptide facilitated axitinib delivery to the posterior ocular section for preventing new blood vessels formation. (Reproduced with permission from Huang et al. [74])

PNPs decorated with transmembrane receptor targeting peptide were developed in the form of an eye drop and administered to the interior of the eye. The functionalized PNPs significantly inhibited angiogenesis in choroidal neovascularization (CNV) as well as in oxygen-induced retinopathy (OIR) disease model (Fig. 4.6) [74]. Besides lipidic and polymeric NPs, metallic or carbon NPs have also shown applicability in ophthalmic drug delivery [75, 76]. Advantages exhibited by the nanocarriers like the controlled release, active targeting, and high encapsulation efficiency make them potential candidate in treating posterior oculopathies using the topical route, and the fabrication of such nanosystems could lead to a new paradigm in drug delivery to the posterior segment of the eye [46, 77].

4.4.2 Liposomes

Liposomes as the ocular drug delivery agent have been explored first in the 1980s for prolonged action. Liposomes are unilamellar or multilamellar vesicular system comprised of phospholipids and amphipathic lipids [78]. Structurally, conventional liposomes have an outer bilayer of lipids that encloses a hydrophilic center.

Verteporfin-loaded liposomes marketed as Visudyne® (Novartis) administered via intravenous route are used for the therapy of CNV, myopic degeneration, and ocular histoplasmosis [79]. Another intravenously administered marketed liposomal product for the treatment of age-related macular degeneration is Photrex® (Miravant) which contains rostaporfin [80]. Research and development in the field of fabrication of liposomes that can be administered topically for the therapy of posterior ocular diseases is very promising.

The topically administered liposomes can improve the permeation of loaded drugs across the ocular tissues and lengthen their residence time [81]. As the drugs remained unexposed to ocular environment because of entrapment within the liposomes, their hydrolysis is prevented, resulting in enhanced bioavailability. The ability of liposomes to act as effective drug delivery agent depends on several factors like employed preparation method that will decide attributes such as drug encapsulation efficiency, particle size, zeta potential, and affinity to corneal surface [82]. However, traditional liposomes have some limitations like poor encapsulation efficiency and limited ocular permeability, which can be overcome by coating these liposomes with polymers that are very well compatible with the posterior ocular tissues. Triamcinolone acetonide–loaded chitosan-capped liposomes were prepared and administered topically to the eyes of rats. Around 12.7 μg of drug was detected in the vitreous humor of the dissected right eye of rats that were treated with chitosan-capped liposomes for 15 days, whereas the presence of drug was not observed in the left vitreous humor that was treated with pure drug. The prepared chitosan-capped liposomes can be seen as the potential nanocarrier system for the therapy of posterior oculopathies as they were able to effectively penetrate the cornea and get accumulated in the vitreous humor [83].

4.4.3 Niosomes

Niosomes are stable bilayer nanocarriers made up of nonionic surfactants, which can encapsulate both polar and nonpolar molecules. They are basically preferred over liposomes as they are comparatively inexpensive, stable, and have less chance of oxidative degradation [84, 85]. It is favored in ocular drug delivery as it can protect the therapeutics from metabolic enzymes and ensure site-specific release in a constant manner. Due to its lipidic nature, this system has the potential to improve the absorption and decrease the systemic drainage of drugs, thereby resulting in their prolonged retention at the diseased site, which eventually improves their bioavailability than the traditional drug solutions. The advantages offered by niosomes in enhancing the permeability as well as residence in posterior ocular sites in delivering the therapeutics are significant [86, 87].

Niosomes made up of Span 60 and Tween 80 are evaluated for their corneal permeability using porcine cornea. A significant increase in permeability up to 67% than the drug suspension was reported [88]. As per the result of the fluorescent emission, 6-carboxy fluorescein–loaded niosomes administered as the eyedrops were successful in delivering the compound to the posterior section. Niosomes loaded with cyclopentolate can enhance absorption by changing the conjunctival as well as scleral permeability. Additionally, no irritation was reported with the administration of niosomal systems [86]. Hence, niosomes are considered as the potential elastic delivery vesicles for both polar and nonpolar drugs to the posterior segment of the eye via topical route.

4.4.4 Polymeric Micelles

Amphiphilic block copolymers are the building blocks of polymeric micelles known for their core/shell structures, which are characterized by the lipophilic and hydrophilic regions, in the nanometre size range [89]. Advantages like easy fabrication methods, scale-up potential, biocompatible nature, high encapsulation efficiency, and surface functionalization shifted the focus of researchers to polymeric micelles for effectively delivering the drugs in the body [90]. The polymers employed for micelle fabrication must be biodegradable, and the degraded by-products should be safe without any toxicity and inflammations to the ocular tissue. The most used polymers explored for micelles formation are PLA, poly (propylene oxide) (PPO), poly (lactic acid-co-polyethylene glycol), and PCL. Lipophilic drugs can be loaded in the core region of micelles and appear as clear solution, which cannot cause the blurring of vision and inflammation that occasionally occur with conventional topicals [91, 92].

The topically administered micelles must travel through the conjunctiva-scleral or the corneal route to reach the posterior ocular section. When the micelles are administered topically, they can perfuse from the conjunctiva to the sclera through the polar pores of size 30–300 nm, after which they cross choroid, RPE, and retina [91]. The nano size range and hydrophilic nature make the micelles to permeate the cornea and travel via the scleral pores [93]. D-α-tocopheryl polyethylene glycol succinate (TPGS) or TPGS/poloxamer 407 micellar systems were formulated for increasing the solubility of simvastatin and delivering it to the posterior ocular section through scleral tissue penetration. The ex vivo results using the porcine model suggested that TPGS micelles efficiently delivered simvastatin via the conjunctiva/sclera [94]. Dexamethasone-loaded chitosan oligomer-stearic acid Valyl-Valyl-Valine/1–2-Dioleoyl-sn-glycero-3-phospho-ethanolamine (CSO-SA-VVVS:2/DOPE) micelles were formulated for targeting posterior section of the eye using peptide transporter-1 (PepT-1). The micelles showed improved tissue penetration and successfully transported to the posterior section of the eye without undergoing any destruction [95].

4.4.5 Contact Lens

Drug-loaded corneal contact lens as ocular drug delivery system was first used in 1965. The modified contact lens can improve the bioavailability and prolong the retention time for effectively managing the posterior oculopathies. The data reported from the clinical studies have shown the high drug penetration from the drug-loaded soft contact lens as compared to subconjunctival injections, thereby drug-laden lenses have shown improvement in the drug bioavailability and reduction in the untoward effects [96, 97]. Release of drug in a sustained manner results in the achievement of optimal drug concentration at the posterior ocular diseased site. Acuvue® Theravision™, the first drug-eluting contact lens, was produced by Johnson & Johnson for the treatment of itching and inflammation associated with ocular allergies, but this ketotifen-loaded lens was discontinued from the market in June 2023 because it did not achieve the sales expectations due to factors like limited market reach, potential concerns about the effectiveness of the low dose of ketotifen delivered by the lens, and the need for further research to optimize drug delivery technology in contact lenses [98, 99]. Another drug-laden contact lens was developed, in which dexamethasone-PLGA film was prepared and incorporated in the lens with a total loading amount of drug 1484 ± 118 μg resulting in sustained release of drug to the retina for about 1 week in both anterior and posterior segments. These drug-laden lenses could be explored as a better alternative to eye drops which need to be taken hourly [100].

Moreover, topical contact lenses have also been laden with drug-loaded nanocarriers instead of plain drugs and explored as an effective noninvasive nanodrug delivery system. pHEMA-constructed hydrogels laden with silica shell cross-linked micelles (SSCMs) have been developed for the delivery of dexamethasone acetate. The novel strategy of administering drug-laden micelles-hydrogel contact lenses was observed to be a unique approach for the therapy of both anterior and posterior oculopathy via prolonged drug delivery [101]. The advancements in contact lenses that have grown over time have demonstrated the convenience and practicality for use in posterior eye drug delivery.

4.5 Conclusion

The ocular conditions associated with the posterior section are responsible for issues like blindness in elderly people. The existing therapies for posterior ophthalmopathy are invasive in nature, which have a reliable clinical efficacy, but at the same time, these invasive therapies (intravitreal injection and implants) face challenges like patient incompliance, discomfort, severe complications, and most importantly require professional's assistance and postoperative recovery. Thus, extensive research on noninvasive delivery systems is required for the treatment of posterior section oculopathy, specifically in the aging population. Yet, the development of effective topical formulation for the therapy of posterior oculopathy is extremely challenging, because of the complex physiological barriers of the eye. At the beginning of the 1990s, scientists gave attention to formulating drug delivery systems for sustained release to improve patient compliance and for efficient treatment of eye disorders [102]. With the advent and continuous advancements in the field of nanotechnology, biotechnology as well as biomaterials engineering, the formulation and development of effective topical systems for the therapy of posterior oculopathies are being investigated by researchers significantly.

Most of the ocular topical formulations for the treatment of posterior ophthalmopathy are in their initial phase of development and their availability in market is still seen as a milestone to achieve due to numerous issues like less efficacy, production costs, scalability, as well as regulatory clearance [103]. Translating theoretical research into practical clinical applications for posterior oculopathies presents a significant challenge. This necessitates not only innovative drug delivery system development but also rigorous attention to ensure satisfactory application safety and long-term storage stability.

Acknowledgments The authors are thankful to the Department of Pharmaceuticals (DoP), Ministry of Chemicals and Fertilizers, Government of India, for providing facilities for writing this chapter. NIPER-Raebareli communication number for this manuscript is NIPER-R/Communication/724.

Conflict of Interest The authors report no conflict of interest related to this chapter.

References

1. How the Eyes Work | National Eye Institute. n.d.. https://www.nei.nih.gov/learn-about-eye-health/healthy-vision/how-eyes-work. Accessed 7 Jan 2025.
2. Agemy S, Azam Z, Shrier E. Overview of anterior and posterior segment complications. 2021:87–105. https://doi.org/10.2174/9789811466465121020009.
3. Qi Q, Wei Y, Zhang X, Guan J, Mao S. Challenges and strategies for ocular posterior diseases therapy via non-invasive advanced drug delivery. J Control Release. 2023;361:191–211. https://doi.org/10.1016/J.JCONREL.2023.07.055.
4. Chuang K, Fields MA, Del Priore LV. Potential of gene editing and induced pluripotent stem cells (iPSCs) in treatment of retinal diseases. Yale J Biol Med. 2017;90:635. https://pmc.ncbi.nlm.nih.gov/articles/PMC5733854/. Accessed 25 Feb 2025.
5. Kanoujia J, Tarannum S, Kaurav M, Raina N, Jain K, Gupta M. Insight into recent updates on vaccines development and immunology of monkeypox infection. Curr Treat Options Infect Dis. 2024;16:118–28. https://doi.org/10.1007/S40506-024-00276-Z.
6. Thrimawithana TR, Young S, Bunt CR, Green C, Alany RG. Drug delivery to the posterior segment of the eye. Drug Discov Today. 2011;16:270–7. https://doi.org/10.1016/J.DRUDIS.2010.12.004.
7. Wang Y, Xu X, Gu Y, Cheng Y, Cao F. Recent advance of nanoparticle-based topical drug delivery to the posterior segment of the eye. Expert Opin Drug Deliv. 2018;15:687–701. https://doi.org/10.1080/17425247.2018.1496080.
8. Biswas A, Choudhury AD, Bisen AC, Agrawal S, Sanap SN, Verma SK, Mishra A, Kumar S, Bhatta RS. Trends in formulation approaches for sustained drug delivery to the posterior segment of the eye. AAPS PharmSciTech. 2023;24 https://doi.org/10.1208/S12249-023-02673-X.
9. Moisseiev E, Loewenstein A. Drug delivery to the posterior segment of the eye. Dev Ophthalmol. 2017;58:87–101. https://doi.org/10.1159/000455276.
10. Christoforidis JB, Chang S, Jiang A, Wang J, Cebulla CM. Intravitreal devices for the treatment of vitreous inflammation. Mediat Inflamm. 2012;2012 https://doi.org/10.1155/2012/126463.

11. Fischer C, Bruggemann A, Hager A, Callizo Planas J, Roider J, Hoerauf H. Vascular occlusions following ocular surgical procedures: a clinical observation of vascular complications after ocular surgery. J Ophthalmol. 2017;2017:9120892. https://doi.org/10.1155/2017/9120892.
12. Varela-Fernández R, Díaz-Tomé V, Luaces-Rodríguez A, Conde-Penedo A, García-Otero X, Luzardo-álvarez A, Fernández-Ferreiro A, Otero-Espinar FJ. Drug delivery to the posterior segment of the eye: biopharmaceutic and pharmacokinetic considerations. Pharmaceutics. 2020;12:269. https://doi.org/10.3390/PHARMACEUTICS12030269.
13. Murali A, Krishnakumar S, Subramanian A, Parameswaran S. Bruch's membrane pathology: a mechanistic perspective. Eur J Ophthalmol. 2020;30:1195–206. https://doi.org/10.1177/1120672120919337.
14. The Effect of Age on the Macromolecular Permeability of Human Bruch's Membrane | IOVS | ARVO Journals. n.d.. https://iovs.arvojournals.org/article.aspx?articleid=2122905. Accessed 7 Jan 2025.
15. Drusen in Eyes: Causes, Treatment, and Macular Degeneration. n.d.. https://www.healthline.com/health/drusen. Accessed 7 Jan 2025.
16. O'Leary F, Campbell M. The blood–retina barrier in health and disease. FEBS J. 2023;290:878–91. https://doi.org/10.1111/FEBS.16330.
17. Bachu RD, Chowdhury P, Al-Saedi ZHF, Karla PK, Boddu SHS. Ocular drug delivery barriers—role of nanocarriers in the treatment of anterior segment ocular diseases. Pharmaceutics. 2018;10:28. https://doi.org/10.3390/PHARMACEUTICS10010028.
18. Brinks J, van Dijk EHC, Klaassen I, Schlingemann RO, Kielbasa SM, Emri E, Quax PHA, Bergen AA, Meijer OC, Boon CJF. Exploring the choroidal vascular labyrinth and its molecular and structural roles in health and disease. Prog Retin Eye Res. 2022;87:100994. https://doi.org/10.1016/J.PRETEYERES.2021.100994.
19. Balachandran RK, Barocas VH. Computer modeling of drug delivery to the posterior eye: effect of active transport and loss to choroidal blood flow. Pharm Res. 2008;25:2685–96. https://doi.org/10.1007/S11095-008-9691-3.
20. del Amo EM, Rimpelä AK, Heikkinen E, Kari OK, Ramsay E, Lajunen T, Schmitt M, Pelkonen L, Bhattacharya M, Richardson D, Subrizi A, Turunen T, Reinisalo M, Itkonen J, Toropainen E, Casteleijn M, Kidron H, Antopolsky M, Vellonen KS, Ruponen M, Urtti A. Pharmacokinetic aspects of retinal drug delivery. Prog Retin Eye Res. 2017;57:134–85. https://doi.org/10.1016/J.PRETEYERES.2016.12.001.
21. Toffoletto N, Saramago B, Serro AP. Therapeutic ophthalmic lenses: a review. Pharmaceutics. 2020;13:36. https://doi.org/10.3390/PHARMACEUTICS13010036.
22. Doshi RR, Bakri SJ, Fung AE. Intravitreal injection technique. Semin Ophthalmol. 2011;26:104–13. https://doi.org/10.3109/08820538.2010.541318.
23. Yang Y, Bailey C, Loewenstein A, Massin P. Intravitreal corticosteroids in diabetic macular edema: pharmacokinetic considerations. Retina. 2015;35:2440. https://doi.org/10.1097/IAE.0000000000000726.
24. Anwary M, Kumar P, du Toit LC, Choonara YE, Pillay V. Polymeric, injectable, intravitreal hydrogel devices for posterior segment applications and interventions, Artif cells Nanomed. Biotechnol. 2018;46:1074–81. https://doi.org/10.1080/21691401.2018.1478845.
25. Choonara YE, Pillay V, Danckwerts MP, Carmichael TR, Du Toit LC. A review of implantable intravitreal drug delivery technologies for the treatment of posterior segment eye diseases. J Pharm Sci. 2010;99:2219–39. https://doi.org/10.1002/JPS.21987.
26. Pacella E, Loffredo L, Malvasi M, Battagliola ET, Messineo D, Pacella F, Arrico L. Effects of repeated intravitreal injections of dexamethasone implants on intraocular pressure: a 4-year study. Clin Ophthalmol. 2020;14:3611. https://doi.org/10.2147/OPTH.S265691.
27. Wang J, Jiang A, Joshi M, Christoforidis J. Drug delivery implants in the treatment of vitreous inflammation. Mediat Inflamm. 2013;2013:780634. https://doi.org/10.1155/2013/780634.
28. Merrill PT, Holekamp N, Roth D, Kasper J, Grigorian R. The 0.19-mg fluocinolone acetonide intravitreal implant reduces treatment burden in diabetic macular edema. Am J Ophthalmol. 2023;248:16–23. https://doi.org/10.1016/J.AJO.2022.09.017.

29. Costello MA, Liu J, Chen B, Wang Y, Qin B, Xu X, Li Q, Lynd NA, Zhang F. Drug release mechanisms of high-drug-load, melt-extruded dexamethasone intravitreal implants. Eur J Pharm Biopharm. 2023;187:46–56. https://doi.org/10.1016/J.EJPB.2023.04.003.
30. Nayak K, Misra M. A review on recent drug delivery systems for posterior segment of eye. Biomed Pharmacother. 2018;107:1564–82. https://doi.org/10.1016/J.BIOPHA.2018.08.138.
31. OZURDEX® Resources for Your Practice | OZURDEX® for HCPs. n.d.. https://hcp.ozurdex.com/resources. Accessed 26 Feb 2025.
32. Overview – ILUVIEN. n.d.. https://hcp.iluvien.com/overview/. Accessed 26 Feb 2025.
33. Agban Y, Thakur SS, Mugisho OO, Rupenthal ID. Depot formulations to sustain periocular drug delivery to the posterior eye segment. Drug Discov Today. 2019;24:1458–69. https://doi.org/10.1016/J.DRUDIS.2019.03.023.
34. Waite D, Wang Y, Jones D, Stitt A, Raj Singh TR. Posterior drug delivery via periocular route: challenges and opportunities. Ther Deliv. 2017;8:685–99. https://doi.org/10.4155/TDE-2017-0097.
35. Raghava S, Hammond M, Kompella UB. Periocular routes for retinal drug delivery. Expert Opin Drug Deliv. 2004;1:99–114. https://doi.org/10.1517/17425247.1.1.99.
36. Michels S, Schmidt-Erfurth U. Photodynamic therapy with verteporfin: a new treatment in ophthalmology. Semin Ophthalmol. 2001;16:201–6. https://doi.org/10.1076/SOPH.16.4.201.10298.
37. Gaudana R, Ananthula HK, Parenky A, Mitra AK. Ocular drug delivery. AAPS J. 2010;12:348. https://doi.org/10.1208/S12248-010-9183-3.
38. Maulvi FA, Shetty KH, Desai DT, Shah DO, Willcox MDP. Recent advances in ophthalmic preparations: ocular barriers, dosage forms and routes of administration. Int J Pharm. 2021;608:121105. https://doi.org/10.1016/J.IJPHARM.2021.121105.
39. Löscher M, Seiz C, Hurst J, Schnichels S. Topical drug delivery to the posterior segment of the eye. Pharmaceutics. 2022;14:134. https://doi.org/10.3390/PHARMACEUTICS14010134.
40. Sripetch S, Loftsson T. Topical drug delivery to the posterior segment of the eye: thermodynamic considerations. Int J Pharm. 2021;597:120332. https://doi.org/10.1016/J.IJPHARM.2021.120332.
41. Rodrigues GA, Lutz D, Shen J, Yuan X, Shen H, Cunningham J, Rivers HM. Topical drug delivery to the posterior segment of the eye: addressing the challenge of preclinical to clinical translation. Pharm Res. 2018;35 https://doi.org/10.1007/S11095-018-2519-X.
42. Yavuz B, Kompella UB. Ocular drug delivery. Handb Exp Pharmacol. 2017;242:57–93. https://doi.org/10.1007/164_2016_84.
43. Li LH, Lee JCY, Leung HH, Lam WC, Fu Z, Lo ACY. Lutein supplementation for eye diseases. Nutrients. 2020;12:1–27. https://doi.org/10.3390/NU12061721.
44. Lim C, Kang JK, Jung CE, Sim T, Her J, Kang K, Lee ES, Youn YS, Choi HG, Oh KT. Preparation and characterization of a lutein solid dispersion to improve its solubility and stability. AAPS PharmSciTech. 2021;22:1–9. https://doi.org/10.1208/S12249-021-02036-4/METRICS.
45. Janoria KG, Gunda S, Boddu SHS, Mitra AK. Novel approaches to retinal drug delivery. Expert Opin Drug Deliv. 2007;4:371–88. https://doi.org/10.1517/17425247.4.4.371.
46. Madni A, Rahem MA, Tahir N, Sarfraz M, Jabar A, Rehman M, Kashif PM, Badshah SF, Khan KU, Santos HA. Non-invasive strategies for targeting the posterior segment of eye. Int J Pharm. 2017;530:326–45. https://doi.org/10.1016/J.IJPHARM.2017.07.065.
47. Jain K, Zhong J. Theranostic applications of nanomaterials. Curr Pharm Des. 2022;28:77. https://doi.org/10.2174/138161282802211223150153/CITE/REFWORKS.
48. Souza JG, Dias K, Silva SAM, De Rezende LCD, Rocha EM, Emery FS, Lopez RFV. Transcorneal iontophoresis of dendrimers: PAMAM corneal penetration and dexamethasone delivery. J Control Release. 2015;200:115–24. https://doi.org/10.1016/J.JCONREL.2014.12.037.
49. Bajwa N, Kumar Mehra N, Jain K, Kumar Jain N. Targeted anticancer drug delivery through anthracycline antibiotic bearing functionalized quantum dots. Artif Cells Nanomed Biotechnol. 2016;44(7):1774–82. https://doi.org/10.3109/21691401.2015.1102740.

50. Sahu RC, Suthar T, Kumar D, Singh P, Datusalia AK, Jain K. Novel ligand conjugated poly (propylene imine) dendrimers for brain targeted delivery of tacrine hydrochloride. J Drug Deliv Sci Technology. 2024;92(105336):1–10. https://doi.org/10.1016/j.jddst.2024.105336.
51. Rizwanullah M, Ahmad MZ, Ghoneim MM, Alshehri S, Imam SS, Md S, Alhakamy NA, Jain K, Ahmad J. Receptor-mediated targeted delivery of surface-modifiednanomedicine in breast cancer: recent update and challenges. Pharmaceutics. 2021;13(12):2039. https://doi.org/10.3390/pharmaceutics13122039.
52. Meza-Rios A, Navarro-Partida J, Armendariz-Borunda J, Santos A. Therapies based on nanoparticles for eye drug delivery. Ophthalmol Ther. 2020;9:1. https://doi.org/10.1007/S40123-020-00257-7.
53. Joyson N, Pathak A, Jain K. One platform comparison of polymeric and lipidic nanoparticles for the delivery of amphotericin B. AAPS PharmSciTech. 2023;24:1–16. https://doi.org/10.1208/S12249-023-02672-Y/METRICS.
54. Battaglia L, Serpe L, Foglietta F, Muntoni E, Gallarate M, Del Pozo Rodriguez A, Solinis MA. Application of lipid nanoparticles to ocular drug delivery. Expert Opin Drug Deliv. 2016;13:1743–57. https://doi.org/10.1080/17425247.2016.1201059.
55. Duong VA, Nguyen TTL, Maeng HJ. Preparation of solid lipid nanoparticles and nanostructured lipid carriers for drug delivery and the effects of preparation parameters of solvent injection method. Molecules. 2020;25:4781. https://doi.org/10.3390/MOLECULES25204781.
56. Tahara K, Karasawa K, Onodera R, Takeuchi H. Feasibility of drug delivery to the eye's posterior segment by topical instillation of PLGA nanoparticles, Asian. J Pharm Sci. 2017;12:394–9. https://doi.org/10.1016/J.AJPS.2017.03.002.
57. Bonilla L, Espina M, Severino P, Cano A, Ettcheto M, Camins A, García ML, Souto EB, Sánchez-López E. Lipid nanoparticles for the posterior eye segment. Pharmaceutics. 2021;14:90. https://doi.org/10.3390/PHARMACEUTICS14010090.
58. Baig MS, Karade SK, Ahmad A, Khan MA, Haque A, Webster TJ, Faiyazuddin M, Al-Qahtani NH. Lipid-based nanoparticles: innovations in ocular drug delivery. Front Mol Biosci. 2024;11:1421959. https://doi.org/10.3389/FMOLB.2024.1421959.
59. de Oliveira IF, Barbosa EJ, Peters MCC, Henostroza MAB, Yukuyama MN, dos Santos Neto E, Löbenberg R, Bou-Chacra N. Cutting-edge advances in therapy for the posterior segment of the eye: solid lipid nanoparticles and nanostructured lipid carriers. Int J Pharm. 2020;589:119831. https://doi.org/10.1016/J.IJPHARM.2020.119831.
60. Gugleva V, Andonova V. Recent progress of solid lipid nanoparticles and nanostructured lipid carriers as ocular drug delivery platforms. Pharmaceuticals. 2023;16:474. https://doi.org/10.3390/PH16030474/S1.
61. Dobreva M, Stefanov S, Andonova V. Natural lipids as structural components of solid lipid nanoparticles and nanostructured lipid carriers for topical delivery. Curr Pharm Des. 2020;26:4524–35. https://doi.org/10.2174/1381612826666200514221649.
62. Yadav M, Schiavone N, Guzman-Aranguez A, Giansanti F, Papucci L, Perez de Lara MJ, Singh M, Kaur IP. Atorvastatin-loaded solid lipid nanoparticles as eye drops: proposed treatment option for age-related macular degeneration (AMD). Drug Deliv Transl Res. 2020;10:919. https://doi.org/10.1007/S13346-020-00733-4.
63. Gordillo-Galeano A, Mora-Huertas CE. Solid lipid nanoparticles and nanostructured lipid carriers: a review emphasizing on particle structure and drug release. Eur J Pharm Biopharm. 2018;133:285–308. https://doi.org/10.1016/J.EJPB.2018.10.017.
64. Singh AK, Kumar M, Upadhyay PK. Nanostructured lipid carriers mediated drug delivery to posterior segment of eye and their in-vivo successes. Curr Pharm Biotechnol. 2023;25:713–23. https://doi.org/10.2174/1389201025666230907145019.
65. Araújo J, Nikolic S, Egea MA, Souto EB, Garcia ML. Nanostructured lipid carriers for triamcinolone acetonide delivery to the posterior segment of the eye. Colloids Surf B Biointerfaces. 2011;88:150–7. https://doi.org/10.1016/J.COLSURFB.2011.06.025.
66. Puglia C, Santonocito D, Romeo G, Intagliata S, Romano GL, Strettoi E, Novelli E, Ostacolo C, Campiglia P, Sommella EM, Pignatello R, Bucolo C. Lipid nanoparticles traverse non-

corneal path to reach the posterior eye segment: in vivo evidence. Molecules. 2021;26:4673. https://doi.org/10.3390/MOLECULES26154673.
67. Tsung TH, Tsai YC, Lee HP, Chen YH, Lu DW. Biodegradable polymer-based drug-delivery systems for ocular diseases. Int J Mol Sci. 2023;24:12976. https://doi.org/10.3390/IJMS241612976.
68. Nguyen DD, Luo LJ, Yang CJ, Lai JY. Highly retina-permeating and long-acting resveratrol/metformin nanotherapeutics for enhanced treatment of macular degeneration. ACS Nano. 2023;17:168–83. https://doi.org/10.1021/ACSNANO.2C05824/ASSET/IMAGES/LARGE/NN2C05824_0007.JPEG.
69. Lynch C, Kondiah PPD, Choonara YE, du Toit LC, Ally N, Pillay V. Advances in biodegradable nano-sized polymer-based ocular drug delivery. Polymers (Basel). 2019;11:1371. https://doi.org/10.3390/POLYM11081371.
70. Mehra P, Tarannum S, Jain K. Polymer blends and additives. In: Polymers for oral drug delivery technologies. Elsevier; 2025. p. 787–809. https://doi.org/10.1016/B978-0-443-13774-7.00020-7.
71. Mansuri S, Kesharwani P, Jain K, Tekade RK, Jain NK. Mucoadhesion: a promising approach in drug delivery system. React Funct Polym. 2016;100:151–72. https://doi.org/10.1016/J.REACTFUNCTPOLYM.2016.01.011.
72. Swetledge S, Jung JP, Carter R, Sabliov C. Distribution of polymeric nanoparticles in the eye: implications in ocular disease therapy. J Nanobiotechnol. 2021;19:10. https://doi.org/10.1186/S12951-020-00745-9.
73. Esteruelas G, Halbaut L, García-Torra V, Espina M, Cano A, Ettcheto M, Camins A, Souto EB, Luisa García M, Sánchez-López E. Development and optimization of Riluzole-loaded biodegradable nanoparticles incorporated in a mucoadhesive in situ gel for the posterior eye segment. Int J Pharm. 2022;612:121379. https://doi.org/10.1016/J.IJPHARM.2021.121379.
74. Huang Y, Zhu Y, Cai D, Guo Q, Wang J, Lei L, Li X, Shi S. Penetrating-peptide-mediated non-invasive Axitinib delivery for anti-neovascularisation. J Control Release. 2022;347:449–59. https://doi.org/10.1016/J.JCONREL.2022.05.009.
75. Anand A, Jian HJ, Huang HH, Hean LE, Li YJ, Lai JY, Da Chou H, Kang YC, Wu WC, Lai CC, Huang CC, Chang HT. Anti-angiogenic carbon nanovesicles loaded with bevacizumab for the treatment of age-related macular degeneration. Carbon NY. 2023;201:362–70. https://doi.org/10.1016/J.CARBON.2022.09.045.
76. Apaolaza PS, Busch M, Asin-Prieto E, Peynshaert K, Rathod R, Remaut K, Dünker N, Göpferich A. Hyaluronic acid coating of gold nanoparticles for intraocular drug delivery: evaluation of the surface properties and effect on their distribution. Exp Eye Res. 2020;198:108151. https://doi.org/10.1016/J.EXER.2020.108151.
77. Jain K, Kumar Mehra N, Jain NK. Nanotechnology in drug delivery: safety and toxicity issues. Curr Pharm Des 2015;21(29):4252–61. https://doi.org/10.2174/1381612821666150901103208
78. Gu Y, Xu C, Wang Y, Zhou X, Fang L, Cao F. Multifunctional nanocomposites based on liposomes and layered double hydroxides conjugated with glycylsarcosine for efficient topical drug delivery to the posterior segment of the eye. Mol Pharm. 2019;16:2845–57. https://doi.org/10.1021/ACS.MOLPHARMACEUT.8B01136/ASSET/IMAGES/LARGE/MP-2018-011367_0003.JPEG.
79. Bressler NM. Guidelines for using verteporfin (Visudyne) in photodynamic therapy for choroidal neovascularization due to age-related macular degeneration and other causes: update. Retina. 2005;25:119–34. https://doi.org/10.1097/00006982-200502000-00002.
80. Rostaporfin – Drug Targets, Indications, Patents – Synapse. n.d.. https://synapse.patsnap.com/drug/6cfe5403ddeb4f1ab2c2bc41fcfc9da4. Accessed 9 Jan 2025.
81. Mishra GP, Bagui M, Tamboli V, Mitra AK. Recent applications of liposomes in ophthalmic drug delivery. J Drug Deliv. 2011;2011:863734. https://doi.org/10.1155/2011/863734.
82. Davis BM, Normando EM, Guo L, Turner LA, Nizari S, O'Shea P, Moss SE, Somavarapu S, Cordeiro MF. Topical delivery of avastin to the posterior segment of the eye in vivo

using annexin a5-associated liposomes. Small. 2014;10:1575–84. https://doi.org/10.1002/SMLL.201303433.
83. Khalil M, Hashmi U, Riaz R, Rukh Abbas S. Chitosan coated liposomes (CCL) containing triamcinolone acetonide for sustained delivery: a potential topical treatment for posterior segment diseases. Int J Biol Macromol. 2020;143:483–91. https://doi.org/10.1016/J.IJBIOMAC.2019.10.256.
84. Akhter MH, Ahmad I, Alshahrani MY, Al-Harbi AI, Khalilullah H, Afzal O, Altamimi ASA, Najib Ullah SNM, Ojha A, Karim S. Drug delivery challenges and current progress in nanocarrier-based ocular therapeutic system. Gels. 2022;8:82. https://doi.org/10.3390/GELS8020082.
85. Chen S, Hanning S, Falconer J, Locke M, Wen J. Recent advances in non-ionic surfactant vesicles (niosomes): fabrication, characterization, pharmaceutical and cosmetic applications. Eur J Pharm Biopharm. 2019;144:18–39. https://doi.org/10.1016/J.EJPB.2019.08.015.
86. Sahoo RK, Biswas N, Guha A, Sahoo N, Kuotsu K. Nonionic surfactant vesicles in ocular delivery: innovative approaches and perspectives. Biomed Res Int. 2014;2014:263604. https://doi.org/10.1155/2014/263604.
87. Owodeha-Ashaka K, Ilomuanya MO, Iyire A. Evaluation of sonication on stability-indicating properties of optimized pilocarpine hydrochloride-loaded niosomes in ocular drug delivery. Prog Biomater. 2021;10:207. https://doi.org/10.1007/S40204-021-00164-5.
88. Kakkar S, Kaur IP. Spanlastics—a novel nanovesicular carrier system for ocular delivery. Int J Pharm. 2011;413:202–10. https://doi.org/10.1016/J.IJPHARM.2011.04.027.
89. Mandal A, Bisht R, Rupenthal ID, Mitra AK. Polymeric micelles for ocular drug delivery: from structural frameworks to recent preclinical studies. J Control Release. 2017;248:96–116. https://doi.org/10.1016/J.JCONREL.2017.01.012.
90. Pignatello R, Corsaro R, Bonaccorso A, Zingale E, Carbone C, Musumeci T. Soluplus® polymeric nanomicelles improve solubility of BCS-class II drugs. Drug Deliv Transl Res. 2022;12:1991. https://doi.org/10.1007/S13346-022-01182-X.
91. Grimaudo MA, Pescina S, Padula C, Santi P, Concheiro A, Alvarez-Lorenzo C, Nicoli S. Topical application of polymeric nanomicelles in ophthalmology: a review on research efforts for the noninvasive delivery of ocular therapeutics. Expert Opin Drug Deliv. 2019;16:397–413. https://doi.org/10.1080/17425247.2019.1597848.
92. Gote V, Ansong M, Pal D. Prodrugs and nanomicelles to overcome ocular barriers for drug penetration. Expert Opin Drug Metab Toxicol. 2020;16:885–906. https://doi.org/10.1080/17425255.2020.1803278.
93. Signorini S, Delledonne A, Pescina S, Bianchera A, Sissa C, Vivero-Lopez M, Alvarez-Lorenzo C, Santi P, Padula C, Nicoli S. A sterilizable platform based on crosslinked xanthan gum for controlled-release of polymeric micelles: ocular application for the delivery of neuroprotective compounds to the posterior eye segment. Int J Pharm. 2024;657:124141. https://doi.org/10.1016/J.IJPHARM.2024.124141.
94. Pescina S, Sonvico F, Clementino A, Padula C, Santi P, Nicoli S. Preliminary investigation on simvastatin-loaded polymeric micelles in view of the treatment of the back of the eye. Pharmaceutics. 2021;13:855. https://doi.org/10.3390/PHARMACEUTICS13060855/S1.
95. Sun L, Zhang M, Shi Y, Fang L, Cao F. Rational design of mixed nanomicelle eye drops with structural integrity investigation. Acta Biomater. 2022;141:164–77. https://doi.org/10.1016/J.ACTBIO.2022.01.014.
96. Kim TY, Lee GH, Mun J, Cheong S, Choi I, Kim H, Hahn SK. Smart contact lens systems for ocular drug delivery and therapy. Adv Drug Deliv Rev. 2023;196:114817. https://doi.org/10.1016/J.ADDR.2023.114817.
97. Xu J, Xue Y, Hu G, Lin T, Gou J, Yin T, He H, Zhang Y, Tang X. A comprehensive review on contact lens for ophthalmic drug delivery. J Control Release. 2018;281:97–118. https://doi.org/10.1016/J.JCONREL.2018.05.020.
98. Schultz C, Breaux J, Schentag J, Morck D. Drug delivery to the posterior segment of the eye through hydrogel contact lenses. Clin Exp Optom. 2011;94:212–8. https://doi.org/10.1111/J.1444-0938.2010.00553.X.

99. Ono J, Toshida H. Use of ketotifen fumarate-eluting daily disposable soft contact lens in management of ocular allergy: literature review and report of two cases. Cureus. 2022;14:e27093. https://doi.org/10.7759/CUREUS.27093.
100. Ross AE, Bengani LC, Tulsan R, Maidana DE, Salvador-Culla B, Kobashi H, Kolovou PE, Zhai H, Taghizadeh K, Kuang L, Mehta M, Vavvas DG, Kohane DS, Ciolino JB. Topical sustained drug delivery to the retina with a drug-eluting contact lens. Biomaterials. 2019;217:119285. https://doi.org/10.1016/J.BIOMATERIALS.2019.119285.
101. Lu C, Yoganathan RB, Kociolek M, Allen C. Hydrogel containing silica shell cross-linked micelles for ocular drug delivery. J Pharm Sci. 2013;102:627–37. https://doi.org/10.1002/JPS.23390.
102. Wu Y, Li X, Fu X, Huang X, Zhang S, Zhao N, Ma X, Saiding Q, Yang M, Tao W, Zhou X, Huang J. Innovative nanotechnology in drug delivery systems for advanced treatment of posterior segment ocular diseases. Adv Sci. 2024;11:2403399. https://doi.org/10.1002/ADVS.202403399.
103. Kaur IP, Kakkar S. Nanotherapy for posterior eye diseases. J Control Release. 2014;193:100–12. https://doi.org/10.1016/J.JCONREL.2014.05.031.

Fixed-Dose Combination Approach in Therapeutic Delivery to the Eye

5

Sakshi Soni, Vandana Soni, and Sushil K. Kashaw

Abstract

Fixed-dose combination (FDC) therapy offers a promising approach to enhance ocular drug delivery by combining multiple therapeutic agents into a single formulation. This strategy can significantly improve patient compliance, reduce dosing frequency, and optimize therapeutic outcomes. This chapter delves into the principles and advantages of FDC therapy for ocular diseases. We explore the various formulation strategies, including topical, intravitreal, and ocular inserts, and discuss the challenges and considerations associated with their development. Additionally, we highlight recent advancements in FDC therapy for ocular diseases such as glaucoma, age-related macular degeneration, and dry eye disease. By understanding the potential benefits and limitations of FDC therapy, researchers and clinicians can work towards developing innovative and effective treatment options for ocular disorders.

Keywords

Fixed-dose combination · Ocular drug delivery · Topical delivery · Intravitreal injection · Ocular inserts · Glaucoma · Age-related macular degeneration · Dry eye disease

S. Soni · V. Soni · S. K. Kashaw (✉)
Department of Pharmaceutical Sciences, Dr. Harisingh Gour University, Sagar, Madhya Pradesh, India

N. K. Mehra (ed.), *Complex Ophthalmic Dosage Forms: Advances in Biomedical Applications and Future Perspectives*, https://doi.org/10.1007/978-981-96-6306-4_5

5.1 Introduction

The eye, a highly specialized and delicate organ, presents unique challenges for effective drug delivery due to several anatomical and physiological barriers. The blood-ocular barrier, comprising the blood-aqueous and blood-retinal barriers, limits drug entry from the systemic circulation, while the cornea's impermeability further restricts drug penetration into deeper tissues. Additionally, rapid tear turnover and blinking reduce the retention time of topically applied drugs, leading to poor bioavailability. Traditional systemic routes are often inadequate for treating ocular conditions, as they fail to achieve therapeutic concentrations in the targeted tissues without posing systemic side effects. These limitations necessitate the development of innovative ocular drug delivery systems designed to enhance drug absorption and retention within the eye [4, 47, 62].

Several delivery approaches have been explored to address these challenges. Topical formulations, such as eye drops, gels, and ointments, remain the most common due to their noninvasive nature, although frequent administration is often required. Ocular inserts and in situ gelling systems offer prolonged drug release by adhering to the ocular surface, reducing the need for repeated dosing. Intraocular implants provide sustained drug release over extended periods through surgical placement for chronic or posterior segment diseases. Meanwhile, nanoparticle-based systems have gained attention for their ability to cross physiological barriers and improve drug penetration and retention, particularly for posterior segment therapies [15, 16, 56, 98].

Despite these advancements, ocular drug delivery systems still face numerous obstacles, including poor patient adherence due to complex regimens and the need to balance therapeutic efficacy with safety. Fixed-dose combination (FDC) therapy has emerged as a promising solution, aiming to simplify treatment by combining two or more active ingredients in a single formulation. FDCs offer several advantages, such as improved patient compliance, enhanced therapeutic efficacy through synergistic effects, reduced dosing frequency, and minimized side effects. In particular, FDCs are beneficial in managing chronic ocular diseases like glaucoma, infections, and age-related macular degeneration, where multidrug regimens are often required [20, 44, 67, 91] (Fig. 5.1).

However, the formulation of FDCs for ocular use presents significant challenges, including drug-drug compatibility, stability issues, and ensuring optimal release profiles for each active ingredient. Furthermore, regulatory hurdles and the need for robust preclinical and clinical evaluation strategies complicate the development and approval of these formulations. This chapter delves into the formulation strategies, challenges, and clinical implications associated with ocular FDCs, with a focus on both anterior and posterior segment diseases. It also highlights case studies of successful FDC applications, discusses their limitations, and identifies emerging trends in the field. Ultimately, understanding the evolving landscape of ocular drug delivery and FDC therapy can pave the way for improved therapeutic outcomes and patient quality of life [26, 38, 85].

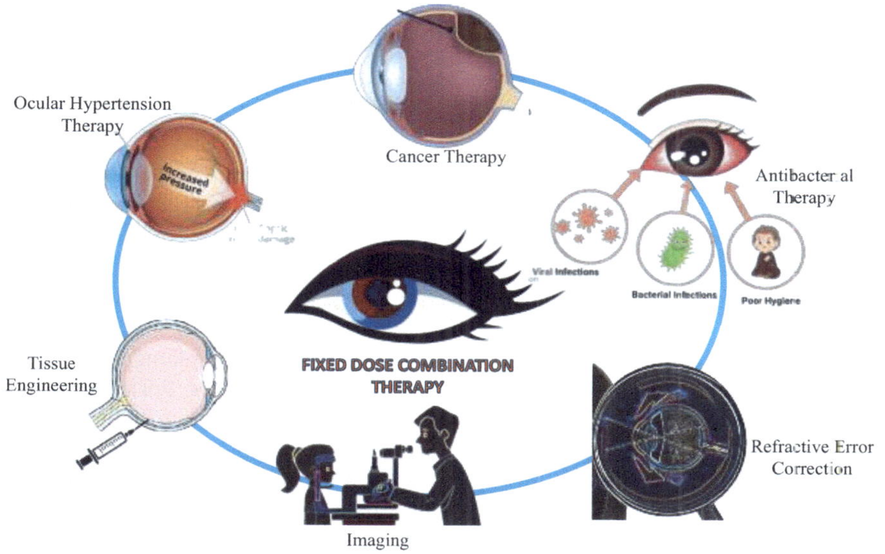

Fig. 5.1 Application of fixed-dose therapy in ocular therapy. (Self-created with biorender.com licensed version)

5.2 FDCs in Ocular Therapy

Fixed-dose combinations (FDCs) in ocular therapy represent an innovative approach to improve treatment outcomes by delivering multiple active pharmaceutical ingredients (APIs) in a single formulation. These formulations can take various forms, including eye drops, ointments, or ocular inserts, designed to simplify medication regimens. The primary goal of FDCs is to enhance patient compliance, reduce the burden of frequent dosing, and achieve synergistic therapeutic effects, making them particularly beneficial in managing complex or chronic ocular conditions such as glaucoma, infections, or uveitis [41, 46, 57].

An FDC consists of two or more APIs combined in a fixed ratio within one pharmaceutical form, ensuring that each dose contains a consistent amount of all active components. In ocular therapy, FDCs help patients manage complex treatment regimens, reducing the need for multiple medications administered separately. FDCs provide a more streamlined and effective therapeutic approach by integrating drugs with different mechanisms of action into a single formulation. These formulations are particularly useful in conditions requiring polytherapy, such as glaucoma, where both intraocular pressure-lowering agents and anti-inflammatory drugs are often needed [27, 50, 54].

5.2.1 Advantages of FDCs in Ocular Therapy

Improved Patient Compliance One of the most significant benefits of FDCs is their ability to simplify treatment regimens, which is essential for conditions requiring long-term therapy. Consolidating multiple medications into a single dose reduces dosing frequency, making it easier for patients to adhere to their prescribed treatment plan. This is especially important for elderly patients, who may struggle with administering multiple eye drops throughout the day. Enhanced adherence to the treatment improves therapeutic outcomes and reduces the likelihood of disease progression [27, 70, 95].

Enhanced Therapeutic Efficacy By combining drugs with complementary mechanisms of action, FDCs can offer synergistic therapeutic benefits. For example, in glaucoma, combining a beta-blocker with a prostaglandin analog results in a more significant reduction in intraocular pressure than either agent alone. Moreover, FDCs can be formulated to optimize drug delivery to ocular tissues, ensuring that each component reaches its target site with minimal loss. This allows for better disease control and minimizes the need for additional medications [27].

Reduced Side Effects FDCs offer the potential to reduce adverse effects by enabling lower doses of individual drugs. When medications with different modes of action are combined, each can be used at sub-therapeutic doses without compromising overall efficacy. Additionally, targeted ocular drug delivery through FDCs minimizes systemic absorption and off-target effects, enhancing the safety profile of the treatment [36].

Cost-Effectiveness FDCs also bring economic advantages by consolidating multiple therapies into a single product, reducing the overall cost of medications. This can be particularly beneficial for patients managing multiple eye conditions or chronic diseases that require lifelong treatment. From a healthcare perspective, FDCs reduce the need for frequent medical visits and lessen the workload on healthcare providers, improving resource utilization [1, 83, 84].

5.2.2 Limitations of FDCs

Limited Flexibility While FDCs simplify treatment, they also introduce limitations in adjusting individual drug doses. Fixed drug ratios may not suit every patient, especially those with varying disease severities or comorbid conditions. If the optimal dose for one drug differs from that in the FDC, it can lead to underdosing or overdosing, resulting in suboptimal therapeutic outcomes or increased side effects [75, 82, 100].

Potential for Drug Interactions Combining multiple APIs into a single formulation increases the risk of drug interactions, which can alter the pharmacokinetics or pharmacodynamics of the individual components. This can lead to unexpected adverse effects or diminished efficacy. Understanding the interaction profiles and optimizing the dosage regimen for each drug in an FDC is crucial to mitigate these risks [42, 49, 87].

Formulation Complexity Developing stable and effective FDC formulations is challenging due to the need for drug compatibility and optimal release profiles. APIs in an FDC must maintain stability without degrading or interacting adversely over time. Controlling the release kinetics of multiple drugs to achieve consistent therapeutic levels is also complicated, particularly in ocular systems where barriers like tear turnover affect drug retention. Ensuring physical and chemical stability during storage further adds to the complexity of formulation development [59, 62].

5.3 FDC Formulations for Ocular Delivery

Fixed-dose combinations (FDCs) offer innovative ways to improve drug delivery to the eye by simplifying treatment regimens and enhancing therapeutic outcomes. Depending on the type of ocular condition and the desired release profile, FDCs can take various forms, including topical formulations, ocular inserts, intraocular implants, and nanoparticle-based systems.

Each formulation presents unique advantages and challenges in achieving optimal drug delivery. Most ophthalmic diseases are chronic, and managing chronic, asymptomatic conditions poses a significant challenge. For example, in early-stage glaucoma, patients often remain unaware of the disease, leading to delayed treatment and suboptimal control of intraocular pressure (IOP). To address a diverse range of ophthalmic conditions, several complex products have been introduced to the market. Table 5.1 provides a list of these products, their manufacturers, dosage forms, active ingredients, and therapeutic indications, based on information from regulatory agencies like the US FDA and EMA. Figure 5.2 illustrates the various routes of drug administration for both anterior and posterior segments of the eye.

Topical FDCs Topical FDCs remain the most popular mode of ocular drug delivery due to their ease of administration, noninvasiveness, and accessibility. They are applied directly to the eye's surface in forms such as eye drops, ointments, or gels. Despite their convenience, these formulations often face challenges with short retention times, poor drug penetration, and patient adherence [3, 40, 52].

Eye Drops Eye drops are widely used for both anterior segment diseases, such as glaucoma or dry eye syndrome, and postoperative care. FDC eye drops combine multiple APIs to target different pathways simultaneously. There are various advantages: easy and noninvasive to administer; flexible dosing allows adjustment of fre-

Table 5.1 List of marketed products of fixed-dose combination for ocular therapy

Product name	Drug combination	Manufacturer	Regulatory approval	Dosage regimen: individual and FDC	MOA	References
Xalacom®	Timolol 0.5%, latanoprost 0.005%	Pfizer, Belgium	EU in 2000	Timolol (b.i.d.), Latanoprost (q.d.), combination (q.d.)	Timolol lowers pressure, latanoprost increases outflow of fluid	[55]
Duotrav®	Timolol 0.5%, travoprost 0.004%	Alcon Laboratories Inc., TX, USA	TGA in 2005, EU in 2010	Timolol (b.i.d.), Travoprost (q.d.), combination (q.d.)	Timolol lowers pressure, travoprost enhances uveoscleral outflow	[43]
Ganfort®	Timolol 0.5%, bimatoprost 0.03%	Allergen Pharm Inc., Westfort, Ireland	EU in 2006	Timolol (b.i.d.), Bimatoprost (q.d.), combination (q.d.)	Timolol lowers fluid formation, bimatoprost increases outflow	[28]
COSOPT®	Timolol 0.5%, dorzolamide 2%	Merck & Co., NJ, USA	FDA in 1998	Timolol (b.i.d.), Dorzolamide (t.i.d.), combination (b.i.d.)	Both drugs reduce fluid secretion	[45]
Extravan®	Timolol 0.5%, travoprost 0.03%	Alcon Laboratories Inc., TX, USA	TGA in 2005	Timolol (b.i.d.), Travoprost (q.d.), combination (q.d.)	Same as duotrav	[7]
COMBIGAN™	Timolol 0.5%, brimonidine 0.2%	Allergen, Irvine, CA, USA	EU in 2006, FDA in late 2006	Timolol (b.i.d.), Brimonidine (t.i.d.), combination (b.i.d.)	Timolol lowers pressure, brimonidine enhances uveoscleral outflow	[69]
Azarga®	Brinzolamide 1%, timolol 0.5%	Novartis, USA	EU in 2008	Brinzolamide (t.i.d.), Timolol (b.i.d.), combination (b.i.d.)	Both reduce fluid formation rate	[45]
SIMBRINZA™	Brinzolamide 1%, brimonidine 0.2%	Alcon Laboratories Inc., TX, USA	FDA in 2013	Brinzolamide (t.i.d.), Brimonidine (t.i.d.), combination (b.i.d.)	Reduces pressure by decreasing fluid formation and enhancing uveoscleral outflow	[24]

ROCKLATAN®	Netarsudil 0.02%, latanoprost 0.005%	Aerie Pharmaceuticals, Irvine, CA, USA	FDA in 2019	Netrasudil (q.d.), Latanoprost (q.d.), combination (q.d.)	Rho kinase inhibitor reduces fluid formation, latanoprost enhances fluid outflow	[14]
PAREMYD®	Hydroxyamphetamine 1%, tropicamide 0.25%	Akorn Pharmaceuticals, Lakeforest, IL, USA	FDA in 1992	Combination (instill 1–2 drops)	Hydroxyamphetamine causes muscle contraction, tropicamide relaxes papillary sphincter	[85]
Cyclomydri®	Cyclopentolate 0.2%, phenylephrine 1%	Alcon Laboratories Inc., TX, USA	FDA in 1975	Combination (instill 1 drop)	Cyclopentolate relaxes muscles, phenylephrine constricts vessels	[85]
AK-poly-bac®	Polymyxin B, bacitracin	Akorn Pharmaceuticals	FDA in 1985	Apply as per severity	Polymyxin B interacts with LPS of bacteria, bacitracin blocks cell wall formation	[97]
Naphcon-A®	Naphazoline HCl 0.025%, pheniramine 0.3%	Alcon Laboratories Inc., TX, USA	FDA in 1994	1–2 drops maximum q.i.d. as per severity	Naphazoline acts as redness reliever, pheniramine is for antihistaminic action	[77]
Visine-A®	Naphazoline HCl 0.025%, pheniramine 0.3%	Johnson & Johnson, US	FDA in 1996	1–2 drops maximum q.i.d. as per severity	Naphazoline acts as redness reliever, pheniramine as antihistamine	[12]
Omidria®	Ketorolac 0.3%, phenylephrine 1%	Omeros Corporation, US	FDA in 2014	Used in cataract surgery	Ketorolac inhibits PG synthesis, phenylephrine acts as vasoconstrictor	[12]

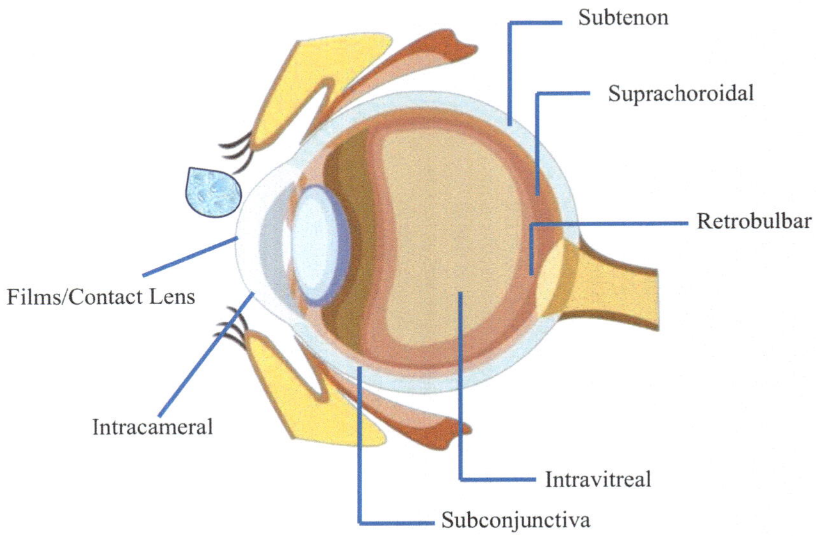

Fig. 5.2 Different ophthalmic routes of administration in the anterior and posterior segments of the eye. (Self-created with biorender.com licensed version)

quency based on the severity of the condition; rapid onset of action due to quick drug absorption by the corneal surface. There are various disadvantages: short residence time on the ocular surface due to tear turnover and blinking; risk of systemic absorption through the nasolacrimal duct, potentially causing side effects; patient compliance issues due to the need for frequent administration [2, 9, 92].

Ointments Ointments are thick, viscous preparations designed to remain on the ocular surface for extended periods, providing a protective layer that prevents evaporation and promotes healing. FDC ointments are often used for treating infections or inflammation. There are various advantages: prolonged retention time on the ocular surface, leading to sustained drug release; suitable for delivering poorly water-soluble drugs; forms a protective barrier, shielding the eye from environmental irritants. There are various disadvantages: blurred vision due to the greasy consistency, limited daytime use; messy and inconvenient application; and risk of irritation, especially if applied incorrectly or in excessive amounts [6, 13, 72].

Gels Gels provide a middle ground between eye drops and ointments, offering the benefits of both rapid drug release and prolonged contact time. These FDC formulations are often used to treat conditions such as dry eye and allergic conjunctivitis. There are various advantages: longer residence time compared to eye drops, reducing the need for frequent administration; less messy than ointments and more comfortable for patients; can be formulated as in situ gelling systems that solidify upon contact with the ocular surface. There are various disadvantages: may cause transient blurred vision after application; less comfortable than eye drops due to their thicker consistency; potential for clumping or uneven drug release if not formulated correctly [23, 60, 74].

Ocular Inserts Ocular inserts are small, solid, or semi-solid devices loaded with drugs that are inserted into the conjunctival sac, where they release medication slowly over time. These FDC systems are particularly useful in chronic conditions requiring sustained drug delivery. There are various advantages provides prolonged and controlled drug release, reduces dosing frequency; improves patient compliance, especially for chronic diseases like glaucoma; bypasses tear turnover and blinking, allowing consistent drug absorption. There are various disadvantages: can cause foreign body sensation or discomfort, reducing patient acceptance; limited drug loading capacity restricts the use of potent APIs; risk of dislodgement or expulsion from the eye, requiring careful monitoring [10, 68].

Intraocular Implants Intraocular implants are surgically placed inside the eye to deliver drugs directly to specific ocular tissues over extended periods. These implants are typically used for long-term treatment of chronic conditions such as uveitis, diabetic macular edema, and age-related macular degeneration. There are various advantages: precise and sustained drug release to the target tissue, improving therapeutic outcomes; minimal systemic absorption, reducing the risk of side effects; and reducing the need for frequent dosing or repeated clinical visits. There are various disadvantages: requires an invasive surgical procedure, which carries risks of complications such as infection or inflammation; limited drug loading capacity may necessitate multiple or large implants; if complications arise, implant removal can be challenging [96].

Nanoparticle-Based FDCs Nanoparticle-based FDCs are emerging as a promising strategy for ocular drug delivery. These formulations use nanoparticles, liposomes, or polymeric systems to encapsulate APIs, improving drug stability, bioavailability, and tissue penetration. FDC nanoparticles can be designed to deliver multiple drugs simultaneously, targeting both the anterior and posterior segments of the eye. There are various advantages: enhanced penetration through ocular barriers, reaching deeper tissues such as the retina; improved drug solubility, especially for hydrophobic drugs; prolonged residence time in ocular tissues, reducing dosing frequency. There are various disadvantages: complex and costly formulation and manufacturing processes; potential toxicity concerns due to the use of synthetic materials or surfactants; regulatory challenges related to safety, stability, and large-scale production (Table 5.2) [58, 63, 86].

5.4 Formulation Considerations for FDCs in Ocular Therapy

Developing fixed-dose combination (FDC) formulations for ocular delivery involves several critical considerations to ensure drug stability, optimal release profiles, bioavailability, and safety. Since the eye is a sensitive organ, every component of the

Table 5.2 Various eye formulations and their outcomes

Formulation type	Example	Study design	References
Eye drops	Timolol/latanoprost (Combigan) Dorzolamide/timolol Brimonidine/brinzolamide UV LUBE eye drops (Chloroquine phosphate 0.03% w/v)	A multicenter, randomized, double-masked, active-controlled clinical trial evaluating the efficacy and safety of the fixed-dose combination of timolol maleate 0.5% and latanoprost 0.005% ophthalmic solution in patients with open-angle glaucoma or ocular hypertension	[11, 33]
Ointments	Tobramycin/dexamethasone (tobradex) Loteprednol etabonate ointment Bacitracin/neomycin/polymyxin B ointment Erythromycin/polymyxin B ointment Dexamethasone/tobramycin ointment Levofloxacin/dexamethasone ointment	A randomized, double-masked, vehicle-controlled clinical trial evaluating the efficacy and safety of tobramycin/dexamethasone ophthalmic ointment 0.3%/0.1% in the treatment of bacterial conjunctivitis	[53, 99]
Gels	Cyclosporine A/hyaluronic acid Brimonidine/ripasudil gel Latanoprost/timolol gel	A randomized, double-masked, vehicle-controlled clinical trial evaluating the efficacy and safety of cyclosporine A 0.05% ophthalmic emulsion in the treatment of dry eye disease	[34, 66]
Ocular inserts	Brimonidine/ripasudil combination Brimonidine/brinzolamide combination Latanoprost/timolol combination	A clinical study evaluating the efficacy and safety of the Ocusert system for the treatment of glaucoma	[64, 89]
Intraocular implants	Bimatoprost/timolol implant Dorzolamide/timolol implant Brimonidine/brinzolamide implant	A clinical trial evaluating the efficacy and safety of an implantable drug delivery system for the treatment of age-related macular degeneration	[19, 48]

(continued)

Table 5.2 (continued)

Formulation type	Example	Study design	References
Nanoparticle-based FDCs	Fluorometholone nanoparticles with bromfenac and levofloxacin Chitosan-coated poly-L-lactide nanoparticles loaded with azithromycin and tretinoin Hydroxypropyl methacrylamide-loaded moxifloxacin nanoparticles Quantum dot-coupled bevacizumab	A preclinical study evaluating the efficacy and safety of nanoparticle-based drug delivery systems for the treatment of ocular infections	[13, 39]

formulation must work harmoniously to maintain therapeutic efficacy while minimizing side effects. Below are key aspects to consider during the formulation of FDCs for ocular applications.

Drug Compatibility and Stability

Ensuring the compatibility and stability of multiple drugs within a single formulation is essential to prevent drug degradation and ensure consistent therapeutic performance. Physical interactions between drugs, such as precipitation or formation of insoluble complexes, can impair drug solubility and bioavailability. In FDC eye drops or suspensions, incompatibility may lead to drug separation or cloudiness, reducing effectiveness and patient acceptance. The chemical stability of each drug component is crucial to prevent degradation through processes like oxidation, hydrolysis, or photolysis. For example, light-sensitive drugs require opaque packaging to prevent photodegradation, while antioxidants may be added to formulations containing oxidation-prone APIs. The interaction between APIs and excipients (e.g., preservatives, surfactants) must be evaluated to ensure long-term stability. Changes in pH, ionic strength, or temperature can trigger degradation, so stability testing under various environmental conditions is essential [93].

Release Kinetics and Drug Release Profiles

Achieving the desired therapeutic effect requires careful design of drug release profiles. Depending on the disease being treated, different release strategies may be employed. In controlled-release FDCs, the formulation releases the drugs gradually over time, providing a steady concentration at the target site. This approach minimizes peak-trough fluctuations and reduces side effects, making it ideal for chronic conditions like glaucoma. Sustained-release formulations extend the release of drugs over hours or days, improving patient compliance by reducing dosing frequency. Ocular inserts and implants are often used to achieve this effect.

Immediate-release FDCs provide rapid drug delivery for acute conditions, such as eye infections or inflammation. These formulations ensure that therapeutic concentrations are achieved quickly, but their effect may be short-lived, necessitating frequent dosing [22, 71, 79].

Particle Size and Distribution

Particle size plays a crucial role in determining the bioavailability, ocular penetration, and consistency of drug release in FDCs. Smaller particles have a larger surface area, promoting faster dissolution and absorption through the cornea. Nanoparticle-based FDCs leverage this principle to enhance drug delivery to deeper ocular tissues, such as the retina. However, excessively small particles may also increase toxicity risks. A narrow particle size distribution ensures consistent drug release and uniformity across the product. Variability in particle size can affect drug stability, absorption, and efficacy, necessitating stringent quality control during manufacturing [61].

Excipients and Their Role in FDC Formulations

Excipients play multiple roles in enhancing the stability, efficacy, and patient acceptability of FDC formulations for ocular use. The selection of excipients must be optimized to ensure compatibility with the APIs and safety for ocular tissues. The vehicle serves as the solvent or carrier for the drugs. Aqueous vehicles are commonly used in eye drops, while oils or emulsions may be used for poorly water-soluble drugs. Preservatives, such as benzalkonium chloride, prevent microbial contamination. However, they must be chosen carefully, as they can cause ocular irritation, especially with long-term use. Buffers maintain the formulation's pH within the physiological range (around 7.4) to prevent irritation and ensure drug stability. Tonicity adjusters like sodium chloride help match the osmotic pressure of the formulation with that of the eye, preventing discomfort or damage to ocular tissues. Increasing the viscosity of the formulation (e.g., using hydroxypropyl methylcellulose) enhances its residence time on the ocular surface, improving drug absorption and reducing dosing frequency. Surfactants enhance the solubility and penetration of hydrophobic drugs, improving bioavailability. Care must be taken to use nonirritating surfactants, as some can disrupt the tear film or damage corneal cells [5, 32].

Sterility and Preservation

Maintaining the sterility of ocular FDC formulations is critical to prevent infections, as the eye is highly vulnerable to microbial contamination. All ophthalmic formulations must be sterile to prevent infections. Sterilization methods such as autoclaving, filtration, or gamma irradiation are employed, depending on the heat and moisture sensitivity of the APIs. Preservatives are added to the multidose containers to prevent microbial growth during use. However, preservative-free formulations may be preferred for patients with sensitive eyes or for post-surgical care, necessitating single-use packaging or advanced delivery systems. Manufacturing ocular

FDCs requires strict adherence to aseptic processes to minimize contamination risks. Facilities must maintain sterile environments, and personnel must follow rigorous hygiene protocols to ensure product safety. Before a fixed-dose combination (FDC) can be advanced to clinical trials, it undergoes rigorous preclinical evaluation to assess its safety, efficacy, and pharmacokinetic properties. Preclinical studies involve both in vitro and in vivo experiments [37].

5.5 In Vitro Studies

Drug release studies evaluate the rate and extent of drug release from the FDC formulation under simulated physiological conditions. Various techniques, such as dissolution testing, dialysis, and diffusion cells, are used to assess the release kinetics. The goal is to optimize the formulation to achieve a desired release profile that matches the therapeutic needs. Compatibility studies assess the physical and chemical stability of the FDC components. These studies involve mixing the drugs and excipients under various conditions, such as different temperatures and pH, and monitoring for any signs of incompatibility, such as precipitation, color change, or loss of potency. The studies evaluate the cellular uptake and potential toxicity of the FDC on ocular cells. Cell lines derived from various ocular tissues, such as corneal epithelial cells, conjunctival cells, and retinal cells, are used to assess the uptake of the drugs, their intracellular distribution, and their effects on cell viability and function [8, 29, 30, 85].

5.6 In Vivo Studies

Pharmacokinetic studies evaluate the absorption, distribution, metabolism, and excretion of the FDC. These studies involve administering the FDC to animal models, such as rabbits or rats, and measuring drug concentrations in various tissues, including the eye, blood, and urine. Pharmacokinetic parameters, such as peak plasma concentration, time to peak concentration, and area under the curve, are determined to assess the drug's systemic exposure and bioavailability. Pharmacodynamic studies evaluate the efficacy of the FDC in producing the desired therapeutic effect. These studies involve measuring the effects of the FDC on various physiological parameters, such as intraocular pressure, inflammation, or angiogenesis. For example, in the case of glaucoma, the efficacy of a FDC can be assessed by measuring its ability to lower intraocular pressure. Safety and toxicity studies are essential to assess the potential adverse effects of the FDC. These studies involve administering the FDC to animals at various doses and monitoring for any signs of toxicity, such as ocular irritation, systemic toxicity, or developmental toxicity. Histopathological examination of ocular tissues can also be performed to identify any structural changes [25, 90].

5.7 Clinical Trials for FDCs

Once an FDC has completed preclinical evaluation, it can be advanced to clinical trials to assess its safety and efficacy in humans [35, 80]. In various phases of clinical trials and meeting regulatory requirements, FDCs can be approved for the market and provide new treatment options for patients with ocular diseases (Table 5.3).

5.8 Future Directions and Challenges in Fixed-Dose Combinations (FDCs) for Ocular Therapy

As ocular diseases become increasingly complex, the development of advanced fixed-dose combination (FDC) therapies is crucial. Future directions in FDC development involve the integration of emerging technologies, addressing regulatory and intellectual property challenges and promoting patient education to ensure proper use. This section discusses these aspects in detail.

5.8.1 Emerging Technologies in FDC Development

Nanotechnology in Ocular FDCs Nanotechnology offers immense potential in delivering FDCs more effectively to the eye. Nanoparticles—such as liposomes, polymeric nanoparticles, and nanoemulsions—can be tailored to target specific ocular tissues. This ensures precise drug delivery to the cornea, retina, or other structures, minimizing drug loss through tear turnover and enhancing penetration through the blood-ocular barrier. This approach offers several advantages, including improved bioavailability by enhancing drug permeation through ocular barriers, reduced systemic absorption, lowering the risk of side effects, and the possibility of combining multiple drugs or APIs in one nanoparticle system for synergistic effects [17, 78].

Biodegradable Polymers for Sustained Drug Delivery Biodegradable polymers, such as polylactic-co-glycolic acid (PLGA), can be used to develop sustained-release FDC systems. These polymers degrade naturally over time, releasing the drugs gradually, reducing dosing frequency, and ensuring prolonged therapeutic effects. This approach is particularly useful for chronic conditions such as glaucoma or diabetic retinopathy, where sustained delivery is essential. By reducing patient burden and minimizing the need for frequent administration, biodegradable polymer-based FDCs can significantly improve adherence to treatment regimens [21].

Gene Therapy with FDC Approaches Gene therapy is an emerging area in ocular disease management, offering a way to treat genetic disorders of the eye. FDC strategies can facilitate the delivery of multiple therapeutic genes simultaneously, addressing complex pathways involved in disease progression. Potential use cases include the treatment of retinal dystrophies by delivering multiple genes to restore

Table 5.3 The list of clinical trials available for fixed-dose combination therapy

NCT number	Study status	Conditions	Interventions	Company
NCT05283395	Completed	Glaucoma	Drug: netarsudil 0.02%/latanoprost 0.005% fixed-dose combination ophthalmic solution	Aerie Pharmaceuticals
NCT04891588	Unknown	Primary open-angle glaucoma, ocular hypertension	Drug: switching the preserved prostaglandin analog-timolol FC to Fixalpost (preservative-free prostaglandin analog-timolol FC)	Centar Zagreb University of Zagreb
NCT01937299	Completed	Ocular hypertension, open-angle glaucoma	Drug: brinzolamide 1%/brimonidine 0.2% ophthalmic suspension Drug: vehicle Drug: travoprost 0.004% ophthalmic solution	Alcon Research
NCT01937312	Completed	Ocular hypertension, open-angle glaucoma	Drug: brinzolamide 1%/brimonidine 0.2% ophthalmic suspension Drug: vehicle Drug: prostaglandin analog	Alcon Research
NCT03150160	Completed	Glaucoma	Drug: brinzolamide 1%/brimonidine 0.2% fixed combination Drug: placebo Drug: travoprost 0.004% ophthalmic solution	Novartis Pharmaceuticals
NCT01327599	Completed	Open-angle glaucoma, ocular hypertension, pigment dispersion glaucoma	Drug: travoprost 0.004% + timolol 0.5% ophthalmic solution	Alcon Research
NCT01978600	Completed	Open-angle glaucoma, ocular hypertension	Drug: brinzolamide 1%/brimonidine 0.2% ophthalmic suspension Drug: timolol maleate 0.5%	Alcon Research

(continued)

Table 5.3 (continued)

NCT number	Study status	Conditions	Interventions	Company
NCT02419508	Completed	Open-angle glaucoma, ocular hypertension	Drug: brinzolamide 1%/brimonidine 0.2% tartrate ophthalmic suspension Drug: brinz/brim vehicle Drug: prostaglandin analog	Alcon Research
NCT03822559	Completed	Open-angle glaucoma, ocular hypertension	Drug: a preservative-free, clear, colorless sterile aqueous ophthalmic solution containing 0.015 mg of tafluprost and 5.0 mg of timolol in 1 mL Drug: a preservative-free, clear, colorless sterile aqueous ophthalmic solution containing 0.015 mg of tafluprost in 1 mL	Santen Pharmaceutical Co., Ltd.
NCT04088019	Recruiting	Tuberculosis; uveitis	Drug: fixed drug combination (FDC) of ATT and oral steroid	Fakultas Kedokteran Universitas Indonesia
NCT00887029	Completed	Open-angle glaucoma, ocular h ypertension	Drug: DuoTrav (travoprost-timolol ophthalmic drops) Drug: Xalacom (latanoprost-timolol ophthalmic drops)	Ophthalmic Consultants Centres, Canada Alcon Research
NCT05987332	Completed	Metastatic uveal melanoma	Drug: IDE196|drug: Crizotinib Drug: pembrolizumab Drug: ipilimumab Drug: nivolumab Drug: dacarbazine	IDEAYA Biosciences
NCT02712008	Completed	Diabetic macular edema	Drug: REGN910–3 Drug: intravitreal aflibercept injection (IAI)	Regeneron Pharmaceuticals Bayer
NCT02713204	Completed	Neovascular age-related macular degeneration	Drug: REGN910-3 Drug: intravitreal aflibercept injection (IAI)	Regeneron Pharmaceuticals

(continued)

Table 5.3 (continued)

NCT number	Study status	Conditions	Interventions	Company
NCT05005637	Enrolling_ by_ invitation	Tuberculous uveitis	Drug: anti-tuberculosis drug Drug: prednisone tablet	Fakultas Kedokteran Universitas Indonesia
NCT01135914	Completed	Diabetic macular edema	D: ranibizumab Procedure: laser	Novartis Pharmaceuticals
NCT05581875	Not_yet_ recruiting	Multiple myeloma	D: belantamab mafodotin-Blmf D: daratumumab D: pomalidomide D: dexamethasone	Hellenic Society of Hematology GlaxoSmithKline

or modify affected pathways. Gene-loaded nanoparticles or viral vectors can ensure localized and controlled release to minimize systemic exposure [31, 73, 94].

5.8.2 Regulatory Hurdles and Intellectual Property Considerations

The development and approval of FDCs involve more regulatory complexities than single-drug formulations. Each component of the combination must demonstrate compatibility, safety, and efficacy individually and in synergy. Furthermore, novel delivery systems—such as nanoparticles or implants—often require additional regulatory scrutiny to ensure they meet safety and quality standards. To address these challenges, pharmaceutical companies can collaborate with regulatory authorities early in the development process to discuss regulatory expectations and potential hurdles. This proactive approach can help to streamline the approval process and minimize delays. Conduct robust preclinical and clinical studies to establish the therapeutic benefits of the FDC. Well-designed and executed studies can provide strong evidence to support regulatory approval [65, 81].

Securing intellectual property rights is critical to incentivize innovation and investment in the development of new FDC technologies. Patents protect the formulation, delivery mechanisms, or novel APIs used in FDCs, helping developers recover research and development costs. However, the overlap of patents on multiple drugs can complicate the commercialization process. To navigate these challenges, strategic patent filing and partnerships with drug manufacturers can be employed. Additionally, licensing agreements can facilitate access to patented technologies while ensuring that innovation is rewarded [51]. Even the most advanced FDCs are only effective if used correctly by patients. Effective patient education is essential to ensure that patients understand how to use the formulation properly and adhere to the prescribed regimen. This is particularly important for chronic ocular conditions where long-term therapy is required. Strategies to improve patient education include providing easy-to-understand instructions and demonstrations on how to administer eye drops or use ocular inserts, as well as educating patients about the importance of adherence to achieve therapeutic outcomes [76].

Noncompliance remains a major barrier to the success of ocular therapies, especially for complex dosing regimens. FDCs simplify treatment by reducing the frequency of administration, but additional strategies are needed to enhance patient adherence. Examples of compliance strategies include simplified dosing regimens, patient support programs, and personalized care. By combining multiple drugs into a single formulation, FDCs can reduce the burden of multiple administrations. Additionally, patient support programs, such as reminders, mobile apps, or refill management systems, can assist patients with chronic therapies. Furthermore, offering tailored treatment plans based on the patient's lifestyle and needs can further enhance adherence [88].

5.9 Conclusion

The future of FDCs in ocular therapy lies at the intersection of emerging technologies, regulatory frameworks, and patient-centered care. Nanotechnology and biodegradable polymers will play pivotal roles in enhancing drug delivery, while gene therapy presents new opportunities to address previously untreatable ocular diseases. Overcoming regulatory and intellectual property challenges will be essential to ensure these innovations reach patients. Additionally, patient education and compliance strategies will remain central to the success of FDC therapies, ensuring that advances in drug delivery translate into meaningful improvements in clinical outcomes. As research progresses, collaborative efforts between scientists, regulatory bodies, and healthcare providers will be key to unlocking the full potential of FDCs in ocular care.

References

1. Adjei K, Ali A. PSS4 a cost effectiveness analysis of latanoprost monotherapy, netarsudil monotherapy and fixed dose combination latanoprost/netarsudil in management of primary open angle glaucoma in USA. Value Health. 2021;24:S220–1.
2. Akhter S, Anwar M, Siddiqui MA, Ahmad I, Ahmad J, Ahmad MZ, Bhatnagar A, Ahmad FJ. Improving the topical ocular pharmacokinetics of an immunosuppressant agent with mucoadhesive nanoemulsions: formulation development, in-vitro and in-vivo studies. Colloids Surf B: Biointerfaces. 2016;148:19–29.
3. Alzahrani A. An investigation into the feasibility of coupling hot-melt extrusion technology with fused deposition modeling 3D printing in different pharmaceutical dosage forms. University: The University of Mississippi; 2022.
4. Arranz-Romera A, Esteban-Pérez S, Garcia-Herranz D, Aragón-Navas A, Bravo-Osuna I, Herrero-Vanrell R. Combination therapy and co-delivery strategies to optimize treatment of posterior segment neurodegenerative diseases. Drug Discov Today. 2019;24:1644–53.
5. Bagul US, Nazirkar MV, Mane AK, Khot SV, Tagalpallewar AA, Kokare CR. Fabrication of architectonic nanosponges for intraocular delivery of brinzolamide: an insight into

QbD driven optimization, in vitro characterization, and pharmacodynamics. Int J Pharm. 2024;650:123746.
6. Bao Q, Newman B, Wang Y, Choi S, Burgess DJ. In vitro and ex vivo correlation of drug release from ophthalmic ointments. J Control Release. 2018;276:93–101.
7. Barnebey HS, Orengo-Nania S, Flowers BE, Samples J, Mallick S, Landry TA, Bergamini MV. The safety and efficacy of travoprost 0.004%/timolol 0.5% fixed combination ophthalmic solution. Am J Ophthalmol. 2005;140(1):e1-1. e1-8
8. Bell D. Combine and conquer: advantages and disadvantages of fixed-dose combination therapy. Diabetes Obes Metab. 2013;15:291–300.
9. Bennett NH, Chinnery HR, Downie LE, Hill LJ, Grover LM. Material, immunological and practical perspectives on eye drop formulation. Adv Funct Mater. 2020;30:1908476.
10. Bhagav P, Trivedi V, Shah D, Chandran S. Sustained release ocular inserts of brimonidine tartrate for better treatment in open-angle glaucoma. Drug Deliv Transl Res. 2011;1:161–74.
11. Bhavsar AS, Bhavsar SG, Jain SM. Evaluation of the effects of chloroquine phosphate eye drops in patients with dry eye syndrome. Int J Biomed Adv Res. 2011;2:198–214.
12. Bielory L. Ocular allergy treatment. Immunol Allergy Clin N Am. 2008;28:189–224.
13. Bisen AC, Dubey A, Agrawal S, Biswas A, Rawat KS, Srivastava S, Bhatta RS. Recent updates on ocular disease management with ophthalmic ointments. Ther Deliv. 2024;15:1–8.
14. Brubaker JW, Teymoorian S, Lewis RA, Usner D, McKee HJ, Ramirez N, Kopczynski CC, Heah T. One year of netarsudil and latanoprost fixed-dose combination for elevated intraocular pressure: phase 3, randomized MERCURY-1 study. Ophthalmol Glaucoma. 2020;3:327–38.
15. Brunaugh AD, Sharma S, Smyth H. Inhaled fixed-dose combination powders for the treatment of respiratory infections. Expert Opin Drug Deliv. 2021;18:1101–15.
16. Bukhary H, Williams GR, Orlu M. Fabrication of electrospun levodopa-carbidopa fixed-dose combinations. Adv Fiber Mater. 2020;2:194–203.
17. Cattelan G, Guerrero Gerbolés A, Foresti R, Pramstaller PP, Rossini A, Miragoli M, Caffarra Malvezzi C. Alginate formulations: current developments in the race for hydrogel-based cardiac regeneration. Front Bioeng Biotechnol. 2020;8:414.
18. Çetin K, Şarkaya K. Drug delivery based on Nanoparticulate systems. Düzce Üniversitesi Bilim ve Teknoloji Dergisi. 2024;12:1993–2015.
19. Chen Y, Han F. Profile of ranibizumab: efficacy and safety for the treatment of wet age-related macular degeneration. Ther Clin Risk Manag. 2012;8:343–51.
20. Cheng K-J, Hsieh C-M, Nepali K, Liou J-P. Ocular disease therapeutics: design and delivery of drugs for diseases of the eye. J Med Chem. 2020;63:10533–93.
21. Costa J, Silva N, Sarmento B, Pintado M. Potential chitosan-coated alginate nanoparticles for ocular delivery of daptomycin. Eur J Clin Microbiol Infect Dis. 2015;34:1255–62.
22. Das S, Suresh PK, Desmukh R. Design of Eudragit RL 100 nanoparticles by nanoprecipitation method for ocular drug delivery. Nanomedicine. 2010;6:318–23.
23. Dol H, Gandhi S, Pardhi D, Vyawahare N. Formulation and evaluation of in situ ophthalmic gel of moxifloxacin hydrochloride. Pharm Innov. 2014;3:60.
24. Greig SL, Deeks E. Brinzolamide/brimonidine: a review of its use in patients with open-angle glaucoma or ocular hypertension. Drugs & Aging. 32;251–60.
25. Feczko T. Polymeric nanotherapeutics acting at special regions of body. J Drug Deliv Sci Technol. 2021;64:102597.
26. Figueirêdo CBM, Nadvorny D, De Medeiros Vieira ACQ, De Medeiros Schver GCR, Sobrinho JLS, Neto PJR, Lee PI, Soares MFDLR. Enhanced delivery of fixed-dose combination of synergistic antichagasic agents posaconazole-benznidazole based on amorphous solid dispersions. Eur J Pharm Sci. 2018;119:208–18.
27. Gambini SRG, De Vico U, Kilian CRR. A one-week course of levofloxacin/dexamethasone eye drops: a review on a new approach in managing patients after cataract surgery. Ophthalmol Ther. 2021;11(1):101–11.
28. Goldberg I, Pina RG, Lanzagorta-Aresti A, Schiffman RM, Liu C, Bejanian M. Bimatoprost 0.03%/timolol 0.5% preservative-free ophthalmic solution versus bimatoprost 0.03%/timolol

0.5% ophthalmic solution (Ganfort) for glaucoma or ocular hypertension: a 12-week randomised controlled trial. Br J Ophthalmol. 2014;98:926–31.
29. Gómez-Aguayo F, Paczka JA, Leñero-Córdova R, Jiménez-Román J, Davila-Villarreal J, Hartleben C, Baiza-Durán L, Olvera-Montaño O, García-Velez F, Muñoz-Villegas P. A phase III randomized clinical trial of a 0.5% timolol+ 0.2% brimonidine+ 2.0% dorzolamide fixed combination, preservative-free ophthalmic solution vs. 0.5% timolol+ 0.2% brimonidine+ 2.0% dorzolamide fixed combination in patients with controlled primary open-angle glaucoma. Ophthalmol Therapy. 2018;7:145–56.
30. Haak T, Meinicke T, Jones R, Weber S, Von Eynatten M, Woerle HJ. Initial combination of linagliptin and metformin in patients with type 2 diabetes: efficacy and safety in a randomised, double-blind 1-year extension study. Int J Clin Pract. 2013;67:1283–93.
31. He J, Fan K, Yan X. Ferritin drug carrier (FDC) for tumor targeting therapy. J Control Release. 2019;311:288–300.
32. Hens B, Corsetti M, Bermejo M, Löbenberg R, González PM, Mitra A, Desai D, Chilukuri DM, Aceituno A. "Development of fixed dose combination products" workshop report: considerations of gastrointestinal physiology and overall development strategy. AAPS J. 2019;21:75.
33. Hollo G, Bozkurt B, Irkec M. Brinzolamide/timolol fixed combination: a new ocular suspension for the treatment of open-angle glaucoma and ocular hypertension. Expert Opin Pharmacother. 2009;10:2015–24.
34. Hussain ME, Khan MKH, Hasan Z, Shammi SA, Quader MA. Efficacy of using topical cyclosporine a (0 05%) eye drops for the treatment of mild to severe dry eye disease. Commun Based Med J. 2021;10:98–104.
35. Imperiale JC, Acosta GB, Sosnik A. Polymer-based carriers for ophthalmic drug delivery. J Control Release. 2018;285:106–41.
36. Javitt M, Novack GD. Advantages and disadvantages of using rho kinase inhibitors to reduce intraocular pressure. Expert Rev Ophthalmol. 2024;19:201–9.
37. Kaarniranta K, Ikäheimo K, Mannermaa E, Ropo A. Pharmacokinetics, efficacy, and safety of the preservative-free fixed combination of tafluprost 0.0015% and timolol 0.5% in healthy volunteers: a phase I comparison vs. the corresponding preservative-free monotherapies. Clin Pharmacokinet. 2016;55:485–94.
38. Karnik I, Youssef AAA, Joshi P, Munnangi SR, Narala S, Varner C, Vemula SK, Majumdar S, Repka M. Formulation development and characterization of dual drug loaded hot-melt extruded inserts for better ocular therapeutic outcomes: sulfacetamide/prednisolone. J Drug Deliv Sci Technol. 2023;84:104558.
39. Khan SU, Ullah M, Saeed S, Saleh EAM, Kassem AF, Arbi FM, Wahab A, Rehman M, Ur Rehman K, Khan D. Nanotherapeutic approaches for transdermal drug delivery systems and their biomedical applications. Eur Polym J. 2024;207:112819.
40. Khopade AJ, Halder A, Patel V, Pateliya B, Burade V. A stable fixed-dose combination of benzalkonium chloride-free latanoprost and once-a-day timolol. RPS Pharm Pharmacol Rep. 2024;3:rqae011.
41. Kikuchi C, Ohno M, Izumo T, Takahashi S, Aoki M, Shimomura H, Kawano Y, Shimada S, Aoyama T. Investigation of approval trends and benefits of new fixed-dose combination drugs in Japan. Ther Innov Regul Sci. 2020;54:200–10.
42. Kim D-W, Weon KY. Pharmaceutical application and development of fixed-dose combination: dosage form review. J Pharm Investig. 2021;51:555–70.
43. Kitazawa Y, Smith P, Sasaki N, Kotake S, Bae K, Iwamoto Y. Travoprost 0.004%/timolol 0.5%-fixed combination with and without benzalkonium chloride: a prospective, randomized, doubled-masked comparison of safety and efficacy. Eye. 2011;25:1161–9.
44. Konstas AG, Schmetterer L, Costa VP, Holló G, Katsanos A, Denis P, Quaranta L, Irkec M, Castejón MA, Teus MA. Current and emerging fixed combination therapies in glaucoma: a safety and tolerability review. Expert Opin Drug Saf. 2020;19:1445–60.

45. Konstas AG, Schmetterer L, Katsanos A, Hutnik CM, Holló G, Quaranta L, Teus MA, Uusitalo H, Pfeiffer N, Katz LJ. Dorzolamide/timolol fixed combination: learning from the past and looking toward the future. Adv Ther. 2021;38:24–51.
46. Kshirsagar NA, Munshi R, Bavdekar SB, Saxena R. Irrational ophthalmic fixed-dose combinations for dry eye syndrome. Indian J Ophthalmol. 2022;70:3687–9.
47. Kumari G, Kumar TS, Zare M, Ramakrishna S. Electrosprayed core-shell nanoparticles for sustained release fixed combination monotherapy in glaucoma treatment. J Drug Del Sci Technol. 2023;88:104923.
48. Kuppermann BD, Patel SS, Boyer DS, Augustin AJ, Freeman WR, Kerr KJ, Guo Q, Schneider S, López FJ, Brimo DDS Gen 1 Study Group. Phase 2 study of the safety and efficacy of brimonidine drug delivery system (Brimo DDS) generation 1 in patients with geographic atrophy secondary to age-related macular degeneration. Retina. 2021;41:144–55.
49. Kushwaha V, Agrawal P, Azmi T, Shukla V, Das A. Fixed dose combination-rationale, uses, advantages, and disadvantages-a brief review. WJPR. 2022;11:628–38.
50. Leary KA, Steibel JP, Harman CD, Anderson AL, Komáromy AM. Safety and efficacy of topically administered netarsudil-latanoprost fixed dose combination (FDC; Rocklatan™) in normal and glaucomatous dogs with ADAMTS10-open-angle glaucoma (ADAMTS10-OAG). Vet Ophthalmol. 2021;24:610–9.
51. Luo Y, Yang L, Feng P, Qiu H, Wu X, Lu S, Zhou M, Xu L, Zhu Y. Pranoprofen nanoparticles with poly (L-lactide)-b-poly (ethylene glycol)-b-poly (L-lactide) as the matrix toward improving ocular anti-inflammation. Front Bioeng Biotechnol. 2020;8:581621.
52. Madaan S. Developing implants for ophthalmic drug delivery and flow modulation. London: UCL (University College London); 2020.
53. Mah FS, Karpecki PM. Review of loteprednol etabonate 0.5%/tobramycin 0.3% in the treatment of blepharokeratoconjunctivitis. Ophthalmol Therapy. 2021;10:859–75.
54. Manu MS, Mehta K, Das M, Balakrishnan S, Rakesh P, Sindhu M, Valamparampil MJ, Neena P, Satyanarayana S. Ocular adverse events in drug sensitive TB patients on daily fixed dose combination anti-TB drugs: a record review study from Kerala, India. Indian J Tuberc. 2020;67:216–21.
55. Miglior S, Grunden J, Kwok K. Efficacy and safety of fixed combinations of latanoprost/timolol and dorzolamide/timolol in open-angle glaucoma or ocular hypertension. Eye. 2010;24:1234–42.
56. Miranda MRH, Dubey A, Gs R, Charyulu RN. Fixed-dose combinations banned in India: is it the right decision? An eye-opening review. Expert Opin Drug Saf. 2019;18:977–85.
57. Morgan-Warren PJ, Morarji JB. Trends in licence approvals for ophthalmic medicines in the United Kingdom. Eye. 2020;34:1856–65.
58. Morrison PW, Porfiryeva NN, Chahal S, Salakhov IA, Lacourt C, Semina II, Moustafine RI, Khutoryanskiy VV. Crown ethers: novel permeability enhancers for ocular drug delivery? Mol Pharm. 2017;14:3528–38.
59. Mueller C, Altenburger U, Mohl S. Challenges for the pharmaceutical technical development of protein coformulations. J Pharm Pharmacol. 2018;70:666–74.
60. Nagargoje S, Phatak A, Bhingare C, Chaudhari S. Formulation and evaluation of ophthalmic delivery of fluconazole from ion activated in situ gelling system. Pharm Lett. 2012;4:1228–35.
61. Nair KL, Vidyanand S, James J, Kumar GV. Pilocarpine-loaded poly (dl-lactic-co-glycolic acid) nanoparticles as potential candidates for controlled drug delivery with enhanced ocular pharmacological response. J Appl Polym Sci. 2012;124:2030–6.
62. Onwuzuligbo CC, Onwuzuligbo AU. Fixed dose combination, a model therapy optimization strategy: a review. GSC Adv Res Rev. 2024;20:036–46.
63. Page S, Khan T, Kühl P, Schwach G, Storch K, Chokshi H. Patient centricity driving formulation innovation: improvements in patient care facilitated by novel therapeutics and drug delivery technologies. Annu Rev Pharmacol Toxicol. 2022;62:341–63.
64. Parveen N, Joshi H. Ocuserts: a novel formulation approach in drug delivery system. Saudi J Med Pharm Sci. 2020;6:420–5.

65. Pawar SN, Edgar KJ. Alginate derivatization: a review of chemistry, properties and applications. Biomaterials. 2012;33:3279–305.
66. Peng W, Jiang X, Zhu L, Li X, Zhou Q, Jie Y, You Z, Wu M, Jin X, Li X. Cyclosporine A (0.05%) ophthalmic gel in the treatment of dry eye disease: a multicenter, randomized, double-masked, phase III, COSMO trial. Drug Des Dev Ther. 2023;16:3183–94.
67. Rahić O, Tucak A, Omerović N, Sirbubalo M, Hindija L, Hadžiabdić J, Vranić E. Novel drug delivery systems fighting glaucoma: formulation obstacles and solutions. Pharmaceutics. 2020;13:28.
68. Rasheed SH, Samanvitha M, Manjunath S. Formulation and in-vitro characterization of erythromycin ocular inserts. Int J Res Pharm Sci Technol. 2020;2:44–50.
69. Razeghinejad MR, Sawchyn AK, Katz LJ. Fixed combinations of dorzolamide-timolol and brimonidine-timolol in the management of glaucoma. Expert Opin Pharmacother. 2010;11:959–68.
70. Rizzo S, Gambini G, De Vico U, Rizzo C, Kilian R. A one-week course of levofloxacin/dexamethasone eye drops: a review on a new approach in managing patients after cataract surgery. Ophthalmol Therapy. 2022;11:101–11.
71. Rupenthal ID, Green CR, Alany RG. Comparison of ion-activated in situ gelling systems for ocular drug delivery. Part 1: physicochemical characterisation and in vitro release. Int J Pharm. 2011;411:69–77.
72. Sanap SN, Bisen AC, Mishra A, Biswas A, Agrawal S, Yadav KS, Krishna A, Chopra S, Mugale MN, Bhatta RS. QbD based antifungal drug-loaded ophthalmic liposomal formulation for the management of fungal keratitis: in vitro, ex vivo and in vivo pharmacokinetic studies. J Drug Deliv Sci Technol. 2022;74:103517.
73. Sarella PNK, Vipparthi AK, Valluri S, Vegi S, Vendi VK. Nanorobotics: pioneering drug delivery and development in pharmaceuticals. 2024; https://doi.org/10.52711/0975-4377.2024.00014.
74. Sawant D, Dandagi PM, Gadad AP. Formulation and evaluation of sparfloxacin emulsomes-loaded thermosensitive in situ gel for ophthalmic delivery. J Sol-Gel Sci Technol. 2016;77:654–65.
75. Shah J, Kotadiya R. A critical review on analytical methods for recently approved FDC drugs: pregabalin and etoricoxib. Crit Rev Anal Chem. 2022;52:1048–68.
76. Shaikh R, Singh TRR, Garland MJ, Woolfson AD, Donnelly RF. Mucoadhesive drug delivery systems. J Pharm Bioallied Sci. 2011;3:89–100.
77. Dockhorn RJ, Duckett TG. Comparison of Naphcon-A and its components (naphazoline and pheniramine) in a provocative model of allergic conjunctivitis. Curr Eye Res. 1994;13(5):319–24.
78. Sharma AK, Arya A, Sahoo PK, Majumdar DK. Overview of biopolymers as carriers of antiphlogistic agents for treatment of diverse ocular inflammations. Mater Sci Eng C. 2016;67:779–91.
79. Siafaka PI, Titopoulou A, Koukaras EN, Kostoglou M, Koutris E, Karavas E, Bikiaris DN. Chitosan derivatives as effective nanocarriers for ocular release of timolol drug. Int J Pharm. 2015;495:249–64.
80. Silva B, São Braz B, Delgado E, Gonçalves L. Colloidal nanosystems with mucoadhesive properties designed for ocular topical delivery. Int J Pharm. 2021;606:120873.
81. Soni V. In silico molecular docking studies of cell-penetrating peptide and doxorubicin toward multiple tumor receptors. Asian J Pharm. 2024;18. https://doi.org/10.22377/ajp.v18i01.5274.
82. Soni S, Namdeo S, Agrawal P, Haider T, Soni V. Novel modalities of delivering herbal medicines for wound healing: a review. Dermatol Rev. 2023;4:194–210.
83. Soni S, Baghel K, Soni ML, Kashaw SK, Soni V. Size-dependent effects of niosomes on the penetration of methotrexate in skin layers. Future J Pharm Sci. 2024;10:48.
84. Soni S, Kori SK, Nema P, Iyer AK, Soni V, Kashaw SK. Cell-penetrating peptides as keys to endosomal escape and intracellular trafficking in nanomedicine delivery. Curr Med Chem. 2024;32(7):1288–312.

85. Srivastava V, Chary PS, Rajana N, Pardhi ER, Singh V, Khatri D, Singh SB, Mehra NK. Complex ophthalmic formulation technologies: advancement and future perspectives. J Drug Deliv Sci Technol. 2023;82:104317.
86. Subashini R. Oral delivery of ascorbic acid stabilized rifampicin nanoparticles for enhanced bioavailability of rifampicin. Tiruchengode: Swamy Vivekanandha College of Pharmacy; 2013.
87. Sun L, Mi K, Hou Y, Hui T, Zhang L, Tao Y, Liu Z, Huang L. Pharmacokinetic and pharmacodynamic drug–drug interactions: research methods and applications. Metabolites. 2023;13:897.
88. Suwannoi P, Chomnawang M, Sarisuta N, Reichl S, Müller-Goymann CC. Development of acyclovir-loaded albumin nanoparticles and improvement of acyclovir permeation across human corneal epithelial T cells. J Ocul Pharmacol Ther. 2017;33:743–52.
89. Tanihara H, Yamamoto T, Aihara M, Koizumi N, Fukushima A, Kawakita K, Kojima S, Nakamura T, Suganami H. Long-term intraocular pressure-lowering efficacy and safety of ripasudil-brimonidine fixed-dose combination for glaucoma and ocular hypertension: a multicentre, open-label, phase 3 study. Graefes Arch Clin Exp Ophthalmol. 2024;262:1–13.
90. Traitel T, Goldbart R, Kost J. Smart polymers for responsive drug-delivery systems. J Biomater Sci Polym Ed. 2008;19:755–67.
91. Tsung T-H, Chen Y-H, Lu D-W. Updates on biodegradable formulations for ocular drug delivery. Pharmaceutics. 2023;15:734.
92. Ubhe A, Oldenkamp H, Wu K. Small molecule topical ophthalmic formulation development–data driven trends & perspectives from commercially available products in the US. J Pharm Sci. 2024;113(10):2997–3011.
93. Vijayendra Swamy S, Nanjawade B, Ravichandran M, Manvi F. Development of sustained ocular drug delivery system for betaxolol. Indian J Pharm Educ Res. 2006;40:165.
94. Vishwakarma M, Agrawal P, Soni S, Tomar S, Haider T, Kashaw SK, Soni V. Cationic nanocarriers: a potential approach for targeting negatively charged cancer cell. Adv Colloid Interf Sci. 2024;327:103160.
95. Wilkins CA, Hamman H, Hamman JH, Steenekamp JH. Fixed-dose combination formulations in solid oral drug therapy: advantages, limitations, and design features. Pharmaceutics. 2024;16:178.
96. Yadav M, Guzman-Aranguez A, De Lara MJP, Singh M, Singh J, Kaur IP. Bimatoprost loaded nanovesicular long-acting sub-conjunctival in-situ gelling implant: in vitro and in vivo evaluation. Mater Sci Eng C. 2019;103:109730.
97. Yolton DP, Haesart S. Anti-infective drugs. Clin Ocul Pharmacol. 2008;11:219–64.
98. Zewail MB, El-Gizawy SA, Osman MA, Haggag YA. Preparation and in vitro characterization of a novel self-nano emulsifying drug delivery system for a fixed-dose combination of candesartan cilexetil and hydrochlorothiazide. J Drug Deliv Sci Technol. 2021;61 102320.
99. Zhang W, Han H, Feng K, Wang X, Du M, Meng X, Liu Y, Huang E, Brant R, Yan H. Is it necessary to use tobramycin-dexamethasone eye ointment prophylactically in eyes at the end of intraocular surgery? BMC Ophthalmol. 2020;20:1–6.
100. Zhuo Y, Yin Z, Ge Z. Attack and defense: adversarial security of data-driven FDC systems. IEEE Trans Industr Inform. 2022;19:5–19.

Role of Preservatives in Ophthalmic Products

6

Sumit Sharma, Sonali Batra, Kanika Nawani, Kritika Sharma, and Pradeep Kumar

Abstract

Preservatives are essential components in ophthalmic products, ensuring their sterility and extending shelf life as they play a critical role in maintaining the microbial integrity of formulations therefore safeguarding patient safety. The selection of the right preservative depends on multiple factors, including the product formulation, duration of use, and patient-specific considerations. Preservatives are indispensable for maintaining the sterility of ophthalmic products, but their potential toxicity necessitates careful selection and evaluation. Benzalkonium chloride being the most frequently used preservative is being frequently reported for its toxic effects. Preservatives that lessen or completely eradicate ocular discomfort are to be a significant advancement for both patients and physicians too.

Keywords

Preservative · Ophthalmic · Glaucoma · Benzalkonium chloride

S. Sharma (✉) · K. Sharma
Delhi Pharmaceutical Sciences and Research University, New Delhi, India

S. Batra
Department of Pharmaceutical Sciences, Indira Gandhi University, Meerpur, Rewari, Haryana, India

K. Nawani
Department of Medical Laboratory Technology and Sciences, Delhi Pharmaceutical Sciences and Research University, New Delhi, India

P. Kumar
Department of Botany, Swami Shraddhanand College, New Delhi, India

N. K. Mehra (ed.), *Complex Ophthalmic Dosage Forms: Advances in Biomedical Applications and Future Perspectives*, https://doi.org/10.1007/978-981-96-6306-4_6

6.1 Introduction

Ophthalmic products are pharmaceutical formulations that are applied in therapeutically effective concentrations in the eye for the management and eradication of eye ailments, such as glaucoma, conjunctivitis, dry eyes, neovascular age-related macular degeneration, diabetic retinopathy, and immune and bacterial keratitis [1, 2]. To address such diseases, patients are frequently put on therapy regimes, the timeframe of which can vary depending on the chronicity of the disease.

The ophthalmic pharmaceutical dispensing systems can be categorized into unit and multidose systems, which streamline the administration and packaging of pharmaceuticals. Under the Unit Dose Dispensing (UDD) drug distribution system used in hospitals, ophthalmic are packaged and provided to patients in single-use quantities during the course of their treatment. The key features of a unit dose system include pharmaceuticals being supplied in amounts not exceeding a 24-hour supply, availability of medications at all times in the patient care area, and packaging in individual, single-use units only [3]. In contrast, a multidose ophthalmic system entails putting several doses of a drug in one container as it is a customized delivery method that provides pharmaceutical use with a framework [4]. This system provides an alternative to the standard prescription method for those individuals who regularly administer medications and for medications that need to be administered over an extended period [5]. Preservatives are added to formulate a stable ophthalmic preparation as their antimicrobial action preserves sterility and extension of shelf life by preventing active drug degradation [6]. The multidose medications as the name suggests require multiple administrations from the same packaging, therefore the incorporation of preservatives in these formulations has been mandated by the US Food and Drug Administration (USFDA) and the US Pharmacopoeia (USP) [7].

Topical administration remains the first-line approach to treating ocular diseases and often requires exposure to different classes of drug formulations and preservatives. The perpetual use of these treatment regimens often leads to dry eye and cornea irritation due to not only the unwanted side effects of active ingredients but preservatives as well depending upon the exposure time. Pharmaceutically preservatives have been classified as: detergent and oxidizing preservatives. Benzalkonium chloride (BAK) which falls under the category of detergent preservative is a quaternary ammonium chloride that is most widely used for the preparation of multidose ophthalmic preparations. It acts by interfering with the lipid generation of the bacterial cell leading to its death [8]. It is frequently used in doses of 0.02–0.04% as a preservative in topical ophthalmic preparations [9]. There is ample evidence linking prolonged use of BAK causes ocular surface toxicity, corneal epithelial breakdown, and increased permeability of the cornea, which includes irritation and dry eyes. According to studies, the threshold concentration at which toxicity occurs has been estimated to be approximately 0.005% [10]. Over time, there has been a notable change in the way preservatives are used in eye products. In 1927, mercury-based chemicals were developed and used as topical ophthalmic treatments and vaccinations frequently in the 1930s, known as thimerosal, also referred to as thiomersal or Merthiolate. Thiosalicylate and ethyl mercury are produced as thimerosal breaks down, and this is what gives it its antibacterial properties [11].

6.2 Ideal Characteristics of Preservatives

Preservatives are highly effective chemical substances for ensuring the stability of pharmaceutical formulations. An ideal preservative should be [12]:

- Stable
- Should not exhibit interaction with other ingredients in the formulation
- Must have antimicrobial activity even at low concentrations
- Nontoxic
- Nonirritating

6.2.1 Classification System of Preservatives

Preservatives when classified based on their chemical nature are classified into two broad categories: natural and synthetic preservatives.

6.2.1.1 Synthetic Preservatives

As the name suggests, synthetic preservatives are synthesized chemically; therefore, they are also known as chemical or artificial preservatives. They exhibit antimicrobial activity even in small quantities. Sorbate and sodium benzoate are some examples. As they can stay fresh for long production time, storage, and use, these preservatives are preferred over natural ones.

6.2.1.2 Natural Preservatives

Natural preservatives are extracted from natural sources. With the increased incidence of chemical preservative resistance, the industry is looking for safer alternatives to artificial preservatives [13]. There is a study that analyzed the essential oil of *Euphrasia rostkoviana* and its antimicrobial effects, highlighting its potential as a natural preservative [14].

6.3 Preservatives Used in Ophthalmic Preparations

6.3.1 Benzalkonium Chloride (BAK)

Benzalkonium is a quaternary ammonium compound and is widely utilized as a preservative in ocular medicine, particularly drops. It helps preserve about 70% of all ocular formulations available [15]. BAK is reported to act through either one or a combination of the following mechanism of action:

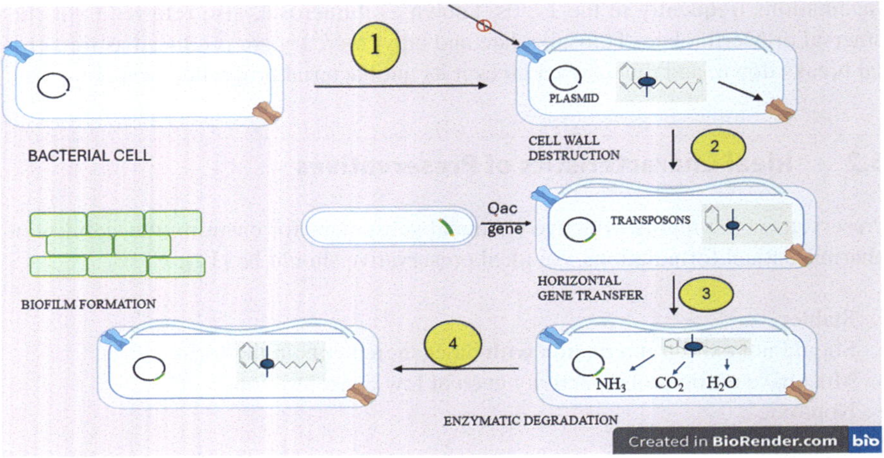

Fig. 6.1 The graphical representation of the mechanism of action of BAK. Self-created with biorwender.com

Cell Wall Destruction BAK is a quaternary ammonium compound as well as a cationic detergent that attaches itself to the negatively charged phospholipid bilayer found in microbial cell membranes [16]. It increases the membrane permeability due to the disruption of the lipid bilayer and eventually causes cell lysis.

Protein Denaturation Once the BAK enters the cell, it binds itself to the bacterial proteins and causes denaturation or unfolding of proteins. Denatured proteins lose their functional activity, which affects vital cellular processes such as enzyme activity, transport mechanism, and structural integrity.

Bacterial Biofilm Formation BAK can penetrate the extracellular polymeric substance matrix of bacterial biofilm and this biofilm matrix, making bacteria within the biofilm more susceptible to BAK [17]. The mechanism of action of BAK is shown in Fig. 6.1.

Bacteria such as *Escherichia coli* consist of an influx and efflux. The efflux pumps our membrane proteins that actively transport BAK out of the bacterial cells. This pumping of BAK out of the cell results in a reduction activity of efflux pump, which ultimately decreases its effectiveness and reduces its intracellular concentration. This efflux pump is often encoded by specific resistance genes such as qac genes, which can be located on a plasmid.

Membrane Alterations *E. coli* can alter the composition of their cell membranes to reduce the affinity of BAK binding. Some bacteria increase the proportion of negatively charged lipids in their membranes, which repels the positively charged BAK molecules. These changes make the membrane less permeable to BAK, thus reducing its disruptive effects.

Biofilm Formation Biofilms provide a protective environment that can shield bacteria from antimicrobial agents. The EPS matrix of biofilms can impede the penetration of BAK, reducing its effectiveness. Bacteria within biofilms are often in a slow-growing or dormant stage, which makes them more resistant to antimicrobials, including BAK.

Enzymatic Degradation Some bacteria produce enzymes that degrade BAK. These enzymes modify the chemical structures of BAK rendering it ineffective. Enzymatic degradation serves as a detoxification mechanism, protecting bacteria from the antimicrobial effects of BAK.

Horizontal Gene Transfer Resistance genes can be transferred between bacteria through horizontal gene transfer mechanisms such as transduction, conjugation, and transformation. These mobile genetic elements such as a plasmid, transposons, and integrons facilitate the spread of resistance traits. Horizontal gene transfer allows for the rapid dissemination of resistance genes within bacterial populations [17].

The use of BAK is currently decreasing as a result of several drawbacks that have been identified throughout time. Following topical drop installation, BAK exhibits deposition qualities in ocular tissue, including the lens, retina, corneal endothelium, and trabecular meshwork. When used frequently or in large amounts, this may cause ocular injury [18].

In 1941, Swan et al. reported that 0.04% BAK was found to be responsible for punctate keratitis in corneal epithelium, whereas its 0.1% demonstrated edema [19]. The toxicity due to BAK has regularly been reported over the last decade. A prior study that compared the toxicity of BAK and other preservatives found that BAK, at 0.01% (w/v), was more harmful than the majority of the other preservatives. The compounds were categorized according to their level of toxicity: thiomersal (0.01%), methylparaben (0.01%), sodium perborate (0.02%), chloro-butanol (0.5%), BAK (0.01%), ethylene diamine tetra acetic acid (EDTA), and salt. The BAK formulation resulted in a lower density of goblet cells when compared to the preservative-free formulation, according to a study that employed human conjunctival goblet cells to simulate the effects of BAK eye drops vs. the preservative-free equivalent. Furthermore, duration and concentration dependencies were observed in the action of BAK on the goblet cells. Higher doses of BAK or longer usage times have been related to damage to the surface of the eyes. In a different study, the safety and effectiveness of 0.5% timolol eye drops and antiglaucoma medication were examined. It was found that patients with glaucoma would benefit from moving from BAK-preserved Timolol drops to the preservative-free version. According to the notion, BAK increases the amount of Timolol that penetrates the cornea and improves the effectiveness of Timolol eye drops. This study demonstrated that the formulation without preservatives lowers intraocular pressure just as well as the formulation containing BAK [18, 20, 21].

6.3.2 Polyquaternium-1 (POLYQUAD®)

In order to minimize the use of BAK and its related side effects, a new quaternary ammonium compound was developed (Alcon Laboratories, Inc., Fort Worth, TX). POLYQUAD® is being employed in contact lens disinfecting solutions [22]. Its use has also been tested for artificial tear formulations [23].

6.3.3 EDTA (Edetate Disodium)

EDTA is a chelating agent and is frequently employed in various cosmeceutical and pharmaceutical formulations such as hair conditioners and face washes. The chemical nature of EDTA includes its ability to bind metals, which is why it is now being used in ophthalmic preparations. EDTA is reported for its action of removing calcified plagues in the superficial cornea in band keratopathy [24].

6.3.4 SofZia™

SofZia™ being the most recent advancement in the field of preservatives contains an active agent named travoprost which after coming in contact with cations gets inactive which is why it induces less cytotoxicity to the cornea as compared to conventional preservatives [25]. Some of the widely used preservatives (Table 6.1) for eye delivery include Benzalkonium chloride (BAK), Polyquaternium-1, Cetrimonium chloride, EDTA, Sodium Perborate, SOS, Thiomersal, Chlorobutanol, and *SofZia*™. The various classes of preservatives have been summarized in Table 6.1.

6.4 Applications of Preservatives

Preservatives are essential for preserving sterility and prolonging the shelf life of multidose formulations of topical ophthalmic medicines because they have significant and essential antibacterial activity [9]. The details of the preservative with storage condition, shelf life in both open and closed containers are summarized in Table 6.2.

6.5 Clinical Challenges and Concerns

Various clinical challenges and concerns can be explained via supported evidence.

Prostaglandin analogs (PGs) are often employed to treat glaucoma, i.e., elevated intraocular pressure. Since PGs efficiently lower intraocular pressure, they may cause conjunctival hyperemia, increased pigmentation of the iris, and potential cytotoxicity as adverse effects. In this case study, four distinct corneal-conjunctival

Table 6.1 Various types of preservatives used in the ocular formulation development

S. No	Chemical class	Characteristics of each class	Advantages	Disadvantages	References
1.	Quaternary ammoniums Benzalkonium chloride (BAK) Cetrimide Polyquaternium-1 Polyquaternium-42 Benzododecinium bromide (BDB) Cetrimonium chloride Streartrimonium chloride	An effective class of cationic surfactants that exhibits a strong affinity for the microbial membrane, allowing them to adhere and destroy the microbial agent's membrane	Highly efficient Prolonged shelf life Stable in solutions	May result in dryness and inflammation Prolonged use may be harmful to corneal cells	[18, 20, 26]
2.	Mercury derivatives Thiomersal or thimerosal Disodium EDTA (disodium-ethylene diamine tetra-acetate) Phenylmercuric nitrate or acetate (0.002% w/v)	These preservatives contain mercury, which was utilized as a preservative in ophthalmic preparation at a low dosage because of its historical germicidal activity and capacity to penetrate the cornea	Strong stability in a solution Effective at small dosages	Potential of allergic reactions Use has declined because of worries about safety Long-term exposure to mercury and its compounds leads to intraocular accumulation of mercury derivatives	[27–29]
3.	Oxidative complexes (soft preservatives) Sodium perborate $NaBO_3$ Sodium chloride SOC (stabilized oxychloro complex) (Pu-rite)	It can function as an antibacterial against diseases caused by bacteria, viruses, and fungi because it can produce free radicals, which are oxidizers	Cost-effective Easily available Preserves isotonicity, ensuring ocular comfort Antimicrobial action with a broad spectrum against bacteria, fungi, and viruses It lowers the possibility of preservative building in the eye by self-degrading into nontoxic by-products such as oxygen, water, and borate	Minimal antibacterial efficacy Adequate antibacterial protection requires conjunction with additional preservatives High amounts may be uncomfortable Possible hazards for the environment posed by borate by-products It requires meticulous formulation to guarantee effectiveness and stability	[28, 30]

(continued)

Table 6.1 (continued)

4.	Amidines (biguanides) SCP (stabilized chlorite peroxide) Polyhexamethylene biguanide hydrochloride (PHMB) Polyaminopropyl biguanide	Chlorhexidine has been used as an ophthalmic preservative for more than 30 years now in addition to this a recent study demonstrates that 0.2% chlorhexidine is a successful sequential combination, particularly for treating fungal infections. Chlorhexidine is the least utilized preservative day because of its compatibility issues, which require it to work more satisfactorily and consistently with all ophthalmic solutions	Use in contact lens solutions Less irritating than BAC Effective at low concentrations	Less effective against some bacterial strains Potential for resistance development Potential for staining and taste alteration	[28, 30]
5.	Phenols Benzyl alcohol Methylparaben Propylparaben Phenol m-cresol	The paraben series was uniform and had antibacterial properties, making it ideal for use against yeasts and molds. Even in minimal concentrations that are effective against bacteria and suitable for preservatives, they may still be regarded as ocular irritation and cause harm to epithelial cells. Antimicrobial activity is exhibited by this group	Wide ranging antimicrobial activity Effective at low concentrations Stable and long-lasting in formulations Frequently used in conjunction with other preservatives	Possibility of toxicity and irritation with extended use May irritate skin and eyes May have carcinogenic effects with prolonged exposure Limited use in some regions due to safety concerns	[31]
6.	Polymyxin Polymyxin B sulfate Sodium silver chloride complex and silver sulfate	About 85 strains of *Pseudomonas aeruginosa* were susceptible to polymyxin B sulfate, indicating its high effectiveness against *P. Aeruginosa*. However, several of those strains acquire resistance to another preservative.	Successful in combating gram-negative bacteria Frequently used in conjunction with other antibiotics to provide broad-spectrum coverage May help lower the likelihood of bacterial resistance	Has a narrow range of spectrum activity (mostly against gram-negative bacteria) May cause toxicity and allergic responses Unsuitable as a preservative on its own	[28]

7.	NIG/EDTA N-hydroxy-methyl-glycinate/disodium-ethylene diamine tetra-acetate	Having a synergistic antimicrobial impact compared to each preservative alone, this kind of combination preserving system	Efficacious against a wide range of bacteria, fungi, and yeast Prolongs the antimicrobial protection period and prolongs the ophthalmic solutions' shelf life	Has the potential to emit formaldehyde, a well-known irritant and sensitizer Causes eye discomfort and hypersensitivity	[32]
8.	SofZia™, multiple component Borate Sorbitol Propylene glycol Zinc	SofZia™ is a unique form of component preservative system that derives its preservative activity from the ionic buffer process	Broad-spectrum antimicrobial effects Decomposes spontaneously into innocuous by-products	Some people may experience eye discomfort. Stability problems in some formulations. Toxicity is low when used adequately Efficacy can be affected by the presence of organic materials	[28]
9.	Alcohols Polyhexamethyl biguanide Chlorobutanol Chloroxylenol Chlorocresol Phenoxyethanol	This class of drugs acts by an antioxidant impact that is combined with stabilizing lens membrane integrity, boosting Na-K ATPase, and increasing membrane sodium pump. These compounds possess exceptional hydrophilic and lipophilic solubility, which allows them to be employed as topical eye preparations. The lipophilic nature of these compounds regulates the drug's trans-corneal penetration	Effective broad-spectrum activity Efficiency in a wide range of activities Frequently used in conjunction with other preservatives	Can be unstable at ambient temperature May irritate the skin	[28]
10.	Carboxylic acid Sorbic acid Hexachlorophene	It has very little antibacterial activity, and the way it inhibits microbial development is by making the microorganism's cell environment more acidic	Effective against a variety of microbes particularly fungus Effective at low pH levels May require a combination of other preservatives for broad-spectrum efficacy Nontoxic and biodegradable	Limited ability to combat some microorganisms Generally well-tolerated Can cause eye irritation at higher concentrations Problems with stability in particular combinations	[28]

Table 6.2 Applications/benefits of preservatives

Name of preservatives	Storage conditions	Applications	Shelf life (opened container)	Shelf life (closed container)	Medication examples	References
Benzalkonium chloride	Keep containers tightly closed in a dry, cool, and well-ventilated place. Temperature-15–30 °C	It is used in eye drops, cleansing solutions for cleansing skin, etc., pain relieving creams, and hand sanitizer	About 1 to 3 months	About 2 to 3 years	Timoptic*, Azopt, Lumigan*, Xalatan	[33]
Polyquaternium-1 (POLYQUAD®)	Temperature - +4 °C, sealed storage, away from moisture	It is used in eye drops, contact lens solutions, skin care products, mouthwashes, etc.	About 1–2 months	About 2–3 years	Tears Naturale II, Opti free* Express disinfecting solutions	[34]
Chloro-butanol	Stored in a polyethylene container, temperature- 2–25 °C	It is used as a detergent preservative in eye drops and other ophthalmic products	About 37 days to 3 months	About 2–3 years	TobraDex* Ointment	[35] [36]
Sodium perborate (GenAqua® and Dequest®; SPB)	Stored in a cool, dry, well-ventilated place, and kept in a tightly closed container, a maximum temperature is less than 40 °C	It is used as an oxidative preservative in lubricating drops	After 1 year	About 36 months	Genteal*	[37]
SofZia™	Store at 2 to 25 °C, at cool room temperature, away from sunlight and heat	Oxidative preservative used in ophthalmic solution formulation	6 weeks or typically one year	About 1–2 years	TravatanZ*	[38]

cell lines are used to examine the cytotoxicity of several PGs ophthalmic solutions. The study used several approaches to ensure reliable and acceptable outcomes.

The cell lines used in this study were BCE C/D-1b (bovine corneal epithelial cells, JCRB-9129, Health Science Research Resource Bank, Osaka, Japan), RC-1 (rabbit corneal epithelium, JCRB-0246; Health Science Research Resource Bank), Chang conjunctiva (human conjunctival cells, ATCC CCL-20.2), and SIRC (rabbit corneal epithelium, CCL-60, American Type Culture Collection [ATCC], Manassas, VA). All cell lines were cultivated according to the standard methods supplied by the providers.

After being taken out of 96-well culture plates, approximately 2×10^4 cells were cultured for two days. After that, several ophthalmic medications were added to the culture media, and the cells were cultured for a further period. Following medication treatments, the cells were incubated for 48 hours in new media that had replaced the culture medium. Neutral red and 3-(4,5-dimethylthiazol-2-yl)-2,5-diphenyl tetrazolium bromide (MTT) tests were used to assess cell viability. While lysosomes in live cells are assessed spectrophotometrically using neutral red labeling, mitochondrial activity and proliferation are measured using the MTT test. The proportion of control cells in the culture medium alone is represented by the treated cells' cell viability.

The cell viability score, which is comparable to the MIC50 (minimum inhibitory concentration of medicine necessary to block the growth of 50% of organisms), was used in the study to compare cell viability. The researchers compared the cell survival of each of the other solutions at each dose with Xalatan, a top PG, using 24 comparisons (two experiments, three exposure durations, and four corneal cell lines). Tapros and 0.01% benzalkonium chloride were also contrasted. 0.0015% Tapros, 0.005% Xalatan, 0.12% unoproston, 0.004% travoprost, 0.004% travoprost, and 0.01% benzalkonium chloride were among the cytotoxicity solutions. The t-test was utilized to do statistical comparisons, and the findings were displayed as the mean ± standard deviation (SD). $P < 0.05$ was regarded as noteworthy.

Four cell lines were examined for cell viability; Travatan Z had the greatest viability rate and Rescula had the lowest [39]. When comparing three dosages of Travatan and Xalatan, the findings were contradictory; after a 10-min treatment, the chloride level was consistently higher, but after a 30-min exposure, there was no change in the level of benzalkonium chloride. Following a 60-min exposure, no change in cell viability was seen between Tapros and benzalkonium chloride. There were no inconsistent results between the MTT and neutral red tests. In general, the MTT results were lower than those produced by the neutral red test, particularly in solutions that were undiluted and twice diluted. There were no significant differences between the four cell lines' outcomes. When Tapros and 0.01% benzalkonium chloride were compared, it was shown that cell viability.

6.6 Preservative-Free Alternatives

The preservative-free formulation is required because all the preservatives currently used in clinics cause irritation, allergies, instability in the tear film, inflammation of the conjunctiva, subconjunctival fibrosis, and deterioration of the corneal surface. This is especially evident in the treatment of long-term ailments like glaucoma with eye drops [39].

Preservative-free eye drops have been shown in studies to lessen glaucoma patients' subjective and objective ocular symptoms. However, the more the preserved drops are used, the more frequently these symptoms occur. Patients with greater daily dosages of BAK and numerous topical glaucoma medicines have lower OSDI and quality of life ratings. Preservative-free drugs come in single-dose vials, which can be troublesome for chronic users and those with underlying tremors or arthritis. They are also more costly and may not be fully covered by insurance [40, 41].

Developing multidose preservative-free bottles (ABAK® and COMOD®) as an alternative to single-dose preservative-free drops has resulted in cost savings and improved usability. Thanks to a bi-functional membrane with antimicrobial qualities, these bottles keep their sterility for up to three months after opening. However, the usage of artificial tears made without preservatives and packaged in multidose bottles was connected to an epidemic of drug-resistant *Pseudomonas aeruginosa* infections in the United States in 2022–2023. The US Centers for Disease Control and Prevention are looking at the epidemic. In the United States, three topical glaucoma drugs are available without preservatives: Zioptan (refrigerated at 2° to 8 °C (36° to 46 °F), Cosopt PF, and Timoptic in Ocudose. The single-use containers containing these sterile liquids should be thrown away after use. These ocular surface irritation and dry eye problems are reduced, and these preservative-free treatments are better tolerated.

Preservative-free formulations of many topical treatments are not commercially accessible, and compounding pharmacies charge more for their preservative-free versions of drugs [42].

Case Study: Comparison of Effects of Preserved and Preservative-Free Antiglaucoma Drugs in Causing Dry Eye

Over 111.8 million persons over 40 worldwide suffer from glaucoma, a chronic and progressive visual neuropathy. The multifactorial ocular ailments referred to as dry eye syndrome (DES) are characterized by pain, blurred vision, and persistent tear films, all of which have the potential to harm the ocular surface. One of the strongest causes of dry eyes is topical antiglaucoma drugs. Nevertheless, the use of preservatives in the manufacturing of these drugs is a problem and exacerbates the symptoms of dry eyes. Preservatives like benzalkonium chloride (BAK) have been shown to produce negative effects such as eye irritation and pain as well as ocular surface toxicity. To reduce possible ocular surface damage while maintaining treatment efficacy, this has prompted the creation of antiglaucoma medication compositions devoid of preservatives. To examine the impact of preserved and preservative-free antiglaucoma medications on inducing dry eyes in glaucoma patients, this study used a randomized controlled trial methodology. The study is to offer insightful information that optimizes the treatment of glaucoma by striking a balance between ocular surface health and therapeutic efficacy. From March to June 2023, this non-randomized controlled experiment was executed at Rawalpindi's Fauji Foundation Hospital. The Med Calc online sample size calculator was applied to estimate the sample size, which included 108 patients in two groups of 54. Adults who were 18 years of age or older and had a verified diagnosis of open-angle glaucoma that required antiglaucoma drug treatment were eligible for participation. All participants gave their informed consent for the trial, which split the patients into two groups. Group B was given the antiglaucoma medication without preservatives, whereas group A was given the medication with preservatives. The study used Schirmer's test and TBUT at baseline to evaluate the effects of the various formulations on ocular health. After taking their prescribed drugs for three months, participants in both groups underwent further testing. By inserting a filter paper sheet into the conjunctival sac between the middle and lateral thirds of the eyelid, the Schirmer 1 Test calculates the total tear flow. The time interval between the final blink and the initial black point or tear film breaking is measured by the TBUT test. By comparing the degree of healing impact (TBUT test and Schirmer test scores) on glaucoma patients induced by each intervention, the efficacy of preservative-free medicine was evaluated. The mean and standard deviation for age, TBUT score (sec), and Schirmer's test (mm) were computed using SPSS version 26. Significant results were defined as a p-value of less than 0.05. The TBUT and Schirmer's tests were the main focus of the study's analysis of individuals with a particular ailment. At the initial visit, the patients' mean TBUT score was 10.90 ± 1.43 s, and their mean age was 34.88 ± 10.22 years. Both groups' TBUT scores dropped after three months, suggesting a decline in the TBUT score. Based on the values, the findings were classified as either normal or abnormal; a score of more than 10 s was deemed normal, and a score of less than 10 s was deemed abnormal. During the first and third appointments, Schirmer's test results were also noted. Group A and group B had mean Schirmer's test scores of 13.77 ± 2.27 mm and 14.51 ± 2.35 mm, respectively. Three months later, group A's mean Schirmer's test scores were 11.94 ± 2.54 mm, whereas group B's were 13.18 ± 1.37 mm. Based on the values, the findings were further divided into two categories: moderate and severe. A score of more than 10 mm was deemed moderate, while a score of less than 10 mm was deemed severe. The effects of preserved and preservative-free antiglaucoma drugs on inducing dry eyes in glaucoma patients were studied in this study. Tear film integrity was evaluated using the tear film stability test (TBUT), where a lower TBUT indicates poor tear film integrity. Participants' ocular health was comparatively stable at the start of the trial, as seen by the same findings of the initial TBUT assessments in both groups. After three months, however, group A's mean TBUT score was 9.68 ± 1.25 s, whereas group B's was 10.12 ± 1.16 s. Preservative-free prostaglandin analogs (PGAs) often have higher levels of preservatives than preserved-prostaglandin analogs (PF-PGAs). Following a 6-month follow-up, TBUT was 5.00 ± 1.88 s while taking tafluprost (NPT) without preservatives. For the next six months, TBUT was 3.60 ± 2.07 s after switching to a preservative that included tafluprost (PT). At

12 weeks after switching to preservative-free tafluprost, TBUT scores climbed from 5.1 ± 2.3 s to 10.1 ± 3.6 s, according to Lee et al., Kim KE et al., and Mohamed et al., who observed that this difference was statistically significant. The Schirmer's test, which gauges tear production, showed intriguing patterns in both groups. Both groups had very comparable Schirmer's test results at baseline. The mean Schirmer's test scores, however, were 13.18 ± 1.37 mm in group B and 11.94 ± 2.54 mm in group A three months following the first visit. This disparity implies that over time, the presence of preservatives may lead to a more noticeable decrease in tear formation. While both preserved and preservative-free drugs may result in lower tear film stability and tear production, the effect seems to be more noticeable in the group receiving the preserved medication, according to the comparison of the two outcome measures—TBUT and Schirmer's Test. According to a study, participants taking preserved antiglaucoma medication had a greater frequency of ocular surface illness than those not using any topical drugs. Formulations without preservatives are required for long-term eye disorders.

The primary cause of membrane integrity disturbance is thought to be BAK, which, when added to eye drops, destabilizes the tear film and results in inflammation, squamous metaplasia, and fibrotic alterations in the conjunctiva. Ocular symptoms brought on by preservatives and antiglaucoma drugs include burning or stinging sensations while using eye drops, dryness, excessive weeping, itching of the eyelids, and a feeling of a foreign object in the eye.

The study does have several drawbacks, though, such as a small sample size and a 4-month time frame. Individual differences in response to treatment and any confounding variables should be addressed, as should the long-term effects of antiglaucoma drugs on dry eyes in bigger, longitudinal studies. The study highlights the detrimental effects of tears and how preservatives in antiglaucoma drugs contribute to dry eye illness, highlighting the detrimental impacts on the Pakistani population. According to the results, antiglaucoma drugs without preservatives should be used in clinical settings, highlighting the need for safer and more efficient therapies [43].

The cytotoxicity of many ophthalmic solutions, with increasing benzalkonium chloride concentrations, was investigated in the study. These solutions included Xalatan, Travatan Z, Rescula, Tapros, and Travatan Z. Apart from Rescula, which was more toxic than Travatan and Tapros because it contained more benzalkonium chloride, cell viability was inversely correlated with this increase. Unlike other PGs, unoproston in Rescula is a digested version of natural PG F2 alpha. It was discovered that the cytotoxicity of conserved tafluprost was similar to that of Xalatan and Travatan. After ten minutes of exposure, the hazardous levels of polysorbate 80 and benzalkonium chloride in Tapros caused a decrease in cell viability. In the trials, it was unclear if tafluprost had a cytoprotective impact on ocular surface cell lines. Two bioassays were employed in the study to assess cell viability under various circumstances. The ten-fold dilutions were impractical in this experimental situation, but the unadulterated and twice-diluted solutions were appropriate for clinically meaningful cytotoxicity assessments. The actual residence time of the administered solution on an ocular surface was reflected by the 10-minute exposure period. Because of their modest survivability rates, differences between medications lost significance after 60 minutes of exposure.

Some of the study's drawbacks were examining four cell lines with two separate assays, assessing just cytotoxicity, and failing to assess kerato-conjunctival cell toxicity directly. Many assessments had to be carried out to learn more about the possible negative effects of PGs in glaucoma patients. Travatan Z without benzalkonium chloride was shown to be the least harmful medication in research employing two cytotoxicity tests on four ocular surface cell lines. The best exposure durations and dilutions for ophthalmic solution cytotoxicity tests were also included in the study. Cell viability in the medications was compared using CVS50, and the recently released preserved Tapros was included in the pharmacological comparison [39].

Case Study 2: A Case Report on Polyquaternium-1 (POLYQUAD® -1)-Associated Dendritiform Keratopathy

PQ-containing eye drops have been demonstrated in vitro to have the ability to reduce cell viability and activate NF-kB and other inflammatory markers, suggesting that PQ-1 may have cytotoxic effects [44]. A polycationic preservative called polyquaternium-1, which is present in contact lens solutions and replacement tears, has been connected in 16 cases to dendritiform keratopathy. The ailment was initially documented in 2016 and was made worse by a rise in usage to once an hour.

The 78-year-old man had a history of glaucoma, aqueous tear deficit, and herpes zoster-related hypesthesia. He also had acute microbial keratitis in his right eye, resulting in a two-week reduction in vision. He had no issues after years of taking Systane, a tear substitution. But two weeks later, his vision got worse. The patient's central corneal dendritiform keratopathy was accompanied by widespread, dispersed, punctate epithelial erosions. He was told to switch to non-preserved tear substitutions and stop using topical Systane. Three weeks later, the keratopathy subsided, and the patient's eyesight improved to 20/100.

The most prevalent preservative in topical ophthalmic drugs, benzalkonium chloride (BAK), is more hazardous to human corneal epithelial cells than PQ-1, a polymeric quaternary ammonium antibacterial agent that is also better tolerated. But in other circumstances, it can cause greater inflammation. An unusual side effect of ocular preservative exposure is dendritiform keratopathy, which may be mediated by immunologic pathways, according to past research. After employing preservative-free tear replacements and stopping Systane, a tear substitute, the patient's reversible dendritiform keratopathy condition improved. One unusual etiology for reversible dendritiform keratopathy is PQ-1 [45].

6.7 Regulatory Guidelines and Standards

Ophthalmic preparations must be sterile to ensure safety as nonsterile products are considered adulterated or misbranded by the FDA. Containers should be sealed tightly to maintain sterility, and eye droppers or dispensers must be sterile and properly labeled with usage instructions and warnings to prevent contamination. The

FDA also requires tamper-resistant packaging for over-the-counter ophthalmic medications. For multidose containers, preservatives that prevent bacterial growth are essential, and packaging must reduce contamination risks.

The European Medicines Agency (EMA) advises that ophthalmic solutions without preservatives should be used for patients who are sensitive, especially infants. While preservatives may be necessary, their concentrations should be minimal, and their safety should be carefully evaluated. The use of mercury-based preservatives like thiomersal is discouraged to minimize environmental mercury exposure.

Ophthalmic product labeling must include clear instructions, warning about contamination risks, and any mercury-containing preservatives used. Manufacturers are urged to adhere to USP standards for safety, efficacy, and quality, ensuring their products meet regulatory requirements. This ongoing oversight ensures that ophthalmic products remain safe and effective for consumers.

6.8 Future Challenges and Outcomes

For many years, there were no other options available to eye physicians, so they had to utilize topical therapies that caused corneal damage and unpleasant symptoms. When physicians had to decide between glaucoma advancement and ocular irritation, which might be rather severe, they naturally went with the less severe option. These days, we have preservatives like sodium perborate and *SofZia*™, which are known as "vanishing preservatives" because they react with the tear film to produce oxygen and water. There are currently or will soon be other choices, such as multiuse bottles that eliminate the need for preservatives and prevent contamination. Although corneal toxicity may not result in blindness, it is nevertheless a significant concern for patients and should not be disregarded for the "greater good" of treating glaucoma. The detrimental effects of the preservatives are expected to impact adherence, which is particularly true for glaucoma, a chronic illness. Preservatives that lessen or completely eradicate ocular discomfort are a significant advancement for both patients and physicians too.

References

1. Patel PB, Shastri D, Shelat P, Shukla A. Ophthalmic drug delivery system: challenges and approaches. Systematic Rev Pharm. 2010;1:113.
2. Hughes PM. Overview of ocular drug delivery and latrogenic ocular cytopathologies. Ophthalmic Drug Delivery Sys. 1993:1–27.
3. Larasati N, Wulandadari CA. Implementation of unit dose dispensing using management information system in pharmacy installation of panti rapih hospital. Jurnal Manajemen Dan Pelayanan Farmasi (J Manag Pharm Pract). 2019;8:153–64.
4. Mertens BJ, Kwint H-F, Bouvy ML, van Marum RJ. Multidose drug dispensing: what does it really involve? Ned Tijdschr Geneeskd. 2017;161:D1770.
5. Johnell K, Fastbom J. Multi-dose drug dispensing and inappropriate drug use: a nationwide register-based study of over 700 000 elderly. Scand J Prim Health Care. 2008;26:86–91.

6. Gomes JAP, Azar DT, Baudouin C, Efron N, Hirayama M, Horwath-Winter J, Kim T, Mehta JS, Messmer EM, Pepose JS, Sangwan VS, Weiner AL, Wilson SE, Wolffsohn JS. TFOS DEWS II iatrogenic report. Ocul Surf. 2017;15:511–38. https://doi.org/10.1016/j.jtos.2017.05.004.
7. Noecker R. Effects of common ophthalmic preservatives on ocular health. Adv Ther. 2001;18:205–15.
8. Freeman PD, Kahook MY. Preservatives in topical ophthalmic medications: historical and clinical perspectives. Expert Rev Ophthalmol. 2009;4:59–64. https://doi.org/10.1586/17469899.4.1.59.
9. Goldstein MH, Silva FQ, Blender N, Tran T, Vantipalli S. Ocular benzalkonium chloride exposure: problems and solutions. Eye. 2022;36:361–8. https://doi.org/10.1038/s41433-021-01668-x.
10. Burstein NL, Klyce SD. Electrophysiologic and morphologic effects of ophthalmic preparations on rabbit cornea epithelium. Invest Ophthalmol Vis Sci. 1977;16:899–911.
11. Clements CJ, McIntyre PB. When science is not enough–a risk/benefit profile of thiomersal-containing vaccines. Expert Opin Drug Saf. 2006;5:17–29.
12. Sasseville D. Hypersensitivity to preservatives. Dermatol Ther. 2004;17:251–63.
13. Jain A, Radiya P, Wadekar R, Limaye S, Pawar C. Natural excipients-an alternative to synthetic excipients: a comprehensive review. Int J Pharm Med Res. 2014;2:123–7.
14. Novy P, Davidova H, Serrano-Rojero CS, Rondevaldova J, Pulkrabek J, Kokoska L. Composition and antimicrobial activity of Euphrasia rostkoviana Hayne essential oil. Evid Based Complement Alternat Med. 2015;2015:734101.
15. Gasset AR. Benzalkonium chloride toxicity to the human cornea. Am J Ophthalmol. 1977;84:169–71.
16. Labbé A, Pauly A, Liang H, Brignole-Baudouin F, Martin C, Warnet J-M, Baudouin C. Comparison of toxicological profiles of Benzalkonium chloride and Polyquaternium-1: an experimental study. J Ocul Pharmacol Ther. 2006;22:267–78. https://doi.org/10.1089/jop.2006.22.267.
17. Beatriz MPP, Ilias T. Benzalkonium chlorides: uses, regulatory status, and microbial resistance. Appl Environ Microbiol. 2019;85:e00377–19. https://doi.org/10.1128/AEM.00377-19.
18. Epstein SP, Ahdoot M, Marcus E, Asbell PA. Comparative toxicity of preservatives on immortalized corneal and conjunctival epithelial cells. J Ocul Pharmacol Ther. 2009;25:113–9. https://doi.org/10.1089/jop.2008.0098.
19. Swan KC. Reactivity of the ocular tissues to wetting agents. Am J Ophthalmol. 1944;27:1118–22.
20. Hedengran A, Begun X, Müllertz O, Mouhammad Z, Vohra R, Bair J, Dartt DA, Cvenkel B, Heegaard S, Petrovski G. Benzalkonium chloride-preserved anti-glaucomatous eye drops and their effect on human conjunctival goblet cells in vitro. Biomed Hub. 2021;6:69–76.
21. Rosin LM, Bell NP. Preservative toxicity in glaucoma medication: clinical evaluation of benzalkonium chloride-free 0.5% timolol eye drops. Clin Ophthalmol. 2013:2131–5.
22. Codling CE, Maillard J-Y, Russell AD. Aspects of the antimicrobial mechanisms of action of a polyquaternium and an amidoamine. J Antimicrob Chemother. 2003;51:1153–8.
23. Ubels J, Clousing D, Van Haitsma T, Hong B-S, Stauffer P, Asgharian B, Meadows D. Pre-clinical investigation of the efficacy of an artificial tear solution containing hydroxypropyl-guar as a gelling agent. Curr Eye Res. 2004;28:437–44.
24. Grant WM. New treatment for calcific corneal opacities. AMA Arch Ophthalmol. 1952;48:681–5.
25. Kahook MY, Noecker RJ. Comparison of corneal and conjunctival changes after dosing of travoprost preserved with sofZia, latanoprost with 0.02% benzalkonium chloride, and preservative-free artificial tears. Cornea. 2008;27:339–43.
26. Sarkar R. Effects of preservatives used in ocular medications on the eye: a comparative review. Ophthalmology J. 2021;6:44–52.
27. Winder AF, Sheraidah GAK, Astbury NJ, Ruben M. Penetration of mercury from ophthalmic preservatives into the human eye. Lancet. 1980;316:237–9.

28. Hameed AR, Al-Shohani ADHH. Preservatives in ophthalmic formulations and the urgent need for preservative-free formulation. Bionatura. 2023;8:1–22.
29. Zhao M, Li Y, Wang Z. Mercury and mercury-containing preparations: history of use, clinical applications, pharmacology, toxicology, and pharmacokinetics in traditional Chinese medicine. Front Pharmacol. 2022;13:807807. https://doi.org/10.5603/OJ.2021.0009
30. Kaur IP, Lal S, Rana C, Kakkar S, Singh H. Ocular preservatives: associated risks and newer options. Cutan Ocul Toxicol. 2009;28:93–103. https://doi.org/10.1080/15569520902995834.
31. Wasito H, Phechkrajang C. Validated visible spectrophotometry for quantitative analysis of pirenoxine in the presence of paraben preservatives in eye drop preparations. In: 2015 4th international conference on instrumentation, communications, information technology, and biomedical engineering (ICICI-BME). IEEE; 2015. p. 327–31.
32. Cristaldi M, Olivieri M, Lupo G, Anfuso CD, Pezzino S, Rusciano D. N-hydroxymethylglycinate with EDTA is an efficient eye drop preservative with very low toxicity: an in vitro comparative study. Cutan Ocul Toxicol. 2018;37:71–6. https://doi.org/10.1080/15569527.2017.1347942.
33. https://www.fishersci.com/us/en/catalog/search/products?keyword=benzalkonium+chloride
34. https://www.lgcstandards.com/
35. Holdsworth DG, Roberts MS, Polack AE. Fate of chlorbutol during storage in polyethylene dropper containers and simulated patient use. J Clin Pharm Ther. 1984;9:29–39. https://doi.org/10.1111/j.1365-2710.1984.tb00916.x.
36. https://go.drugbank.com/drugs/DB11386
37. https://www.wardsci.com/, https://www.solvay.com/
38. https://eyewiki.org/.
39. Ayaki M, Iwasawa A. Cytotoxicity of prostaglandin analog eye drops preserved with benzalkonium chloride in multiple corneoconjunctival cell lines. Clin Ophthalmol. 2010:919–24. https://doi.org/10.2147/opth.s13406.
40. Pisella PJ, Pouliquen P, Baudouin C. Prevalence of ocular symptoms and signs with preserved and preservative free glaucoma medication. Br J Ophthalmol. 2002;86:418. https://doi.org/10.1136/bjo.86.4.418.
41. Jaenen N, Baudouin C, Pouliquen P, Manni G, Figueiredo A, Zeyen T. Ocular symptoms and signs with preserved and preservative-free glaucoma medications. Eur J Ophthalmol. 2007;17:341–9. https://doi.org/10.1177/112067210701700311.
42. Walsh K, Jones L. The use of preservatives in dry eye drops. Clin Ophthalmol. 2019;13:1409–25. https://doi.org/10.2147/OPTH.S211611.
43. Khan FS, Asghar A, Intisar ul haq BR, Nazir T, Nadeem S, Obaid N. Comparison of effects of preserved and preservative free anti-glaucoma drugs in causing dry eyes, vol. 20. Annals of PIMS-Shaheed Zulfiqar Ali Bhutto Medical University; 2024. p. 171–5.
44. https://eyewiki.org/Preservatives_in_Topical_Ophthalmic_Medications
45. Verkade A, Matoba AY. Polyquaternium-1–associated Dendritiform keratopathy. Cornea. 2019;38:502–3.

7 Reverse Engineering in Complex/ Combination Ophthalmic Product

Khushi Sandeep Kachhwaha, Pratiksha Jadhav, Varnna Prashob, Anita Singh, and Yuvraj Singh

Abstract

Pharmaceutical products have been continuously evolving in the field of ophthalmology to address unmet needs of the patients. This has resulted in increasing reliance on dosage forms like emulsions, gels, ointments, and implants, often referred to as complex ophthalmics, owing to the difficulty in their chemistry, manufacturing and quality control. Converting complex ophthalmic products into generic versions is imperative to expand access to these advanced therapies across all segments of society. However, reverse engineering is often riddled with scientific challenges, which require investment of time and money. This chapter scrutinizes reverse engineering in complex ophthalmic drug products, specifying key methodologies crucial for their development and assessment. We begin with deformulation, exploring the systematic breakdown of formulations to identify their components. Following this, in vitro comparative characterization studies are discussed, which are critical for assessing the physicochemical properties of ophthalmic products comparative to reference formulations. Special emphasis is laid on in vitro release testing as a decisive component for assessing drug release rates and potential ocular bioavailability. The importance of clinical endpoint studies is also captured, signifying how these trials estimate the effectiveness and safety of generic formulations in real-world scenario vis-à-vis the already

K. S. Kachhwaha · P. Jadhav · V. Prashob · Y. Singh (✉)
Department of Pharmacy, Birla Institute of Technology and Science, Pilani, Hyderabad Campus, Jawahar Nagar, Kapra Mandal, Medchal District, Hyderabad, Telangana, India
e-mail: yuvraj.singh@hyderabad.bits-pilani.ac.in

A. Singh
Department of Pharmaceutical Sciences, Sir J.C. Bose Technical Campus, Kumaon University, Nainital, Uttarakhand, India

N. K. Mehra (ed.), *Complex Ophthalmic Dosage Forms: Advances in Biomedical Applications and Future Perspectives*, https://doi.org/10.1007/978-981-96-6306-4_7

approved innovator product. The chapter also briefly discusses physiologically based pharmacokinetic (PBPK) models, which envisage drug pharmacokinetic behaviour based on its physicochemical properties and the dosage form in which it is administered, allowing for feasible prediction of absorption and action. This all-inclusive overview aims to elucidate the reverse engineering processes, emphasizing challenges and innovations in ophthalmic drug development and tries to provide a deeper understanding of this critical field.

Keywords

Reverse engineering · Complex ophthalmic dosage forms · In vitro comparative characterization tests · In vitro release studies · Clinical pharmacokinetic study

7.1 Introduction

As per the Food and Drug Administration (FDA), a generic drug is a type of pharmaceutical product that mirrors a previously marketed brand-name drug covering aspects like dosage form, administration method, quality, safety, strength, performance characteristics and their intended use. By comparing these similarities and performing bioequivalence (BE) studies, it can be proved that a generic medication is comparable to a brand-name medication and has the same therapeutic benefit; that is, a generic medication can be served in place of a brand-name medication ensuring comparable effectiveness [1]. Generic products are generally less expensive (80–85% less) when compared to innovator drugs despite being almost similar in nature [2].

However, it is important to acknowledge that there will always be slight, expected variability from the reference listed drug (RLD) or the reference standard (RS) product. The generic product will be approved only if the variability is justified and does not have marked effect on drug's therapeutic index. An investigation by Davit et al. reported that there is very faint distinction in bioavailability, (approximately 3.5%) between the innovator product and the generic version. This amount of difference is clinically accepted [3]. Generic drug approvals are obtained by filing an abbreviated new drug application (ANDA), which allow a generic manufacturer to utilize at least in part, pre-existing data regarding safety and effectiveness accrued by the innovator, during its initial submission, as well as follow on public data generated during real-world product utilization or various research and development activities. Thus, an ANDA spares the generic manufacturer from the tedious and expensive cycle of generating exhaustive original clinical data obligatory for authorization of a full-fledged new drug application (NDA). In the process, the product becomes eligible for accelerated marketing authorization, provided that bioequivalence (BE) with the innovator product is established and there are no patent infringements or violations of market exclusivity. The time and money saved in the approval process is transferred to the product and the customer, allowing the FDA-approved generics to be marketed at substantially lower costs than the RLD [4]. For instance, Loteprednol Etabonate suspension (Lotemax) is sold at an average cost of 300$ per 5 mL bottle whilst its generic variants are sold at an average cost of 60$–100$ per

5 mL bottle [5]. Similarly, Maxidex® (dexamethasone ophthalmic suspension) is sold at 150$ per 5 mL bottle, whereas the generic versions are sold at an average cost of 25$–40$ [6].

Effective navigation of Para IV filing (an artificial litigation under Hatch Waxman Act) grants generic manufacturers with a 180-day generic drug marketing exclusivity window [7]. Generic companies are often at loggerheads over the 180-day exclusivity incentive, vying to be the 'first to file'. This competitive pressure simultaneously compels innovator companies to continuously evolve their product pipeline—introducing safer and more efficacious formulations—to retain market supremacy through lifecycle management and product evergreening strategies..

The interplay of market forces, unmet patient needs and general advancement of pharmaceutical sciences has resulted in emergence of increasingly complex drug products. The word 'complex' has a special connotation, as highlighted in the GDUFA II (Generic drug user fee agreement) commitment letter and is indicative of a pharmaceutical product containing either:

- Complex active pharmaceutical ingredient (API) like naturally sourced ingredients, complex mixture of APIs, variable peptides, oligonucleotides, metal complexes (ferumoxytol), etc. For instance, glatiramer acetate (Copaxone), a heterogenous composite of random peptide-copolymers. Generic version of Copaxone was approved as an ANDA in 2015 (Sandoz) and required submission of fundamental reaction scheme, comparative physicochemical characterization data including amino acid composition, polymerization signatures, biological assay as evidence of sameness versus the innovator [8]. Conjugated estrogens—marketed under the brand name Premarin (a combination of sodium salts of equilin sulfate and estrone sulfate)—Sevelamer carbonate (an anion exchange resin that binds phosphate), botanical drug products such as Crofelemer, and oil-based or oil-derived products like omega-3 acid derivatives, are also classified as complex APIs [9].
- Complex formulations or dosage forms include a wide range of technologies such as colloidal systems (e.g., cyclosporine ophthalmic emulsion, RESTASIS®; thermoreversible microemulsion containing a eutectic blend of lidocaine and prilocaine, Oraquix®), lipid complexes (e.g., ABELCET®, Amphotericin B lipid complex injection), transdermal delivery systems (e.g., Lidocaine patch), liposomal formulations (Doxil®), implantable devices (Etonogestrel implant), extended-release injectables (e.g., BRIXADI™, a FluidCrystal®-based buprenorphine injection for subcutaneous use), polymeric microparticles (Bydureon®), crystalline extended-release injectable suspensions (INVEGA SUSTENNA®, paliperidone palmitate for intramuscular use), amorphous dispersions, and lyophilized formulations for injection (e.g., albumin-bound paclitaxel nanoparticles, Abraxane®).
- Complex routes of delivery: products for localized dermatological delivery. Since ophthalmological products are usually applied to the eye for localized drug delivery, such products are complex [10], for example, ophthalmological products like prednisolone acetate ophthalmic suspension, acyclovir cream, etc. However, FDA does not solely rely on route of delivery to determine product complexity, but also assesses the dosage form to arrive at a conclusion. This

implies that an ophthalmic solution/drop would not be considered a complex product if no additional complexity (complex API, drug device combination, drug-drug combinations) is involved in the product.

- Drug-device combinations: orally inhaled and nasal drug products (OINDPs) like calcitonin nasal spray, dry powder inhalers like salmeterol inhalation powder, metered dose inhalers, autoinjectors like diazepam intramuscular injection for management of status epilepticus, respule-nebulizer combinations, etc. Recently, FDA approved Mydcombi®, a first (and only) in class fixed-dose combination ophthalmic spray containing phenylephrine hydrochloride 2.5% and tropicamide 1%. Mydcombi® leverages the innovator company's proprietary Optejet® device platform, which adds to the layers of complexity (i.e. route of administration + fixed combination of drug substances + innovative device for drug administration).

Complex drug products are critical for curative or palliative management of several life threatening, life debilitating or orphan ailments like cancers, multiple sclerosis, schizophrenia, diabetes mellitus, osteoporosis, etc. For instance, Victoza® (Liraglutide injection, NovoNordisk), a blockbuster GLP-1 analogue, which has revolutionized type II diabetes mellitus therapy by introducing cardioprotective and weight reduction as auxiliary modalities for disease management has annual sales of $1.66 billion (April 2024) and is priced at a premium of ~Rs 5500 per pre-filled syringe (India). Yet the legal and regulatory difficulties induced by 'complexness' of such products in replicating RLD sameness/equivalence often results in little generic competition depriving vital medicine access to the patient.

The well-publicized shortage of 'Doxil®' and platinum-based antineoplastics during the previous decade substantiates the importance of complex generic drugs in the market [11]. Doxil® was a key drug for treatment of certain cancers during the first decade of 2000s (this is true even in 2024). Due to complexity of the liposomal product which included a peculiar drug-loading mechanism, a generic variant could not be developed/approved until 2013 stretching the gap to 18 years since the approval of Doxil® (1995). This exacerbated the deficiencies in global supplies that had arisen since 2011, due to manufacturing issues at Ben Venue Laboratories, the main supplier of Doxil®. The production lines were shut down for almost 2 years. The shortage led to treatment delays and adverse health outcomes and was only normalized by the year 2018, when cohorts of generic manufacturers entered the market after obtaining expediated FDA approvals. Generics at affordable process settled the supply-demand ratio and put an end to the suffering of the patients. This case highlights not only the concerns about the fragility of the pharmaceutical supply chain and its influence but also the importance of the presence of complex generic drugs in the market [12].

To stimulate generic formulations access, the FDA has pushed GDUFA, a user fee programme to fund regulatory science research for streamlined availability of effective and safe generic medicine. GDUFA is carried out by the Office of Research and Standards, Office of Generic Drugs, FDA. The programme is implemented through internal (within the FDA) and external collaborations with industry and academia. Additionally, since 2007, the FDA has continuously strengthened the available pathways for establishing bioequivalence for generic manufacturers by supplementing the existing regulations under 21 CFR 320 and general guidances with dynamic Product-Specific Guidances (PSGs). PSGs describes the agencies current perspectives and anticipated approaches for the development of generic drugs that are therapeutically equivalent (TE) to RLDs. FDA's PSG webpage has more than 2000 searchable entries of which almost 40% are dedicated to complex products [13]. Under GDUFA III commitment letter, FDA aspires to continually develop PSGs for complex products within specified time frames to aid the generic applicants in preparing their ANDA submissions. The FDAs PSG webpage also helps identify the RLD/RS associated with a particular PSG; however, for confirmation of current RS, the applicants should use the Orange Book [14].

The current chapter is centred around reverse engineering of complex ophthalmic products. It is directed towards researchers working in the complex generic R&D domain and will dissect the complex ophthalmic landscape from the 'FDA-approved products' point of view (Table 7.1). Accordingly, we will cover basics of de-formulation, i.e. how to initiate a generic drug product development using a relevant RLD. Thereafter, an overview of in vitro and in vivo characterization techniques has been provided, including various compendial and non-compendial in vitro release testing methods, utilized in proving qualitative and quantitative sameness between RLD and proposed generic. A brief analysis of currently available PSG's for complex ophthalmic has also been tabulated at one place (Table 7.3). The overall objective is to facilitate the pathway of complex generic development for ophthalmic products.

7.2 Complex Ophthalmic Products

The eye is a sophisticated sensory component that acts as preceptory gateway to the world around us. Consequently, any ailment associated with the eyes should be tackled with the utmost urgency, precision, and care as the cost of not doing so could be astronomical to the patient (risk of vision loss). According to an analysis by World Health Organization (WHO), more than two million people deal with vision impairment [78]. Vision impairment could be caused due to underlying pathologies like refractive errors [79], ocular hypertension [80], age related tissue degeneration [81], cataracts [82], diabetic retinopathy [83] or diabetic macular edema [84]. Hence, to maintain eye health, drug products introduced into the eye must be safe and efficacious.

Complexity in ophthalmic dosage forms is not a recent phenomenon as ocular inserts made of gelatin were first launched in the late 1800s. These contained cocaine and were intended to produce prolonged local anaesthesia [85]. The first dissolvable insert was developed in the Soviet Union in the year 1960. After almost a decennium, Ocusert® was introduced by the Alza Corporation, a system which could release pilocarpine at a constant rate for an entire week [86]. The first ophthalmic suspension was prednisolone acetate 1% (Pred Forte,1970) followed by dexamethasone 1% (Maxidex®,1973) [87].

Traditional ophthalmic dosage forms generally encompass solutions, ointments, suspensions and emulsions [88]. Out of these, the most used dosage form are

Table 7.1 A categorical list of complex ophthalmic products approved by the FDA, along with the status of their respective product-specific guidance

Brand name	API/strength	Composition	Indication	NDA holder year of approval	Remarks (Complexity)	Exclusivity/ patent data	Product-specific guidance (PSG)	References
	Ophthalmic suspension/drops							
FML®	Fluorometholone, 0.1%	*Inactive ingredients*: sodium chloride, monobasic sodium phosphate, edetate disodium, hydroxyethyl cellulose, tyloxapol, hydrochloric acid and/or sodium hydroxide, purified water	Inflammation of eye	ABBVIE INC July 28, 1982	Manufacturing and maintaining the stability of ophthalmic suspension is complex	–	Available	[15]
Zylet™	Loteprednol etabonate 0.5% and tobramycin 0.3%	*Inactive ingredients*: edetate disodium, glycerin, povidone, purified water, tyloxapol, benzalkonium chloride 0.01%, sulfuric acid and/or sodium hydroxide	Superficial bacterial ocular infection	BAUSCH AND LOMB INC Dec 14, 2004	Fixed-dose combination	Patent information available	Available	[16]
Ilevro®	Nepafenac 0.3%, topical ophthalmic,1.7 mL in a 4 mL multidose bottle.	*Inactive ingredients*: BKC, CMC, guar gum, carbomer 974P, boric acid, edetate sodium, sodium chloride and sodium hydroxide/hydrochloric acid	Anti-inflammatory action indicated for the treatment of pain and inflammation associated with cataract surgery	HARROW EYE LLC Aug 19, 2005	Manufacturing and maintaining the stability of ophthalmic suspension is complex	–	Available	[17]

Maxitrol®	Dexamethasone, neomycin sulfate, polymyxin B sulfate 0.1%; EQ. 3.5 mg base/mL; 10,000 units/mL	*Inactive ingredients*: hypromellose 2910 0.5%, sodium chloride, polysorbate 20, hydrochloric acid and/or sodium hydroxide, purified water, benzalkonium chloride 0.004%	Steroid responsive inflammatory ocular conditions	HARROW EYE LLC Approved Prior to Jan 1, 1982	Fixed-dose combination	–	Available	
Betoptic S®	Betaxolol hydrochloride 2.5 mg/mL of betaxolol as base	*Inactive ingredients*: carbomer 934P, edetate disodium, hydrochloric acid or sodium hydroxide, mannitol, poly(styrene-divinyl benzene) sulfonic acid, benzalkonium chloride 0.01%. and purified water	Treatment of elevated intraocular pressure (IOP) in patients with chronic open-angle glaucoma or ocular hypertension	NOVARTIS PHARMACEUTICALS CORP Dec 29, 1989	In-situ formulation	–	Available	[18]
Simbrinza™	Brinzolamide (10 mg/mL) and brimonidine tartrate (2 mg/mL)	*Inactive ingredients*: mannitol, carbomer 974P, sodium, chloride, tyloxapol, boric acid, propylene glycol, sodium hydroxide and/or HCl	Reduction of elevated intraocular pressure inpatients with open-angle glaucoma or ocular hypertension	ALCON LABORATORIES INC Apr 19, 2013 Patent information available	Fixed-dose combination	Patent information available	Available	[19]
Azopt®	Brinzolamide10 mg	*Inactive ingredients*: mannitol, carbomer 974P, sodium chloride, tyloxapol, edentate disodium benzalkonium chloride 0.005% (preservative), sodium hydroxide and/or hydrochloric acid to adjust pH and purified water, USP	Reduction of elevated intraocular pressure inpatients with open-angle glaucoma or ocular hypertension	SANDOZ INC Apr 1, 1998	Manufacturing and maintaining the stability of ophthalmic suspension is complex	–	Available	[20]

(continued)

Table 7.1 (continued)

Brand name	API/strength	Composition	Indication	NDA holder year of approval	Remarks (Complexity)	Exclusivity/ patent data	Product-specific guidance (PSG)	References
	Ophthalmic gel							
Iheezo™	Chloroprocaine hydrochloride3% contains 26 mg of chloroprocaine per gram of gel	*Inactive ingredients*: hydroxyethyl cellulose (HEC), hydrochloric acid, water for injection	Ocular surface anaesthesia	HARROW EYE LLC Sept 27, 2022 Exclusivity: 09/27/2025 Patent validity: 09/15/2038	Sterile, single-patient-use ophthalmic gel. Preparation, manufacturing and maintaining the stability of ophthalmic gel is complex. *Does not contain an antimicrobial preservative*	Exclusivity: 09/27/2025 Patent: 09/15/2038	Available	[21]
Lotemax®	Loteprednol etabonate: contains 5 mg/g of loteprednol etabonate	*Inactive ingredients:* boric acid, edetate disodium dihydrate, glycerin, polycarbophil, propylene glycol, sodium chloride, tyloxapol, water for injection, and sodium hydroxide to adjust to a pH of between 6 and 7	Steroid responsive inflammatory ocular conditions	BAUSCH AND LOMB Sept 28, 2012	Formulating, manufacturing and maintaining the stability of ophthalmic gel is complex	–	Available	[22]
Timoptic-XE®	Timolol maleate: each mL ofTimoptic-XE0.25% contains 2.5 mg of timolol; also available as 0.5% strength	*Inactive ingredients*: gellan gum, tromethamine, mannitol, and water for injection. Preservative: benzododecinium bromide	Treatment of elevated intraocular pressure inpatients with ocular hypertension or open-angle glaucoma	BAUSCH AND LOMB INC Nov 4, 1993	Gel forming/drops. In-situ gelling formulation	–	Available	[23]

Zirgan	Ganciclovir 0.15%	*Inactive ingredients*: carbomer, mannitol, benzalkonium chloride (preservative), sodium hydroxide, WFI	Antiviral, acute herpetic keratitis	BAUSCH AND LOMB Sept 15, 2009	Formulating, manufacturing and maintaining the stability of ophthalmic gel is complex	–	Available	[24]
Akten®	Lidocaine hydrochloride 3.5% (35 mg/mL)	*Inactive ingredients*: hypromellose, NaCl/HCl, WFI	Local anaesthetic for ocular surface anaesthesia	THEA PHARMA INC Oct 7, 2008	Formulating, manufacturing and maintaining the stability of ophthalmic gel is complex	–	Available	[25]
	Ophthalmic solutions							
Phospholine iodide®	Echothiophate iodide 0.125%	*Inactive ingredients*: potassium acetate, sodium hydroxide or acetic acid chlorobutanol (chloral derivative), mannitol, boric acid, and sodium phosphate	Reduction of elevated IOP	FERA PHARMACEUTICALS LLC Approved Prior to Jan 1, 1982	For solution, freeze-dried product. Manufacturing is complex	–	Not available	[26]
VisionBlue®	Trypan blue, 0.06%	*Inactive ingredients*: sodium mono-hydrogen orthophosphate, sodium di-hydrogen orthophosphate. Sodium chloride, and water for injection	Helps in ophthalmic surgery by staining the anterior capsule of the lens	DORC INTERNATIONAL BVDec 16, 2004	Prefilled syringe	–	Not available	[27]
Opcon A®	PheniramineMaleate 0.315%,Naphazoline HCl 0.02675%	*Inactive ingredients*: boric acid, edetate sodium, hypromellose, edetate disodium, benzalkonium chloride, purified water, sodium borate, sodium chloride and hydrochloric acid	Temporary relief from itching and redness caused due to pollen, ragweed	BAUSCH AND LOMB Jun 8, 1994	Injectable solution and fixed-dose combination	–	Not available	[28]

(continued)

Table 7.1 (continued)

Brand name	API/strength	Composition	Indication	NDA holder year of approval	Remarks (Complexity)	Exclusivity/ patent data	Product-specific guidance (PSG)	References
TissueBlue™	Brilliant Blue G 0.025%	*Inactive ingredients:* polyethylene glycol and buffered sodium chloride solution (sodium chloride, sodium phosphate dibasic dihydrate, sodium phosphate monobasic dihydrate, water for injection). Phosphoric acid and/or sodium hydroxide	In ophthalmic surgery by staining the anterior capsule of the lens	DUTCH OPHTHALMIC RESEARCH CENTER INTERNATIONAL BV Dec 20, 2019	Prefilled syringe	Exclusivity: NCE-12/20/2024	Available	[29]
Photrexa®	Riboflavin 5′- phosphate0.146%	*Inactive ingredients:* dibasic sodium phosphate, dextran, monobasic sodium phosphate, sodium chloride, and water for injection	For progressive keratoconus and corneal ectasia following refractive surgery	GLAUKOS CORP Apr 15, 2016	Prefilled syringe	–	Not available	[30]
Cosopt®	Dorzolamide hydrochloride20 mg/mL (2%), timolol maleate5 mg/mL (0.5%)	*Inactive ingredients*: sodium citrate, hydroxyethyl cellulose, sodium hydroxide, mannitol, and water for injection. Benzalkonium chloride	Treatment of elevated intraocular pressure inpatients with ocular hypertension or open-angle glaucoma	THEA PHARMA INC Apr 7, 1998	Fixed-dose combination	–	Not available	[31]
Rocklatan®	Netarsudildimesylate 0.2 mg/mL (0.02%) and latanoprost0.05 mg/mL (0.005%)	*Inactive ingredients*: boric acid, mannitol, sodium hydroxide, benzalkonium chloride and water for injection	Treatment of elevated intraocular pressure in patients with ocular hypertension or open-angle glaucoma	ALCON LABS INC Mar 12, 2019	Fixed-dose combination	Patent information available	Available	[32]

	Ophthalmic ointments							
Lumi-Sporyn	Neomycin andpolymyxin B sulfates and acitracin zinc Eq. 400 units/ gm; Eq. 3.5 mg base/ gm; Eq 10,000 units/ gm	*Inactive ingredient*: petrolatum	Conjunctivitis, keratitis and keratoconjunctivitis, blepharitis and blepharoconjunctivitis	CASPER PHARMA LLC Approved Prior to Jan 1, 1982	Fixed-dose combination	–	Not available	[33]
Ciloxan®	Ciprofloxacin hydrochlorideEq. 0.3% base	*Inactive ingredients*: mineral oil and white petrolatum	Bacterial conjunctivitis	SANDOZ INC Mar 30, 1998	Manufacturing and maintaining stability of ointment is complex	–	Available	[34]
Maxitrol®	Neomycin and polymyxin B sulfates and dexamethasone 0.1%; Eq. 3.5 mg base/gm; 10,000 units/g	*Inactive ingredients*: anhydrous liquid lanolin and white petrolatum. Preservatives: methylparaben, propylparaben	Steroid responsive inflammatory ocular conditions	SANDOZ INC Approved Prior to Jan 1, 1982	Fixed-dose combination	–	Not available	[35]
Tobradex®	Tobramycin0.3% (3 mg) and dexamethasone 0.1% (1 mg)	*Inactive ingredients*: mineral oil and white petrolatum, chlorobutanol	Steroid responsive inflammatory ocular conditions	NOVARTIS PHARMACEUTICALS CORP Sept 28, 1988	Fixed-dose combination	–	Available	[36]
Lotemax®	Loteprednol etabonate 0.5%	*Inactive ingredients*: mineral oil USP, white petrolatum	Treatment of postoperative inflammation and pain following ocular surgery	BAUSCH AND LOMB INC Apr 15, 2011	Manufacturing and maintaining stability of ointment is complex	–	Available	[37]

(continucd)

Table 7.1 (continued)

Brand name	API/strength	Composition	Indication	NDA holder year of approval	Remarks (Complexity)	Exclusivity/ patent data	Product-specific guidance (PSG)	References
Tobrex®	Tobramycin 0.3%	*Inactive ingredients*: chlorobutanol, mineral oil, white petrolatum	External infection of eye	NOVARTIS PHARMACEUTICALS CORP Approved Prior to Jan 1, 1982	Manufacturing and maintaining stability of ointment is complex	–	Available	[38]
	Ophthalmic emulsions							
Restasis Multidose®	Cyclosporine 0.05%	*Inactive ingredients*: glycerin, castor oil, polysorbate 80, carbomer copolymer type A, purified water and sodium hydroxide	Increase tear production inpatients whose tear production is presumed to be suppressed due to ocular inflammation associated with keratoconjunctivitis sicca	ABBVIE INC Oct 27, 2016	Manufacturing, to get the desired globule size and maintaining stability is complex for emulsion	Patent information available	Available	[39]
Durezol	Difluprednate 0.05%	*Inactive ingredients*: boric acid, castor oil, glycerin, polysorbate 80, water for injection, sodium acetate, edetate disodium, sodium hydroxide. Preservative: sorbic acid	Treatment of inflammation and pain associated with ocular surgery. The treatment of endogenous anterior uveitis	SANDOZ INC Jun 23, 2008	Manufacturing, to get the desired globule size and maintaining stability is complex for emulsion	–	Available	[40]

Xelpros®	Latanoprost 0.005%	*Inactive ingredients*: castor oil, sodium borate, boric acid, propylene glycol, edetate disodium, polyoxyl 15 hydroxy stearate, sodium hydroxide, hydrochloric acid, and water for injection	Elevated intraocular pressure inpatients with open-angle glaucoma or ocular hypertension	SUN PHARMACEUTICAL INDUSTRIES LTD Sept 12, 2018	Emulsion	Patent expiration for DP-09/12/2029	Available	[41]
	Ophthalmic inserts and implants							
Durysta®	Bimatoprost 10 mcg	*Inactive ingredients:* (D,L-lactide), poly (D,L-lactide-co-glycolide), poly (D,L-lactide) acid end and polyethylene glycol 3350	Reduction of intraocular pressure (IOP) inpatients with open angle glaucoma (OAG) or ocular hypertension	ABBVIE INC Mar 4, 2020	Implant	Patent information available	Available	[42]
Dextenza®	Dexamethasone 0.4 mg	*Inactive ingredients*: 4-arm polyethylene glycol (PEG)N hydroxysuccinimidyl glutarate (20 K), trilysine acetate, N-hydroxy succinimide-fluorescein, sodium phosphate dibasic, sodium phosphate monobasic, water for injection	The treatment of ocular inflammation and pain following ophthalmic surgery	OCULAR THERAPEUTIX INC Nov 30, 2018	Insert-manufacturing process is complex	Exclusivity-10/7/2024 Patent information available	Not available	[43]
Ozurdex™	Dexamethasone 0.7 mg	OZURDEX™ is an intravitreal implant containing 0.7 mg (700 µg) dexamethasone in the NOVADUR™ solid polymer drug delivery system	Macular oedema	ABBVIE INC Jun 17, 2009	Intravitreal implant	–	Not available	[44]

(continued)

Table 7.1 (continued)

Brand name	API/strength	Composition	Indication	NDA holder year of approval	Remarks (Complexity)	Exclusivity/ patent data	Product-specific guidance (PSG)	References
Yutiq™	Fluocinolone acetonide 0.18 mg	*Inactive ingredients:* polyimide tube, polyvinyl alcohol, silicone adhesive and water for injection	Treatment of chronic non-infectious uveitis affecting the posterior segment of the eye	ALIMERA SCIENCES INC Oct 12, 2018	Complex dosage form (long-acting injectable)/ complex drug-device combination (intravitreal applicator)/Complex route of delivery–36-month sustained-release drug delivery system	Patent information available	Not available	[45]
Retisert®	Fluocinolone acetonide 0.59 mg	*Inactive ingredients*: microcrystalline cellulose, polyvinyl alcohol and magnesium stearate	Treatment of chronic non-infectious uveitis affecting the posterior segment of the eye	BAUSCH AND LOMB INC Apr 8, 2005	Intravitreal implant	–	Not available	[46]
Iluvien®	Fluocinolone acetonide 0.19 mg	*Inactive ingredients:* polyimide tube, polyvinyl alcohol, silicone adhesive and water for injection	Treatment of diabetic macular oedema	ALIMERA SCIENCES INC Sept 26, 2014	Intravitreal implant	Patent expiration 08/12/2027	Not available	[47]
	Spray metered							
Mydcombi®	Phenylephrine hydrochloride 2.5%, tropicamide 1%	*Inactive ingredients:* Hydrochloric acid and/or sodium hydroxide, water for injection; Preservative: benzalkonium chloride	Indicated to induce mydriasis for diagnostic procedures and in conditions where short term pupil dilation is desired.	EYENOVIA INC May 5, 2023	Spray metered ophthalmic product	Exclusivity NP-05/05/2026 Patent information available	Not available	[48]

The listed inactives (excipients) in the product label are often the starting point of deciphering the formulation solutions, which account for 62.4% of all ophthalmic products, while ointments account for 17.4% and suspensions account for 20% [89]. The simple solution (eye drops) are most widely used because they are easy to administer and are patient compliant. Moreover, from a chemistry manufacturing and controls (CMC) standpoint, solution dosage forms are easier to fabricate, requiring fewer excipients and lesser unit operations. Suspensions and emulsions are considered when the active ingredient is hydrophobic with minimum aqueous solubility [90]. However, it is a task to ensure elegance, stability, sterility and safety of suspensions, emulsions and semi-solid dosage forms. The number of characterization tests and regulatory requirements also increase significantly [91]. Ophthalmic formulations have several limitations like unavoidable drug loss during instillation, inefficient corneal permeability, drainage through nasolacrimal drainage duct, reflex blinking and tear turnover [92]. To overcome these disadvantages, the most appropriate option is to use complex ophthalmic products. They are designed in such a way that they may contain certain complex ingredients, complex formulation design, delivery systems or drug device combinations [93]. These formulations overcome the limitations of conventional ophthalmic products and offer several advantages, including improved patient compliance, reduced dosing frequency, enhanced delivery efficiency, minimized side effects, and slower drug clearance. As a result, they enable extended ocular residence time, site-specific drug delivery, prolonged contact time, and increased ocular bioavailability [94].

As per a 2018 survey on marketed ophthalmic generics within the USA, following statistics was reported: ophthalmic solutions (market cap of 17.9 billion $, more than (>) 100 RLDs, >50% approved generics), emulsions (4.4 billion $, 4 RLDs, 25% approved generics), suspension (1.9 billion $, ~22 RLDs, 23% approved generics), ointments (730 million $, ~15 RLDs, 30% approved generics). These data suggest that pharmaceutical firms have been very successful in reverse engineering ophthalmic solutions. But there have been fewer approvals for complex ophthalmic generics [95].

The trend is also supported by the fact that for solutions, where manufacturing conditions or processing steps do not impact the properties of the end desired product (critical quality attributes (CQAs)) "the in vivo bioavailability or bioequivalence of the drug product may be self-evident" as per 21 CFR320.22(b). However, for more complex products where manufacturing conditions, processing steps or excipient choice could impact the properties of the end product, the bioavailability (BA) or bioequivalence (BE) may be shown by various in vivo (local pharmacokinetic, pharmacodynamics, comparative clinical endpoint study) and in vitro methods.

Each of the listed bioequivalence pathways carries inherent limitations, primarily because ophthalmic drugs act locally, and direct drug quantification in ocular tissues at multiple time points is both prohibitively expensive and ethically challenging. Furthermore, the limited volume of aqueous humour available for analysis necessitates sparse sampling across larger cohorts, which introduces significant inter-subject variability and complicates study outcomes. Consequently, to measure the bioavailability of a drug product or establish the bioequivalence of specific drug products, the FDA requires in vivo or in vitro testing, or both, as per CFR 320.24(a).

Substantive physiologic-based pharmacokinetic modelling (PBPK) data is also encouraged to integrate physicochemical drug product/substance properties to existing in vitro/vivo data using preclinical/human physiology. Table 7.2 captures the most recent approved ophthalmic generics in the USA.

7.3 NDA and ANDA

In terms of marketing authorization, generic approvals (obtained through ANDA applications) are 20-fold more frequent than new drug approvals (obtained through NDA application) [96]. An Abbreviated New Drug Application (ANDA) can only be filed for drug products that are no longer under patent protection or regulatory exclusivity. The applicant must provide evidence that the proposed generic drug is an exact replicate of the innovator product—also known as the Reference Listed Drug (RLD)—which has already been approved as safe and effective by the regulatory authority. The primary objective is to establish equivalence to the RLD through a bioequivalence (BE) assessment. A pharmaceutically equivalent (PE) generic, when supported by appropriate BE data, is deemed therapeutically equivalent (TE) to the RLD [97].The difference between RLD and RS is highlighted in Fig. 7.1.

The inactive ingredients, route of administration, dosage form and strength of the generic drug must all be the same in terms of quality (Q1) and quantity (Q2). Generally, a product meant for ocular use must have the same inactive components as the RLD in the same concentration. If an ophthalmic drug product has a preservative, buffer, substance to adjust tonicity or thickening agent that is different from the RLD [as allowed by the chemistry, manufacturing, and controls (CMC) regulations for ANDAs, 21 CFR 314.94(a)(9)(iv)], the applicant must describe the differences and provide evidence that they have no bearing on the proposed drug product's safety or effectiveness. Therefore, the FDA has decided that, any qualitative (Q1) or quantitative (Q2) deviations from the RLD should be accompanied by an appropriate in vivo BE study or studies, even though 21 CFR 314.94(a)(9)(iv) provides a similar allowance to parenteral products [17].

Q1 Sameness Evaluation is the qualitative sameness of the inactive ingredients, that is, correctly identifying the inactive ingredients and providing information regarding the grade and chemistry of each inactive ingredients.

Q2 Sameness Evaluation is the demonstration of the quantitative sameness, that is, correctly determining the quantity or concentration of the inactive ingredients. The FDA accepts a difference of +/−5% of the generic drug from the innovator drug [98].

Q3 Sameness Evaluation is done to confirm the physicochemical similarity of the generic drug with the RLD. Even if a product is Q1/Q2 similar, there could be dissimilarities in the arrangement of drug substance within the dosage form, which may affect product performance. Often labelled as 'totality of evidence approach to bioequivalence', the studied parameters may include comparative appearance, globule size distribution, rheological behaviour, pH, polymorphic

Table 7.2 Select list of 'FDA-approved' complex ophthalmic generics. The approvals have been classified as per dosage form. This information has been mined from, Orange Book website and is subject to change with time

Generic name	API strength	Indication	ANDA holder	Year of approval	PSG
Ophthalmic suspension/drops					
FlurometHalone	Flurometholone 0.1%	Inflammation of eye	AMNEAL EU LTD Exclusivity 07/07/2024	Jan 9, 2024	Available
Dexasporin	Dexamethasone, neomycin sulfate, polymyxin B sulfate 0.1%; Eq. 3.5 mg base/ml; 10,000 units/ml	Steroid-responsive inflammatory ocular conditions	BAUSCH AND LOMB PHARMACEU TICALS INC.	Sept 13, 1995	Available
Brinzolamide	Brinzolamide 1%	Reduction of elevated intraocular pressure in patients with open-angle glaucoma or ocular hypertension	PADAGIS US LLC Exclusivity 04/07/2024	July 28, 2023	Available
Ophthalmic gel					
Loteprednol etabonate	Loteprednol etabonate 0.5%	Steroid responsive inflammatory ocular conditions	HIKMA PHARMACEUTICALS USAINC	Nov 29, 2023	Available
Timolol Maleate	Timolol maleate Eq. 0.5% base	Treatment of elevated intraocular pressure in patients with ocular hypertension or open-angle glaucoma	EUGIA PHARMA SPECIALITIES LTD	Mar 23, 2023	Available
Ophthalmic solutions					
Phenaramine maleate, naphazoline HCl	Phenaramine maleate-0.315%; naphazoline HCl 0.02675%	Temporary relief from itching and redness caused due to pollen, ragweed	ALTAIRE PHARMACEUTICALS INC	Sept 27, 2010	Not available
Dorzolamide hydrochloride, timolol maleate	Dorzolamide 20 mg/mL (2%) and timolol 5 mg/mL (0.5%)	Treatment of elevated intraocular pressure in patients with ocular hypertension or open-angle glaucoma	INGENUS PHARMACEUTICALS LLC	May 1, 2023	Not available

(continued)

Table 7.2 (continued)

Generic name	API strength	Indication	ANDA holder	Year of approval	PSG
Ophthalmic ointments					
Neomycin and polymyxin B sulfates and bacitracin zinc	Neomycin and polymyxin B sulfates and bacitracin zinc Eq. 400 units/gm; Eq. 3.5 mg base/gm; Eq. 10,000 units/gm	Conjunctivitis, keratitis and keratoconjunctivitis, blepharitis and blepharoconjunctivitis	BAUSCH AND LOMB PHARMACEUTICALS INC	Oct 30, 1995	Not available
Dexamethasone, neomycin and polymyxin B sulfates	Dexamethasone, neomycin and polymyxin B sulfates 0.1%; Eq. 3.5 mg base/gm; 10,000 units/g	Steroid-responsive inflammatory ocular conditions	BAUSCH AND LOMB PHARMACEUTICALS INC	July 25, 1994	Not available
Ophthalmic emulsions					
Cyclosporine	Cyclosporine 0.05%	Increase tear production inpatients whose tear production is presumed to be suppressed due to ocular inflammation associated with keratoconjunctivitis sicca	SAPTALIS PHARMACEUTICALS LLC	July 5, 2024	Available
Difluprednate	Difluprednate 0.05%	Treatment of inflammation and pain associated with ocular surgery, The treatment of endogenous anterior uveitis	EPIC PHARMA LLC	May 14, 2024	Available

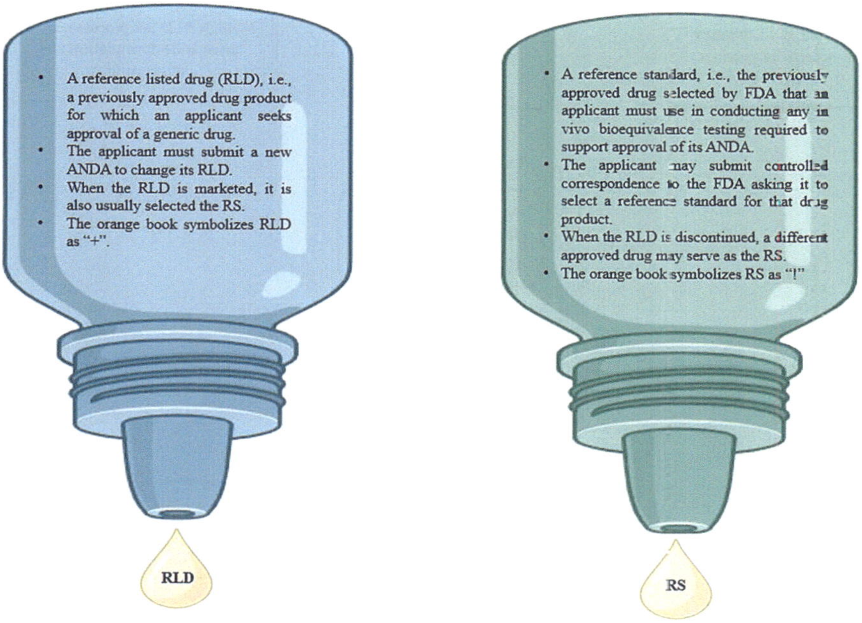

Fig. 7.1 Difference between RLD and RS. Per FDA, the major difference between an RS and an RLD is its application during the development and submission process of a generic drug product. Self-creation

form of the drug substance or a critical excipient, crystalline habit of API, drug release profile, surface tension, specific gravity, particle size, etc. According to the FDA's guideline titled 'Guidance for Industry: Bioequivalence Studies with Pharmacokinetic Endpoints for Orally Inhaled Drug Products', if a complex generic formulation demonstrates Q1/Q2 sameness (i.e., qualitative and quantitative equivalence), comparable physicochemical properties, and a statistically similar in vitro drug release profile, in vivo bioequivalence (BE) testing may be waived. This is based on the rationale that the concentration of the active ingredient at the target site is unlikely to differ significantly from that of the RLD [99].

To establish Q1/Q2 sameness with the Reference Listed Drug (RLD), it is essential to identify the critical quality attributes (CQAs) of the RLD. This process involves a detailed assessment of product characteristics influenced by the manufacturing process and critical process parameters, raw material attributes (including the grade and source of critical excipients), and formulation development steps The exercise is driven by prior knowledge—acquired through reverse engineering of earlier innovator products—as well as data mining from scientific literature and patents, analysis of product labels, review of regulatory documents such as the 'Summary Basis of Approval' available on the FDA's product webpage, and advanced analytical characterization of the procured RLD (Fig. 7.2). The Q1/Q2 sameness evaluation is expanded in later sections.

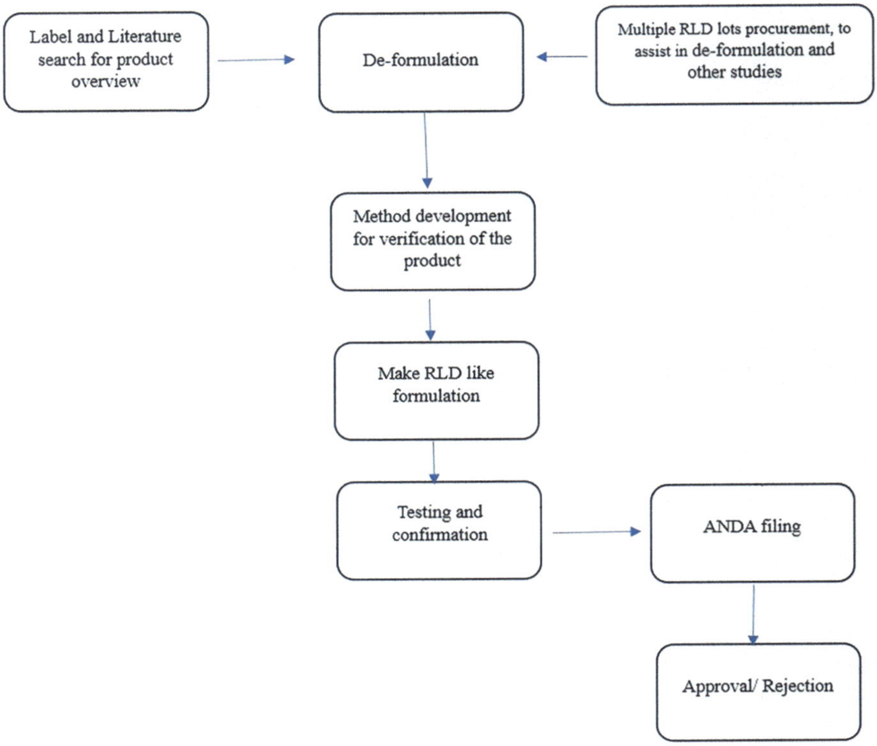

Fig. 7.2 Process of developing a generic formulation and establishing Q1/Q2/Q3 sameness. Self-creation

7.4 Deformulation

To produce a generic product with therapeutic and pharmaceutical sameness, scientists deformulate the existing innovator product. Multiple RLD batches are procured and tested analytically to find the closest possible 'recipe' for the generic product [100]. It requires extensive literature research, and in some cases, controlled correspondence may be filed to the FDA for further information. De-formulation is also called re-engineering or reverse engineering because it aims to find both the quantity and function of the inactive ingredients used. De-formulation may also be used to find the replacement of banned or obsolete components. Along with this, comparative de-formulation helps in understanding the reason behind batch-to-batch variation.

7.4.1 De-formulation Challenges That Can Impact the Q1/Q2 of a Product

Complex products deploy multiple complex components and complex manufacturing procedures. Hence, establishing a reliable, validated analytical method for the estimation and evaluation of formulation components and grades can be challenging. Some common challenges faced are-non-uniform distribution of the API/excipient. If the drugs particle size is overly diminutive, the analytical method or instrument may not be sensitive enough to identify the API dynamics and quantity [101]. For example, in a study, the concentration of furosemide that is initially measured using UV-visible spectroscopy changes and reduces significantly during the formulation of nanocrystals because the particle size of the crystals becomes too small for the standard UV-spectroscopy to detect accurately [102].

Complexities in the API/excipient solid-state characterization can further muddle the picture. Situations like possibility of existing in metastable form or susceptibility to solid-state transitions are deeply problematic to decipher. If a formulation contains both crystalline and amorphous ingredients, the dominating ingredient interferes with the signal from others, and it becomes challenging to find the nature of the minor ingredients. For example, solid dispersion of griseofulvin, which is highly crystalline, loses this characteristic tendency when distributed homogenously in amorphous polymers like polyvinylpyrrolidone (PVP), polyethylene glycol, hydroxypropyl methylcellulose and Eudragits [103]. Some complex formulations may also contain 'two' or more crystalline APIs, which interfere with each other during the analytical testing [104]. Finally, establishment of Q1/Q2/Q3 sameness requires demonstration of statistical insignificance of the tested parameter over multiple lots of exhibit batches with fully validated analytical methods.

A list of ophthalmic products that are considered 'complex' as per the current regulatory understanding of the authors are listed in Table 7.1 along with their label disclosed ingredients and indications. The type and amount of excipient is found by preliminary de-formulation studies, using the FDA label or package insert of the RLD product or by literature survey. Multiple trial and error batches can be attempted and tested for their closeness to the innovator product.

7.5 Evaluation of Generic Ophthalmic Products

For the purpose of developing a generic version of complex ophthalmic drug product, it is obligatory to corroborate that the RLD and test product ought to be the same. In order to achieve Q1/Q2 sameness, the excipients in the test product must be identical to those in the Reference Listed Drug (RLD), both qualitatively (Q1) and quantitatively (Q2). No deviation in excipient type or concentration is permitted, as even minor changes may lead to local irritation or altered drug performance [95]. In order to showcase the Q1 sameness, the chemistry and related characterization data should be presented by the applicant. Any variances in the excipients amongst the test and RLD must not be more than 5%, which is stated as Q2 sameness. The applicant should itemize all the inactive ingredients, excluding those used as needed (q.s.) and the quantity in terms of w/w% or mg/ml. If at all, any test product does not meet these norms, then it is not cogitated as Q1/Q2 sameness. Generally, FDA has construed that Q2 sameness for a product will be a concentration that is within the 95–105% of RLD concentration. The percentage is calculated as per formula given in Eq. (7.1).

$$\%\text{difference in inactive ingredient} = \frac{(\text{Test} - \text{Reference})}{\text{Reference}} * 100\% \tag{7.1}$$

In cases where the test product matches the innovator product in terms of Q1/Q2, several PSGs outline one or more of the studies listed in Fig. 7.3 [105].

Since solutions are the simplest form of formulations and have direct absorption (no dissolution), there is no obligation to perform clinical or comparative in vitro studies (if the generic has the same composition as the brand product). That is bioequivalence is undeniable in ophthalmic solutions as attributes of final product remain unaffected by the manufacturing process [106]. A clinical endpoint bioequivalence study is mandated in ophthalmic products that do not qualify as Q1/Q2 equivalent. The rationale behind this is that any variations in inactive ingredients can significantly impact the safety and efficacy of the product.

For non-solution formulations like suspensions and emulsions, evaluation often involves two pathways: (1) either execution of two in vitro studies with additional comparative characterization studies or (2) one in vivo bioequivalence study with clinical endpoints. In this context, in vitro studies can be carried out through either of the following approaches: analysis of drug particle size and drug particle size distribution (suspensions)/globule size distribution (emulsions) and comparative

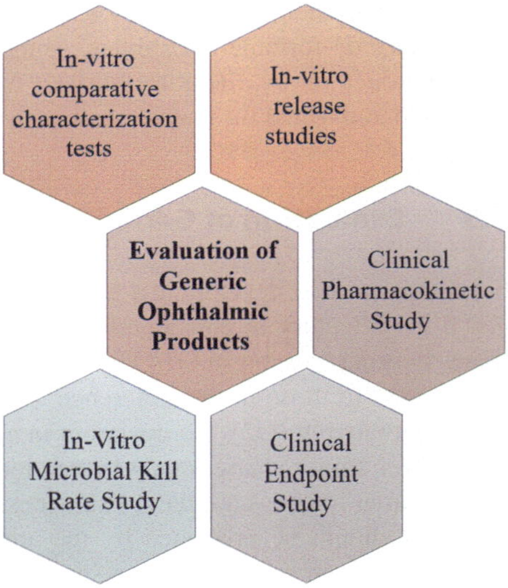

Fig. 7.3 Complimentary studies which are used to supplement the Q1/Q2 sameness in an ANDA application for obtainment of marketing authorization by the FDA. Self-creation

in vitro drug release testing. Evaluation of semisolid drug products like gels and ointments often call for one in vitro study with additional comparative characterization studies or in vivo bioequivalence study. Here, the in vitro studies entail comparative in vitro drug release testing [13].

A formulation is classified as Q1/Q2 non-equivalent when the formulation contains non-identical excipients or if there is a difference exceeding 5% in any inactive component compared to the RLD. In such scenarios, it is advisable to request a meeting with the FDA's Office of Generic Drugs under the Pre-ANDA Meeting Program prior to initiating bioequivalence studies. Table 7.3 contains a summary of currently available product-specific guidance for various ophthalmic dosage forms issued by the FDA. This guidance (though not mandatory) can be safely followed to understand the exact pathway for navigating the Q1/Q2/Q3/in vivo studies quagmire for obtaining generic approval for various dosage forms of complex ophthalmic products. The following subsections throw light on select studies that are mentioned in Table 7.3 and are routinely utilized to showcase the pharmaceutical and therapeutic equivalence of an intended generic to a RS /RLD.

7.5.1 In Vitro Comparative Characterization Studies

Stated below are universal tests, also known as Q3 or physical and chemical similarity tests for ophthalmic formulations. It involves comparing the physiochemical attributes of the test product as well as inactive ingredients present in the RLD. The results of these tests of the generic formulation should be comparable with that of the RLD product to establish the sameness of the product. The results are required for the ANDA application and further the approval of the generic drug.

7.5.1.1 Identification Test

Identification tests are necessary to validate the presence of the required API and excipients in the formulation. The identity of API is crucial to ensure that the formulation is safe and efficacious. These tests ensure that the product contains ingredients as mentioned in the label. The tests should be sensitive enough to differentiate between nearly identical compounds. They should be drug product/drug substance specific. Some common identification methods are UV-visible spectroscopy, IR spectroscopy, high-performance liquid chromatography (HPLC), thin layer chromatography (TLC), mass spectrometry, nuclear magnetic resonance (NMR), etc. [107].

7.5.1.2 Appearance

Appearance plays a major role in patient acceptability. Ophthalmic formulations are generally required to be clear and transparent. They cannot have a characteristic colour because that would limit the vision of the patient. However, appearance can vary for different dosage forms. Solutions and suspensions are generally transparent, whereas ointments, gels and emulsions can be cloudy to opaque. Viscosity, nature of ointment base, presence of bubbles/foam and particle size significantly impact the appearance of the formulation. A general thumb rule is that the intended

Table 7.3 Summaries of draft or finalized PSGs for various ophthalmic dosage forms

Active ingredient dosage form	RLD or RS number, type of PSG, date recommended	Recommendations detailed in PSGs towards generic product development	Reference
Ointment			
Acyclovir	202408 Draft 08/20/2021	In vitro option Q1/Q2 sameness with RLD. BE may be established on comparative in vitro testing of three exhibit batches of both the test product and designated RS product including following characterization: Appearance, polymorphic form of acyclovir, acidity and alkalinity of the extracted ointment base, drug particle size and size distribution, rheological properties including yield stress and viscosity. The applicant should characterize viscosity over a range of shear rates. In vitro drug release tests of acyclovir from the test and RS products with detailed information on development and validation of a proposed in vitro drug release testing method should be provided.	[49]
Bacitracin	061212 Draft 10/04/2016	In vitro option Acceptable comparative physicochemical characterization of the test and RLD formulations. Q1/Q2 sameness with RLD. The comparative study should be performed on at least three exhibit batches of both test and reference products and should include: Comparative appearance, acidity and alkalinity of the extracted ointment base, comparative rheological properties including yield stress and viscosity. The applicant should characterize viscosity over a range of shear rates. Comparative drug particle size and size distribution. Full profiles of the particle size distribution should also be submitted for all samples tested. Comparative in vitro drug release rates of bacitracin from the test and RLD formulations. The methodology used for in vitro drug release testing should be able to discriminate the effect of process variability in the production of the test formulation.	[50]

Ciprofloxacin hydrochloride	020369 Draft 09/13/2018	Option I: In vitro studies Q1/Q2 sameness with RLD. The comparative study should be performed on at least three exhibit batches of both test and reference products and should include: Solid-state form of ciprofloxacin hydrochloride, appearance, acidity and alkalinity of the extracted ointment base, drug particle size and size distribution, rheological properties including yield stress and viscosity. The applicant should characterize viscosity over a range of shear rates. Full profiles of the particle size distribution should also be submitted for all samples tested. Acceptable comparative in vitro drug release rates of ciprofloxacin from the test and RS formulations. The methodology used for in vitro drug release testing should be able to discriminate the effect of process variability in the production of the test formulation. Option II: In vivo clinical endpoint study	[51]
Dexamethasone; Tobramycin ointment	050616 Draft 02/15/2024	Option I: One in vitro bioequivalence (BE) study with supportive characterization studies. To demonstrate BE by this option, the test product should be qualitatively (Q1) and quantitatively (Q2) the same as the RLD, and in vitro drug release testing (IVRT) of dexamethasone and tobramycin should be performed on three batches of both test and RS products using at least 12 units from each batch. Additional comments: The IVRT method study should include information on the method development and validation to detect potential formulation differences and capture the complete release profile of dexamethasone and tobramycin from the test and RS products. A prospective applicant may use the same method or different methods for IVRT of dexamethasone and tobramycin. BE based on comparative analysis of release profiles using an appropriate statistical method. Supportive characterization studies: Comparative physicochemical characterization of the test product and RS product should be performed on at least three batches of both the test product and RS product and should include appearance, specific gravity, acidity and alkalinity of the extracted ointment base; crystallinity and crystal habit of dexamethasone; particle size and size distribution of dexamethasone; particle size and size distribution of tobramycin; and rheological properties including yield stress and viscosity. Viscosity should be characterized over a range of shear rates. Option II: One in vitro BE study + one in vivo BE study with pharmacokinetic (PK) endpoints	[52]

(continued)

Table 7.3 (continued)

Active ingredient dosage form	RLD or RS number, type of PSG, date recommended	Recommendations detailed in PSGs towards generic product development	Reference
Erythromycin ointment	062447 Draft 10/04/2016	Option I: In vitro studies Q1/Q2 sameness with RLD. The comparative study should be performed on at least three exhibit batches of both test and reference products and should include: Comparative solid-state form of erythromycin, appearance, acidity and alkalinity of the extracted ointment base, rheological properties including yield stress and viscosity. The applicant should characterize viscosity over a range of shear rates. Comparative drug particle size and size distribution. Full profiles of the particle size distribution should also be submitted for all samples tested. Comparative in vitro drug release rates of erythromycin from the test and RLD formulations. The methodology used for in vitro drug release testing should be able to discriminate the effect of process variability in the production of the test formulation.	[53]
Loteprednol etabonate ointment	200,738 Draft 02/15/2024	Option I: One in vitro BE study with supportive comparative characterization studies. To demonstrate BE by this option, the test product should be qualitatively (Q1) and quantitatively (Q2) the same as the RLD. One in vitro BE study: IVRT of loteprednol etabonate should be performed on three batches of both test and RS products using at least 12 units from each batch. The IVRT method study should include information on the method development and validation to detect potential formulation differences and capture the complete release profile of loteprednol etabonate from the test and RS formulations. A prospective applicant may use the same method or different methods for IVRT of loteprednol etabonate. *BE is established using* comparative analysis of release profiles using an appropriate statistical method. Comparative characterization studies: Comparative physicochemical characterization of the test and RS products. The comparative study should be performed on at least three batches of both the test4 and RS products and should include: appearance, acidity and alkalinity of the extracted ointment base, rheological properties including yield stress and viscosity. The applicant should characterize viscosity over a range of shear rates. Drug particle size and size distribution. Option II: BE study with PK endpoints. Design: single-dose, crossover or parallel design, in vivo in aqueous humour.	[54]

Tobramycin ointment	050555 Draft 05/15/2019	In vitro *study:* To qualify for the in vitro option for this drug product *all of the following criteria* should be met: Q1/Q2 sameness between test and RLD. Acceptable comparative physicochemical characterizations of the test and RS products. The comparative study should be performed on at least three exhibit batches of both the test and RS products and should include: Acceptable comparative in vitro drug release of tobramycin from the test and RS products. The methodology used for in vitro drug release testing should be able to discriminate the effect of process variability in the production of the test formulation. Comparative appearance, acidity and alkalinity of the extracted ointment base, polymorphic form of tobramycin rheological properties including yield stress and viscosity. The applicant should characterize viscosity over a range of shear rates. Comparative drug particle size and size distribution. The particle size distribution should be compared using *population bioequivalence (PBE)* (95% upper confidence bound) based on D50 and SPAN [i.e. (D90-D10)/D50)]. The applicant should provide no fewer than ten datasets from three different batches of both the test and RS products for PBE analysis. Full profiles of the particle distribution should also be submitted for all samples tested. Refer to the Guidance on Budesonide inhalation suspension for additional information regarding PBE.	[55]
Gel			
Ganciclovir gel	022211 Draft 05/16/2024	One in vitro BE study with supportive characterization studies To demonstrate BE by this option, the test product should be qualitatively (Q1) and quantitatively (Q2) the same as the RLD. One in vitro BE study: Comparative IVRT of ganciclovir should be performed on three batches each of the test and RS products using at least 12 units from each batch. The IVRT method study should include information on the method development and validation to detect potential formulation differences and capture the complete release profile of ganciclovir. BE based on comparative analysis of release profiles should be established using an appropriate statistical method (e.g. model independent approach using similarity factor (f2)). Supportive Characterization Studies should be performed on at least three batches each of the test and RS products and should include appearance, pH, specific gravity, osmolality and rheological properties, including yield stress and viscosity. The applicant should characterize viscosity over a range of shear rates. Additional information: The RLD is presented in a tube with a dropper tip. The tube with a dropper tip is the device constituent part. FDA recommends that prospective applicants examine the size and shape, the external critical design attributes, and the external operating principles of the RLD device when designing the test device. An ANDA for this product should include complete comparative user interface analyses.	[56]

(continued)

Table 7.3 (continued)

Active ingredient dosage form	RLD or RS number, type of PSG, date recommended	Recommendations detailed in PSGs towards generic product development	Reference
Lidocaine hydrochloride gel	022221 Draft 05/16/2024	One in vitro BE study with supportive characterization studies. To demonstrate BE by this option, the test product should be qualitatively (Q1) and quantitatively (Q2) the same as the RLD. One in vitro BE study: Comparative IVRT of lidocaine hydrochloride should be performed on three batches of both test (T) and RS products using at least 12 units from each batch Comparative characterization studies: They should be performed on at least three batches of both the T and RS products and should include appearance, pH, specific gravity, osmolality and rheological properties, including yield stress and viscosity. The applicant should characterize viscosity over a range of shear rates. Additional information: The RLD is presented both in a bottle with a dropper tip and in a tube with a dropper tip. The bottle with a dropper tip or the tube with a dropper tip are the device constituent parts. FDA recommends that prospective applicants examine the size and shape, the external critical design attributes and the external operating principles of the RLD device when designing the T device. An ANDA for this product should include complete comparative user interface analyses.	[57]
Chloroprocaine hydrochloride gel	216,227 Draft 08/21/2023	One in vitro BE study with supportive characterization studies. Q1/Q2 sameness with RLD. The comparative study should be performed on at least three exhibit batches1 of both test and reference products and should include: Type of study: Comparative IVRT for chloroprocaine hydrochloride from test and RS products. A properly developed and validated method that can detect potential formulation differences and capture the complete release profile of chloroprocaine hydrochloride should be provided. *Equivalence* in chloroprocaine hydrochloride release should be established using a proper statistical method from three batches of Test and RS products using a model independent similarity (f2) factor approach. Supportive studies should include acceptable comparative physicochemical characterization of the test and the reference with respect to appearance, pH, specific gravity, osmolality, rheological properties including yield stress and viscosity. The applicant should characterize viscosity over a range of shear rates. Additional information: The RLD is presented in a vial with a dropper tip. The vial with dropper tip is the device constituent part. The FDA recommends that prospective applicants examine the size and shape, the external critical design attributes, and the external operating principles of the RLD device when designing the test device. An ANDA for this product should include complete comparative user interface analyses.	[58]

Loteprednol etabonate gel	202872 Draft 09/16/2019	Option I: In vitro To qualify for the in vitro option for this drug product all of the following criteria should be met: The test and RLD formulations are qualitatively (Q1) and quantitatively (Q2) the same (Q1/Q2). Acceptable comparative physicochemical characterizations of the test and RS products. The comparative study should be performed on at least three exhibit batches of both the test and RS products and should include: Comparative crystalline habit of loteprednol etabonate. Comparative appearance, pH, specific gravity and osmolality. Comparative soluble fraction of loteprednol etabonate in the final drug product. Comparable rheological properties including yield stress and viscosity. The applicant should characterize viscosity over a range of shear rates. Comparative drug particle size distribution. The particle size distribution should be compared using PBE (95% upper confidence bound) based on D50 and SPAN [i.e. (D90-D10)/D50]. The applicant should provide no fewer than ten datasets from three different batches of both the test and reference products for PBE analysis. Full profiles of the particle size distributions should also be submitted for all samples tested. Please refer to the Guidance on Budesonide inhalation suspension for additional information regarding PBE. Acceptable comparative in vitro drug release of loteprednol etabonate from the test and RS formulations. Option II: In vivo BE study with PK endpoints	[59]
Suspension/drops			
Besifloxacin hydrochloride suspension/drops	022308 Draft 03/02/2020	In vitro option Q1/Q2 sameness with RLD. Acceptable comparative in vitro drug release of besifloxacin from the test and RS formulations along with following supportive data: Comparable crystalline habit of besifloxacin. Comparable appearance, pH, specific gravity, osmolality, surface tension and viscosity profile Comparable soluble fraction of besifloxacin in the final drug product. Comparable drug particle size distribution. The particle size distribution should be compared using PBE (95% upper confidence bound) based on D50 and SPAN [i.e. (D90-D10)/D50]. The applicant should provide no fewer than ten datasets from three different batches of both the Test and reference products for PBE analysis. Full profiles of the particle size distributions should also be submitted for all samples tested. Refer to the Guidance on Budesonide inhalation suspension for additional information regarding PBE.	[60]

(continued)

Table 7.3 (continued)

Active ingredient dosage form	RLD or RS number, type of PSG, date recommended	Recommendations detailed in PSGs towards generic product development	Reference
Betaxolol hydrochloride suspension/drops	019845 Draft 05/15/2019	In vivo option BE based on clinical endpoint (95% CI). A BE study with a clinical endpoint in the treatment of open-angle glaucoma and ocular hypertension comparing the test product versus the RLD, each applied as one drop in both eyes two times daily at approximately 8:00 a.m., and 8:00 p.m. for 42 days.	[61]
Brimonidine tartrate; brinzolamide suspension/drops	204251 Draft 05/15/2019	In vivo option One study; BE study with clinical endpoint. Recommended primary endpoint is the mean difference in IOP of both eyes between the two treatment groups; To establish BE, the limits of each two-sided 95% confidence interval of the treatment difference (test—reference) for mean IOP of both eyes (continuous variable) at all four follow-up points.	[62]
Brimonidine tartrate; brinzolamide suspension/drops	204251 Draft 05/15/2019	In vivo option: BE study with clinical endpoint. Recommended primary endpoint is the mean difference in IOP of both eyes between the two treatment groups; To establish BE, the limits of each two-sided 95% confidence interval of the treatment difference (test—reference) for mean IOP of both eyes (continuous variable) at all four follow-up points.	[62]

Dexamethasone; neomycin sulfate; polymyxin B sulfate suspension/drops	050023 Draft 02/15/2024	Option I: In vitro BE studies with supportive comparative characterization studies, i.e. the test product should be qualitatively (Q1) and quantitatively (Q2) the same as the RLD. Comparative IVRT of dexamethasone should be performed on three batches of both test and RS products using at least 12 units from each batch. The IVRT method study should include information on the method development and validation to detect potential formulation differences and capture the complete release profile of dexamethasone. BE based on comparative analysis of release profiles should be established using an appropriate statistical method (e.g. model independent approach using similarity factor (f2)). Drug particle size and particle size distribution of dexamethasone. This in vitro BE study would be done on three batches of both test and RS) products. The sample preparation method and selected particle sizing methodology should be adequately optimized and validated to demonstrate the adequacy of the selected method in accurately and reliably identifying and measuring the size of the drug particles. Prospective applicant should perform size characterization at different dilution conditions as part of method development to demonstrate the impact of dilution. Full particle size distribution profiles representative of all test product and RS product batches tested should be submitted as supporting information. *PBE* analysis of the D50 and SPAN. Prospective applicants should provide no less than ten datasets from three batches each of the test and RS products to be used in the PBE analysis. For additional information on PBE statistical analysis, refer to the most recent version of the FDA product-specific guidance on Budesonide Inhalation Suspension (NDA 020929). Supportive characterization studies: Comparative physicochemical characterization of the test product and RS product. The comparative study should be performed on at least three batches of both the test product and RS product and should include crystalline habit of dexamethasone, appearance, pH, specific gravity, osmolality, surface tension, viscosity, soluble fraction of dexamethasone in the final drug product. Quality assessment: For quality-related recommendations for supporting drug product development, refer to the most recent version of the FDA guidance for industry on quality considerations for topical ophthalmic drug products. Option II: In vivo BE study with clinical endpoints in aqueous humour.	[63]

(continued)

Table 7.3 (continued)

Active ingredient dosage form	RLD or RS number, type of PSG, date recommended	Recommendations detailed in PSGs towards generic product development	Reference
Dexamethasone; tobramycin suspension/drops	050818 Draft 02/15/2024	Option I: Two in vitro BE studies with supportive comparative characterization studies. To demonstrate BE by this option, the test product should be qualitatively (Q1) and quantitatively (Q2) the same as the RLD. Drug particle size and particle size distribution of dexamethasone. This in vitro BE study will be performed on three batches of both test and RS products. The sample preparation method and selected particle sizing methodology should be adequately optimized and validated to demonstrate the adequacy of the selected method in accurately and reliably identifying and measuring the size of the drug particles. Prospective applicant should perform size characterization at different dilution conditions as part of method development to demonstrate the impact of dilution. Full particle size distribution profiles representative of all test product and RS product batches tested should be submitted as supporting information. Parameters to measure: D50 and SPAN [(D90-D10)/D50]. *PBE* analysis of the D50 and SPAN. Prospective applicants should provide no less than ten datasets from three batches each of the test and RS products to be used in the PBE analysis. For additional information on PBE statistical analysis, refer to the most recent version of the FDA product-specific guidance on Budesonide Inhalation Suspension (NDA 020929). Comparative IVRT of dexamethasone: Should be performed on three batches of both test and RS products using at least 12 units from each batch. The IVRT method study should include information on the method development and validation to detect potential formulation differences and capture the complete release profile of dexamethasone from the test and RS products. BE based on comparative analysis of release profiles should be established using an appropriate statistical method (e.g. model independent approach using similarity factor (f2)). Comparative characterization studies should be conducted on at least three batches each of the test product and the Reference Standard (RS) product. These studies should assess parameters such as the crystalline habit of dexamethasone, appearance, pH, specific gravity, osmolality, surface tension, viscosity, and the soluble fraction of dexamethasone in the final drug product. Additional information: The RLD is presented in a bottle with a dropper tip. The bottle with dropper tip is the device constituent part. FDA recommends that prospective applicants examine the size and shape, the external critical design attributes, and the external operating principles of the RLD device when designing the test device. An ANDA for this product should include complete comparative user interface assessment analyses. Option II: One in vivo BE study with PK endpoints in aqueous humour	[52]

Fluorometholone suspension/ Drops	019216 Draft 06/03/2020	Option I:In vitro option To qualify for the in vitro option for this drug product all of the following criteria should be met. The test and RLD formulations are qualitatively (Q1) and quantitatively (Q2) the same (Q1/Q2). Acceptable comparative physicochemical characterization of the test and RS products. The comparative study should be performed on at least three batches of both the test and RS products and should include: Comparable appearance, pH, specific gravity, osmolality, surface tension, buffer capacity and viscosity. Comparable soluble fraction of fluorometholone acetate in the final drug product. Comparable drug particle size distribution. The particle size distribution should be compared using PBE (95% upper confidence bound) based on D50 and SPAN [i.e. (D90-D10)/D50]. The applicant should provide no fewer than ten datasets from three different batches of both the test and reference products for PBE analysis. Full profiles of the particle size distributions should also be submitted for all samples tested. Please refer to the Guidance on Budesonide inhalation suspension for additional information regarding PBE.	[64]
Rimexolone suspension/drops	020474 Draft 06/16/2016	Option I: In vivo study. Type of study: BE study with PK endpoints. Design: single-dose, crossover or parallel design in vivo in aqueous humour. Additional recommendation pertains to development of an In vitro dissolution test method for this drug product for stability and quality control purposes. The specifications of the IVRT will be determined upon review of the ANDA.	[65]
Prednisolone acetate suspension/drops	017011 Draft 05/15/2019	Option I: In vitro studies To qualify for the in vitro option all of the following criteria should be met. The test and RLD formulations are qualitatively and quantitatively the same (Q1/Q2). Acceptable comparative physicochemical characterization of the test and RS products. on at least three batches of both the test and RS products and should include: Comparable appearance, pH, specific gravity, osmolality, surface tension, buffer capacity, and viscosity. Comparable soluble fraction of prednisolone acetate in the final drug product. Comparable drug particle size distribution. The particle size distribution should be compared using PBE (95% upper confidence bound) based on D50 and SPAN [i.e. (D90-D10)/ D50)]. The applicant should provide no fewer than ten datasets from three different batches of both the test and reference products for PBE analysis. Full profiles of the particle size distributions should also be submitted for all samples tested. Acceptable comparative in vitro drug release of prednisolone acetate from the test and RS formulations. The methodology used for in vitro drug release testing should be able to discriminate the effect of process variability in the production of the test formulation. Option II: In vivo BE study with PK endpoints in aqueous humour.	[66]

(continued)

Table 7.3 (continued)

Active ingredient dosage form	RLD or RS number, type of PSG, date recommended	Recommendations detailed in PSGs towards generic product development	Reference
Nepafenac suspension/drops 0.1%	021862 Draft 12/22/2016	Option I: In vitro studies. To qualify for the in vitro option for nepafenac ophthalmic suspension all of the following criteria should be met: The test and RLD formulations are qualitatively and quantitatively the same (Q1/Q2). Acceptable comparative in vitro drug release of nepafenac from the test and RLD formulations. The methodology used for in vitro drug release testing should be able to discriminate the effect of process and variability in the production of the test formulation. The characterization study should be performed on at least three exhibit batches of both the test and RLD products, and should include: Comparative appearance, pH, specific gravity, osmolality, surface tension, buffer capacity, and viscosity as a function of applied shear. Viscosity measurement should also be conducted at the tear pH. Comparative re-dispersibility (time required to re-disperse the formulation and sedimentation time). Comparative soluble fraction of nepafenac in the final drug product. Comparative unit dose content (one drop per unit dose). Provide data for the amount of unit dose (one drop) with assay for nepafenac from a minimum of ten units from three batches each of the test and reference products. The unit dose content should be compared using PBE (95% upper confidence bound). Comparative drug particle size distribution based on D50 and SPAN [(D90-D10)/D50]. The applicant should provide no fewer than ten datasets from three different batches each of the test and reference products for the PBE analysis. Full profiles of the particle size distributions should also be submitted for all samples tested. Dissolution test method and sampling times: The dissolution information for this drug product can be found on the FDA-Recommended Dissolution Methods website Conduct comparative dissolution testing on 12 dosage units each of all strengths of the test and reference products. Specifications will be determined upon review of the ANDA). Option II: Pharmacokinetic BE studies with PK endpoints in aqueous humour. Option III: Comparative BE clinical endpoint study.	[67]

Loteprednol etabonate; tobramycin suspension/drops	050804 Draft 02/15/2024	Option I:Two in vitro BE studies with supportive comparative characterization studies. To demonstrate BE by this option, the test product should be qualitatively (Q1) and quantitatively (Q2) the same as the RLD. Drug particle size and particle size distribution of loteprednol etabonate: In vitro BE study on three batches of both test and RS products. The sample preparation method and selected particle sizing methodology should be adequately optimized and validated to demonstrate the adequacy of the selected method in accurately and reliably identifying and measuring the size of the drug particles. Prospective applicant should perform size characterization at different dilution conditions as part of method development to demonstrate the impact of dilution. Full particle size distribution profiles representative of all test product and RS product batches tested should be submitted as supporting information. Parameters to measure: D50 and SPAN [(D90-D10)/D50]. PBE analysis of the D50 and SPAN. Prospective applicants should provide no less than ten datasets from three batches each of the test and RS products to be used in the PBE analysis. For additional information on PBE statistical analysis, refer to the most recent version of the FDA product-specific guidance on Budesonide Inhalation Suspension (NDA 020929). Comparative IVRT of loteprednol etabonate should be performed on three batches of both test and RS products using at least 12 units from each batch. The IVRT method study should include information on the method development and validation to detect potential formulation differences and capture the complete release profile of loteprednol etabonate. BE based on comparative analysis of release profiles should be established using an appropriate statistical method (e.g. model independent approach using similarity factor (f2)). Comparative characterization studies: Comparative physicochemical characterization of the test product and RS product. The comparative study should be performed on at least three batches of both the test product4 and RS product and should include appearance, pH, specific gravity, osmolality, surface tension, viscosity and soluble fraction of loteprednol etabonate in the final drug product.	[68]

(continued)

Table 7.3 (continued)

Active ingredient dosage form	RLD or RS number, type of PSG, date recommended	Recommendations detailed in PSGs towards generic product development	Reference
Emulsion			
Cyclosporine emulsion	214,965 Draft 08/02/2022	Option I: Two in vitro BE studies with supportive comparative characterization studies. Q1/Q2 sameness with RLD. Acceptable comparative physicochemical characterization of the test and RS product. The comparative study should be performed on at least three exhibit batches of both test and reference products and should include: Zeta potential, viscosity profile as a function of applied shear, pH, osmolality, surface tension and cyclosporine distribution in different phases within the formulation. Comparative globule size distribution of test and RS products. Dynamic light scattering method is recommended to measure the globule size distributions of test and RS formulations. Comparable globule size distribution profiles (intensity-based histograms) upon serial dilution should be provided. Non-ionic electrolyte-free dilution media are recommended. Information on the instrument, mode (if applicable), dilution medium and dilution level should be provided. Equivalence between test and RS products with respect to globule size distribution should be demonstrated by a suitable method proposed by the sponsor. PBE based on D50 and SPAN (alternatively harmonic intensity weighted average particle diameter and polydispersity index derived from cumulant analysis of the intensity size distribution) for the globule size distribution only (the other parameters do not require PBE analysis). Applicants should provide no less than ten datasets from three batches each of the Test and Reference products to be used in the PBE analysis. Please refer to the most recent version of the FDA product-specific guidance on Budesonide Inhalation Suspension (NDA 020929) for additional information regarding PBE. Acceptable comparative in vitro drug release study of cyclosporine from the test and RS products. The methodology used for in vitro drug release testing should be discriminative of process variabilities in the production of the test product. Option II: One in vivo BE study with clinical endpoints.	[69]

Difluprednate emulsion	022212 Draft 02/14/2017	Option I: In vitro option. The test and RLD formulations are qualitatively and quantitatively the same (Q1/Q2). Acceptable comparative physicochemical characterization of the test and RLD formulations should be performed on at least three exhibit lots of both test and reference products. Parameters to measure: Globule size distribution, viscosity profile as a function of applied shear, pH, zeta potential, osmolality and surface tension. Sponsors should also submit information on the drug distribution in different phases within the formulation. PBE based on D50 and SPAN (alternatively harmonic intensity weighted average particle diameter and polydispersity index derived from cumulant analysis of the intensity size distribution) for the globule size distribution only (the other parameters do not require PBE analysis). The applicants should provide no less than ten datasets from three batches each of the test and reference products to be used in the PBE analysis. Sponsors should compare the size parameter upon serial dilution (if applicable) of the Test and Reference products, and provide histograms of size distribution data of each diluted sample. Acceptable comparative in vitro drug release rate tests of difluprednate from the test and Reference formulations. The methodology used for in vitro drug release testing. Option II: In vivo option Analytes to measure (in appropriate biological fluid): Difluprednate and its active metabolite, 6α, 9-difluoroprednisolone 17-butyrate (DFB), in aqueous humour.	[70]
Latanoprost emulsion	206,185 Draft 06/03/2020	One study: BE study with clinical endpoint	[71]
Implant			
Bimatoprost implant	211911 Draft 05/18/2023	Option I: One comparative clinical endpoint BE study with additional information. Additional information: Dissolution test method and sampling times: Conduct comparative dissolution testing on 12 dosage units each of all strengths of the test and reference products. Specifications will be determined upon review of the ANDA. FDA recommends that prospective applicants examine the size and shape, the external critical design attributes, and the external operating principles of the RLD device when designing the test (T) devices including: Size of the biodegradable implant, preloaded, sterile, single-use applicator. Needle features: silicone coating, gauge, length and implant retention mechanism. An ANDA for this product should include complete comparative user interface assessment analyses.	[72]

(continued)

Table 7.3 (continued)

Active ingredient dosage form	RLD or RS number, type of PSG, date recommended	Recommendations detailed in PSGs towards generic product development	Reference
Solution/Drop			
Bromfenac sodium solution/drops	021664 Draft 10/19/2017	Option I. Waiver: To qualify for a waiver of the in vivo BE study requirement, a generic bromfenac sodium ophthalmic solution product must be qualitatively (Q1) and quantitatively (Q2) the same as the RLD. Bromfenac sodium ophthalmic solution products should have comparable physicochemical properties to the RS including but not limited to pH, specific gravity, buffer capacity, osmolality and viscosity, if applicable. Comparative analysis should be performed on three exhibit batches, if available, of both test and RS products. An in vivo BE study is requested for any bromfenac sodium ophthalmic solution product that has a different inactive ingredient from the RLD, a difference of more than 5% in the amount of any inactive ingredient compared to that of the RLD, or differences in comparative physicochemical characterization data.	[73]
Lotilaner solution/drops	217,603 Draft 05/16/2024	To qualify for a waiver from submitting an in vivo BE study on the basis that BE is self-evident under 21 CFR 320.22(b)(1), a generic lotilaner ophthalmic solution/drops product should be qualitatively (Q1) and quantitatively (Q2) the same as the RLD. An applicant may seek approval of a drug product that differs from the RLD in preservative, buffer, substance to adjust tonicity or thickening agent provided that the applicant identifies and characterizes the differences and provides information demonstrating that the differences do not affect the safety or efficacy of the proposed drug product. Additional information: The RLD is presented in a bottle with a dropper tip. The bottle with dropper tip is the device constituent part. FDA recommends that prospective applicants examine the size and shape, the external critical design attributes, and the external operating principles of the RLD device when designing the test device. An ANDA for this product should include complete comparative user interface assessment analyses.	[74]

Timolol maleate solution, gel forming/drops	020330 Draft 11/21/2019	Option I: In vitro studies. Following criteria should be met: The test and RLD formulations are qualitatively (Q1) and quantitatively (Q2) the same (Q1/Q2). Acceptable comparative physicochemical characterization of the test and RS products including pH, specific gravity, buffer capacity, osmolality and viscosity. Comparative studies should be performed on at least three batches of both the test product and RS acceptable comparative rheological properties of the test and RS gels formed using physiologically relevant media and dilution. This can include measuring the gel strength and/or the stress-strain profile including yield point. Testing should be performed on at least three batches of both the test and RS products. Gels should form using simulated tear fluid.	[75]
Brilliant blue G Solution	209,569 Draft 12/17/2021	An in vivo BE study is requested for any brilliant blue G (0.025%) ophthalmic solution product that has a different inactive ingredient from the RLD, a difference of more than 5% in the amount of any inactive ingredient compared to that of the RLD or differences in comparative physicochemical characterization data.	[76]
Brimonidine tartrate; timolol maleate solution/ drops	021398 Draft 09/13/2018	To qualify for a waiver of the in vivo BE study requirement, a brimonidinetartrate; timolol maleate ophthalmic solution product must be qualitatively (Q1) and quantitatively (Q2) the same as the RLD including but not limited to pH, specific gravity, buffer capacity, osmolality and viscosity, if applicable. Comparative analysis should be performed on three exhibit batches, if available, of both test and RS products. An in vivo BE study is requested for any brimonidine tartrate; timolol maleate ophthalmic solution product that has a different inactive ingredient from the RLD, a difference of more than 5% in the amount of any inactive ingredient compared to that of the RLD or differences in comparative physicochemical characterization data.	[77]

Please note that the content of the table is subject to continuous updating and should be cross checked with the latest versions of PSGs issues by the FDA generic formulation should be the same as innovator product in terms of appearance [108].

7.5.1.3 Viscosity

The viscosity of the formulation plays a pivotal role in patient compliance, retention time, product stability and product administration. Solutions being less viscous can be easily instilled into the eye. The immediate spreading of the eye drops solution/suspension over the corneal surface reduces the need to blink multiple times and hence is more patient compliant. However, low viscosity will also lead to higher amount of nasolacrimal drainage that will lead to wastage of the drug and hence low bioavailability. Ideally, viscosity of an ophthalmic formulation should be high at low shear rates, that is when the eyes are open, to prevent the drainage of the formulation from the surface of the eye; and low at high shear conditions such as blinking to minimize friction and discomfort. Therefore, the formulations should be shear-thinning in nature [109]. Some ocular formulations, especially when the drug is potent, aim to attain prolonged retention time and frequency of administration. This could be achieved by increasing the viscosity of the formulation like in cases of gels, ointments and suspensions. Viscosity can be measured using capillary methods <911>, rotational methods <912> or rolling ball methods <913> as mentioned in the USP. In industry, the in-situ gelling properties of a formulation are tested using a rheometer or vial invert method.

7.5.1.4 Surface Tension

Surface tension plays a critical role in the performance of ophthalmic formulations, influencing both safety and efficacy. From a safety perspective, the surface tension of the formulation should closely match that of tear fluid to minimize ocular discomfort. In terms of efficacy, surface tension affects various aspects—from the initial spreading and retention of the formulation on the ocular surface to the subsequent permeation of the active pharmaceutical ingredient through the mucosal membrane of the eye [12]. Surface tension governs the shape of eye drops allowing them to have a volume ranging from 20 to 100 μL. If the surface tension of the formulation is low, it will form un-evenly shaped drops, which will affect the dose consistency [110]. The stability and administration of the formulation are also affected if the surface tension is very high because it could lead to the formation of foam in the container. Surface tension is crucial when it comes to facilitating good spreadibility of the formulation of the surface of the eye. Ideal surface tension will lead to more coverage of the surface in an even manner, allowing more drug to come in contact and hence more permeation. Therefore, surface tension directly affects the bioavailability of the drug. The presence of surfactants and certain solvents, viscosity, pH and temperature directly impact surface tension. Capillary rise method, drop-weight method, ring method, drop shape analysis and maximum bubble pressure method are some conventional methods used to calculate surface tension [111]. Sophisticated equipment like goniometer is used in the industry to measure the same. The surface tension of commercial eye drops such as Thera® Tears Lubricant Eye Drops, Refresh

Liquigel®, and Refresh Optive® Lubricant ranges from approximately 65 to 70 dynes/cm, aligning closely with the surface tension of natural tear fluid [112].

7.5.1.5 Drug Particle Size and Particle Size Distribution

The physical properties, manufacturing processability, and quality attributes of any dosage form containing drug particles—such as granules, suspensions, or ointments—are inherently influenced by particle size and particle size distribution of the finished drug product [113]. Non-micronized particles in ophthalmic formulations can lead to abrasion of soft ocular tissues and reflex tearing. Therefore, particle size less than 10 μm is a prerequisite when dealing with a suspension. On the other hand, no guidance hints at the permissible limit of particle size for semi-solid dosage forms; therefore, matching the dimensions of particles present in the RLD seems to be the most plausible way forward. While numerous techniques exist, polarized light microscopy is prime choice for determining 'particle size distribution' of dispersed phase in ointments because it is a non-destructive technique and does not trigger any physiochemical changes in the dosage form [113]. Estimation of particle size distribution (suspensions, gels, ointments) or globule size distribution (emulsions) is carried out using population bioequivalence (PBE) statistical method. D_{50} and SPAN[$(D_{90}-D_{10})/D_{50}$] are the essential metrics of PBE, whilst D_{10} and D_{90} act as supportive parameters. Additionally, earth mover's distance may be deployed when particle size distribution is not monomodal. SPAN is preferred over polydispersity index (PDI) to bring the relative difference in test and reference products, as the derived values for PDI are very small and the allowable margin is just 10% of 0.1 making it much more difficult to pass [114].

7.5.1.6 Osmolarity

The osmolarity of tear fluid lining the cornea of a healthy eye is estimated to be 290–310 mOsm/kg [115]. Formulations are expected to match the osmolarity of the eye to avoid any discomfort to the patient. Hypertonic solutions could cause dryness and irritation, whereas hypotonic solutions that have low osmolarity could disrupt the corneal membrane leading to a stinging sensation.

Generally, hypotonic solutions are better tolerated in comparison to hypertonic solutions. Osmolarity also affects the drug absorption from the corneal surface because some osmotic agents act as permeation and absorption enhancers. It may sometimes be important to ensure that the osmolarity of a product is maintained throughout their shelf life as it may be linked to DS stability. For example, chloramphenicol solution can only be stable in hypertonic state because its co-solvency depends on the PEG 3000 and glycerol [116]. The osmolarity and concentration of these co-solvents impact the solvency and hence stability of the API. Extractables and leachables can also highly impact the osmolarity of the formulation and hence are to be considered whilst developing the generic version of a complex ophthalmic RLD. Vapour pressure osmometer and freezing point depression osmometer are generally used to measure osmolarity [117].

7.5.1.7 Assay

Assay is critical quality attribute of the formulation because it has a direct impact on the safety and efficacy of the formulation. Assay testing ensures that right amount of drug is present in the formulation. Over-dosing could lead to safety issues, and under-dosing leads to efficacy issues. Regular assay testing done at different time points helps in deciding the shelf life of the product. It also ensures batch to batch uniformity. Assay is generally measured using chromatography and related techniques or microbial assay [118].

7.5.1.8 Impurities

Organic or inorganic impurities might already be there in the drug substance/excipients or may arise during the processing of the drug product. If the impurity level crosses the specified limit as mentioned in USP <232>,<233>, and ICH Q3A, B, D, it could have a deleterious impact on the safety of the formulation. Hence, impurity testing is to be done at different stability time points to identify/quantify the degradation products and pathways. According to the threshold detected, the impurities are classified into 'identification', 'qualification' and 'reporting'. This is an important criterion that has to be submitted for regulatory approval and has been further elaborated in ICH Q3 guidelines [119]. Prednisolone acetate suspension, a synthetic glucocorticoid steroid for inflammation, has eight known impurities that can be detected using reversed phase HPLC technique [120]. This impurity profile would have to be matched both qualitatively and quantitatively with RLD in generic product development.

7.5.1.9 Uniformity of Dosage forms

This study is done to check the equivalence of amount of drug and excipient present, in every batch of the drug product that is manufactured. It includes testing for both the content of the drug substance and the mass of the formulation. These tests are done to ensure consistent efficacy and safety. The type and proportion of ingredients used, the manufacturing process, the packaging material and environmental factors like stability and formulation factors like viscosity have a bearing on the uniformity of the dosage form. The tests are conducted either by weight variation or content uniformity. For semi-solid dosage forms like ointments and gels, samples are collected from different parts of the tube usually, the beginning, the middle and the end, which are then analysed for the amount of active ingredient present. More information on uniformity of dosage forms can be found in USP <905> [118].

7.5.1.10 Specific Gravity

Specific gravity measures the comparative density of drug product to that of water. This parameter is an indicator of formulation stability, especially emulsions and suspensions. For instance, in the case of emulsion if the dispersed phase density is lesser than that of continuous phase, there are chances that the droplets can rise to the surface thus leading to instability. In case of suspensions, the particles have higher specific gravity and tend to settle faster, thus leading to sedimentation and affecting uniformity. The specific gravity ensures that the consistency of a formulation is in close proximity to that of the RLD. If at all any change is the specific gravity occurs, it suggests that there are disparities in the concentration of API or excipients that will ultimately impact the safety and efficacy of the product [121].

7.5.1.11 Particulate and Foreign Matter

Particulates in topical ophthalmic products have the potential to cause certain amounts of iritis and uveitis. They may also block Schlemm's canal, obstructing aqueous humour drainage and causing acute glaucoma. Electronic or microscopic particle analysis is done by dissolving ophthalmic products in suitable, particle-free solvent. USP chapter <789> can be referred for further information [122].

7.5.1.12 Pump Delivery

This test is done to check the performance of ophthalmic formulations (sprays) intended to be delivered by a metered pump. The dimensions of each pump and the weight of spray delivered by each pump should be compared to the RLD. This guarantees pump-to-pump reproducibility and batch-to-batch consistency. Ideally, the weight of the spray delivered should be within 15% of the desired weight, and their mean weight should be within 10% of their target weight. The pump delivery and its acceptance criteria should be present in the drug product specification [123, 124].

7.5.1.13 Spray Content Uniformity (SCU)

This test gives information about the overall the performance of the pump. The uniformity of content of the therapeutic agent present in the droplets of each spray should be tested for multiple sprays from the starting to completion of use of a single container, among various containers and batches of the drug product. Generally, two sprays per determination are used. For the test to give accurate and comparable results for multiple batches, the controls for actuation parameters like actuation force and stroke length should be set. Both the actuation quantity and the percentage of the label claim should be used to describe the amount of drug delivered [123, 124]. *The following are the acceptance criteria:*

- Amount of API per determination for batch 1 should be within 100 ± 20 percent of label claim for more than 2–20 determinations from ten containers.
- For batch 2, none of the determinations should be outside 100 ± 25 percent of the label claim.
- For batch 3, the mean for each of the beginning and end determinations should be within 100 ± 15 percent of label claim.
- If the product does not meet the above acceptance criteria, an additional 20 containers should be sampled for second-tier testing.
- *Acceptance criteria for tier 2*: For batch 1, the amount of API per determination is within 100 ± 20 percent of the label claim for more than 6 of all 60 determinations; for batch 2, all the 60 determinations should be within 100 ± 25 percent of label claim; for batch 3, the means for each of the beginning and end determinations should be within 85–115 percent of label claim [123].

7.5.1.14 Spray Pattern and Plume Geometry

This is a characteristic feature of pump-based products and is influenced by the performance of the pump. The factors that could impact the plume geometry and spray pattern are shape and size of the nozzle, the size of the metering chamber, the pump design and the physiochemical characteristics of the formulation. This test is to be done on a timely basis as a quality control for the drug release from an ophthalmic-metered spray device. The controls, such as the number of sprays per pattern, the orientation and position of the collection surface with respect to the nozzle, the distance of spray between the nozzle and the surface, and the visualization procedure, should be set and specified in order to obtain comparable results for multiple batches. The acceptance criteria include the size and shape of the pattern. The acceptance testing for plume geometry and spray pattern of incoming pumps can substitute the release testing of drug product if justified. The acceptance criteria for the same should be present in the drug product specification [125].

7.5.1.15 Sterility

Ophthalmic dosage forms are supposed to be sterile in nature and must adhere to the specifications of sterility tests described in USP chapter <71>. The ingredients employed in the formulation should be sterile or low bioburden grade. If at all the ingredients in the formulation cannot be regularly sterilized, then components which adhere to the sterility standards must be employed and the process/manufacturing should be carried out aseptically. Usage of sealed container closure system should be employed to guarantee sterility [126].

7.5.1.16 Bacterial Endotoxins

Ophthalmic dosage form should be formulated in a way that the bacterial endotoxin is within limit. The bacterial endotoxin test discussed in USP chapter <85> should be used to testify the endotoxin level. This test employs the use of Limulus amoebocyte lysate reagent and to quantify bacterial endotoxin. For ophthalmic irrigation products, the bacterial endotoxin limit is not more than (NMT) 0.5 EU/mL, while for injectable or implantable ophthalmic dosage forms, the limit is NMT 2 EU per dose per eye. Bacterial endotoxin testing is not mandatory for topically applied ophthalmic products [126].

7.5.1.17 Extractable and Leachable

Extractable and leachable impurities are derived from the manufacturing systems or container closure systems (CCS) or that may or may not react with DS or DP components. They can be organic or inorganic and can severely impact the drug's potency, safety, efficacy and stability. Hence, it is important to identify and qualify them. Extractable are substances/chemical species released, extracted or pulled out of the CCS/packaging material under stress conditions like pH, temperature, moisture, pressure using techniques such as solvent extraction. Leachables are chemical species that migrate into the formulation from the container closure system (CCS) or packaging materials during the product's shelf life. These substances are typically monitored as part of stability studies to ensure product safety and integrity. Components of the formulation like plasticizers, sequestrants, metal catalysts, antioxidants, polymer degradation products and dyes could be a source of leachable [127]. USP chapters <661.1> and <661.2> provide a basic understanding of extraction profile required for routine qualification. The testing for extractables and leachable is described in USP chapters <1663>, <1664>, <1665> and <1664.1> [128].

7.5.1.18 Stability Studies

Drug product manufacturers are obligated to implement a stability evaluation programme and leverage the obtained results to understand appropriate storage conditions and hence the expiration dates (21 CFR211.166). The stability studies are to be conducted in its primary packaging/container closure system marketed. Stability studies must be conducted using the primary packaging or container closure system intended for marketing. These studies provide critical information on the compatibility of the formulation with the device, the presence of leachables, the compatibility of excipients with each other and with the API, as well as the performance of the delivery device over the product's shelf life. The following stability recommendations should guide the development of stability testing programme of complex ophthalmic generic products [129].

7.5.1.18.1 Container Orientation During Storage

Applicant seeking ANDA approval must position primary stability batches in both an upright position and an inverted or horizontal position, with data from both setups included in original submission. The worst-case orientation identified in this comparison should be used as standard orientation for future routine stability tests once the product is approved. These orientational permutations are carried out to ensure product quality by subjecting it to the most challenging conditions [130].

7.5.1.18.2 Water Loss

For ophthalmic products contained in semipermeable CCS, it is recommended that both applicants and manufacturers should perform water loss test. The purpose of this test is to understand the moisture transmission properties of CCS and the protective capabilities of secondary packaging.

7.5.1.18.3 Freeze/Thaw Study for Emulsions and Suspensions

Conducting a one-time freeze/thaw thermal cycling study is crucial for applicants to determine the effects of temperature fluctuations during shipping and handling of ophthalmic formulations like suspensions and emulsions. The FDA recommends: three cycles, switching between freezing (−20 °C to 0 °C) and ambient conditions (25 °C to 35 °C) over a total minimum duration of 3 days. Applicants need to provide rationale in case of using alternative conditions for their choices [131].

7.5.1.18.4 In-Use Stability Studies

These studies are utilized to set expiration dates and substantiate labelling claims regarding appropriate storage conditions that might alter after the drug product is opened, such as temperature or light exposure (see 21CFR 211.166, 21 CFR 211.137 (b)).

7.5.1.18.5 Testing of Container Closure System of Ophthalmic Products

The CCS of pharmaceutical products is one of the most crucial factors governing the quality of the product along with establishing a market identity. Its quality should be at par with the quality of the product it withholds. The container ensures the safety of the product by ensuring sterility and stability of the formulation. The containers for ophthalmic products are designed to allow maximum patient convenience while minimizing the risk of contamination during dispensing and therefore ensures safe application. The application for ANDA or NDA has a dedicated section for the description and evaluation of the container closure system used. Most widely used container closure systems for ophthalmic use are high-density polyethylene (HDPE) containers, used to deliver solutions, suspensions or emulsions. The tests are listed in Fig. 7.4.

7.5.2 In Vitro Release Testing (IVRT)

Even if complex ophthalmic products are compositionally equivalent, they may still show significant differences in physicochemical properties. These differences can include variations in rheological behavior and in vitro drug release, particularly when the source or grade of excipients is not the same. Typically, IVRT for bioequivalence assessment is just one element of a comprehensive evidence-based strategy. In order to meet the criteria for the in vitro option for bioequivalence as per CFR 320.24 (b) (6), three criteria must be fulfilled: (1) Q1/Q2 sameness, (2) their acceptable comparative physiochemical characteristics and [3] the in vitro drug release rates from both test and reference product must be comparable.

Therefore, IVRT is designed to verify whether the envisioned generic product has a release rate, which resembles the RLD. When formulations are nearly identical, IVRT gives conclusive evidence that the drug release profiles are similar. For the formulations which display variations, IVRT employed with in vivo testing

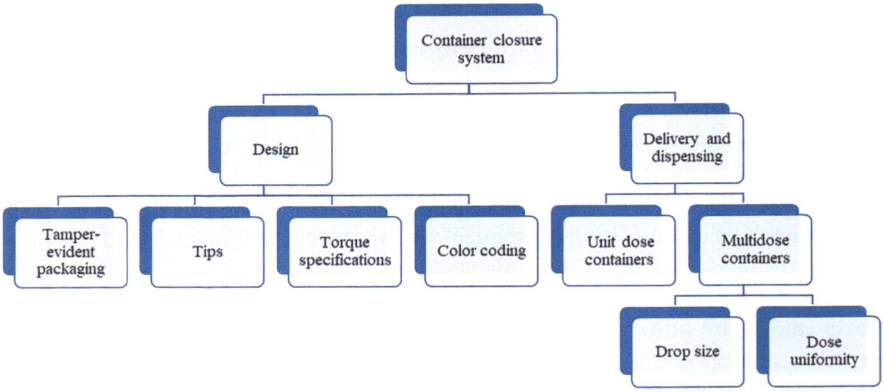

Fig. 7.4 Classification of tests associated with container closure systems

affirms their overall similarity. Once validated, IVRT can serve as viable quality control tool and provide reliable means to verify post approval manufacturing changes. Utilizing a 'Design of experimental (DoE)' model, applicants can obtain insights into how manufacturing processes and attributes of input material can exert influence on drug release properties and obtain clues towards product optimization [132].

The IVRT of ophthalmic ointments is usually done using USP type 2 apparatus equipped with an immersion cell-Model-A and USP type 4 with semi-solid adapter. The USP type 2 apparatus is modified into ointment-holding cells with mini-paddles that can hold lower volumes of ointments. The USP apparatus 4 with trans-cap cell method can also be used for the same [133], as exemplified by Mekjaruskul et al. who investigated drug release from dexamethasone ointment using USP dissolution apparatus 4 (flow-through cell) with a semisolid adapter [134].

Modified USP 1 apparatus with novel two-sided semisolid adapter has also been used to investigate IVRT for ophthalmic ointment. The equipment entails a bottom cap which is fitted with a membrane and a spacer kept on top. The reservoir between spacer and membrane is filled with ointment, followed by a second membrane positioned and closed with a top cap. The experimental conditions such as stirring speed and media temperature should be adjusted to ensure complete drug release. This setup can be modified to study unidirectional drug release by deploying an impermeable polyethylene membrane on one side [134]. Mekjaruskul et al. found that USP type 1 apparatus with novel two-sided ointment adapter facilitates greater surface area for ointment and release medium to interact [135]. A vertical Franz diffusion cell apparatus can also be used to access drug release from ophthalmic formulations as per USP <1724>.

In recent times, especially tailored dissolution assemblies have been fitted with in situ fibre optics which quantitate the extent of drug released in real time. The equipment design varies depending on the type of probe and detector used [136]. For instance, brinzolamide ophthalmic suspension has been subjected to membrane-free in vitro dissolution testing in an assembly fitted with an in-situ UV fibre optic (μDISS profiler™). Briefly, the dissolution assembly contained a mini-bath platform to control the temperature (34 ± 0.5 °C) and dissolution media (simulated tear fluid (STF), pH 7.4, 15 mL) and agitation (125 ± 5 rpm) in cylindrical sample cells. The selected temperature mimicked the prevalent temperature of the eye. The detection probe assembly was placed in the centre of the vessel, with the tip positioned 2 cm above the bottom of the vessel. Spectrum acquisition from the signal being supplied by the probes was started prior to introduction of test sample in the cells containing prewarmed dissolution media. This novel dissolution method was completed within minutes with conditions mimicking low temperature, high shear (due to blinking) and non-sink conditions of the eye. The dissolution method could additionally discriminate between batches exhibiting changes in material attributes like suspension's PSD [137].

Developing IVRT methods for ophthalmic emulsions is challenging. Commonly used compendial dissolution methods or Franz diffusion cell cannot be used effectively for emulsions, as such methods are not sensitive enough to detect trivial

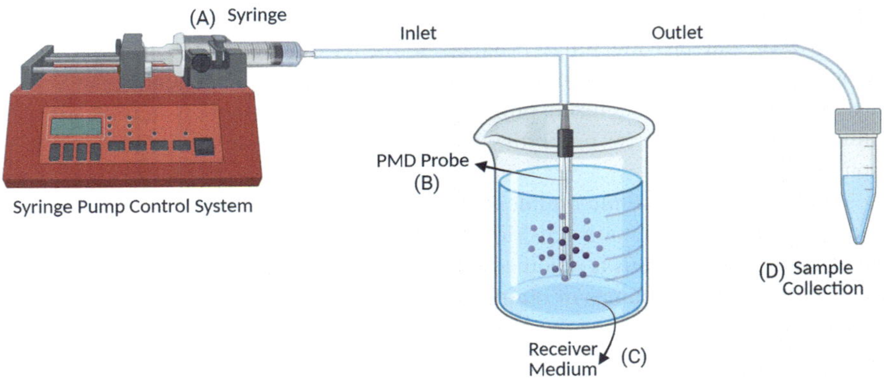

Fig. 7.5 Diagrammatic representation of pulsatile microanalysis (PMD); the inlet syringe pump (**a**) carries the formulation and pumps into the PMD probe window (**b**), the drug diffuses from the probe window to receiver m (**c**) consisting of receiver medium. Further, the sample is collected from sample collection window (**d**) and receiver window for analysis of differential drug concentration in free form and in oil dispersed phase

changes in concentration over infinitesimal time periods like minutes or seconds that are of potential interest due to the brief retention time of emulsions. Also, traditional methods lack the ability to mimic the abrupt temperature change that occurs immediately after the emulsion comes in contact with the cornea. Additionally, for emulsions, onset of free-drug release is driven by passive diffusion. Passive diffusion is governed by several factors like surface area and concentration gradient which are difficult to replicate in an in vitro setting. Pulsatile microanalysis (PMD) can address some of these deficiencies associated with conventional IVRT methods. It is especially useful in comprehending the drug-release behaviour of micro/nano emulsions and ophthalmic gels. The setup for PMD IVRT is shown in Fig. 7.5, and it involves an inlet syringe pump (A), a PMD probe window (B), a receiver window (C), a sample collection outlet (D) and impermeable tubes. The PMD probe consists of a probe window, which is a porous tubing segment. The probes are usually made of cellulosic dialysis membrane having a molecular weight cut-off of 18 kDa (can vary). The pores of the probe window are so small that they hinder the passage of globules or micelles, and hence, only unbound solubilized drug can pass through the probe. This probe is connected to a pulsatile syringe inlet on one side and a sample collection outlet on the other using an impermeable tubing. The probe is fully dipped in a receiver medium which is maintained at a constant temperature of 20–35 °C (relevant to eye temperature). The dissolution medium, which could consist of water, polysorbate 80 and a very low concentration of the drug product, is stirred continuously. The formulation is loaded in the syringe pump and pumped into the probe window and stopped at chosen resting times. It is then ejected out for sample collection. When the emulsion is in the probe window, the dissolved drug diffuses out of the probe window into the receiver medium. Samples are to be collected from both the receiving medium and the sample collection outlet to understand the concentration of drug in free dissolved state and in the oil phase over a range of time. This also gives information about the partition properties of the drug used in the emulsion. Each pulse from the syringe is desired to have a specific volume of emulsion (15 μL) and a flowrate (~20 μL/min). Ideal resting times after discarding the first 2 pulses and combining the end 2 pulses would be 0, 20, 40, 60, 90, 120, 180, 300 and 600 s. The study is to be performed using sample replicates of 3, 6 or 9 for producing statistically valid results [138].

The majority of available IVRT methods like dialysis and Franz cell diffusion are chiefly aimed at quantifying the cumulative amount of drug released with secondary attention accorded to drug release rate. These methods rely on diffusion process, which has certain limitations. For instance, ophthalmic products with rapid drug release profiles may not be pliant with introduction of an artificial diffusion rate limiting membrane (as with dialysis), which tends to prolong the total duration required to obtain complete drug release. Moreover, diffusion-based methods are not sensitive enough to detect minor changes produced by modifications in manufacturing process [139].

USP apparatus 3,4 and 7 have been explored for use in in vitro studies of ophthalmic implants. For instance, USP apparatus 3 (designated as reciprocating cylinder) has been repurposed to investigate the release of ciprofloxacin from PLGA/PLA implants. In this method, implants were loaded in cylindrical holder from which approximately 30 μL sampling was done at pre-fixed time intervals. Further, samples were tested using HPLC technique to measure ciprofloxacin levels. Drug release from this method was akin to that observed in micro dialysis system [140]. USP type 4 apparatus has been utilized by Shen and Burgess to determine the dexamethasone release from PLGA microsphere embedded in a PVA hydrogel implant. In this approach, implants were placed in apparatus' cells, which were then flushed with phosphate buffer (pH 7.4 supplemented with sodium azide). Sampling was done at regular intervals and tested using an HPLC method [116].

Stein and colleagues have conducted in vitro release studies on Ozurdex® implant (indicated to address non-infectious uveitis and macular oedema) using USP apparatus 7 (400-DS Dissolution Apparatus 7). Briefly, dissolution cells were filled with Ringer solution of pH 7.4 maintained at 37 °C. Each cell was conditioned with a meshed basket holder that dipped in dissolution media at a rate of 20 times per minute. It was observed that drug release from different implants varied a lot, likely due to fragmentation of implant causing sudden bursts of drug release. Moreover, the broken fragments adhered to the base, thereby partially clogging the perforated base. While other USP dissolution methods mainly deal with diffusion and swelling, USP apparatus 7 also provides the possibility of simulating drug release by erosion [141].

Given the absence of standard methods for in vitro dissolution testing of implantable ocular devices, various innovative methods have also been employed in research labs to elucidate drug release from complex ophthalmic products. Static diffusion method is an uncomplicated approach, in which implants are immersed in suitable dissolution medium, which is kept at constant temperature with or without stirring. At set time points, drug release is monitored by manually retrieving the implant or sampling the medium using an appropriate analytical technique [142]. Static methods can be employed to examine the release of drug from in-situ depot forming implants and solid implants using appropriate release media adjusted to different volumes according to required sink conditions for respective drug. Balasubramaniam et al. highlighted some of the issues associated with such simplistic approaches, as they could not account for uneven swelling of implant, protein instability, inability to mimic natural environment of vitreous humour accurately, while studying indomethacin release from ophthalmic implant constructed using HPMC and sodium alginate [143].

The agar diffusion method for dissolution testing involves preparing a sterile gel by pouring a solution of either a polysaccharide, such as agar, or a protein, such as collagen, into a petri dish. To test the implant, a circular portion of the gel is excised from the central area, and the implant is placed in the resulting void. Additional agar or collagen is then poured over the implant to seal it in position. After solidification, implants are taken out at specified time intervals to quantify drug diffused by analysing the gel. One benefit of this method is its adaptability to study drug release in highly viscous settings reminiscent of intravitreal fluids. Nevertheless, this method does not capture true ocular physiological conditions [144].

Drug molecules encounter different barriers in the eye depending on the location of implant. Thus, incorporation of membranes in in vitro release assays may offer a way to simulate these barriers and get results in better alignment with in vivo findings. Chennamaneni et al. developed a novel bio-erodible dexamethasone implant by compressing drug-loaded PLGA microspheres to achieve synchronic delivery to anterior as well as to posterior segment of eye. To account for the extent of drug released with respect to time, the authors loaded the implant in a cellulosic dialysis bag (10 kDa molecular weight cut off (MWCO)) filled with 1 mL balanced salt solution (BSS, pH 7.4). The sealed bag was submerged in 100 mL drug releasing medium (BSS, pH 7.4) under constant stirring at 37 °C. Sampling was done at predetermined intervals and further analysed using LC/MS/MS technique [145]. Takeaki Kawashima and colleagues have tested a model of a sclera-implantable membrane-based capsule to provide sustained release of ciliary neurotrophic factor, brain-derived neurotrophic factor (BDNF) or other protein-based drugs to retina. To establish proof of concept, a transscleral device was constructed from a chamber composed of dimethacrylate, polyethylene glycol (PEG) and collagen mix. Each drug formulation contained tiny collagen beads loaded with BDNF and was sealed with collagen membrane of various concentrations. Capsules were placed in transwell cell filled with 40 μL PBS, and the whole system was incubated in 1 ml of PBS at 37 °C [146]..

In order to develop long-acting ophthalmic preparations, there should be evidence that drug candidates exhibit prolonged half-lives when subjected to aqueous flow. In an effort to achieve this, Awwad et al. created an in vitro model called the PK-eye model. This model was fabricated to gauge the clearance of therapeutic protein and dissolution profile of therapeutic suspensions and implants in vitreous compartment of the eye [147]. Built with two compartments, whose dimensions and volumes were similar to that of human eye, the device featured a dialysis membrane

(12–14 kDa MWCO) as a barrier to solubilized proteins in posterior cavity. Emulation of posterior compartment of eye was achieved by using a melange of PBS and simulated vitreous humour (made up of agar and hyaluronic acid with a dynamic velocity of 0.6 Pa.s). Further, the device was equipped with injection port and aqueous inlet port in vitreous cavity in addition to single outlet port in anterior chamber. Results suggested that model was capable of discriminating the release profiles of protein drugs (ranibizumab) and poorly soluble drugs (triamcinolone acetonide) [148].

To incorporate the impact of eye fluid flux on the direction of drug diffusion from an implant, a specialized assembly which combines the vitreous model and eye movement system has been developed. As a part of this experiment, a PCL (polycaprolactone)-based implant, filled with dexamethasone was tested. The implant was placed into vitreous model (VM) filled with simulated vitreous humour, i.e. Ringer's buffered solution and polyacrylamide gel (PAAG) [149]. In the eye movement system model (EyeMoS) setup, six VMs were organized in a horizontal row, and eye movements were simulated for up to 24 h. For sampling from release medium, devices were disengaged, and implant was removed and reinserted after completion of this process [150].

7.5.3 Clinical Endpoint Studies

For conditions like ocular hypertension and open-angle glaucoma, it is recommended to demonstrate bioequivalence (BE) using clinical end point studies. Draft guidance on dorzolamide hydrochloride suggest this type of study [151].Briefly, the subjects should have the condition in both eyes. One drop of the test product in addition to reference product is to be applied thrice daily for 6 weeks. Primary endpoint is regarded as the mean difference in the intra-ocular pressure (IOP) of both eyes between the two treatment groups at four time points—pre-dose, 2 h, 2 weeks and 6 weeks post dose, respectively. BE can be established if the difference in mean IOP between the test and reference products, along with the 95% confidence interval, remains within ±1.0 mmHg at all four time points [152]. Readers are directed towards Table 7.3 for further case studies.

For conditions associated with pain and inflammation due to cataract surgery, a different clinical end point study is suggested, referenced in the PSG of Nepafenac suspension (Table 7.3) [153]. Briefly, it is advised to have a vehicle-based placebo control group to show that both the test product and RLD are efficacious. It also helps in establishing the fact that the study is sensitive enough to recognize small differences between products. A drop of either the test product, vehicle control, or RLD is instilled into the operative eye three times daily, starting one day before cataract surgery and continuing through the two-week postoperative period (days 1 to 14). This makes the treatment duration of 16 days. The number of subjects with a significant cure on day 14 is marked as an endpoint and is defined by assigning a score of 0 for aqueous cells, 0 for aqueous flare and a score of NMT 3 for pain. BE can be established if the difference of primary endpoints of test and reference product with 90% confidence interval are within −0.20 to +0.20 for dichotomous variables. Dichotomous variables are variables whose result/outcomes are divided into two distinct categories or outcomes like Yes/No or Pass/Fail. Using the modified intent to treat test population (mITT) and last observation carried forward (LOCF), the test and RLD should both be statistically superior to the placebo/vehicle control ($p < 0.05$) for the primary endpoint in order to determine the study sensitivity [152].

7.5.4 Pharmacokinetic (PK) Study in Tear Fluid/ Aqueous Humour

These studies are required when in vitro studies cannot be completely relied upon for estimating true performance of the drug product. The proceeding steps are performed during in vivo clinical pharmacokinetic testing of complex ophthalmic products. If feasible, analyte of interest may be imaged using methods like fluorescence or radiolabelling. In other cases, traditional bioanalytical approaches like drug extraction followed by chromatographic analysis are deployed.

The subjects are selected from a cohort of healthy or diseased volunteers suffering with the ailment on which the test product is intended to act. Additionally, the subject is to be assessed properly to know the difference between before and after results. This is followed by treatment of the subjects with test and reference ophthalmic products, followed by collection of pharmacokinetic samples from tissue/fluid of interest. Tear samples can be accessed using various methods like Schirmer test [154], tear meniscus measurement [155], microcapillary tubes [156], punctal plugs [157] or lacrimal punctum occlusion [158]. Certain studies necessitate collection of aqueous humour via anterior chamber paracentesis, which requires patient anesthetization. Briefly, a fine 30G or higher sterile needle is positioned into the aqueous chamber of the eye through the corneal limbus. A small volume of 0.1–0.2 mL is aspirated using the syringe. After removing the needle slowly, gentle pressure is applied on the eye to avoid any residual leakage [159], [160].

Vitreous humour is sampled using pars plana vitrectomy or needle aspiration technique. In pars plana vitrectomy, a very small and fine incision is made in the pars plana, and using a vitrector or cutter, a small amount of sample is aspirated. In needle aspiration technique, a fine needle of 27G or finer is inserted via the pars plana into the vitreous chamber and sample is aspirated out.

Additionally, blood or plasma samples may be simultaneously collected to check the systemic exposure of the drug after administration as an ophthalmic [161]. After topical administration, the blood samples are to be collected from appropriate veins. The amount of drug present in the whole blood or treated plasma/serum is then measured after treating the collected sample using appropriate bioanalytical methods.

7.5.5 Microbial Kill Rate Study

If the ophthalmic product houses an antibiotic, it is advisable to do additional testing to assess bio equivalency. Microbial kill rate study is an in vitro method used to scrutinize the antibiotic activity of a combination product and if the active and inactive ingredients of the product have an impact on the same. This study is recommended by FDA for drugs like tobramycin in combination products is posted in form of a draft guidance [162, 163]. The aim is to check the antimicrobial activity of the test and reference products against all the listed microorganisms in the Preservative Effectiveness Test (USP <51>) and in the 'Indications' section of the RLD label. The study should be performed with at least 12 replicates per kill rate study of each organism to illustrate BE between the test and the reference products. The test, reference and the negative control are compared here. Following incubation, each test organism or product's population is recommended to be ascertained through tallying the surviving colonies at predetermined intervals. Following this phase, the kill rates are determined and compared for the same organism at same timepoints. The time points are generally set as 0, 7.5, 15, 30 and 60 mins. To establish the 90% confidence interval for the test/reference ratios of average kill rate at each sampling stage, two one-sided test procedures are recommended [152].

7.5.6 Physiologically Based Pharmacokinetic Models

It is well established that a proposed generic drug product must demonstrate its bioequivalence to RLD in order to gain approval. However, conventional in vivo bioequivalence methodologies mentioned above are often unethical, impractical and cost-prohibitive for locally acting drug products like complex topical ophthalmic generics. When performing local pharmacokinetic (PK) studies of ophthalmic products, researchers encounter substantial challenges like scarcity of tissue samples, especially aqueous humour for comprehending drug' distribution and effects. Additionally, sparse sampling often produces limited data leading to considerable variability that can cloud the interpretation of results. Consequently, to achieve statistical significance, large population size is required which complicates the study design and escalates the overall cost of the study. Alongside PK studies, comparative clinical endpoint studies also face unique challenges. Patient outcomes are the major determining factors of therapeutic efficacy in these kinds of studies. However, the severity, of a patient's underlying conditions can skew results, hindering the accurate assessment of the drug's therapeutic efficacy. Furthermore, variability in individual patient responses adds to the intricacy of evaluation.

To address these barriers, mathematical modelling approaches such as physiologically based pharmacokinetic (PBPK) modelling were introduced [164]. PBPK integrates information from biopharmaceutic properties of the drug and human physiology to anticipate the expected drug concentrations in local tissues. This approach may streamline evaluation of critical drug attributes in assessing BE during development of generic products. The Office of Generic Drugs, FDA, has recently launched scientific research programmes to expediate the development and evaluation of generic products through application of model-integrated BE alternative approaches [165].Thus far, Ocular Compartmental Absorption and Transit model (OCAT™) to predict ocular exposure in humans has been validated for ophthalmic solutions. However, researchers have been working hard to extend the utility of PBPK to other ophthalmic dosage forms as well. Le Medy et al. have highlighted the effectiveness of an ocular PBPK (OPBPK) model to envision human exposure following the use of a besifloxacin ophthalmic suspension. The OPBPK model took factors like nasolacrimal drainage, ocular absorption, the dissolution of particles in tears and distribution within rabbit eye into consideration. This research could predict the implication of ophthalmic suspension CQA's like particle size and viscosity on the dynamics of API retention and dissolution in the pre-corneal region, which influences the overall absorption of ocular drugs. Therefore, this study bolsters the OCAT model by considering both particle size and viscosity as crucial elements that impact ocular exposure [166]. In a different study, the same group tried to illustrate the application of the OCAT™ within Gastroplus® to characterize the drug pharmacokinetic parameters in rabbits using ophthalmic ointment formulation. It involved analysis of aqueous humour concentrations ensuing administration of multiple ointment formulations in rabbits through a newly developed OCAT™ model for fluorometholone. In an effort to describe the rate at which API transitions from the ointment into the tears, the model incorporated various parameters such as a fitted application time surface area (SA) and a fitted Higuchi release constant. Furthermore, parameter sensitivity analysis in Gastroplus® was used to understand the impact of application time, SA and Higuchi rate constant on aqueous humour concentration. It was noted that application time had a significant effect on T_{max} of drug in aqueous humour, whereas Higuchi release constant and application SA had considerable effects on maximum concentration of drug in aqueous humour. This preliminary model represented a foundational step to describe the interplay of drug properties and eye physiology as well as its outcome on in vivo ocular drug pharmacokinetics in rabbit [167]. PBPK modelling is seen as a way to generate more accurate outcomes for diverse formulations at an accelerated rate. Yet, there are many obstacles hindering its success:

- A significant challenge is the lack of direct eye tissue concentrations for model validations.
- There is limited understanding on drug and melanin binding in the eye tissue.
- Knowledge about the presence of metabolizing enzymes in ocular tissues is limited.
- Existing models focus on emphasize passive permeation in ocular tissues neglecting the role of active transport [168].

7.5.7 In Vitro and In Vivo Correlation Development for Ophthalmics

The concept of IVIVC dates back to 1950s, when scientists made attempts to establish a relationship between in vitro dissolution studies of oral formulations with their corresponding in vivo behaviour using mathematical models [169]. Although FDA's IVIVC guidance on extended release dosage forms has been available since 1997, no regulatory framework exists for complex non-oral formulations to this day. Despite extensive efforts, the development of IVIVC for complex ophthalmics has been especially challenging due to scarcity of in vivo data and utilization of non-compendial dissolution methods for several ophthalmic dosage forms [170]. From an anatomical and physiological point of view, eye has complex structure which is dynamic in nature which complicates the development of IVIVC for ophthalmic products. Mimicking precorneal barriers like tear drainage, tear dilution, continuous clearance of dosage forms, lacrimation, and tear turnover hinders the development of a reliable in vitro predictive model. Moreover, depending on the disease, the target of drug delivery also changes like topical formulations for ocular surface and implants/inserts for anterior and vitreous chamber of the eye which again demands for different IVIVCs [171].

Designing an in vitro dissolution test for ocular implants is also fraught with challenge of mimicking the in vivo environment of eye. Among various drug delivery routes, replication of eye's drug distribution and permeation mechanism is complicated. Apart from anatomical and physiological impediments, factors such as ageing and age-related macular degeneration can cause the vitreous humour to liquify, which consequently has a major influence on product performance and ultimately results in discrepancy among individuals [172]. Implants fabricated using biodegradable polymers conventionally show biphasic or multiphasic release profiles, which are governed by the kind of biodegradable polymer used, and the geometry of the implant, etc. For instance, PLGA exhibits triphasic drug release behaviour, i.e. burst release, diffusion and bulk erosion [173]. Gesine Schliecker et al. attempted to investigate the correlation at multiple levels between in vitro drug release profiles and in vivo pharmacokinetic parameters of three Buserelin implants. To gain insight into in vitro drug release curves, theoretical models like Higuchi and Korsmeyer-Peppas were applied [174]. However, basic standalone statistical model can't be relied on to simulate/predict in vivo data as pharmacokinetic parameters are difficult to acquire in the long-acting intraocular implants owing to confounding factors like protein binding, precorneal fluid drainage and drug metabolism, which are essential for deconvolution to establish IVIVC [175].

Additional challenges arise due to difficulty in real-time screening of intraocular implants are owing to the amount of time required for the studies. To speed up the evaluation, accelerated testing is often used [176]. However, operational conditions such as elevated temperature or extreme pH used in these studies may result in a drug release profile that differs from what is observed under real-time testing conditions. These differences can be precipitated by variable factors like degradation of drug, mobility acquisition by the polymer due to glass transition followed by change in the consistency of formulations under such conditions. [177]. Attainment of a level A IVIVC hinges on detailed pharmacokinetic data. This data often requires serial drug concentration measurement, but due to ethical concerns and practical difficulty, researchers cannot do regular ocular sampling [178]. Thus, IVIVC for ophthalmic formulations is in a continuous state of flux and requires extensive validation before deployment as a substantive tool in ANDA dossiers to support waiver of in vivo testing.

7.6 Complex Ophthalmic Dosage Forms: Quality Considerations

A broad outline of quality tests for ophthalmic dosage forms that are useful in showcasing the sameness of RLD and intended generic dosage forms is discussed in the following sub-sections and is summarized in Fig. 7.6.

7.6.1 Ophthalmic Solutions/Drops

Solutions are thermodynamically stable, homogenous mixture of a solute in a solvent. The formula of ophthalmic solution is chosen to match the isotonicity, surface tension and pH of the eye. Solutions could be termed 'complex' if they involve complex manufacturing techniques, complex administration, complex formulation or are aimed to target complex ocular conditions. It is a must for these solutions to be sterile and stable to prevent contamination and degradation. These ophthalmic solutions may be antibacterial, antiviral, anti-inflammatory, lubricating, anti-glaucoma or diagnostic in nature. Commonly used excipients may include buffering agents like potassium acetate, boric acid, sodium phosphate (dibasic/monobasic), sodium chloride and sodium phosphate monobasic dihydrate. Tonicity adjusters like mannitol, sodium chloride, dextran and glycerin. Preservatives like chlorobutanol, benzalkonium chloride. pH modifiers like sodium hydroxide, acetic acid, phosphoric acid and HCl. Viscosity enhancer like polyethylene glycol, dextran and hydroxyethyl cellulose [179].

For fixed-dose or multi-drug solutions, it is essential to ensure precise addition of the active pharmaceutical ingredient(s) and maintain content uniformity throughout each unit and across the entire batch to guarantee the safe and effective use of the drug product. Preservatives can sometimes cause allergic reactions or irritation. Therefore, preservative-free solutions have come to the market. It is possible to make such formulations by ensuring sterility during manufacture and packaging and further relying on the container closure system for sterility. The product could be packed in either single-use containers or multi-use containers. For multi-use preservative free containers, airless pumps, disposable dropper systems and sometimes prefilled syringes can be used [180].

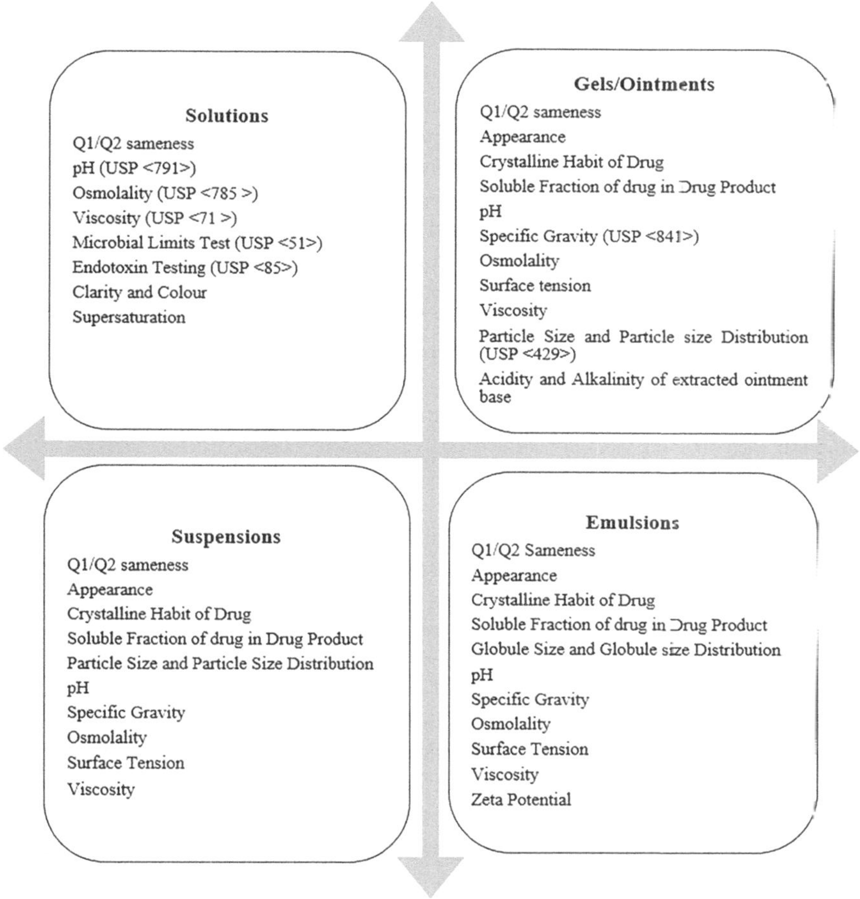

Fig. 7.6 Quality tests for ophthalmic dosage forms; these are the comparative in vitro characterization studies which are carried for the purpose of developing a generic version of complex ophthalmic drug product

Test for reconstitution is specific for powders or concentrates that need to be diluted with a vehicle/solvent before being administered. These drugs might not be stable enough to be marketed in a solution form and hence are lyophilized or solidified. The type and amount of diluent and the concentration of the final solution are mentioned in the label. It is necessary for the powder and diluent to be sterile. All measures should be taken to ensure that the final solution is sterile. After reconstitution, the solution should be visually checked for any insoluble particulate matter. The reconstitution time is contemplated as the principle factor [181].

7.6.2 Suspensions

Ophthalmic suspensions are complex colloidal systems that are formulated to deliver API to the ocular tissues. They comprise of finely divided drug particles dispersed in a liquid medium. These preparations are incorporated in treating eye conditions such as ocular infections and imbalance in the intraocular pressure. The physical and chemical traits of the suspension and the necessity to maintain the physical and chemical attributes to develop a stable suspension contribute to its complexity. Particle size and particle size distribution (PSD) are the most crucial factors affecting the effectiveness of suspensions. Ophthalmic suspensions must comply with quality tests listed in USP <771> Ophthalmic Products-Quality tests. Whether the drug substance (DS) is micronized or acicular or unmicronized will have a bearing on the drug product development and performance [182]. Consequently, a three-tier particle size limit in DS specification is formalized and followed.

Typical studies during the product development phase include comparative physicochemical testing of proposed generic vis-à-vis RLD in terms of appearance, assay, pH, purity, osmolality, redispersiblity, suspendibility, elemental impurities (USP <232>, ICH Q3D), antioxidant or preservative level if required, antimicrobial effectiveness of lowest preservative levels attained during shelf life of RLD, effect of centrifugation, temperature ramping, freeze-thaw cycling, photostability testing, container closure system (CCS) characterization (integrity testing, tamper evident features, child protection packaging, USP <661>, <671>, <87>/<88>, extractables/leachables (USP <1663 and <1664>). Studies on formulation development could further include excipient compatibility, identification of critical material attributes (CMAs) and critical process parameters (CPPs), order of addition (DS versus excipients), hold time study, etc. Studies on grade of excipient deployed could include establishing for instance molecular weight and degree of polymerization of viscosity modifiers like guar gum, carbomers, cellulosic derivatives, PLGA, etc. The amount of carbon atoms in a preservative/surfactant like benzalkonium chloride (C12 vs C16) or the hydrate/dihydrate or anhydrous form of a buffering agent/chelator like EDTA. Excipient specification should include CQA's that can impact the drug product CQA. Potential destabilization mechanisms such as particle size growth due to Ostwald ripening, particle agglomeration, caking, and sedimentation should also be evaluated through accelerated stability testing.

In-process tests during scaled-up manufacturing generally include hold-time testing, homogeneity testing of bulk and packaged product (assessing assay uniformity in samples collected from the top, middle, and bottom of the manufacturing vessel, as well as at the beginning, middle, and end of the filling process), and evaluation of particle size distribution using a three-tier limit, where the maximum cut-off is below 10 μm, as specified in USP 789.

The shelf life and release specification of complex ophthalmic suspension will include similar quality control studies as conducted during the product development stage of the proposed generic. Acceptance criteria for the finalized specification sheet should conform to the data generated by testing of multiple lots of RLD (both in its initial and the terminal phase of shelf life). The stability studies should be carried out in the full packaging configuration and orientations (erect, inverted, supine) utilizing all the stability stations prescribed in ICH Q1A (R2).

Redispersability/Resuspendibility: Resuspendibility of pharmaceutical suspension is envisioned as a critical factor for assuring reliable dosing and desired pharmacological action. It states the capability of the suspended particles to reappear to a state of uniform distribution after settling [183]. The major aspects of suspension which distresses the resuspend ability are particle size, dispersed medium viscosity, density and use of suspending agents/stabilizers. There is no pharmacopeial method for examining the resuspendability but can be iteratively examined by filling the formulation in a sealed cylinder, and subjecting it to repeated fixed rotations and calculating the time required for obtaining a homogenous dispersion [184].

Amount of API in the dispersed medium: The API of the suspension is generally in the form of insoluble particles. However, in some cases, according to the partition coefficient of the drug, the drug could solubilize itself in the dispersion medium. The amount of API both in the dispersed phase and the dispersion medium should be checked in the assay testing of the product over different time points [185].

7.6.3 Emulsions

Emulsions are biphasic colloidal systems in which an oil phase and water phase are intimately dispersed owing to the presence of an emulsifier, providing a nuanced mechanism for drug delivery on the ocular surface. They are generally formulated if the API has poor aqueous solubility [186]. The emulsion could be oil in water or water in oil. The oil phase is generally made of non-polar liquids like vegetable oils, castor oil or medium chain triglycerides. The choice of oil impacts the emulsion's stability, viscosity, drug release, and drug permeation. The stability of emulsions is directly impacted by the globule size and globule size distribution, emulsifier concentration, and physicochemical properties of aqueous and oily phase. Challenges like difficulty to manufacture/process and transient stability on storage (avoiding phase separation) make these formulations complex in nature.

The first step during the development phase is to characterize the API and the inactive ingredients. The probable list of excipients can be ascertained by going through the label and patents of the RLD product. The properties of the API should be comparable to that of the RLD product. The physical, chemical and biological properties of both the API and the excipients should be studied. Investigable physical properties include appearance, polymorphic forms of API and various excipients, solubility of API in various solvents and its thermal properties. The chemical properties include structural analysis using analytical techniques like NMR, FTIR and mass spectroscopy. Further, the assay and purity of the API should be assessed during all phases of development. Attention should be paid to the order of addition of ingredients of the formulation. It is necessary to study the interaction of excipients with each other and with the API at different time points in different conditions.

Once API and excipient distribution are understood, the impact of process variables on drug release must also be delineated.

After formulation of preliminary batches, they are subjected to the Q1/Q2 comparative sameness assessment. The tests should cover all the phases of the emulsion: the oil phase, the aqueous phase and the micellar phase that is formed due to certain excipients like surfactants. In vitro tests of the generic drug should be carried out in conformance with the RLD. These in vitro tests include appearance, assay, pH, purity, osmolality, globule size and globule size distribution, surface tension, crystalline habit of drug, soluble fraction of drug in drug product, rheological properties, specific gravity, viscosity profile as a function of shear, elemental impurities (USP <232>, ICH Q3D), antioxidant or preservative level if required, antimicrobial effectiveness of lowest preservative levels attained during shelf life of RLD, effect of centrifugation, temperature ramping, freeze thaw cycling, photostability testing, container closure system (CCS) characterization (integrity testing, tamper evident features, child protection packaging, USP <661>, <671>, <87>/<88>, extractables/leachables (USP <1663> and <1664>).

Globule size and globule size distribution (GSD) are designated as a CQAs for emulsions because they governs the ocular bioavailability and hence the performance of the emulsion. Larger droplets could be easily expelled during blinking [187]. Globule size and globule size distribution also impact coalescence. High-pressure homogenizer [188], ultrasonicator [189] or microfluidic techniques can be used to control these CQAs. Techniques like dynamic light scattering (DLS) [190], scanning electron microscopy (SEM) [191], Cryo-TEM (transmission electron microscopy) [192], AFM (atomic force microscopy) [193], laser diffraction [194], nanoparticle tracking analysis [195], etc. can be used to monitor these CQAs. Once a GSD for any emulsion sample is established, appropriate statistical methods are needed to compare it to the RLD. Common statistical methods are histogram descriptors, such as 10th, 50th and 90th percentile (i.e. D10, D50 and D90), which have historically been used to compare monodisperse populations. However, these methods are not as accurate when applied to polydisperse samples. To address this, FDA scientists have employed the earth mover's distance (EMD) approach for more effective comparison of polydisperse GSD profiles [196].

The pH of the RLD is generally chosen according to the route of administration, the stability of the drug and the type of emulsifier used, and thus needs to be replicated tightly in the generic version. pH affects the charge of droplets and hence the stability of the emulsion [197]. Adjustment of pH before or after the mixing the aqueous and oil phase also impacts the CQAs of the emulsion, especially the globule size. It is therefore sensible to perform an iterative series of experiments (DoE-based trials) to develop an understanding of relation between critical material attributes (including critical input materials like DS, excipient/stabilizers grade or their category) and their impact on CQAs like pH and GSD. Additionally, fluctuation in temperature during the manufacturing of a formulation can led to changes in the viscosity, precipitation of contents and even phase separation, and should be investigated by subjecting the product to temperature cycling.

Emulsifying agents reduce interfacial tension and hence prevent coalescence [198]. Their type, proportion and order of mixing of emulsifying agents and surfactants impact the globule size and turbidity of an emulsion. Additional emulsion forming excipients include anti-creaming agents, electrolytes and preservatives. Creaming is the upward movement of dispersed globules. Creaming can further lead to phase separation. Stabilizers and viscosity modifiers can avoid creaming by forming a barrier around the globules or increasing the viscosity, respectively [199]. Further, addition of electrolytes or salts in the form of buffering agents can influence the charge of the globules and hence the stability of the emulsion.

It is important to focus on the amount of preservative in the oil phase and aqueous phase. Specific types of preservatives and antioxidants are to be used according to the relative distribution in various phases of the emulsion as preservatives can partition through the oil phase and enter the aqueous phase which may lead to contamination of the oil globule. To prevent this, attention should be given to the partition coefficient, solubility and the amount of preservative to be used for each phase. Emulsions are also subject to Ostwald ripening, which is growth of large droplets at the expense of the smaller ones. This process involves both thermodynamic and transport effects. Interfacial properties have a direct impact on Ostwald ripening. The right type and ratio of emulsifier/surfactant can mitigate Ostwald ripening [198].

Select in process, shelf life and release pharmacopeial tests discussed with respect to ophthalmic suspensions are also applicable to emulsions during scale up manufacturing and testing. The characterization metrics for emulsions are sometimes difficult to apply in practice, as has been exemplified in case of Restasis® eye drops (cyclosporin emulsion 0.5%) by Allergan. Restasis® is a complex, heterogenous, multiphase product. Though an FDA draft guidance for industry on generic cyclosporin exists, yet there are loopholes in it (voiced by the innovator), which pose dilemma to the generic manufacturers in proving bioequivalence with respect to RLD [200].

Cyclosporin's bioavailability is highly dependent on its distribution across the phases of emulsion, which could be very challenging for generic manufacturers to detect and replicate. For instance, in a study on a different ophthalmic emulsion, distribution of API (difluprednate) in various phases (aqueous micellar, microemulsion and emulsion) of intact and diluted ophthalmic emulsion has been attempted using ^{19}F NMR [201]. However, this non-invasive method needs validation and also experimentation on non-fluoride API molecules like cyclosporine before extrapolation. Moreover, traditional statistical metrics such as population bioequivalence analysis (PBE) are inadequate for demonstrating in vitro bioequivalence in the case of Restasis®, due to its nature as a polydisperse emulsion. Consequently, the Product-Specific Guidance (PSG) recommends the use of a more rigorous approach, such as earth mover's distance (EMD). However, notable gaps in the assessment still remain. EMD is sensitive towards data grouping, group distance, number of samples and replicates. So, to avoid misleading results, there should be clear guidelines to ensure valid data processing, which is missing at the moment.

The in vitro methods selected for drug release must distinguish how variations in globule size distribution affect drug release from emulsions. However, current analytical methods are indirect and incapable of detecting the impact of subtle differences between GSD of a Restasis® and a tentative generic product trying to replicate Restasis®. Additionally, method should focus on quantification of free drug release from excipients like castor oil and polysorbate 80. Though globule size is a CQA for emulsion, still there are other parameters which govern the stability of emulsion, and the information pertaining to those parameters is non-existent. This further add to the vagueness on how close the test product and RLD need to be one another to claim pharmaceutic and therapeutic equivalence.

7.6.4 Gels

Ophthalmic gels are semisolid sterile preparations designed for instillation into the space between the eyelid and the eyeball; the active pharmaceutical ingredient is present either in solubilized or suspended form. Development of ophthalmic gel requires incorporation of viscosity building polymers in ophthalmic solutions, which subsequently result in increased precorneal contact time of drugs and decreased drainage [202]. Gels consists of a three-dimensional cross-linked polymers or colloidal network enveloped in fluid. Ophthalmic gels are of two types: (a) gel eye drops are simple viscous formulation that remains unaltered after administration, b) in-situ gels can be described as aqueous solutions/suspensions/emulsions, which transition to gel on contact with tear fluid due to pH, temperature, ion-based gelling mechanisms. Some commonly used excipients in gels are [203] *agents* like gellan gum, carbomers, hydroxyethyl cellulose, hypromellose cellulose, polycarbophil, etc.

One of the primary factors contributing to the complexity of gels is achieving the right balance of viscosity and gelation and use of stimuli-sensitive polymers. In general, the list of inactive ingredients like gelling agents and viscosity modifiers can be learnt from the package insert or label of the RLD product, and the excipient's grade and their physical state should match that of the RLD.

The grade of the excipient has a huge impact on the drug product quality attributes. A low-melting-grade material may melt under accelerated stability testing and affect the physicochemical stability of the product. Consequently, prior to drug product formulation, an excipient compatibility study involving a binary mixture is desirable to guarantee the stability of the product. In most cases, it is assumed that the excipients employed in the RLD are compatible with the drug material. In order to prevent drug degradation, the formulator should be aware that various vendors or grades may have different contaminants, or the excipients may have been processed by the innovator using a proprietary method before incorporation in the formulation.

Since viscosity impacts the retention time and ocular bioavailability, it is advised to present data of appropriate rheometric data and in vitro corneal permeation study to compare the flux of the test and RLD formulations. To study the tendency of Oswald ripening, a temperature cycling study can be done during the drug product development phase with cycles ranging from room temperature to 40 °C. In case of emulsion type gels, the test products are usually put through the following alternating cycles of freeze-thaw: 24 h at −20 °C followed by a 24-h thaw at room temperature, 24 h at −20 °C followed by a 24-h thaw at room temperature, and 72 h at −20 °C followed by a 24-h thaw at room temperature [204]. After the completion of these cycles, the test product should be tested for stability in terms of appearance, solubility, absence of precipitation and crystallization, particle size of drug crystals (if any), viscosity, pH, package integrity, etc. The study is to be done in the original packaging that is decided to be marketed and samples are to be collected from the starting, middle and ending of the packaging container. The results of the antimicrobial preservative study and antioxidants content study must be similar to that of the RLD product. Studies should be done according to USP <51> [205].

During scaled-up manufacturing, appropriate quality attributes of the RLD should be recognized and mimicked. Some information about the process parameters and manufacturing ingredients can be gained from the innovator patents and published literature. The formulated test product should be physically and chemically comparable to the RLD in multiple aspects. These include appearance, assay, pH, purity, osmolality, globule size and its distribution, surface tension, crystalline habit of the drug, soluble drug fraction in the final product, rheological properties, specific gravity, and viscosity profile as a function of shear. Additional parameters to be evaluated include elemental impurities (as per USP <232> and ICH Q3D), levels of antioxidants or preservatives if required, antimicrobial effectiveness based on the lowest preservative concentration maintained during the RLD's shelf life, and stress responses such as centrifugation, temperature ramping, freeze–thaw cycling, and photostability. The container closure system (CCS) must also be characterized through integrity testing, tamper-evident features, childresistant packaging, and evaluations outlined in USP <661>, <671>, <87>/<88>, and extractables and leachables studies according to USP 1663> and <1664. The in-process tests, shelf life and release specification tests are similar to those mentioned for suspensions and emulsions.

7.6.5 Ointments

These are semisolid preparations that could be emulsions or suspensions that are to be applied on the surface of the eye. Rapid blinking leads to even distribution on the corneal surface [206]. Some commonly used excipients in ointments are [207] petrolatum, anhydrous liquid lanolin, mineral oil and preservatives like chlorobutanol, benzalkonium chloride, EDTA, thimerosal, sorbic acid and potassium sorbate. The bases have a melting point close to human body temperature, so they can melt and spread easily with minimum shear. The properties of the base are reflected as the property of the ointment. A majority of ointment based products have incredibly low (less than 1%) amounts of API [208]. Consequently, the impact of ointment base on pharmacokinetic and dynamic properties of the drug load is significant, providing all the more reason for greater scrutiny of the ointment base used in RLD.

Major characteristic of the ointment base such as alkalinity and acidity will have a stern impression on the safety and efficacy of the eye ointment. Ophthalmic preparations are presumed to be of the pH analogous to that of the pH of the eye so as to prevent irritation and any damage to the ocular tissue. The 'acidity and alkalinity' of the extracted ointment base can lead to reduced effectiveness of the active pharmaceutical ingredient as it can intrude with the solubility, stability and absorption of the API; also, alkalinity and acidity can be the prime cause for disruption of tear fluid and can cause dryness of the eye [209, 210]. Organoleptic properties like texture, grittiness, smell of the base, optimal concentration of viscosity agents, use of appropriate type and ratio of preservatives and concentration and type of permeation enhancers could greatly affect the properties of the ointment and thus should be effectively replicated [211, 212].

It is necessary to understand the impact of not only the type and proportion of excipients but also the process parameters during manufacturing of ointments. These could include temperature, heating/cooling rates, mixing methods/ impeller speeds, mixing time, etc. Even slight variance in these process parameters could lead to differences in the quality and performance of the generic product when compared to the RLD [208].

The experimental batches arrived at after the initial filtration exercises discussed above are subjected to the Q1/Q2 comparative testing. The in-process tests, shelf life and release specification tests are similar to those mentioned for gels.

7.6.6 Inserts and Implants

Inserts and implants serve as huge advancements in the field of ophthalmic. The insert/implant degrades gradually without any irritation to the patient. This produces sustained drug release which helps in maintaining therapeutic levels of drug for extended time. Inserts are small devices that are instilled in the cul-de-sac, allowing sustained and extended release of drug directly on the ocular surface. Hydrogel inserts and silicon-based inserts are commonly used in ocular delivery. The hydrogel inserts swell and slowly release the medication, whereas the silicon inserts are used when longer residence time is desired [213].

Implants, unlike inserts, are surgically placed inside or on the surface of the eye and are integrated into the anatomy of the eye. They could function in both therapeutic and restorative way. Intraocular lenses, especially used in cataract and refractive correction surgery, are the most widely used implants. Retinal/subretinal implants [214] are placed on the retina for vision correction, glaucoma implants and micro-stents are used to drain excess aqueous humour, and *intra-vitreal* implants

[215] are placed in the vitreous humour for extended release of drugs usually intended to act on the retina. Varying molecular weight grades of poly (lactic-co-glycolic) acid (PLGA), polylactide (PLA) and macrogols (PEG) are commonly used polymers for manufacturing inserts. PEG can also be used as a plasticizer or co-solvent. For the sake of simplicity, implants or inserts has been used interchangeably in the current text.

The developmental evaluation of a generic implant includes, but is not limited to, the following assessments: molecular weight of the polymer, transparency and refraction studies (in the case of lenses), extractable and leachable testing, mechanical testing of the implant, thermal analysis of both the polymer and the active pharmaceutical ingredient, implant structure and dimensions, porosity and internal pore structure analysis, drug–excipient compatibility testing, forced degradation studies, freeze–thaw testing, and drug release studies at elevated temperatures to accelerate polymer degradation and ensure complete drug release [216–218]. In select cases, generation of additional preclinical data (intraocular pressure measurement/ tonometry/ relevant data) may help development of PBPK models, which may strengthen ANDA submission by developing better process and product understanding.

Mark A. et al. from the University of Texas have reverse engineered an Ozurdex intravitreal implant. The marketed implant is composed of matrix of 50:50 PLGA with the corticosteroid dexamethasone embedded in it for sustained-release to the posterior segment of the eye. Briefly, the list of excipients and the process of manufacturing (hot melt extrusion) of the implant were identified using the product label and patent indicative of the RLD. Analytical method like x-ray diffraction, gel permeation chromatography, scanning electron microscopy, polarized light microscopy, thermal analysis, ATR-FTIR, etc. were used to characterize the API and excipients during the pre-formulation studies. Jet milling was used to decrease the particle size and mix the API with excipients followed by two-stage hot-melt extrusion. The extruded matrix was cut into dimensions similar to that of the RLD. Structural analysis of the implant was done using SEM, the porosity was measured using X-ray microcomputed tomography (micro-CT), and the assay for drug content was estimated by HPLC. The drug release profile was then generated in normal saline solution (37 °C) stirred at 100 rpm in an incubator shaker. Samples were collected and analysed using HPLC. The implant followed a triphasic drug release profile similar to the innovator product. The reverse engineered implants also conformed to the specifications of RLD and had similar properties like appearance, roughness, dimensions, etc. [219].

7.6.6.1 Polymers Employed in the Construction of Implants/Inserts

Evaluating the morphology, category of implant, grade and chemical composition of polymer employed in implant construction is critical as drug release from inserts and implants is directly impacted by the polymer used and the dimensions and structure of the implant (matrix/reservoir). A short note on polymeric biomaterials and their characteristic features which require critical investigation during complex generic development are discussed next [220].

Polyethylene Glycol: It is hydrophilic polymer; it has the ability to increase the biocompatibility and solubility of therapeutics, which are incorporated into it. Dextenza® is an ocular implant comprised of 4-arm PEG, which functions to concoct the formulation and controls drug release. PEG undergoes hydrolytic erosion to release dexamethasone from the implant [221]. ReSure®, an ophthalmic sealant, is an in situ-forming hydrogel based on polyethylene glycol, where the PEG component is supplied in a lyophilized form. This product is mainly used for intraoperative management of clear corneal incision. [222]. Syfovre™ consists of pentadecapeptides, which are covalently bound to PEG. Evidence suggests that PEGylation enhances the half-life of peptide API, thus making it longer acting [223].

*Polyvinyl Alcohol (PVA):*As a water soluble and biodegradable polymer, PVA improves the chemical stability and enhance chemical resistance. PVA can be obtained in diverse hydration rates and molecular weight based on the synthesis method used [224]. Retisert® consists of an outer layer made up of PVA, which has the potential to release fluocinolone acetonide in a controlled manner. Iluvien® is an intravitreal implant comprising PVA and polyimide, enabling the sustain release of fluocinolone acetonide for 36 months.

Polymethacrylates: Polymethyl methacrylates (PMMA) have been approved by FDA for intraocular administration. They are biocompatible, possess good mechanical strength and provide good clarity once processed [225, 226]. This polymer was used as early as in 1936 to manufacture contact lenses [227].

Polyesters: Polyglycolic acid (PGA), polylactic acid (PLA), polylactic and coglycolic acid are FDA-approved polymers for ocular delivery system. A noteworthy advantage of these polymer is the tuneability of biodegradation according to the proportion of monomer fused in the polymer [228]. PGA is hydrophilic in nature; hence, it is vulnerable to hydrolysis, and thus degradation is faster when compared to PLA. It is reported that the biodegradation rate of PLA is 2 years. The end product of biodegradation of both PLA and PLGA is non-toxic in nature [229]. Combination of PGA and PLA by a process of copolymerization results in the formation of PLGA, the proportion of monomer used and the groups that are present at the end of the polymer is responsible for altering the biodegradation profile. The role of PLGA in ocular drug delivery is to facilitate the control release of the drug. PLGA properties can be reformed by adjusting the proportion of PLA and PGA, thus fine-tuning the time taken for degradation, hydrophobicity and degrees of crystallinity [230]. Ozurdex® discussed previously is an intravitreal implant, where PLGA (50:50) is used for the extended release of the dexamethasone [231, 232]. Another noteworthy example is Durysta®, a PLGA-based implant for delivering Bimatoprost in a controlled manner for 4–6 months [233].

7.6.6.2 Challenges of Working with PLGA

In order to develop a generic drug product, Q1 and Q2 sameness has to be achieved [234]. Sterile long-acting injectables containing PLGA have a very complex manufacturing process. It has been estimated that generic drug product manufacturers of PLGA-based long-acting injectables may require about 2000–5000 units of RLD, which may pose a significant logistical challenge [235]. Each PLGA-based formulation is different in its own way. It is designed to meet specific characteristics of a formulation such as particle size, rate of controlled release, drug loading capacity, etc. [236] Hence, without having proper understanding of the drug and polymer, and the anticipated product attribute, it is problematic to achieve favoured outcomes. Development of IVIVC for PLGA-based products is also a challenge [237], which is exacerbated by absence of compendial methods for evaluating in vitro release of PLGA-based complex generics [238].

Constraints due to physicochemical properties of PLGA: The proportion of lactic acid and glycolic acid, molecular weight, molecular weight distribution, polymer construction whether linear or star shaped, crystallinity, end caps on polymer, hydration rate, inherent viscosity, and glass transition temperature have an impact on the release mechanism and biodegradation of the polymer [238]. Research has shown that PLGA comprising of 50:50 lactide to glycolide ratio attains quickest degradation rates [228].

Manufacturing process variations: Any alteration in the manufacturing process can give rise to change in the release profile of the drug. For instance, if a generic company employs 'double emulsification' process instead of coacervation technique employed by the innovator, the release of drug will be affected as the 'double emulsification' process results in the production of porous particles, whereas latter results in solid non-porous structures. The manufacturing steps employed should also be as similar to that of the innovator company, with process parameters such as temperature and manufacturing time carefully set to that of innovator [239].

7.7 Concluding Insights

The exploration of complex ophthalmic formulations has advanced ocular drug delivery, addressing the limitations of conventional dosage form, such as unavoidable drug loss during instillation, inefficient corneal permeability, drainage through nasolacrimal drainage duct, reflex blinking and tear turnover. However, this complexity in ophthalmic formulations has been a bane in genericization strategies, as exemplified by skewness in generic options available for solutions versus suspensions, emulsions, gels, ointments or implants. FDA has been addressing this issue in collaboration with relevant industrial and academic stakeholders by conducting regulatory research programmes. This has resulted in knowledge-based decision-making and enrichment of guidance documents that can be adopted by pharmaceutical firms attempting to reverse engineer complex ophthalmic formulations. Significant research has resulted in the utilization of various in vitro approaches as surrogate for in vivo testing in select scenarios. Establishment of Q1/Q2 sameness needs to be often accompanied by totality of evidence approaches (Q3). However, in the ophthalmic space, there are still several innovative products which are less clearly understood, and filing of a generic application is only possible by harnessing the in vivo approach. Thus, significant work needs to be done in case of developing methods that are universally applicable to ophthalmic dosage forms.

This chapter emphasizes approaches which are obligatory to conclude the Q1/Q2 sameness. A fleeting overview was provided for various quality considerations that may be implemented whilst attempting complex generic version of an existing RLD from scratch. In summary, the ongoing development of complex ophthalmic formulations embodies a transformative dive in ocular therapeutics, with the prospective to significantly ameliorate patient care and treatment efficacy, provided that the prevailing challenges are scientifically addressed through collaborative research and innovation.

References

1. https://www.fda.gov/drugs/frequently-asked-questions-popular-topics/generic-drugs-questions-answers
2. Generic Medicines: Solutions for a Sustainable Drug Market? | Applied Health Economics and Health Policy [Internet]. [cited 2024 Oct 30]. Available from: https://link.springer.com/article/10.1007/s40258-013-0043-z
3. Davit BM, Nwakama PE, Buehler GJ, Conner DP, Haidar SH, Patel DT, et al. Comparing generic and innovator drugs: a review of 12 years of bioequivalence data from the United States Food and Drug Administration. Ann Pharmacother. 2009;43(10):1583–97.
4. Tatham AJ. The use of generic medications for glaucoma. J Ophthalmol. 2020;2020:1651265.
5. DailyMed – LOTEMAX – loteprednol etabonate suspension/ drops [Internet]. [cited 2024 Oct 30]. Available from: https://dailymed.nlm.nih.gov/dailymed/lookup.cfm?setid=f37d74ac-5f9c-40a8-b43b-35148fc66bbe
6. GoodRx [Internet]. [cited 2024 Oct 30]. Dexamethasone (Maxidex): Uses, Side Effects, Dosage & Reviews. Available from: https://www.goodrx.com/maxidex/what-is
7. An International Guide to Patent Case Management for Judges [Internet]. [cited 2024 Oct 30]. Available from: https://www.wipo.int/patent-judicial-guide/
8. Komlosh A, Weinstein V, Loupe P, Hasson T, Timan B, Konya A, et al. Physicochemical and biological examination of two glatiramer acetate products. Biomedicines. 2019;7(3):49.
9. Kleindl PA, Xiong J, Hewarathna A, Mozziconacci O, Nariya MK, Fisher AC, et al. The botanical drug substance crofelemer as a model system for comparative characterization of complex mixture drugs. J Pharm Sci. 2017;106(11):3242–56.
10. download.pdf [Internet]. [cited 2024 Oct 30]. Available from: https://www.fda.gov/media/157675/download
11. Alpert A, Jacobson M. Impact of oncology drug shortages on chemotherapy treatment. Clin Pharmacol Ther. 2019;106(2):415.
12. Ibrahim SS. The role of surface active agents in ophthalmic drug delivery: a comprehensive review. J Pharm Sci. 2019;108(6):1923–33.
13. Research C for DE and. Product-Specific Guidances for Generic Drug Development. FDA [Internet]. 2023 Nov 21 [cited 2024 Oct 30]; Available from: https://www.fda.gov/drugs/guidances-drugs/product-specific-guidances-generic-drug-development
14. Orange Book: Approved Drug Products with Therapeutic Equivalence Evaluations [Internet]. [cited 2024 Oct 30]. Available from: https://www.accessdata.fda.gov/scripts/cder/ob/index.cfm
15. 016851s063lbl.pdf [Internet]. [cited 2024 Dec 23]. Available from: https://www.accessdata.fda.gov/drugsatfda_docs/label/2013/016851s063lbl.pdf
16. 050804s026lbl.pdf [Internet]. [cited 2024 Dec 23]. Available from: https://www.accessdata.fda.gov/drugsatfda_docs/label/2021/050804s026lbl.pdf
17. 203491s001lbl.pdf [Internet]. [cited 2024 Dec 23]. Available from: https://www.accessdata.fda.gov/drugsatfda_docs/label/2012/203491s001lbl.pdf
18. 019845s030lbl.pdf [Internet]. [cited 2024 Dec 23]. Available from: https://www.accessdata.fda.gov/drugsatfda_docs/label/2021/019845s030lbl.pdf
19. 204251s012lbl.pdf [Internet]. [cited 2024 Dec 23]. Available from: https://www.accessdata.fda.gov/drugsatfda_docs/label/2023/204251s012lbl.pdf

20. 020816s025lbl.pdf [Internet]. [cited 2024 Dec 23]. Available from: https://www.accessdata.fda.gov/drugsatfda_docs/label/2023/020816s025lbl.pdf
21. 216227Orig1s000Corrected_lbl.pdf [Internet]. [cited 2024 Dec 23]. Available from: https://www.accessdata.fda.gov/drugsatfda_docs/label/2023/216227Orig1s000Corrected_lbl.pdf
22. 202872s002lbl.pdf [Internet]. [cited 2024 Dec 23]. Available from: https://www.accessdata.fda.gov/drugsatfda_docs/label/2018/202872s002lbl.pdf
23. 020330s032lbl.pdf [Internet]. [cited 2024 Dec 23]. Available from: https://www.accessdata.fda.gov/drugsatfda_docs/label/2018/020330s032lbl.pdf
24. 022211lbl.pdf [Internet]. [cited 2024 Dec 23]. Available from: https://www.accessdata.fda.gov/drugsatfda_docs/label/2009/022211lbl.pdf
25. 022221s008lbl.pdf [Internet]. [cited 2024 Dec 23]. Available from: https://www.accessdata.fda.gov/drugsatfda_docs/label/2022/022221s008lbl.pdf
26. 011963s052lbl.pdf [Internet]. [cited 2024 Dec 23]. Available from: https://www.accessdata.fda.gov/drugsatfda_docs/label/2021/011963s052lbl.pdf
27. 021670s011lbl.pdf [Internet]. [cited 2024 Dec 23]. Available from: https://www.accessdata.fda.gov/drugsatfda_docs/label/2023/021670s011lbl.pdf
28. 020065Orig1s033lbl.pdf [Internet]. [cited 2024 Dec 23]. Available from: https://www.accessdata.fda.gov/drugsatfda_docs/label/2022/020065Orig1s033lbl.pdf
29. 209569Orig1s004lbl.pdf [Internet]. [cited 2024 Dec 23]. Available from: https://www.accessdata.fda.gov/drugsatfda_docs/label/2022/209569Orig1s004lbl.pdf
30. 203324s006lbl.pdf [Internet]. [cited 2024 Dec 23]. Available from: https://www.accessdata.fda.gov/drugsatfda_docs/label/2019/203324s006lbl.pdf
31. 020869s042lbl.pdf [Internet]. [cited 2024 Dec 23]. Available from: https://www.accessdata.fda.gov/drugsatfda_docs/label/2020/020869s042lbl.pdf
32. 208259s006lbl.pdf [Internet]. [cited 2024 Dec 23]. Available from: https://www.accessdata.fda.gov/drugsatfda_docs/label/2020/208259s006lbl.pdf
33. 050417s012lbl.pdf [Internet]. [cited 2024 Dec 23]. Available from: https://www.accessdata.fda.gov/drugsatfda_docs/label/2020/050417s012lbl.pdf
34. 020369s022lbl.pdf [Internet]. [cited 2024 Dec 23]. Available from: https://www.accessdata.fda.gov/drugsatfda_docs/label/2017/020369s022lbl.pdf
35. 050065s065lbl.pdf [Internet]. [cited 2024 Dec 23]. Available from: https://www.accessdata.fda.gov/drugsatfda_docs/label/2021/050065s065lbl.pdf
36. 050616s040s041lbl.pdf [Internet]. [cited 2024 Dec 23]. Available from: https://www.accessdata.fda.gov/drugsatfda_docs/label/2021/050616s040s041lbl.pdf
37. 200738s003lbl.pdf [Internet]. [cited 2024 Dec 23]. Available from: https://www.accessdata.fda.gov/drugsatfda_docs/label/2020/200738s003lbl.pdf
38. 050555s038lbl.pdf [Internet]. [cited 2024 Dec 23]. Available from: https://www.accessdata.fda.gov/drugsatfda_docs/label/2021/050555s038lbl.pdf
39. 050790s027lbl.pdf [Internet]. [cited 2024 Dec 23]. Available from: https://www.accessdata.fda.gov/drugsatfda_docs/label/2017/050790s027lbl.pdf
40. 022212s016lbl.pdf [Internet]. [cited 2024 Dec 23]. Available from: https://www.accessdata.fda.gov/drugsatfda_docs/label/2016/022212s016lbl.pdf
41. 206185s019lbl.pdf [Internet]. [cited 2024 Dec 23]. Available from: https://www.accessdata.fda.gov/drugsatfda_docs/label/2024/206185s019lbl.pdf
42. 211911s002lbl.pdf [Internet]. [cited 2024 Dec 23]. Available from: https://www.accessdata.fda.gov/drugsatfda_docs/label/2024/211911s002lbl.pdf
43. 208742s007lbl.pdf [Internet]. [cited 2024 Dec 23]. Available from: https://www.accessdata.fda.gov/drugsatfda_docs/label/2021/208742s007lbl.pdf
44. 022315Orig1s016lbl.pdf [Internet]. [cited 2024 Dec 23]. Available from: https://www.accessdata.fda.gov/drugsatfda_docs/label/2022/022315Orig1s016lbl.pdf
45. 210331Orig1s000ltr.pdf [Internet]. [cited 2024 Dec 23]. Available from: https://www.accessdata.fda.gov/drugsatfda_docs/appletter/2018/210331Orig1s000ltr.pdf
46. 021737s019.pdf [Internet]. [cited 2024 Dec 23]. Available from: https://www.accessdata.fda.gov/drugsatfda_docs/label/2011/021737s019.pdf

47. 201923s002lbl.pdf [Internet]. [cited 2024 Dec 23]. Available from: https://www.accessdata.fda.gov/drugsatfda_docs/label/2017/201923s002lbl.pdf
48. 215352s000lbl.pdf [Internet]. [cited 2024 Dec 23]. Available from: https://www.accessdata.fda.gov/drugsatfda_docs/label/2023/215352s000lbl.pdf
49. Acyclovir Topical Ointment.
50. Bacitracin_ophthalmic ointment_RLD61212_RV09–16.pdf [Internet]. [cited 2024 Oct 31]. Available from: https://www.accessdata.fda.gov/drugsatfda_docs/psg/Bacitracin_ophthalmic%20ointment_RLD61212_RV09-16.pdf
51. Ciprofloxacin hydrochloride.
52. PSG_050592.pdf [Internet]. [cited 2024 Oct 31]. Available from: https://www.accessdata.fda.gov/drugsatfda_docs/psg/PSG_050592.pdf
53. Erythromycin_ophthalmic ointment_RLD 062447_RV09-16.pdf [Internet]. [cited 2024 Oct 31]. Available from: https://www.accessdata.fda.gov/drugsatfda_docs/psg/Erythromycin_ophthalmic%20ointment_RLD%20062447_RV09-16.pdf
54. PSG_200738.pdf [Internet]. [cited 2024 Oct 31]. Available from: https://www.accessdata.fda.gov/drugsatfda_docs/psg/PSG_200738.pdf
55. Tobramycin ophthalmic ointment 0.3 NDA 050555 PSG Page RC May 2019.pdf [Internet]. [cited 2024 Oct 31]. Available from: https://www.accessdata.fda.gov/drugsatfda_docs/psg/Tobramycin%20ophthalmic%20ointment%200.3%20NDA%20050555%20PSG%20Page%20RC%20May%202019.pdf
56. PSG_022211.pdf [Internet]. [cited 2024 Oct 31]. Available from: https://www.accessdata.fda.gov/drugsatfda_docs/psg/PSG_022211.pdf
57. PSG_022221.pdf [Internet]. [cited 2024 Oct 31]. Available from: https://www.accessdata.fda.gov/drugsatfda_docs/psg/PSG_022221.pdf
58. PSG_216227.pdf [Internet]. [cited 2024 Oct 31]. Available from: https://www.accessdata.fda.gov/drugsatfda_docs/psg/PSG_216227.pdf
59. PSG_208219.pdf [Internet]. [cited 2024 Oct 31]. Available from: https://www.accessdata.fda.gov/drugsatfda_docs/psg/PSG_208219.pdf
60. PSG_022308.pdf [Internet]. [cited 2024 Oct 31]. Available from: https://www.accessdata.fda.gov/drugsatfda_docs/psg/PSG_022308.pdf
61. PDF [Internet]. [cited 2024 Oct 31]. Available from: https://www.accessdata.fda.gov/drugsatfda_docs/psg/BETAXOLOL%20HYDROCHLORIDE%20ophthalmic%20suspension-drops%20NDA%20019845%20PSG%20Page%20RV%20May%202019.pdf
62. Brimonidine tartrate; Brinzolamide.
63. PSG_050023.pdf [Internet]. [cited 2024 Oct 31]. Available from: https://www.accessdata.fda.gov/drugsatfda_docs/psg/PSG_050023.pdf
64. PSG_019216.pdf [Internet]. [cited 2024 Oct 31]. Available from: https://www.accessdata.fda.gov/drugsatfda_docs/psg/PSG_019216.pdf
65. Rimexolone_ophthalmic suspension_RLD 020474_RV06-16.pdf [Internet]. [cited 2024 Oct 31]. Available from: https://www.accessdata.fda.gov/drugsatfda_docs/psg/Rimexolone_ophthalmic%20suspension_RLD%20020474_RV06-16.pdf
66. FDA/CDER/"Mahadevan C. Prednisolone acetate.
67. Nepafenac_ophthalmic suspension 0.1_RLD 021862_RV12-16.pdf [Internet]. [cited 2024 Oct 31]. Available from: https://www.accessdata.fda.gov/drugsatfda_docs/psg/Nepafenac_ophthalmic%20suspension%200.1_RLD%20021862_RV12-16.pdf
68. PSG_050804.pdf [Internet]. [cited 2024 Oct 31]. Available from: https://www.accessdata.fda.gov/drugsatfda_docs/psg/PSG_050804.pdf
69. Cyclosporine Ophthalmic Emulsion.
70. Difluprednate_ophthalmic emulsion_RLD 022212_RV02-17.pdf [Internet]. [cited 2024 Oct 31]. Available from: https://www.accessdata.fda.gov/drugsatfda_docs/psg/Difluprednate_ophthalmic%20emulsion_RLD%20022212_RV02-17.pdf
71. PSG_206185.pdf [Internet]. [cited 2024 Oct 31]. Available from: https://www.accessdata.fda.gov/drugsatfda_docs/psg/PSG_206185.pdf
72. Bimatoprost Ophthalmic Implant.
73. FDA/CDER/"Mahadevan C. Bromfenac sodium.

74. Lotilaner Ophthalmic Solution/Drops.
75. PSG_020330.pdf [Internet]. [cited 2024 Oct 31]. Available from: https://www.accessdata.fda.gov/drugsatfda_docs/psg/PSG_020330.pdf
76. Brilliant Blue G Ophthalmic Solution.
77. FDA/CDER/"Mahadevan C. Brimonidine tartrate; Timolol maleate.
78. Vision impairment and blindness [Internet]. [cited 2024 Oct 30]. Available from: https://www.who.int/news-room/fact-sheets/detail/blindness-and-visual-impairment
79. Ärzteblatt DÄG Redaktion Deutsches. Deutsches Ärzteblatt. [cited 2024 Oct 30]. Refractive errors (14.10.2016). Available from: https://www.aerzteblatt.de/int/archive/article?id=182948
80. Argus WA. Ocular hypertension and central corneal thickness. Ophthalmology. 1995;102(12):1810–2.
81. Age-related macular degeneration and the aging eye | CIA [Internet]. [cited 2024 Oct 30]. Available from: https://www.dovepress.com/age-related-macular-degeneration-and-the-aging-eye-peer-reviewed-fulltext-article-CIA
82. Cataract | Nature Reviews Disease Primers [Internet]. [cited 2024 Oct 30]. Available from: https://www.nature.com/articles/nrdp201514
83. Stitt AW, Curtis TM, Chen M, Medina RJ, McKay GJ, Jenkins A, et al. The progress in understanding and treatment of diabetic retinopathy. Prog Retin Eye Res. 2016;(51):156–86.
84. Diabetic Macular Edema | Ophthalmologica | Karger Publishers [Internet]. [cited 2024 Oct 30]. Available from: https://karger.com/oph/article-abstract/227/Suppl.%201/21/255407/Diabetic-Macular-Edema?redirectedFrom=fulltext
85. Calles JA, Bermúdez J, Vallés E, Allemandi D, Palma S. Polymers in ophthalmology. In: Puoci F, editor. Advanced polymers in medicine [Internet]. Cham: Springer; 2015 [cited 2024 Oct 30]. pp 147–76. Available from: https://doi.org/10.1007/978-3-319-12478-0_6.
86. CRSToday [Internet]. [cited 2024 Oct 30]. Drop the Drops. Available from: https://crstoday.com/articles/2019-june/drop-the-drops
87. Durezol® (Difluprednate Ophthalmic Emulsion 0.05%) Compared with Pred Forte® 1% Ophthalmic Suspension in the Treatment of Endogenous Anterior Uveitis | Journal of Ocular Pharmacology and Therapeutics [Internet]. [cited 2024 Oct 30]. Available from: https://www.liebertpub.com/doi/10.1089/jop.2010.0059
88. Le Bourlais C, Acar L, Zia H, Sado PA, Needham T, Leverge R. Ophthalmic drug delivery systems—recent advances. Prog Retin Eye Res. 1998;17(1):33–58.
89. Ocular Drug Delivery Barriers—Role of Nanocarriers in the Treatment of Anterior Segment Ocular Diseases [Internet]. [cited 2024 Oct 30]. Available from: https://www.mdpi.com/1999-4923/10/1/28
90. Ophthalmic emulsions and suspensions: Cutaneous and Ocular Toxicology: Vol 20, No 2–3 – Get Access [Internet]. [cited 2024 Oct 30]. Available from: https://www.tandfonline.com/doi/full/10.1081/CUS-120001857
91. Roland I, Piel G, Delattre L, Evrard B. Systematic characterization of oil-in-water emulsions for formulation design. Int J Pharm. 2003;263(1):85–94.
92. Topical Drug Delivery to the Posterior Segment of the Eye: Addressing the Challenge of Preclinical to Clinical Translation | Pharmaceutical Research [Internet]. [cited 2024 Oct 30]. Available from: https://link.springer.com/article/10.1007/s11095-018-2519-x
93. Recent Advances in Ophthalmic Drug Delivery: Therapeutic Delivery: Vol 1, No 3 – Get Access [Internet]. [cited 2024 Oct 30]. Available from: https://www.tandfonline.com/doi/full/10.4155/TDE.10.40
94. Ophthalmic Drug Dosage Forms: Characterisation and Research Methods – Baranowski – 2014 – The Scientific World Journal – Wiley Online Library [Internet]. [cited 2024 Oct 30]. Available from: https://onlinelibrary.wiley.com/doi/10.1155/2014/861904
95. Regulating Generic Ophthalmologic Drug Bioequivalence—Envisioning Accessibility for Patients | Journal of Ocular Pharmacology and Therapeutics [Internet]. [cited 2024 Oct 30]. Available from: https://www.liebertpub.com/doi/abs/10.1089/jop.2020.0041
96. Gupta R, Bollyky TJ, Cohen M, Ross JS, Kesselheim AS. Affordability and availability of off-patent drugs in the United States—the case for importing from abroad: observational study. BMJ. 2018;19(360):k831.

97. Raw AS, Lionberger R, Yu LX. Pharmaceutical equivalence by design for generic drugs: modified-release products. Pharm Res. 2011;28(7):1445–53.
98. Generic drug approval: a US perspective: current medical research and opinion: Vol 27, No 3 – Get Access [Internet]. [cited 2024 Sept 30]. Available from: https://www.tandfonline.com/doi/full/10.1185/03007995.2010.548374
99. Bioavailability and bioequivalence studies submitted in NDAs or INDs — general considerations. 2014.
100. Aspects for developing and processing solid forms – Gruss – 2021 – Methods and principles in medicinal chemistry – Wiley Online Library [Internet]. [cited 2024 Oct 30]. Available from: https://onlinelibrary.wiley.com/doi/10.1002/9783527823048.ch1
101. Takechi-Haraya Y, Ohgita T, Demizu Y, Saito H, Izutsu K, Sakai-Kato K. Current status and challenges of analytical methods for evaluation of size and surface modification of nanoparticle-based drug formulations. AAPS PharmSciTech. 2022;23(5):150.
102. Gulsun T, Borna SE, Vural I, Sahin S. Preparation and characterization of furosemide nanosuspensions. J Drug Deliv Sci Technol. 2018;45:93–100.
103. Wani RJ, Sharma P, Zhong HA, Chauhan H. Preparation and characterization of griseofulvin solid dispersions. ASSAY Drug Dev Technol. 2020;18(3):109–18.
104. Bharate SS, Bharate SB, Bajaj AN. Interactions and incompatibilities of pharmaceutical excipients with active pharmaceutical ingredients: a comprehensive review. Int J Pharm Excip [Internet]. 2016 Nov 23 [cited 2024 Oct 30];1(3). Available from: https://jefc.scholasticahq.com/article/1131-interactions-and-incompatibilities-of-pharmaceutical-excipients-with-active-pharmaceutical-ingredients-a-comprehensive-review
105. Xiaohui_(Jeff)_Jiang_GRxBiosims2019.pdf [Internet]. [cited 2024 Oct 30]. Available from: https://accessiblemeds.org/sites/default/files/2019-11/Xiaohui_%28Jeff%29_Jiang_GRxBiosims2019.pdf
106. Chambers WA. Ophthalmic Generics—Are They Really the Same? Ophthalmology 2012 119(6):1095–1096.
107. Uddin MS, Mamun AA, Kabir MT, Setu JR, Zaman S, Begum Y, et al. Quality control tests for ophthalmic pharmaceuticals: pharmacopoeial standards and specifications. J Adv Med Pharm Sci. 2017;14:1–17.
108. Shapiro LJ, McCormick WC, Angorn RA. Implications of drug product appearance on generic drug acceptance. Clin Res Pract Drug Regul Aff. 1986;4(1):43–58.
109. Toropainen E, Fraser-Miller SJ, Novakovic D, Del Amo EM, Vellonen KS, Ruponen M, et al. Biopharmaceutics of topical ophthalmic suspensions: importance of viscosity and particle size in ocular absorption of indomethacin. Pharmaceutics. 2021;13(4):452.
110. Fathi Azarbayjani A, Jouyban A, Chan SY. Impact of surface tension in pharmaceutical sciences. J Pharm Pharm Sci Publ Can Soc Pharm Sci Soc Can Sci Pharm. 2009;12(2):218–28.
111. Yasueda S, Inada K, Matsuhisa K, Terayama H, Ohtori A. Evaluation of ophthalmic suspensions using surface tension. Eur J Pharm Biopharm. 2004;(2):377–82.
112. Han K, Woghiren OE, Priefer R. Surface tension examination of various liquid oral, nasal, and ophthalmic dosage forms. Chem Cent J. 2016;10(1):31.
113. Bao Q, Burgess DJ. Perspectives on physicochemical and in-vitro profiling of ophthalmic ointments. Pharm Res. 2018;35(12):234.
114. Liu M. In-vitro Population Bioequivalence (PBE) Parameters for Particle Size Distribution (PSD).
115. Wu Y, Wang C, Wang X, Mou Y, Yuan K, Huang X, et al. Advances in dry eye disease examination techniques. Front Med [Internet]. 2022 Jan 25 [cited 2024 Oct 30];8. Available from: https://www.frontiersin.org/journals/medicine/articles/10.3389/fmed.2021.826530/full
116. Salih T, El-Mahdi I. The physical, chemical, and microbiological stability of chloramphenicol ophthalmic solution. Libyan Int Med Univ J. 2018;3:42.
117. Mazet R, Yaméogo JBG, Wouessidjewe D, Choisnard L, Gèze A. Recent advances in the design of topical ophthalmic delivery systems in the treatment of ocular surface inflammation and their biopharmaceutical evaluation. Pharmaceutics. 2020;12(6):570.

118. Hudson-Curtis B, Novick S. Assessing content uniformity. In: Zhang L, editor. Nonclinical statistics for pharmaceutical and biotechnology industries [Internet]. Cham: Springer; 2016 [cited 2024 Dec 16]. p. 631–651. Available from: https://doi.org/10.1007/978-3-319-23558-5_24
119. Research C for DE and. ANDAs: Impurities in Drug Products [Internet]. FDA; 2020 [cited 2024 Oct 30]. Available from: https://www.fda.gov/regulatory-information/search-fda-guidance-documents/andas-impurities-drug-products
120. Marley A, Stalcup AM, Connolly D. Development and validation of a new stability indicating reversed phase liquid chromatographic method for the determination of prednisolone acetate and impurities in an ophthalmic suspension. J Pharm Biomed Anal. 2015;102:261–6.
121. Liquid ophthalmic drug products: physicochemical properties, formulations, and manufacturing considerations | SpringerLink [Internet]. [cited 2024 Sept 30]. Available from: https://link.springer.com/chapter/10.1007/978-3-030-76367-1_11
122. <788> Particulate matter in injections.
123. Nasal Spray and Inhalation Solution, Suspension, and spray Drug Products – Chemistry, Manufacturing, and Controls Documentation.
124. usp31nf26s1_c601, General Chapters: <601> AEROSOLS, NASAL SPRAYS, METERED-DOSE INHALERS, AND DRY POWDER INHALERS [Internet]. [cited 2024 Sept 30]. Available from: http://www.uspbpep.com/usp31/v31261/usp31nf26s1_c601.asp
125. Baxter S, Myatt B, Stein S, Parkinson A, Chambers F, Doub B, et al. Spray Pattern and Plume Geometry Testing and Methodology: An IPAC-RS Working Group Overview. AAPS PharmSciTech. 2022 May 18;23(5):145.
126. Microbiological Considerations for Ophthalmic Products: Sterility, Endotoxin Limits, and Preservatives | SpringerLink [Internet]. [cited 2024 Sept 30]. Available from: https://link.springer.com/chapter/10.1007/978-3-030-76367-1_9
127. Approaches to the Evaluation of Extractables and Leachables in Tobacco Product Application Review.
128. ⟨1663⟩ Assessment of Extractables Associated with Pharmaceutical Packaging/Delivery Systems [Internet]. [cited 2024 Sept 30]. Available from: https://doi.usp.org/USPNF/USPNF_M7126_03_01.html
129. ANDAs: Stability Testing of Drug Substances and Products.
130. ANDAs: Stability Testing of Drug Substances and Products: Questions and Answers.
131. Guidance for Industry #5 – Drug Stability Guidelines.
132. Session III: Novel In-vitro Release Testing for Complex Formulations.
133. Bao Q, Newman B, Wang Y, Choi S, Burgess DJ. In-vitro and ex vivo correlation of drug release from ophthalmic ointments. J Control Release Off J Control Release Soc. 2018;4(276):93.
134. Impact of membranes on in-vitro release assessment: a case study using dexamethasone | AAPS PharmSciTech [Internet]. [cited 2024 Sept 30]. Available from: https://link.springer.com/article/10.1208/s12249-020-01874-y
135. Impact of apparatus and adapter on in-vitro drug release of ophthalmic semisolid drug products | Pharmaceutical Research [Internet]. [cited 2024 Sept 30]. Available from: https://link.springer.com/article/10.1007/s11095-023-03586-x
136. Nir I, Lu X. In situ UV fiber optics for dissolution testing—what, why, and where we are after 30 years. Dissolution Technol. 2018;25(3):70–7.
137. Vo A, Feng X, Patel D, Mohammad A, Patel M, Zheng J, et al. In-vitro physicochemical characterization and dissolution of brinzolamide ophthalmic suspensions with similar composition. Int J Pharm. 2020;588:119761.
138. Bellantone RA, Shah KB, Patel PG, Kaplan M, Xu X, Li V, et al. Cyclosporine release and distribution in ophthalmic emulsions determined by pulsatile microdialysis. Int J Pharm. 2022;615:121521.
139. Patel D, Zhang Y, Dong Y, Qu H, Kozak D, Ashraf M, et al. Adaptive perfusion: an in-vitro release test (IVRT) for complex drug products. J Controlled Release. 2021;333:65–75
140. Dash AK, Haney PW, Garavalia MJ. Development of an in-vitro dissolution method using microdialysis sampling technique for implantable drug delivery systems. J Pharm Sci. 1999;88(10):1036–40.

141. Influence of the test method on in-vitro drug release from intravitreal model implants containing dexamethasone or fluorescein sodium in poly (d,l-lactide-co-glycolide) or polycaprolactone – PubMed [Internet]. [cited 2024 Sept 30]. Available from: https://pubmed.ncbi.nlm.nih.gov/29490233/
142. In-situ forming PLGA implants for intraocular dexamethasone delivery – PubMed [Internet]. [cited 2024 Sept 30]. Available from: https://pubmed.ncbi.nlm.nih.gov/29981408/
143. Evaluation of protein drug stability with vitreous humor in a novel ex-vivo intraocular model – PubMed [Internet]. [cited 2024 Sept 30]. Available from: https://pubmed.ncbi.nlm.nih.gov/26032291/
144. Allababidi S, Shah JC. Kinetics and mechanism of release from glyceryl monostearate-based implants: evaluation of release in a gel simulating in-vivo implantation. J Pharm Sci. 1998;87(6):738–44.
145. Chennamaneni SR, Mamalis C, Archer B, Oakey Z, Ambati BK. Development of a novel bioerodible dexamethasone implant for uveitis and postoperative cataract inflammation. J Control Release Off J Control Release Soc. 2013;167(1):53–9.
146. A scalable controlled-release device for transscleral drug delivery to the retina – PubMed [Internet]. [cited 2024 Sept 30]. Available from: https://pubmed.ncbi.nlm.nih.gov/21112628/
147. Awwad S, Lockwood A, Brocchini S, Khaw PT. The PK-eye: a novel in-vitro ocular flow model for use in preclinical drug development. J Pharm Sci. 2015;104(10):3330–42.
148. Artificial vitreous replacements – PubMed [Internet]. [cited 2024 Sept 30]. Available from: https://pubmed.ncbi.nlm.nih.gov/12652023/
149. Loch C, Nagel S, Guthoff R, Seidlitz A, Weitschies W. The vitreous model – a new in-vitro test method simulating the vitreous body. Biomed Tech (Berl). 2012 Sept 6;57 Suppl 1:/j/bmte.2012.57.issue-s1-O/bmt-2012-4106/bmt-2012-4106.xml.
150. Simulation of drug distribution in the vitreous body after local drug application into intact vitreous body and in progress of posterior vitreous detachment – PubMed [Internet]. [cited 2024 Sept 30]. Available from: https://pubmed.ncbi.nlm.nih.gov/24311438/
151. Dorzolamide hydrochloride. 2017;
152. Choi SH, Lionberger RA. Clinical, Pharmacokinetic, and In-vitro Studies to Support Bioequivalence of Ophthalmic Drug Products. AAPS J. 2016;18(4):1032–8.
153. Nepafenac_ophthalmic suspension 0.1_RLD 021862_RV12–16.pdf [Internet]. [cited 2024 Sept 30]. Available from: https://www.accessdata.fda.gov/drugsatfda_docs/psg/Nepafenac_ophthalmic%20suspension%200.1_RLD%20021862_RV12-16.pdf
154. Brott NR, Zeppieri M, Ronquillo Y. Schirmer Test. In: StatPearls [Internet]. Treasure Island (FL): StatPearls Publishing; 2024 [cited 2024 Oct 30]. Available from: http://www.ncbi.nlm.nih.gov/books/NBK559159/
155. Tear meniscus dimensions in tear dysfunction and their correlation with clinical parameters – PubMed [Internet]. [cited 2024 Sept 30]. Available from: https://pubmed.ncbi.nlm.nih.gov/24315297/
156. Tear fluid collection methods: Review of current techniques – PubMed [Internet]. [cited 2024 Sept 30]. Available from: https://pubmed.ncbi.nlm.nih.gov/33631970/
157. Boldin I, Klein A, Haller-Schober EM, Horwath-Winter J. Long-term follow-up of punctal and proximal canalicular stenoses after silicone punctal plug treatment in dry eye patients. Am J Ophthalmol. 2008;146(6):968–972.e1.
158. Yen MT, Pflugfelder SC, Feuer WJ. The effect of punctal occlusion on tear production, tear clearance, and ocular surface sensation in normal subjects. Am J Ophthalmol. 2001;131(3):314–23.
159. Monitoring ocular drug therapy by analysis of aqueous samples – PubMed [Internet]. [cited 2024 Sept 30]. Available from: https://pubmed.ncbi.nlm.nih.gov/19700195/
160. Aqueous humor and plasma concentrations of ciprofloxacin and moxifloxacin following topical ocular administration in ophthalmologically normal horses. American Journal of Veterinary Research. 71(5)nternet]. [cited 2024 Sept 30]. Available from: https://avmajournals.avma.org/view/journals/ajvr/71/5/ajvr.71.5.564.xml

161. Khokhlov AL, Yaichkov II, Korsakov MK, Shetnev AA, Volkhin NN, Petukhov SS. Development of quantification methods of a new selective carbonic anhydrase II inhibitor in plasma and blood and study of the pharmacokinetics of its ophthalmic suspension in rats. Res Results Pharmacol. 2023;9(4):53–64.
162. Dexamethasone Tobramycin_ophthalmic suspension NDA 50818 RV Feb 2019.pdf [Internet]. [cited 2024 Sept 30]. Available from: https://www.accessdata.fda.gov/drugsatfda_docs/psg/Dexamethasone%20Tobramycin_ophthalmic%20suspension%20NDA%2050818%20RV%20Feb%202019.pdf
163. PSG_050804.pdf [Internet]. [cited 2024 Sept 30]. Available from: https://www.accessdata.fda.gov/drugsatfda_docs/psg/PSG_050804.pdf
164. Zhao L, Seo P, Lionberger R. Current scientific considerations to verify physiologically-based pharmacokinetic models and their implications for locally acting products. CPT Pharmacomet Syst Pharmacol. 2019;8(6):347.
165. Tan ML, Chandran S, Jereb R, Alam K, Bies R, Kozak D, et al. Mechanistic modeling of ophthalmic, nasal, injectable, and implant generic drug products: a workshop summary report. CPT Pharmacomet Syst Pharmacol. 2023;12(5):631–8.
166. Le Merdy M, Spires J, Tan ML, Zhao L, Lukacova V. Clinical ocular exposure extrapolation for a complex ophthalmic suspension using physiologically based pharmacokinetic modeling and simulation. Pharmaceutics. 2024;16(7):914.
167. (PDF) Ocular Physiologically Based Pharmacokinetic Modeling for Ointment Formulations [Internet]. [cited 2024 Oct 31]. Available from: https://www.researchgate.net/publication/346535230_Ocular_Physiologically_Based_Pharmacokinetic_Modeling_for_Ointment_Formulations
168. Wang W, Ouyang D. Opportunities and challenges of physiologically based pharmacokinetic modeling in drug delivery. Drug Discov Today. 2022;27(8):2100–20.
169. Nelson E. Solution rate of theophylline salts and effects from oral administration. J Am Pharm Assoc Am Pharm Assoc. 1957;46(10):607–14.
170. Chilukuri DM, Sunkara G, Young D, editors. Pharmaceutical product development: in-vitro-in-vivo correlation. New York: Informa Healthcare; 2007. 204 p. (Drugs and the pharmaceutical sciences).
171. Design and evaluation of polymeric ocular drug delivery system – PubMed [Internet]. [cited 2024 Sept 30]. Available from: https://pubmed.ncbi.nlm.nih.gov/19721251/
172. Shen J, Burgess DJ. In-vitro–in-vivo correlation for complex non-oral drug products: where do we stand? J Controlled Release. 2015;10(219):644–51.
173. Stewart SA, Domínguez-Robles J, Donnelly RF, Larrañeta E. Implantable polymeric drug delivery devices: classification, manufacture, materials, and clinical applications. Polymers. 2018;10(12):1379.
174. In-vitro and in-vivo correlation of buserelin release from biodegradable implants using statistical moment analysis – ScienceDirect [Internet]. [cited 2024 Sept 30]. Available from: https://www.sciencedirect.com/science/article/pii/S0168365903004346?via%3Dihub
175. A comprehensive insight on ocular pharmacokinetics | Drug Delivery and Translational Research [Internet]. [cited 2024 Sept 30]. Available from: https://link.springer.com/article/10.1007/s13346-016-0339-2
176. Elevated temperature accelerated release testing of PLGA microspheres – ScienceDirect [Internet]. [cited 2024 Sept 30]. Available from: https://www.sciencedirect.com/science/article/pii/S0168365906000848?via%3Dihub
177. Accelerated in-vitro release testing methods for extended-release parenteral dosage forms | Journal of Pharmacy and Pharmacology | Oxford Academic [Internet]. [cited 2024 Sept 30]. Available from: https://academic.oup.com/jpp/article-abstract/64/7/986/6135307?redirectedFrom=fulltext&login=false
178. Use of Pd Biomarkers to Drive Dose Selection And Early Clinical Decision Making: Bioanalysis: Vol 4, No 20 – Get Access [Internet]. [cited 2024 Sept 30]. Available from: https://www.tandfonline.com/doi/full/10.4155/bio.12.224

179. Drug Excipients | Bentham Science [Internet]. [cited 2024 Oct 31]. Available from: https://www.eurekaselect.com/article/2681
180. Yoon S, Regn M. Packaging Development: Multi-Dose Container Closure for Preservative Free Products, Extractable/Leachables from Packaging, New Technologies. In: Neervannan S, Kompella UB, editors. Ophthalmic Product Development: From Bench to Bedside [Internet]. Cham: Springer; 2021 [cited 2024 Oct 31]. p. 229–45. Available from: https://doi.org/10.1007/978-3-030-76367-1_10
181. Adeleke OA, Hayeshi RK, Davids H. Development and evaluation of a reconstitutable dry suspension containing isoniazid for flexible pediatric dosing. Pharmaceutics. 2020;12(3):286.
182. Pargaonkar N, Lvov YM, Li N, Steenekamp JH, de Villiers MM. Controlled release of dexamethasone from microcapsules produced by polyelectrolyte layer-by-layer nanoassembly. Pharm Res. 2005;22(5):826–35.
183. Deicke A, Süverkrüp R. Dose uniformity and redispersibility of pharmaceutical suspensions 2: assessment of three commercial erythromycin ethyl succinate oral liquids. Eur J Pharm Biopharm. 2000;49(1):73–8.
184. Alarie H, Roullin VG, Leclair G. Development of a safe and versatile suspension vehicle for pediatric use: Formulation development. Int J Pharm. 2019;569:118552.
185. Impact of formulation parameters on in-vitro release from long-acting injectable suspensions | AAPS J [Internet]. [cited 2024 Sept 30]. Available from: https://link.springer.com/article/10.1208/s12248-021-00566-0
186. Bibette J, Calderon FL, Poulin P. Emulsions: basic principles. Rep Prog Phys. 1999;62(6):969.
187. Ophthalmic emulsions and suspensions. J Toxicol Cutan Ocul Toxicol 20(2–3) Get Access [Internet]. [cited 2024 Oct 31]. Available from: https://www.tandfonline.com/doi/full/10.1081/CUS-120001857
188. High-pressure homogenization as a process for emulsion formation – Schultz – 2004 – Chemical Engineering & Technology – Wiley Online Library [Internet]. [cited 2024 Oct 31]. Available from: https://onlinelibrary.wiley.com/doi/10.1002/ceat.200406111
189. Taha A, Ahmed E, Ismaiel A, Ashokkumar M, Xu X, Pan S, et al. Ultrasonic emulsification: an overview on the preparation of different emulsifiers-stabilized emulsions. Trends Food Sci Technol. 2020;105:363–77.
190. Dynamic light scattering (DLS) – ScienceDirect [Internet]. [cited 2024 Oct 31]. Available from: https://www.sciencedirect.com/science/article/abs/pii/B9780128141823000109?via%3Dihub
191. Limage S, Schmitt M, Vincent-Bonnieu S, Dominici C, Antoni M. Characterization of solid-stabilized water/oil emulsions by scanning electron microscopy. Colloids Surf Physicochem Eng Asp. 2010;365(1):154–61.
192. Spernath L, Regev O, Levi-Kalisman Y, Magdassi S. Phase transitions in O/W lauryl acrylate emulsions during phase inversion, studied by light microscopy and cryo-TEM. Colloids Surf Physicochem Eng Asp. 2009;332(1):19–25.
193. Gillies G, Prestidge CA. Interaction forces, deformation and nano-rheology of emulsion droplets as determined by colloid probe AFM. Adv Colloid Interface Sci. 2004;108–109:197–205.
194. Driscoll DF, Etzler F, Barber TA, Nehne J, Niemann W, Bistrian BR. Physicochemical assessments of parenteral lipid emulsions: light obscuration versus laser diffraction. Int J Pharm. 2001;219(1):21–37.
195. Chan MY, Dowling QM, Sivananthan SJ, Kramer RM. Particle sizing of nanoparticle adjuvant formulations by Dynamic Light Scattering (DLS) and Nanoparticle Tracking Analysis (NTA). In: Fox CB, editor. Vaccine adjuvants: methods and protocols [Internet]. New York: Springer; 2017 [cited 2024 Oct 31]. pp 239–52. Available from: https://doi.org/10.1007/978-1-4939-6445-1_17
196. Equivalence testing of complex particle size distribution profiles based on earth mover's distance | AAPS J [Internet]. [cited 2024 Oct 31]. Available from: https://link.springer.com/article/10.1208/s12248-018-0212-y
197. Chen CM, Lu CH, Chang CH, Yang YM, Maa JR. Influence of pH on the stability of oil-in-water emulsions stabilized by a splittable surfactant. Colloids Surf Physicochem Eng Asp. 2000;170(2):173–9.

198. Ravera F, Dziza K, Santini E, Cristofolini L, Liggieri L. Emulsification and emulsion stability: The role of the interfacial properties. Adv Colloid Interface Sci. 2021;288:102344.
199. Robins MM. Emulsions — creaming phenomena. Curr Opin Colloid Interface Sci. 2000;5(5):265–72.
200. Cyclosporine_ophthalmic emulsion_RLD 050790_RV09–16.pdf [Internet]. [cited 2024 Oct 31]. Available from: https://www.accessdata.fda.gov/drugsatfda_docs/psg/Cyclosporine_ophthalmic%20emulsion_RLD%20050790_RV09-16.pdf
201. Multiphase drug distribution and exchange in oil-in-water nanoemulsion revealed by high-resolution 19F qNMR | Molecular Pharmaceutics [Internet]. [cited 2024 Oct 31]. Available from: https://pubs.acs.org/doi/abs/10.1021/acs.molpharmaceut.2c00025
202. Scherer GW. Structure and properties of gels. Cem Concr Res. 1999;29(8):1149–57.
203. Shah H, Jain A, Laghate G, Prabhudesai D. Chapter 32 – Pharmaceutical excipients. In: Adejare A, editor. Remington (Twenty-third Edition) [Internet]. Academic; 2021 [cited 2024 Sept 30]. p. 633–43. Available from: https://www.sciencedirect.com/science/article/pii/B9780128200070000325
204. Generic Development of Topical Dermatologic Products: Formulation Development, Process Development, and Testing of Topical Dermatologic Products | AAPS J [Internet]. [cited 2024 Oct 31]. Available from: https://link.springer.com/article/10.1208/s12248-012-9411-0
205. Generic Development of Topical Dermatologic Products: Formulation Development, Process Development, and Testing of Topical Dermatologic Products | The AAPS Journal [Internet]. [cited 2024 Oct 31]. Available from: https://link.springer.com/article/10.1208/s12248-012-9411-0
206. Robin JS, Ellis PP. Ophthalmic ointments. Surv Ophthalmol. 1978;22(5):335–40.
207. van Heugten AJP, de Vries WS, Markesteijn MMA, Pieters RJ, Vromans H. The role of excipients in the stability of triamcinolone acetonide in ointments. AAPS PharmSciTech. 2018;19(3):1448–53.
208. Influence of Manufacturing Process Variables on the Properties of Ophthalmic Ointments of Tobramycin | Pharmaceutical Research [Internet]. [cited 2024 Oct 31]. Available from: https://link.springer.com/article/10.1007/s11095-018-2462-x
209. Formulation and process factors influencing product quality and in-vitro performance of ophthalmic ointments – PubMed [Internet]. [cited 2024 Oct 31]. Available from: https://pubmed.ncbi.nlm.nih.gov/26231106/
210. Physicochemical attributes and dissolution testing of ophthalmic ointments – PubMed [Internet]. [cited 2024 Oct 31]. Available from: https://pubmed.ncbi.nlm.nih.gov/28344172/
211. Rational design of topical semi-solid dosage forms-how far are we? [Internet]. [cited 2024 Oct 31]. Available from: https://www.mdpi.com/1999-4923/15/7/1822
212. Ophthalmic preparations – ScienceDirect [Internet]. [cited 2024 Oct 31]. Available from: https://www.sciencedirect.com/science/article/abs/pii/B9780128200070000283
213. Jervis L. A summary of recent advances in ocular inserts and implants. J Bioequivalence Bioavailab. 2016 Jan 1:09.
214. Winter 2004 – Volume 44 – Issue 1: International ophthalmology clinics [Internet]. [cited 2024 Sept 30]. Available from: https://journals.lww.com/internat-ophthalmology/toc/2004/04410
215. Lambiase A, Abdolrahimzadeh S, Solmaz SM. An update on intravitreal implants in use for eye disorders. Drugs Today. 2014;50(3):239.
216. Short BG. Safety evaluation of ocular drug delivery formulations: techniques and practical considerations. Toxicol Pathol. 2008;36(1):49–62.
217. Mittal S, Miranda O. Recent Advancemnts in Biodegradable Ocular Implants. http://www.eurekaselect.com [Internet]. [cited 2024 Sept 30.]; Available from: https://www.eurekaselect.com/article/83328
218. Shen J, Lu GW, Hughes P. Targeted ocular drug delivery with pharmacokinetic/pharmacodynamic considerations. Pharm Res. 2018;35(11):217.
219. Costello MA, Liu J, Wang Y, Qin B, Xu X, Li Q, et al. Reverse engineering the Ozurdex dexamethasone intravitreal implant. Int J Pharm. 2023;634:122625.

220. Gote V, Sikder S, Sicotte J, Pal D. Ocular drug delivery: present innovations and future challenges. J Pharmacol Exp Ther. 2019;370(3):602–24.
221. Allyn MM, Luo RH, Hellwarth EB, Swindle-Reilly KE. Considerations for polymers used in ocular drug delivery. Front Med. 2021;8:787644.
222. P130004b.pdf [Internet]. [cited 2024 Sept 30]. Available from: https://www.accessdata.fda.gov/cdrh_docs/pdf13/P130004b.pdf
223. Shatz W, Hass PE, Peer N, Paluch MT, Blanchette C, Han G, et al. Identification and characterization of an octameric PEG-protein conjugate system for intravitreal long-acting delivery to the back of the eye. PloS One. 2019;14(6):e0218613.
224. Teodorescu M, Bercea M, Morariu S. Biomaterials of PVA and PVP in medical and pharmaceutical applications: Perspectives and challenges. Biotechnol Adv. 2019;37(1):109–31.
225. Activity and mechanism of antimicrobial peptide-mimetic amphiphilic polymethacrylate derivatives [Internet]. [cited 2024 Sept 30]. Available from: https://www.mdpi.com/2073-4360/3/3/1512
226. Full article: a review of the properties and applications of poly (methyl methacrylate) (PMMA) [Internet]. [cited 2024 Sep 30]. Available from: https://www.tandfonline.com/doi/full/10.1080/15583724.2015.1031377
227. Fan X, Torres-Luna C, Azadi M, Domszy R, Hu N, Yang A, et al. Evaluation of commercial soft contact lenses for ocular drug delivery: a review. Acta Biomater. 2020;115:60–74.
228. Makadia HK, Siegel SJ. Poly lactic-co-glycolic acid (PLGA) as biodegradable controlled drug delivery carrier. Polymers. 2011;3(3):1377–97.
229. Physical and mechanical properties of PLA, and their functions in widespread applications — a comprehensive review – ScienceDirect [Internet]. [cited 2024 Sept 30]. Available from: https://www.sciencedirect.com/science/article/pii/S0169409X16302058
230. Regulation of the ocular cell/tissue response by implantable biomaterials and drug delivery systems [Internet]. [cited 2024 Sept 30]. Available from: https://www.mdpi.com/2306-5354/7/3/65
231. Cao Y, Samy KE, Bernards DA, Desai TA. Recent advances in intraocular sustained-release drug delivery devices. Drug Discov Today. 2019;24(8):1694–700.
232. Lavecchia A. Deep learning in drug discovery: opportunities, challenges and future prospects. Drug Discov Today. 2019;24(10):2017–32.
233. Bimatoprost implant: first approval | drugs & aging [Internet]. [cited 2024 Sept 30]. Available from: https://link.springer.com/article/10.1007/s40266-020-00769-8
234. Garner J, Skidmore S, Park H, Park K, Choi S, Wang Y. Beyond Q1/Q2: the impact of manufacturing conditions and test methods on drug release from PLGA-based microparticle depot formulations. J Pharm Sci. 2018;107(1):353–61.
235. Statement from FDA Commissioner Scott Gottlieb, M.D., on new agency actions to further deter 'gaming' of the generic drug approval process by the use of citizen petitions | FDA [Internet]. [cited 2024 Oct 31]. Available from: https://www.fda.gov/news-events/press-announcements/statement-fda-commissioner-scott-gottlieb-md-new-agency-actions-further-deter-gaming-generic-drug
236. Effect of particle size, polydispersity and polymer degradation on progesterone release from PLGA microparticles: Experimental and mathematical modeling – ScienceDirect [Internet]. [cited 2024 Sept 30]. Available from: https://www.sciencedirect.com/science/article/pii/S037851731731133X?via%3Dihub
237. Current approaches for in-vitro drug release study of long acting parenteral formulations | Bentham Science [Internet]. [cited 2024 Sept 30]. Available from: https://www.eurekaselect.com/article/65076
238. Shen J, Choi S, Qu W, Wang Y, Burgess DJ. In-vitro-in-vivo correlation of parenteral risperidone polymeric microspheres. J Controlled Release. 2015;28(218):2–12.
239. Challenges and complications of poly(lactic-co-glycolic acid)-based long-acting drug product development [Internet]. [cited 2024 Sept 30]. Available from: https://www.mdpi.com/1999-4923/14/3/614

Role of Quality by Design (QbD) in Ophthalmic Products

8

Akhila Reddy, Syed Nazrin Ruhina Rahman, Ishika Raju Kamdi, and Arvind Gulbake

Abstract

The drug delivery to the eye undergoes challenges because of its unique anatomical and typical physiological characteristics. The conventional methods which are delivered through oral and systemic routes often lead to significant side effects, so application through the topical route is the most efficient and suitable approach to ocular formulations. Nonetheless, because of the eye's inherent shape, the retention of topical formulations decreases, leading to more frequent dosing, which can lead to undesirable side effects. To enhance the capacity for drug retention, a novel delivery system presents a promising strategy. However, the excipients and solvents could exhibit toxic effects. So, it is important to construct the safety profile in the initial steps by having a thorough understanding of the regulatory aspects and technical challenges. Various reports showed that the nanoformulation development by the quality by design (QbD) approach displayed a higher success rate by reducing the significant loss. It is a risk and knowledge-based systematic tool that ensures an understanding of the material and the process variables of final formulation. The key elements and design tools help in pre-defining and optimizing the final lead formulations with limited experimental trials. This approach focuses on improving the process quality which influences the manufacturing efficiency of the ophthalmic products.

A. Reddy · I. R. Kamdi · A. Gulbake (✉)
Department of Pharmaceutics, National Institute of Pharmaceutical Education and Research, Guwahati, Assam, India
e-mail: arvind@niperguwahati.in

S. N. R. Rahman
Department of Pharmaceutics, Royal School of Pharmacy, The Assam Royal Global University, Guwahati, Assam, India

N. K. Mehra (ed.), *Complex Ophthalmic Dosage Forms: Advances in Biomedical Applications and Future Perspectives*, https://doi.org/10.1007/978-981-96-6306-4_8

Keywords

Quality by design · Design of experiment · Key elements in QbD · Regulatory aspects

Abbreviations

%DL	Drug loading
%EE	Entrapment efficiency
ANDA	Abbreviated new drug applications
ANOVA	Analysis of variance
BBD	Box Behnken design
CCD	Central composite design
CMA	Critical materials attributes
CPP	Critical process parameters
CQA	Critical quality attributes
DoE	Design of experiment
EMA	European medical agency
FbD	Formulation by design
FD	Full factorial design
FFD	Fractional factorial design
FMEA	Failure mode and effect analysis
FTA	Fault tree analysis
ICH	International Council for Harmonisation
PAT	Process analytical technology
PBD	Placket Burman design
PD	Particle size
PHA	Preliminary hazard analyses
QbD	Quality-by-design
QbT	Quality-by-testing
QRM	Quality risk management
QTPP	Quality targeted product profile
RAM	Risk assessment matrix
USFDA	United States Food and Drug Administration
ZP	Zeta potential

8.1 Introduction

The eye is a distinctive and intricate organ in the body, due to its physiological and anatomical nature. The structure of ocular tissues complicates drug delivery to the target site(s), and the efficacy of conventional dosage forms, including ointment, suspension and eye drops, is limited due to eye's inherent characteristics that include reflex blinking and tear turnover after administration. In addition, delivering to the posterior part is another challenge due to presence of various barriers like static, dynamic, and metabolic factors. This led to shifting to an alternate method with improved physicochemical properties to ameliorate the challenges associated with therapeutic molecules such as novel biomaterials including nanoparticles, liposomes, the nanosized particles overcome the significant hurdles without causing damage to the ocular area. Also, these biomaterials show an increase in the ocular bioavailability of the drugs by protecting them from degradation. But, excipients and solvents which are used in the preparation may show toxicities, so it is an important to construct the safety profile in the initial step by having a thorough understanding of the regulatory aspects and technical challenges, followed by the optimization of critical and process parameters involved in the formulation preparation [1].

Optimization of the process parameter, critical quality attributes and composition prior to finalization of any ophthalmic product is a crucial step, an incomplete understanding can lead to undesirable results and perplexed to identify the factors that have a significant impact on the final product. Therefore, the critical process and the material impact on formulation have been illustrated using a scientific approach. Literature suggests that the nanoformulation development by employing quality by design (QbD) and design of experimentation (DoE) approach displayed a higher success rate by reducing the significant loss. It is a risk and knowledge-based systematic tool that ensures an understanding of the material and the process variables of final formulation. This does not have any specific regulatory consideration for nanoformulation preparation, but the United States Food and Drug Administration (USFDA) and European Medical Agency (EMA) promote the QbD principles from the initial product development stage. To start with any development, it is crucial to pre-define the quality characteristics followed by quality targeted product profile (QTPP), critical quality attributes (CMA), critical material attributes (CMA), and critical process parameters of the product. Additionally, screening and optimizing design, by selecting a suitable design space is crucial to obtaining the final lead formulation [2].

The ophthalmic formulation showed a continuous impact on both stability and therapeutic efficacy. The QbD approach focused on achieving outstanding quality of the product by improving the processing capacity, which later positively influenced manufacturing efficiency. This chapter highlights the fundamental principles of QbD and outlines the key elements and design of experiment (DoE). This chapter also provides a summary of case studies related to different ocular delivery systems using the DoE approach [1].

8.2 Principle of QbD

The Q8 guidelines of ICH described, quality by design (QbD) as a systematic approach in pharmaceutical product development, that primarily gives an information and understanding of the process and product, in developing the final quality

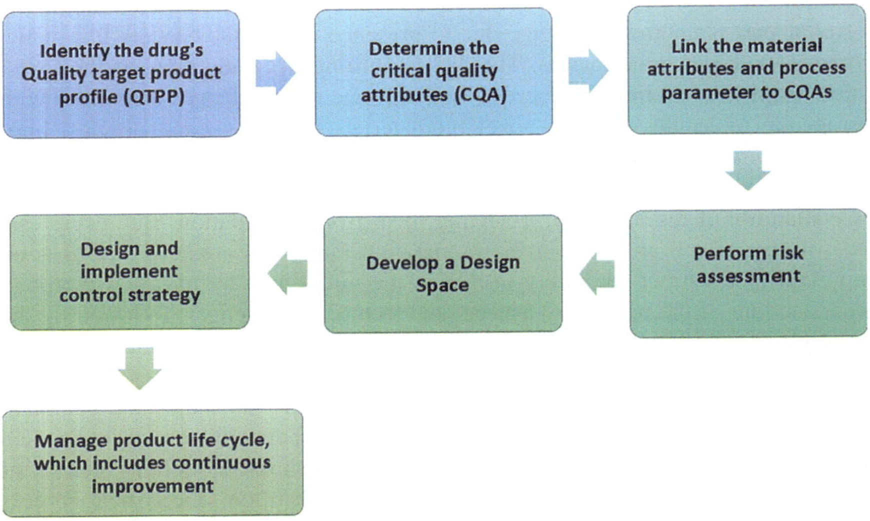

Fig. 8.1 Steps involved in pharmaceutical product development by using QbD approach. (Self-creation)

product [3]. This definition aligns with the FDA's current drug quality system concept, which states that quality cannot be tested into products; it should be built-in, and the steps involved in the product development by QbD approach are illustrated in Fig. 8.1 [2].

8.3 QbD Evolution in the Pharmaceutical Field

QbD is a concept that has been around for a while on a global scale. During the 1970s, an American Engineer, Juran J.M., developed the concept of QbD in his renowned book entitled "Juran on Quality by Design." This book was subsequently embraced by various technology-based sectors, including aerospace, automobile, aviation, and telecommunications, all of which are involved in developing superior-quality services and products for humankind benefits. Woodcock states that a high-quality drug product is both free of contamination and offers the therapeutic benefits promised on the label to the end-users/patients [4]. The concept of QbD was later incorporated by healthcare sectors and particularly implemented by the manufacturers of medical devices in the 1990s. In the twenty-first century, there is an increasing focus on awareness related to the quality of pharmaceutical products. In the early 2000s, the USFDA announced the idea/concept of QbD as a component of its efforts to update pharmaceutical production. The guidelines of the USFDA on QbD gave pharmaceutical companies a structure to follow when creating their products and processes to achieve specific quality goals. Since then, the QbD has transformed into a thorough strategy covering every step of drug development and manufacturing, from designing the product to producing it commercially [5]. In 2005, the USFDA established a culture of quality by publishing a document entitled "Quality Systems Approach to Pharmaceutical CGMP Regulations" for the pharmaceutical industries. This document emphasizes the critical parameters involved in the chemistry and production/manufacturing as well as the safety and efficacy of the pharmaceutical dosage forms. It also demonstrates the scientific basis related to the demonstration and submission of the product, along with an understanding of the process involved in the development of the drug product. The QbD concept allows flexibility in regulatory supervision to improve efficiency in product development before and after submission in the post-approval stages. Moreover, this systemic approach tries to make the development/process easier for scientific innovation and ongoing improvement across the product lifecycle [6].

8.4 Regulatory Aspects of the QbD Concept

The implementation of QbD has spread to various industrial settings, influencing both thinking and operations. Nowadays, globally pharmaceutical companies and regulatory agencies are working together to adopt QbD principles to enhance the quality of pharmaceutical products for the well-being of patients. QbD is a systematic approach that can be used in every aspect of pharmaceutical product development. The federal regulatory agencies/bodies, including EMA, ICH, and USFDA, are consistently promoting the adoption and implementation of QbD principles by pharmaceutical companies to fully utilize its comprehensive benefits in the development of their process and products [7]. The USFDA highlighted that the refinement in the pharmaceutical product quality may not be secure by investigations only; hence, a QbD approach in pharmaceutical product development is highly recommended by USFDA [8]. The regulatory bodies' advice on the implementation of QbD mainly emphasizes the significance of developing a thorough comprehension of the pharmaceutical process and product, recognizing, and managing crucial features of the process development, and implementing risk-based methods to guarantee product quality [9].

In January 2013, the USFDA announced the implementation of QbD into the Abbreviated New Drug Applications (ANDA) submitted for approval of generic drug products in the USA. Even though pharmaceutical companies initially resisted the implementation of QbD, an official person from the USFDA revealed at the recent International Forum Process Analytical Chemistry meeting that there has been a consistent rise in utilization of QbD concept in recent ANDA submissions [10]. Similarly, from February 2018, the EMA also required QbD implementation in the approval of generic drugs in European countries. In addition, numerous regulatory bodies around the world—including CFDA, China; Health Canada; MHRA, the United Kingdom; Health Authority, Chile; MCC, South Africa; SFDA Saudi Arabia; and TGA, Australia—have also supported the ICH guidelines (i.e., Q8, Q9, and Q10) focusing on creating reliable and good quality pharmaceutical products. This has caused a fundamental change in the implementation of QbD methods, eliminating the necessity of quality-by-testing (QbT) at each phase of product development [5]. The QbD-implemented understanding of the process not only streamlines product development but also plays an essential role in the technology transfer. The regulatory bodies need to align the regulatory requirements and understanding throughout their departments. Good communication between the industry and regulatory bodies is necessary to resolve challenges and concerns related to QbD implementation in pharmaceutical companies. Moreover, as quality is embedded in every step of the process development, regulatory bodies feel more at ease when granting drug applications.

8.5 Key Elements of QbD

ICH Q8 Several elements of QbD are addressed in pharmaceutical development that help in establishing the essential framework in product development along with the catalysts. Fig. 8.2 shows a representation of the key elements of QbD [2].

According to the ICH Q8 guideline, the QTPP defines as a "Prospective summary of quality characteristics of a drug product that ideally achieve to ensure the desired quality" [11]. It is a fundamental tool of the QbD and the initial step for design and development of pharmaceutical products and their related manufacturing processes (Table 8.1) [12]. It ought to be presented in tabular form, containing comprehensive details about the dosage form, administration route, delivery systems, container closure system, dosage strength, characteristics influencing pharmacokinetic properties (like dissolution) and standards for product quality (identification, assay, content uniformity, purity, sterility, stability, and drug in vitro release). QTPP delineation should also take regulatory restrictions, patient compliance, and product performance into account [13].

The drug product must have some quality attributes to continuously provide therapeutic advantages which stated on label. QTPP assists formulation scientists in establishing objectives while sustaining an efficient and focused formulation effort. QTPP has been linked with the identity, dosage form, assay, purity, and stability of label. Accordingly, the sterility, pH, viscosity, and drug release profile of an ophthalmic medication would be typical QTPP (Fig. 8.3) [12].

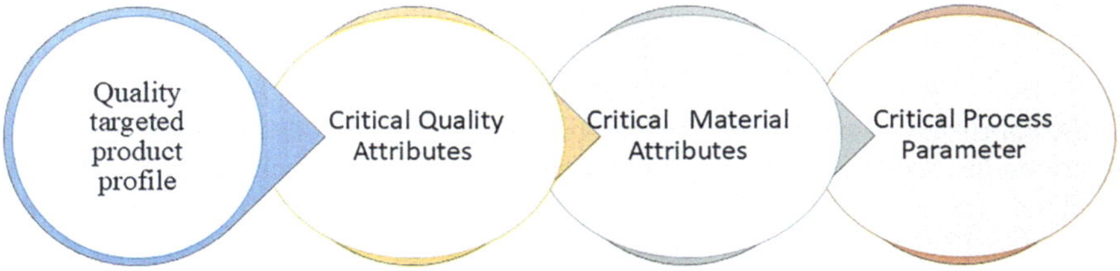

Fig. 8.2 Key elements in QbD. (Self-creation)

Table 8.1 Quality target product profile (QTPP) with their target and justification involved in the development of the ophthalmic product

Target	QTPP	Justification
Sustained release delivery	Dosage form	For improved shelf life, stability, patient compliance, and efficacy with reduced frequency in dosing
Ocular route	Route of administration	Non-invasive, easy to use, less adverse effects with local action
Solutions, gels, and solids	Physical appearance	Appearance of the final formulation and easy-to-administer
Particle size (PS) Zeta potential (ZP) %Entrapment efficiency(%EE) Drug release	Physicochemical characterization	Helps in better absorption and permeation Identifies the product stability Impacts maximum drug loading Impacts pharmacokinetic drug release
The formulation pH suitable to physiological pH	pH	Impacts drug release and permeation
Quality	Assay of drug substances and preservative	Validates the efficacy of the drug and omits the bacterial and microbial growth in the formulation
Stability	Shelf-life and storage condition	Impacts the product quality
The physical stability and clinical performances	Viscosity	Increased viscosity improves contact time between the formulation and ocular tissues, which results in improved BA of the drug
ADME	Pharmacokinetics	Necessitates for attaining the desired efficacy
Primary and secondary packaging materials	Container and closure system	Improves the drugs' shelf life

8.6 Critical Quality Attributes (CQAs)

Identifying appropriate CQAs is subsequent step after determining QTPP. To finalized the desired level of product quality, the physical characteristics (polymorphism, aqueous solubility, and size distribution/morphology), chemical attributes (pKa, oxidative stability, photolytic), and biological/microbiological properties of a material, including final drug productsthat must fall within a optimum range which further ensure the ultimate product quality [14]. Drug products, intermediates (in process materials), and source materials (drug and excipients) are typically linked to CQAs. The qualities of a drug product that are essential to its effectiveness are known as its CQAs, or tailored quality, safety, and efficacy (Table 8.2) [12].

It demonstrates that CQAs are QTPP sections that may be changed by altering formulation or process variables. For example, QTPP may contain extra quality features of drug products not included in CQA since they will not change during the drug development process, including strength and dosage form. However, since they could be modified by formulation or process variables, QTPP features like

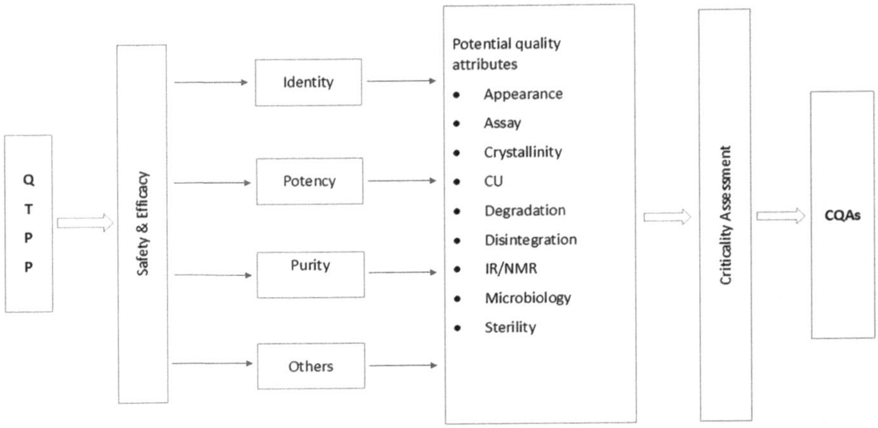

Fig. 8.3 Quality target product profile (QTPP) to critical quality attributes (CQAs) involved in ophthalmic product development. (Self-creation)

Table 8.2 Critical quality attributes (CQAs) involved in ophthalmic product development

Critical quality attributes (CQAs)	Justification
Drug content	The drug content is important for any pharmaceutical product to attain maximum drug concentration in targeted site
Particle size	The particle size should be within the mentioned range for easy penetration through the corneal layers
Size distribution	The PDI assures uniform delivery of drug
% Entrapment efficiency (EE)	The maximum drug entrapment helps in better drug release from the carrier
% Drug release	Prolonged drug release is required at the targeted site for an extended period
Viscosity	The viscosity attributes in increasing the drug bioavailability and permeation for an extended period
pH	A pH of 7.4 helps in better drug release and permeation
Osmolality	The osmolality value of the ophthalmic products should be within the ocular comfort range
Sterile	Product sterility must be considered to avoid ocular infection, which is caused by the contamination of the prepared ophthalmic products

assay, drug content uniformity, drug dissolution, and drug permeation will also be included in CQA [12].

CQAs have been identified utilizing risk assessment by ICH guideline Q9. Making these risk assessments required prior product knowledge, such as the accumulated laboratory, nonclinical, and clinical experience with a particular product-quality attribute [12]. These may incorporate relevant details from related compounds and citations from published works. This data justifies connecting the CQA to the efficacy and safety of products. Frequent characteristics of drug substances include their appearance, particle size, morphological shapes, residual solvents, water content, inorganic impurities, organic impurities, residue on ignition, heavy metals, and assay-based parameters [13].

8.7 Critical Material Attributes (CMAs)

Critical material attributes (CMAs) are the physical, chemical, biological, or microbiological traits or features of a loading material. The CMAs should limit to an appropriate range and distribution for assuring the desired characteristics of a drug substance, excipient, or in-process material [13]. The physicochemical, biopharmaceutical, or microbiological characteristics of the input materials are considered in CMAs; these attributes enhance the end product's quality by ensuring that it remains within the right range. The initial category of variables that can affect CQA variability is CMAs, which have a connection with the therapeutic product's composition. When ophthalmic drug products are being produced, common CMAs are the kind and concentration of the drug, polymer, cosurfactant, and surfactant [14].

8.8 Critical Process Parameter (CPP)

Several process variables known as critical process parameters (CPPs) should be watched to confirm that the procedure provides the intended superiority. These factors' variability affects the product's CQA. They might include process state variables (temperature, homogenization cycle, and homogenization pressure) or input operational parameters (speed and flow rate) for creating an ophthalmic formulation. The multidomain signal design space is defined by the mathematical correlations between the critical material attributes (CMAs) and CPPs with the CQAs that have been demonstrated to ensure safety, quality, and efficacy [15].

Every aspect, such as facilities, machinery, QTPP, manufacturing variables and materials, should be considered while designing an ideal manufacturing process. For example, temperature, mechanical energy input, type of mixer, and mixing/homogenization frequency are the three main factors that affect the production of semisolid formulations. To achieve an identical quality, it is essential to recognize and carefully monitor the process parameters that utilize these related aspects [16].

8.9 Risk Assessment

As a step in the QRM process, risk assessment is critical to the expansion of QbD. When undertaking a risk assessment, the CQA is given priority to monitor the significance of reaching the established goal. Because the CQA is vital to the product's safety and effectiveness, it is given primary importance. The CMAs and CPPs are connected to the CQAs in this process. Risk assessment can be done with a variety of qualitative and quantitative methods. During the risk assessment phase, brainstorming is an important technique followed by road maps, FMEA, fault tree analysis (FTA), fishbone/Ishikawa diagrams, hazard and operability analyses, important control points, and preliminary hazard analyses (PHA) are some other techniques. The Ishikawa fishbone diagram, road maps, and FTA are used for qualitative risk identification; FMEA and PHA are utilized as quantitative tools for risk categorization [17]. Figure 8.4 represents the various crucial elements involved in the development of formulation by using QbD approach.

The methodical development of collecting evidence to facilitate a risk preference made during a risk management process is known as risk assessment. It entails identifying risks and analyzing and assessing the risks that come with being exposed to such risks. It is the first phase in the quality risk management process; risk control and risk review are the other two. Decision-making for risk reduction and/or acceptance is a part of risk control. Reducing risk to a manageable level is the aim of risk control. To include fresh information and experience, the output and outcomes of the risk-managing process should be reassessed at the end. Risk communication, or the exchange of evidence about risk and risk managing among any number of

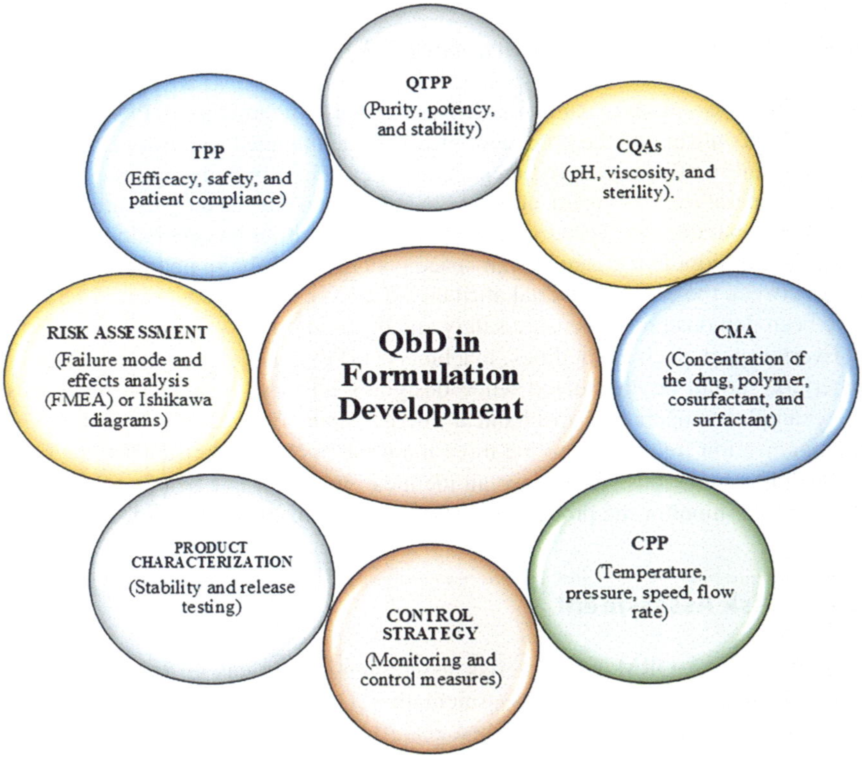

Fig. 8.4 Key elements involved in the formulation development by QbD approach. (Self-creation)

parties (such as industry and patients, regulators and industry, or within a company), should be continuous during the risk management process. The contents may include details about hazards to quality that are present, varied, probabilistic, severe, acceptable, controlled, treated, or have other characteristics [13]. Risk assessment consists of three components: risk identification, risk analysis, and risk evaluation.

Risk Identification This is the systematic use of information to identify sources of harm (hazards) related to the risk question or problem description. This step can include gathering and analyzing historical data, conducting theoretical analysis, incorporating informed opinions, and considering stakeholder concerns. By thoroughly understanding these potential sources of harm, organizations can proactively address issues before they escalate.

Risk Analysis This involves the process of estimating the risk associated with the identified hazards. During risk analysis, the likelihood and potential impact of each hazard are evaluated to understand the overall risk profile. This step is crucial for determining which risks require more immediate attention and resources for mitigation.

Risk Evaluation In this phase, the estimated risks are compared to specified risk criteria using either a quantitative or qualitative scale. The goal is to ascertain the significance of the risks and to decide on the appropriate risk control, treatment, detectability, or other aspects of risks to quality. Risk evaluation helps in prioritizing risks based on their severity and the effectiveness of existing or proposed control measures.

By following these steps, organizations can create a structured approach to risk management that enhances the quality and safety of their products and processes (Table 8.3), and the Ishikawa bone diagram identifies the problems in the system (Fig. 8.5). This comprehensive methodology ensures that risks are identified, analyzed, and evaluated systematically, leading to more informed decision-making and better outcomes [13].

8.10 Importance of the QbD Concept in Ophthalmic Formulations

The potential of QbD to improve the efficiency of a process, and the quality of a product makes it crucial to the pharmaceutical industry. It assists pharmaceutical companies in systematically developing their products and processes to achieve the expected quality characteristics. Through the manufacturing process's critical features identification and control, QbD can minimize the risk of product variability and failure which ultimately enhances the product efficacy as well as the safety of the patients [18]. Furthermore, it promotes ongoing enhancements and creativity in pharmaceutical production through the promotion of cutting-edge technologies and

Table 8.3. Risk assessment matrix as per the regulatory requirements

Dependent variables	Independent variables						
	Drug concentration	Excipient concentration	Solvent	Solvent flow rate	Stirring time	Stirring speed	Temp
PS	Medium	Medium	High	Low	Low	High	Medium
PDI	Medium	Low	Low	Low	Low	Low	Low
ZP	Medium	Medium	Low	Low	Low	Low	Medium
% EE	High	High	Low	Low	High	High	Low
% DL	High	Medium	Low	Low	Medium	High	Low

Self-creation
Abbreviation: *PS* particle size, *ZP* zeta potential, *PDI* particle size distribution, *%EE* %entrapment efficiency, *%DL* %drug loading

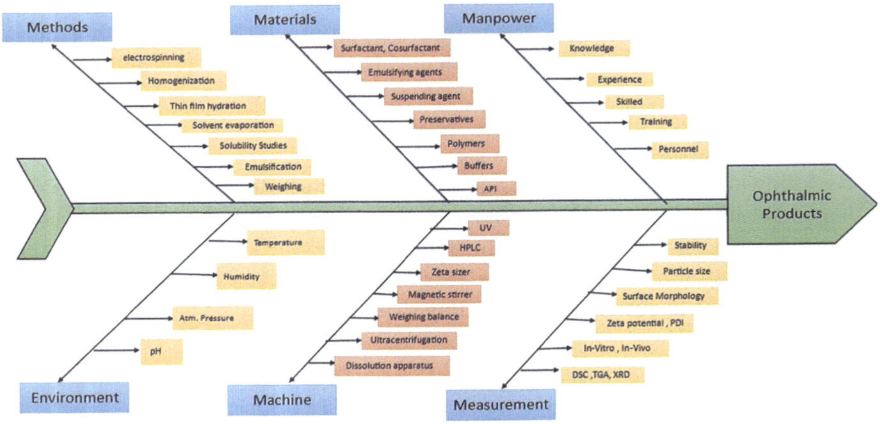

Fig. 8.5 Influence of Ishikawa fishbone diagram on ophthalmic product development. (Self-creation)

analytical tools [19, 20]. There are mainly three important stacks of QbD from different categories, namely, regulatory agencies, end users, and pharmaceutical industries. Where the QbD triangle places the patients at the top, focusing on patients' health as the final goal, then, subsequently the pharmaceutical industries and at last regulators (Fig. 8.6) [21]. The main goal of incorporating QbD principles into the product development process is to improve the health of the patients. QbD is a structured method in pharmaceutical formulation development which begins with pre-defined goals, focusing on product comprehension and process control using scientific concepts along with quality risk management (QRM). QbD leads to improved quality outcomes beyond what consumers request, resulting in a decrease in defects at a very small scale. In contrast, two more sponsors also receive substantial advantages, with the pharmaceutical industry gaining better comprehension of their products and processes, reducing failure of batches, improving control of the scale-up process, and ultimately increasing their return on investment, resulting in saving of cost [22]. Moreover, QbD-assisted development helps in identifying and managing risks effectively, enhancing yield, and cutting down the time for development, costs, and efforts. It also decreases testing and post-approval submissions and improves batch consistency along with confidence in the validation of the process. This approach results in faster market entry, minimizes product rejects and recalls from the market, and addresses drug shortage challenges to ensure greater availability of drug products in the market. Additionally, the QbD approach not only helps to improve the regulatory flexibility for pharmaceutical product development but also decreases the number of submissions related to post-approval changes [23]. Currently, different QbD methodologies such as "design of experiment" (DoE), "formulation by design" (FbD), process analytical technology (PAT), etc., are extensively applied in diverse scientific fields like analytical chemistry, cosmetics, food technology, development of the pharmaceutical product, etc. The adoption of QbD approach over conventional/traditional approach by the pharmaceutical industry gains several benefits such as the risk of error removal due to the conventional method, reduced costs and time spent. The QbD based astounding adaptability process, enhanced both the production and quality of products, assist in fulfilling the regulatorynecessities, improves robustness, enhances the sponsor's transparency by grasping the control strategy for the pharmaceutical products, reduces the post-approval changes, conserves valuable resources by conducting tests in real-time only, outstanding profits from investments, minimizes discrepancies among the batches, streamlines the process of scaling up, validating, and bringing to market in a clear, logical, and anticipated manner, minimizes the risk of facing compliance actions, expensive penalties, and recalls of drug products, etc. [23]. Some of the ophthalmic formulations using QbD approach is summarised in Table 8.4.

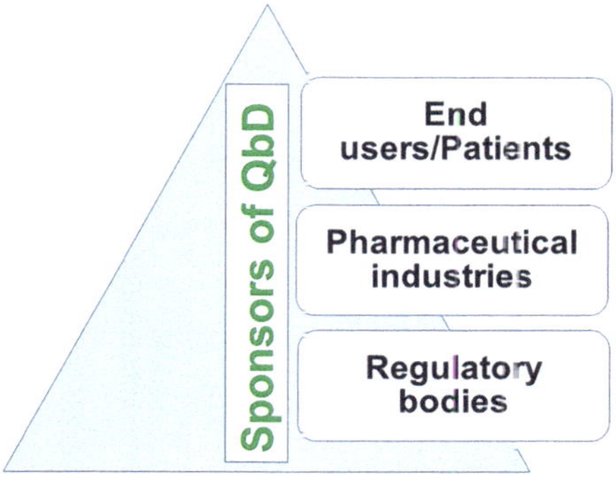

Fig. 8.6 Integration of three important sponsors in the QbD triangle. (Self-creation)

Table 8.4 Summary of the recently utilized QbD approach in ophthalmic formulation development

S. no.	Disease	Drug	Formulation	Independent variables	Dependent variables	Statistical tool	References
1	Inflammatory diseases	Resveratrol	Mucoadhesive lecithin/chitosan nanoparticles	Lecithin concentration, lecithin ratio	Size, PDI, ZP, %EE	Central composite design	[2]
2	Glaucoma	Thymoquinone	Thymoquinone-laden nanoemulsion	Oil concentration S mix concentration (Surfactant to co-surfactant ratio) Stirring speed, flow rate	Size, PDI, ZP% drug content	Box-Behnken design	[35]
3	Ocular inflammation	Dexamethasone and tobramycin	In situ gel for the treatment ocular inflammation	Poloxamer 407 concentration (%) HPMC K4M concentration (%) (X2).	Gel strength, gelation temperature, mucoadhesion index, cumulative percentage drug release for DSP	Central composite design	[36]
4	Fungal keratitis	Voriconazole	Voriconazole loaded lipidic nanoparticles	Lipid ratio, concentration of surfactant	Size, PDI, ZP, %EE	Central composite design	[37]
5	Ocular diseased	Ibuprofen	NSAID-loaded nanostructured lipid carriers	Lipid concentration, ratio, and concentration of surfactant	Size, PDI, ZP, %EE		
6	Inflammatory diseases	Nepafenac	Nepafenac-loaded in situ gel	Concentration of gelrite, concentration of methocel E 15 hypromellose premium LV	Drug release and viscosity of NP	Factorial design	[38]
7	Inflammatory diseases	Loteprednol etabonate	Loteprednol etabonate–loaded in situ ophthalmic gel	Lipid content, surfactant content, co-surfactant content	Size, PDI, ZP, %EE	Central composite design	[39]

S. no.	Disease	Drug	Formulation	Independent variables	Dependent variables	Statistical tool	References
8	Glaucoma	Brinzolamide	Architectonic nanosponges for intraocular delivery of brinzolamide	Weight of ethyl cellulose, concentration of PVA volume of DCM	Size, PDI, ZP, %EE, practical yield	Box-Behnken design	[24]
9	Glaucoma	Dorzolamide hydrochloride, timolol maleate	Viscous eye drops of dorzolamide hydrochloride and timolol maleate	Concentration of HPMC K15, concentration of CP 934	Formulation viscosity Mucoadhesion index, % drug release of dual drugs	Full factorial design	[40]
10	Glaucoma	Brimonidine tartrate	Cationic-charged liposomes	Concentration of lipid, cholesterol, and drug content	Size, PDI, ZP, %EE	Box-Behnken design	[41]
11	Glaucoma	Bimatoprost	Bimatoprost-loaded lipidic nanoformulation	Concentration of lipid, cholesterol and drug content	Size, PDI, ZP, %EE, release kinetics	Central composite design	[42]
12	Ocular diseased	Flurbiprofen, Brimonidine	Dual drug–loaded thermosensitive ocular in situ gel	Concentration of poloxamer 407 HPMC and chitosan	Viscosity, pH, gelation temperature	Central composite design	[43]
13	Fungal keratitis	Posaconazole	In situ gel of posaconazole ophthalmic nanosuspension	Concentration of gellan gum and drug homogenization pressure and number of cycles	Formulation viscosity, mucoadhesion index, % drug release, %EE	Box-Behnken design	[44]
14	Ocular inflammation	Diosmin	Lipid nanoparticles	Solid lipid concentration, liquid lipid concentration, surfactant concentration, solid lipid type.	Size, PDI, ZP, %EE	Box-Behnken design	[45]

(continued)

Table 8.4 (continued)

S. no.	Disease	Drug	Formulation	Independent variables	Dependent variables	Statistical tool	References
15	Keratitis	Fluconazole	Drug-loaded liposome	Lipid amount, sonication time, hydration time	Size, PDI, ZP, %EE	Box Behnken design	[46]
16	Ocular diseased	Chloramphenicol	Thermosensitive in situ gel loaded with chitosan (CS) nanoparticles	Concentrations of CS and triphosphate (TPP)	Size, PDI, ZP, %EE, drug loading	Central composite design	[47]
17	Ocular mycosis	Voriconazole	Voriconazole ternary micellar systems	Weight ratio of Pluronic® P123 and F68	Size, PDI, ZP, %EE, drug loading	D-optimal design	[48]
18	Glaucoma	Methazolamide	Methazolamide liposomal in situ gel	Concentration of cholesterol and drug	Size, PDI, ZP, %EE	Factorial design	[49]
19	Ocular diseased	Ciprofloxacin	Ciprofloxacin-loaded poly(2-ethyl-2-oxazoline) (PSLN) formulation	Amount of drug surfactant concentration and polymer ratio	Size, %EE	Box Behnken design	[50]
20	Glaucoma	Dorzolamide	Nanoparticles of chitosan oligosaccharide and dextran sulfate	Concentration of oligosaccharide and dextran sulfate	Size, PDI, ZP, %EE	Box Behnken design	[51]
21	Fungal keratitis	Itraconazole	Thermosensitive and mucoadhesive in situ ocular gel	Concentration of pluronic® F127, HPMC K100	Gelation temperature, mucoadhesion strength, pH, drug content, viscosity	Central composite design	[52]
22	Ocular diseased	Benzo-dodecenium bromide	In situ gel for ocular of benzo-dodecenium for ocular treatment	Concentration of gellan gum, carbopol 934P and benzo-dodecenium bromide	Viscosity, pH, mucoadhesive strength, permeability coefficient, sustained release of the drug	Box-Behnken design	[53]

S. no.	Disease	Drug	Formulation	Independent variables	Dependent variables	Statistical tool	References
23	Bacterial keratitis	Moxifloxacin	Nanoparticles laden in situ gel	Poloxamer 188 and poloxamer 407 concentration	Gelation temperature	Factorial design	[54]
24	Human cytomegalovirus (CMV) retinitis	Ganciclovir	Ganciclovir nanosuspension-loaded detachable microneedles patch	Concentration of PVA and SLS	Size, PDI, ZP, %EE	Central composite design	[55]
25	Glaucoma	Brinzolamide	Brinzolamide ocular inserts via hot-melt extrusion	Concentration of Metolose® SR 90SH-100000SR and Kolliphor® P 407	%EE, drug release	Factorial design	[56]
26	Glaucoma	Dorzolamide hydrochloride, timolol maleate	Timolol maleate and dorzolamide hydrochloride in injectable liposomes	Amount of lipid and pressure of high-pressure homogenizer	Size, PDI, %EE, release kinetics	Full factorial design	[57]
27	Glaucoma	Brimonidine tartrate	Self-nano emulsifying drug delivery system of brimonidine tartrate	Concentration of oil, surfactant and co-surfactant	Size, PDI, ZP, %EE, self-emulsification	Box-Behnken design	[58]
28	Glaucoma	Methazolamide	Cationic nanostructured lipid carriers	Type of oil, concentration of lipid, oil and polysorbate	Size, PDI, ZP, %EE, drug loading	Full factorial design	[59]

8.11 Design of Experiment (DoE) Approach in Product Development

Design of experiment (DoE) is the statistical approach used by industries to optimize product and process performance during fabrication. This emphasizes in planning and predicting the errors with optimal experimental trials.

General Aspects in DoE

DOE applications involve various steps, like study objectives, critical factors, responses, statistical analysis, and mathematical models. Identifying a problem is the crucial step before implementing the experimental model. To identify this, the objectives of experiment are outlined and considered before implementing an experimental design. Consequently, the effect can be defined by creating a rational relationship between the input and output variables, considering the prior experimental or expertise knowledge on similar types of experiments is represented in fishbone diagram, which reflects causes and effects, such as manpower, method, environment, and materials.

8.12 Screening

In the case of a new system or process, identifying the significant factors that impact the response is challenging. Many factors are involved in the manufacturing process, and the experimenters lack previous knowledge of factors that significantly influence formulation preparation. So, it is an important to perform screening before starting the experiment to minimize or eliminate the unwanted factors. The primary designs used for screening are full factorial design, Taguchi design, fractional factorial design, and Plackett-Burman design [24].

8.12.1 Full Factorial Design

This design is used for developing the factors that are independent of one another. The components are often denoted as A, B, and C, whereas the levels are assigned as -1 and $+1$ and if any intermediate level is present, it is denoted as 0. This design encompasses all potential combinations of the factors and levels [25]. This study explored the formulation of mucoadhesive microemulsions loaded with clotrimazole, an antifungal agent which aimed at treating ocular diseases. A full factorial design (FD) was employed to assess the impact of various formulation variables, including the surfactant type, surfactant and co-surfactant ratio, and percentages of both mixtures. The oleic acid was selected based on the solubility of CLZ, while the surfactant and co-surfactant, along with their ratios, were selected for their emulsifying ability and limited turbidity, even with a high volume of water. The selected components, including Cremophor: oleic acid, Transcutol HP: Cremophor EL, and water (20%, 70%, and 10% w/v), resulted in a particle size of 229.1 ± 0.989 nm, a polydispersity index (PDI) of 0.5 ± 0.0095, and a zeta potential of −33.3 ± 0.9 mV. Additionally, in vitro drug release for this formulation was 37.6%, while the free drug exhibited a release of 83.2% over 6 h. The chitosan-coated microemulsion demonstrated the potential for effective drug delivery of clotrimazole in the ocular diseases [26]. Another investigation was conducted utilizing flexosomes containing tolnaftate (TOL), an antifungal agent. This study employed a 3 × 2^2 full factorial design, using variables such as the amount of stearyl amine, hydration volume, and the type of edge activator (20 mg, 5 mL, and Cremophor RH 40, respectively). The findings showed excellent outcomes, including an excellent entrapment efficiency of 66% and a particle size of 154 ± 29.11 nm. Additionally, the ex vivo and in vivo studies demonstrated that the optimal flexosomal formulation exhibited a promising effect in delivering the TOL [27].

8.12.2 Fractional Factorial Design

Fractional factorial design (FFD) is suitable when the number of factors in the experimental design is between 3 and 7. It is often half or one-fourth of the fractional factorial design. However, this approach has limitations such as confounding and misidentification of the interaction or main result. Thus, resolution serves to assess interactions and effects without confounding. The ambiguity in the result increases if the design resolution is less and vice versa. This design is mainly used to examine the factors that most significantly influence critical quality [28].

8.12.3 Plackett-Burman Design

The PBD are used when main effect of experimental design is highly confounded with the two-factor interaction. This is an economical method and interaction between them is highly negligible [26]. Rathod et al. studied the NSAID-loaded nanostructured lipid carrier (NLC) for ocular treatment. In this, seven factors such as lipid and surfactant concentration, the concentration of lipids, ultrasonication time and frequency, homogenizer time, and speed were displayed across minimum and maximum values with 12 experimental runs and determined their effect on the PS, PDI, and ZP. Subsequently, the in vitro release from the ibuprofen (IBU) NLC showed sustained release and also resulted in a robust and stable ophthalmic formulation with good physicochemical properties [29].

8.12.4 Taguchi Design

This serves as a main and the objective function for optimization. It starts with the minimum two levels and four runs without the requirement of any center points. This design completely relies on orthogonal arrays [1]. Ramzan et al. investigated the development of liposomes loaded with 5-fluorouracil (5-FU) for treating the

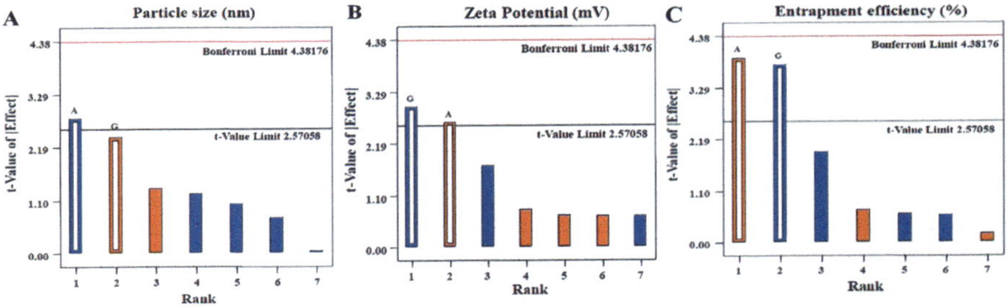

Fig. 8.7 Pareto chart (**a**) particle size, (**b**) zeta potential, (**c**) entrapment efficiency. (Reprinted with permission from Ramzan et al. [30] (figure 1) Copyright © 2025 Elsevier B.V)

proliferation of fibroblast. This design was employed for several formulation trials, and the resulting formulation was analyzed for size, PDI, and ZP. In the present study seven factors, with two levels (indicating lower and higher) were selected. The significant results derived from the Pareto chart (Fig. 8.7) were subsequently confirmed and chosen for additional optimization trials [30].

8.13 Optimization Design

This design is used for optimizing the factors that are significantly affecting the experimental process. The optimization design helps in defining the one model responses which are complex in nature, in this only few factors which are significantly affecting with maximum runs that are selected either from the screening design or based on previous knowledge, preliminary trails and experiences. Examples of this design includes the central composite design, Box Behnken design, Mixture design and Optimal design [1].

8.13.1 Box-Behnken Design

This design is developed by Box and Behnken; it is a family of three-level, two-factor designs that minimize the number of runs for optimization, distinguishing it from complete factorial designs. It primarily facilitates the capture of linear effects; nevertheless, it is limited to certain factors and responses and may not account for higher-order interactions that are very important in particular processes [28]. Bagyl et al. developed brinzolamide (BRN) nanosponges (NS) incorporated into an in situ gel for glaucoma treatment using a solvent evaporation method. The formulation was optimized through Box Behnken design (BBD), where independent factors such as ethyl cellulose, BRN, and dichloromethane volume were studied across 17 runs and 3 levels. This approach allowed for the investigation of interactions among independent factors and their quadratic effects on dependent variables, resulting in a particle size of 114 ± 2.32 nm, a PDI of 0.11 ± 0.01, and an entrapment efficiency of 83.12 ± 1.2%. The optimized formulation demonstrated significant potential in all evaluation studies compared to the marketed product Brinzox® [24]. In this study, Farooq et al. fabricated the bevacizumab (BEV) and dexamethasone (DEX) co-loaded liposomal carrier for ocular delivery. The formulation was initially developed using the thin film hydration method and subsequently optimized through BBD with the help of Mintab software. The concentration of 1,2-dipalimitoyl-Sn-glycero-3-phosphocholine (DPPC), cholesterol, and DEX was used as independent variables. The optimized values of the liposomes included the particle size of 139 ± 2 nm, PDI of 0.03 ± 0.01, zeta potential of −2 ± 0.7 mV, and entrapment efficiencies of 97 ± 0.5% for BEV and 26 ± 0.5% for DEX. Furthermore, the final optimized formulation demonstrated a prolonged release profile when compared to the free form [31].

8.13.2 Central Composite Design

CCD is used when higher order and sequential experimentation is necessary. This design contains information about factorial designs and often used in response surface methodologies. The CCD allows in determining the curvature due to its centre points, in this design each factor will have five levels [28]. Ramzan et al. developed chitosan 5-FU liposomes to treat ocular disease. In this investigation, the factors that significantly affected the screening design were subsequently utilized for optimization. Based on the outcome from the pareto chart (Fig. 8.7), the prime factors such as lipid conc. and the sonication cycle were selected. Subsequently, they assessed the optimization runs to examine the desirability factors, with the desirability range established between 0 and 1 in this design. Values close to zero did not fit the model, whereas values near unity or 1 suggest a good fit within the model and detailed understanding about the study was illustrated in Fig. 8.8. Subsequently, the assessment of the in vivo performance was conducted, comparing the plain drug solution, liposomal formulation, and liposomal loaded gel. The gel exhibited a higher pharmacokinetic profile in comparison to other two formulations [30].

Said et al. conducted an optimization and evaluation of mucoadhesive cubosomes containing voriconazole to enhance ocular bioavailability. This study investigated by incorporating the different concentrations of pluronic 127 and monoolein. The final formulation was achieved by 15% monoolein and 1.2% pluronics along with the other variables. The optimized formulation showed a particle size of 160 nm along with a drug loading of 0.81%. Subsequently, this enhanced formulation underwent chitosan coating to extend the pre-corneal residence duration. The formulation coated with chitosan displayed more effective mucoadhesive properties, as well as safety and biocompatibility. Additionally, it demonstrated an

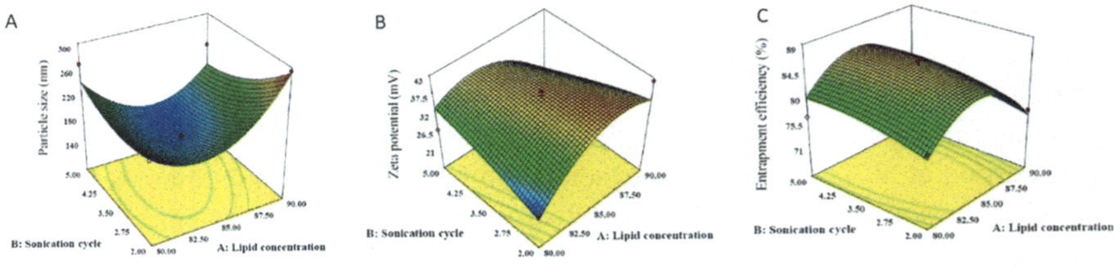

Fig. 8.8 3D surface counter plots using a central composite design: (**a**) particle size, (**b**) zeta potential, and (**c**) entrapment efficiency. (Reprinted with permission from Ramzan et al. [30]. Copyright © 2025 Elsevier B.V.)

improved pharmacokinetic profile when compared to the voriconazole suspension [32]. Saha et al. optimized the lecithin/chitosan NPs for ocular treatment. The initial risk factors were assessed by using the fishbone diagram followed by the risk assessment by RAM. The screening studies were performed on CQAs for CMA and CPPs. Later the significant factors were optimized using CCD. The design showed an optimized desired value of 0.913 with the PS of 163.3 nm with 0.254 and 46.6 mV of PDI and ZP (Fig. 8.9). Later from the extensive characterization studies, it was found that the %EE was 97.03% with the maximum release and release kinetics of 96.87 and 0.9897, respectively, within the predicted range at an alpha level of ±0.05 [2].

8.13.3 Optimal Design

The optimality of the design is completely dependent upon the statistical model, and it is analyzed against statistical criteria; however, identifying the appropriate design for factors and models requires an understanding of both practical knowledge and statistical theory during the experimental design process. The optimum designs include several categories such as A-optimal, D-optimal, and I-optimal. This design contains three levels for a single factor [20]. In this study, Eldeeb et al. formulated the cubosomes for ocular delivery of brimonidine tartrate to improve ocular bioavailability; the final cubosomal formulation was prepared by d-optimal design. Also, this formulation showed the 9.1 folds increase in the residence time compared to the marketed formulation alphagen® P [32].

8.13.4 Mixture Design

It is used, when the proportions of different independent factors blend, and the sum of all factors must be 100%. This design can be of many types such as simple centroid design, simple lattice design and optimal design. The optimal design is the most commonly used mixture design [1]. The study by Park et al. dealt with the

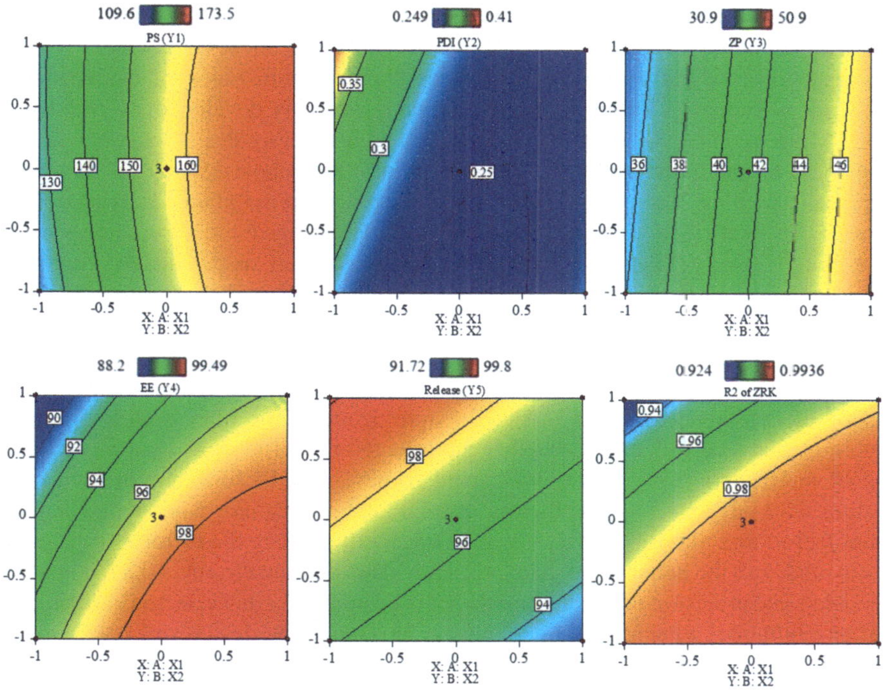

Fig. 8.9 Counter plots of quality attributes. (Reprinted with permission from Saha et al. [2]. Copyrights © 2021 Elsevier B.V)

fabrication and evaluation of dequafosol tetrasodium hot melted extruded inserts, using a design of experiments approach for dry eye disease. The independent variables included hydroxypropyl cellulose (L-HPC), hydroxy methylcellulose acetate succinate (HPMCAS-HG), and polyethylene glycol-6000 (PEG-6000). These three variables were randomized and mixed in proportions to achieve sustained drug release. The formulation was optimized through mixture design and demonstrated stable drug content and surface pH over 3 months. Furthermore, the in vitro corneal permeation study showed enhanced bioavailability and reduced dosing frequency compared to conventional dosage forms. Consequently, the authors concluded that these inserts hold significant promise for improving patient compliance through sustained release effects [33].

8.14 Utilization of Statistical Tools Under the QbD

DoE is one of the most crucial statistical tools of QbD, which utilizes various mathematical/statistical terminologies to optimize the factors involved in the development of process and formulation. The optimization process comprises determining the best possible combination or conditions for an experiment and including all influencing factors to achieve optimal results. One of the most widely used mathematical/statistical software tools under DoE for the optimization of independent/dependent factors is Design-Expert® software, which is developed by Stat-Ease Inc., Minneapolis, USA (https://www.statease.com/software/design-expert/). In the software, research directions are categorized into three choices, such as characterization, screening, and optimization of factors based on the planned experimental design. The statistical tools are now being uniquely utilized in diverse areas within the drug delivery field. Statistical terminologies, such as adequate precision, adjusted R^2, analysis of variance (ANOVA), *f*-value, predicted R^2, *p*-value, PRESS, R^2, regression coefficient, and standard deviation, are used to interpret the influence of factors involved in process and formulation development on the quality of the final product. The *f*-value, *p*-value, and ANOVA are analyzed to check the significance of the selected design model for optimization. The regression coefficient value is analyzed to identify whether the selected model is showing a positive or negative effect. The R^2, adjusted R^2, standard deviation, PRESS, and predicted R^2 values are analyzed to see whether the correlation between independent and dependent factors is good or bad. The signal-to-noise ratio among the factors/variables is measured with the help of Adequate precision value obtained from a selected model. An adequate precision value of more than 4 is desirable for any selected experimental design/model, which ultimately helps in navigating the design space within the model to get the optimized formula [9].

8.15 Challenges Faced in Implementing the QbD Concept

QbD is a vital concept that promotes comprehension of processes, which is essential in ensuring the better performance of the process and the quality of products. The regulatory bodies prioritize quality within the products. The quality checking involves testing of final product which leads to a challenge in the utilization of QbD concept. Furthermore, several testing techniques currently in use are significantly restricted in today's context of mass production [4]. The mentioned few points below indicate the challenges faced in implementing the QbD concept by pharmaceutical companies in their process and product development stage [34].

1. The main reason and significant obstacle to implementing the QbD concept is the insufficient comprehension of the pharmaceutical process.
2. The pharmaceutical companies mainly focus more on the final product rather than on understanding the scientific process involved in the development and manufacturing of the products.
3. There is still an unresolved challenge regarding the coordination and agreement between USFDA review, compliance sectors, and field inspectors on how to implement the QbD concept in the pharmaceutical process and product development.

4. Most pharmaceutical companies believe that there is a necessity for clearer instructions on how to implement the QbD concept into real-time practice.
5. Pharmaceutical companies require clear explanations from the regulatory bodies regarding the terminologies involved in the QbD concept, guidelines for choosing and eliminating critical quality attributes/parameters, acceptable techniques/methodologies, standards for replacing methods used in analysis, and measures for assessing control effectiveness.
6. Enhanced collaboration among various sectors such as process development, production, packaging, and quality control and assurance within the pharmaceutical companies is required to execute QbD in an organization successfully.
7. Additionally, pharmaceutical companies are more concerned that implementation of the QbD concept may delay the approval process of submitted applications or result in the submission of needless information to regulatory bodies, which can potentially hinder the process of application approval.

8.16 Conclusion and Future Perspectives

The global market for ophthalmic drugs was estimated at 37.82 billion USD and is projected to grow at an annual rate of 8.34% from 2024 to 2030. The growth of markets is driven by the increasing prevalence of eye disorders, causing research and development teams and industries to heighten their initiatives in the production of ophthalmic. Nonetheless, the key players are focusing on launching novel formulations with enhanced drug delivery systems to reduce the limitations present with the currently existing marketed formulations. The development of ocular formulation with QbD approach helps in identifying the individual factor's effect on the final product. Furthermore, this helps in risk assessment and control, which is lacking in the routine ocular drug development process. QbD is a systematic tool that consists of various key elements like QTPP, CQA, CMA, CPP, and DoE. Also, QbD is an economical and efficient method than other techniques. It helps in identifying the root cause with reduced variables and improves understanding of the process design and product development. Subsequently, this approach helps in identifying the suitable experimental design and factors that are significantly affecting the formulation development. The manufacturers can effectively meet the regulatory requirements, reducing the risk of compliance issues through QbD principles. Thus, the DoE tools in QbD resolve the gap between the pilot-scale production and the commercial development of ophthalmic-based nanocarrier formulations.

Acknowledgments Dr. Gulbake would like to thank Anusandhan National Research Foundation (EEQ/2022/000950 SERB), New Delhi, India, for extending facilities to write the book chapter.

References

1. Politis SN, Colombo P, Colombo G, Dimitrios MR. Design of experiments (DoE) in pharmaceutical development. Drug Dev Ind Pharm. 2017;43:889–901.
2. Saha M, Saha DR, Ulhosna T, Sharker SM, Shohag MH, Islam MS, Ray SK, Rahman GMS, Reza HM. QbD based development of resveratrol-loaded mucoadhesive lecithin/chitosan nanoparticles for prolonged ocular drug delivery. J Drug Deliv Sci Technol. 2021;63:102480.
3. Kalyane D, Raval N, Polaka S, Tekade RK. Chapter 1 – Quality by design as an emerging concept in the development of pharmaceuticals. In: Tekade RK, editor. The future of pharmaceutical product development and research. London: Academic Press; 2020. p. 1–25.
4. Woodcock J. The concept of pharmaceutical quality. Am Pharm Rev. 2004;7:10–5.
5. Beg S, Hasnain MS, Rahman M, Swain S. Chapter 1 – Introduction to quality by design (QbD): fundamentals, principles, and applications. In: Beg S, Hasnain MS, editors. Pharmaceutical quality by design. London: Academic Press; 2019. p. 1–17.
6. Comoglu T, Ozyilmaz ED. Chapter 10 – Pharmaceutical product development: a "quality by design" (QbD) approach. In: Nayak AK, Sen KK, editors. Dosage forms, formulation developments and regulations. London: Academic Press; 2024. p. 285–310.
7. Benetti AA, Benetti C. Regulatory aspects of quality by design. In: Jain NK, Bajwa N, editors. Introduction to quality by design (QbD): from theory to practice. Singapore: Springer Nature; 2024. p. 35–57.
8. Mishra V, Thakur S, Patil A, Shukla A. Quality by design (QbD) approaches in current pharmaceutical set-up. Expert Opin Drug Deliv. 2018;15:737–58.
9. Rahman SNR, Agarwal N, Goswami A, Sree A, Jala A, Venuganti A, Deka A, Borkar RM, Singh V, Das D, Shunmugaperumal T. Studies on spray dried topical ophthalmic emulsions containing cyclosporin A (0.05% w/w): systematic optimization, in vitro preclinical toxicity and in vivo assessments. Drug Deliv Transl Res. 2023;13:1654–74.
10. Jain S. Quality by design (QBD): a comprehensive understanding of implementation and challenges in pharmaceuticals development, international. J Pharm Pharm Sci. 2014;6:29–35.
11. U.S. Food and Drug Administration. Guidance for industry: Q8 (R2) pharmaceutical development. Silver Spring: Center for Drug Evaluation and Research; 2009.
12. Gandhi A, Roy C. Quality by design (QbD) in pharmaceutical industry: tools, perspectives and challenges. PharmaTutor. 2016;4:12–20.
13. Zhang L, Mao S. Application of quality by design in the current drug development. Asian J Pharm Sci. 2017;12:1–8.
14. Liu M, Cheng G, Tang Z, Zhou L, Wan X, Ding G. Flame retardancy performance and mechanism of polyvinyl alcohol films grafted amino acid ionic liquids with high transparency and excellent flexibility. Polym Degrad Stab. 2022;205:110133.
15. Beg S, Hasnain MS. Pharmaceutical quality by design: principles and applications. London: Academic Press; 2019.
16. Namjoshi S, Dabbaghi M, Roberts MS, Grice JE, Mohammed Y. Quality by design: development of the quality target product profile (QTPP) for semisolid topical products. Pharmaceutics. 2020;12:287.
17. Kalyane D, Raval N, Polaka S, Tekade RK. Quality by design as an emerging concept in the development of pharmaceuticals. In: The future of pharmaceutical product development and research. London: Elsevier; 2020. p. 1–25.
18. Nunavath RS, Singh MT, Jain A, Chakma M, Arivuselvam R, Azeeze M. Quality by design in pharmaceuticals: a review of its impact on regulatory compliance and product quality. Drug Res (Stuttg). 2024;74:18–23.
19. Rahman SNR, Shunmugaperumal T. Formulation development of oil-in-water nanosized emulsions. In Oil-in-water nanosized emulsions for drug delivery and targeting. John Wiley & Sons, Inc., 111 River Street, Hoboken, NJ 07030, USA, 2021;19–68.
20. Aru P, Gulhane M, Katekar V, Deshmukh S. Quality by design (QbD) in pharmaceutical development: a comprehensive review. GSC Biol Pharm Sci. 2024;26:328–40.

21. Valeru C. Pharmaceutical product development: quality by design (QbD). UPI J Pharm Med Health Sci. 2024:11–6.
22. Khan A, Naquvi KJ, Haider MF, Khan MA. Quality by design- newer technique for pharmaceutical product development. Intell Pharm. 2024;2:122–9.
23. Shunmugaperumal T. Oil-in-water nanosized emulsions for drug delivery and targeting. Hoboken: Wiley; 2020.
24. Bagul US, Nazirkar MV, Mane AK, Khot SV, Tagalpallewar AA, Kokare CR. Fabrication of architectonic nanosponges for intraocular delivery of Brinzolamide: an insight into QbD driven optimization, in vitro characterization, and pharmacodynamics. Int J Pharm. 2024;650:123746.
25. Birla D, Khandale N, Bashir B, Shahbaz Alam M, Vishwas S, Gupta G, Dureja H, Kumbhar PS, Disouza J, Patravale V, Veiga F, Paiva-Santos AC, Pillappan R, Paudel KR, Goh BH, Singh M, Dua K, Singh SK. Application of quality by design in optimization of nanoformulations: principle, perspectives and practices. Drug Deliv Transl Res. 2024;15:798–830.
26. Al-mahallawi AM, Ahmed D, Hassan M, El-Setouhy DA. Enhanced ocular delivery of clotrimazole via loading into mucoadhesive microemulsion system: in vitro characterization and in vivo assessment. J Drug Deliv Sci Technol. 2021;64:102561.
27. Aziz D, Mohamed S, Tayel S, Makhlouf A. Flexosomes as a promising nanoplatform for enhancing tolnaftate ocular delivery: formulation, in vitro characterization, statistical optimization, ex vivo and microbial in vivo studies. Int J Pharm. 2023;646:123471.
28. Waghule T, Dabholkar N, Gorantla S, Rapalli VK, Saha RN, Singhvi G. Quality by design (QbD) in the formulation and optimization of liquid crystalline nanoparticles (LCNPs): a risk based industrial approach. Biomed Pharmacother. 2021;141:111940.
29. Rathod VR, Shah DA, Dave RH. Systematic implementation of quality-by-design (QbD) to develop NSAID-loaded nanostructured lipid carriers for ocular application: preformulation screening studies and statistical hybrid-design for optimization of variables. Drug Dev Ind Pharm. 2020;46:443–55.
30. Ramzan M, Khan T, Siddique MUM, Shahid M. Biocompatible cationic 5-fluorouracil loaded elastic liposomes for ocular delivery: in vitro, ex vivo, and in vivo evaluation. J Drug Deliv Sci Technol. 2024;101:106278.
31. Farooq U, O'Reilly NJ, Ahmed Z, Gasco P, Raghu Raj Singh T, Behl G, Fitzhenry L, McLoughlin P. Design of liposomal nanocarriers with a potential for combined dexamethasone and bevacizumab delivery to the eye. Int J Pharm. 2024;654:123958.
32. Said M, Aboelwafa AA, Elshafeey AH, Elsayed I. Central composite optimization of ocular mucoadhesive cubosomes for enhanced bioavailability and controlled delivery of voriconazole. J Drug Deliv Sci Technol. 2021;61:102075.
33. Jin Park Y, Wuk Kim D. Development and evaluation of hot-melt–extruded diquafosol tetrasodium formulations for ophthalmic inserts: a design of experiments approach. Int J Pharm. 2024;659:124249.
34. Rahman SNR, Pawde DM, Katari O, Hmingthansanga V, Shunmugaperumal T. Systematic optimization, in vitro drug release, and preliminary nonclinical toxicity assessment of non-phospholipid-based topical ophthalmic emulsions containing 0.05 or 0.1% w/w cyclosporin A for dry-eye syndrome management. AAPS PharmSciTech. 2019;21:36.
35. Srivastava V, Pardhi ER, Yadav R, Singh V, Khatri DK, Mehra NK. QbD-driven thymoquinone laden nanoemulsion for glaucoma management: In vitro, ex vivo, and pre-clinical evaluation. J Drug Deliv Sci Technol. 2024;94:105493.
36. Patel N, Thakkar V, Metalia V, Baldaniya L, Gandhi T, Gohel M. Formulation and development of ophthalmic in situ gel for the treatment ocular inflammation and infection using application of quality by design concept. Drug Dev Ind Pharm. 2016;42:1406–23.
37. Patel A, Dharamsi A. Voriconazole loaded lipidic nanoparticles for ophthalmic delivery: development using QbD combined with risk-based approach. Curr Nanomed (Formerly: Recent Pat on Nanomed). 2023;13:56–69.
38. Wagh DG, Shahi SR, Bora S, Dabir PD, Deore SV. QBD approach in formulation and evaluation of gelrite based in situ ophthalmic gel of nepafenac. World J Pharm Res. 2016;6:638–54.

39. Priya S, Srividya G, Chandra NS, Singh PP, Saha RN, Sathe P, Nirmal J, Singhvi G. Loteprednol etabonate loaded lyotropic liquid crystalline nanoparticles in-situ ophthalmic gel: Qbd driven optimization and in-vitro, ex-vivo evidence of sustained precorneal residence time. J Drug Deliv Sci Technol. 2023;89:105081.
40. Shah PA, Gevariya NR, Christian JR, Patel KG, Thakkar VT, Gohel MC, Gandhi TR. Science based development of viscous eye drop of dorzolamide hydrochloride and timolol maleate using full factorial design. Pharm Methods. 2018;9:69–78.
41. Soni PK, Saini T. Formulation design and optimization of cationic-charged liposomes of brimonidine tartrate for effective ocular drug delivery by design of experiment (DoE) approach. Drug Dev Ind Pharm. 2021;47:1847–66.
42. Verma P, Singh RK, Wadhwa A, Prabhakar B, Yadav KS. Bimatoprost-loaded lipidic nanoformulation development using quality by design: liposomes versus solid lipid nanoparticles in intraocular pressure reduction. Nanomedicine. 2023;18:1815–37.
43. Polat HK, Arslan A, Ünal S, Haydar MK, Aytekin E, Gözcü S, Karakuyu NF, Mokhtare B. Formulation development of dual drug-loaded thermosensitive ocular in situ gel using factorial design. J Pharm Innov. 2023;18:768–88.
44. Khare P, Chogale MM, Kakade P, Patravale VB. Gellan gum–based in situ gelling ophthalmic nanosuspension of Posaconazole. Drug Deliv Transl Res. 2022;12:2920–35.
45. Zingale E, Rizzo S, Bonaccorso A, Consoli V, Vanella L, Musumeci T, Spadaro A, Pignatello R. Optimization of lipid nanoparticles by response surface methodology to improve the ocular delivery of diosmin: characterization and in-vitro anti-inflammatory assessment. Pharmaceutics. 2022;14:1961.
46. Sanap SN, Bisen AC, Mishra A, Biswas A, Agrawal S, Yadav KS, Krishna A, Chopra S, Mugale MN, Bhatta RS. QbD based antifungal drug-loaded ophthalmic liposomal formulation for the management of fungal keratitis: in vitro, ex vivo and in vivo pharmacokinetic studies. J Drug Deliv Sci Technol. 2022;74:103517.
47. Huang C, Peng L, Xu X, Lu Y, Wang X, Lan Z, Chen J, Zhou Y. Preparation and characteristics of a thermosensitive in situ gel loaded with chitosan nanoparticles for optimal ocular delivery of chloramphenicol. J Drug Deliv Sci Technol. 2023;89:104962.
48. Fahmy AM, Hassan M, El-Setouhy DA, Tayel SA, Al-mahallawi AM. Voriconazole ternary micellar systems for the treatment of ocular mycosis: statistical optimization and in vivo evaluation. J Pharm Sci. 2021;110:2130–8.
49. Londhe VY, Sharma S. Formulation, characterization, optimization and in-vivo evaluation of methazolamide liposomal in-situ gel for treating glaucoma. J Drug Deliv Sci Technol. 2022;67:102951.
50. Onugwu AL, Attama AA, Nnamani PO, Onugwu SO, Onuigbo EB, Khutoryanskiy VV. Development and optimization of solid lipid nanoparticles coated with chitosan and poly(2-ethyl-2-oxazoline) for ocular drug delivery of ciprofloxacin. J Drug Deliv Sci Technol. 2022;74:103527.
51. Pardeshi SR, More MP, Pardeshi CV, Chaudhari PJ, Gholap AD, Patil A, Patil PB, Naik JB. Novel crosslinked nanoparticles of chitosan oligosaccharide and dextran sulfate for ocular administration of dorzolamide against glaucoma. J Drug Deliv Sci Technol. 2023;86:104719.
52. Permana AD, Utami RN, Layadi P, Himawan A, Juniarti N, Anjani QK, Utomo E, Mardikasari SA, Arjuna A, Donnelly RF. Thermosensitive and mucoadhesive in situ ocular gel for effective local delivery and antifungal activity of itraconazole nanocrystal in the treatment of fungal keratitis. Int J Pharm. 2021;602:120623.
53. Ranch KM, Maulvi FA, Naik MJ, Koli AR, Parikh RK, Shah DO. Optimization of a novel in situ gel for sustained ocular drug delivery using Box-Behnken design: in vitro, ex vivo, in vivo and human studies. Int J Pharm. 2019;554:264–75.
54. Upadhyay Shivam U, Chavan Siddhi K, Gajjar Devarshi U, Upadhyay Umeshkumar M, Patel Jayvadan K. Nanoparticles laden in situ gel for sustained drug release after topical ocular administration. J Drug Deliv Sci Technol. 2020;57:101736.
55. Suriyaamporn P, Pornpitchanarong C, Pamornpathomkul B, Patrojanasophon P, Rojanarata T, Opanasopit P, Ngawhirunpat T. Ganciclovir nanosuspension-loaded detachable microneе-

dles patch for enhanced drug delivery to posterior eye segment. J Drug Deliv Sci Technol. 2023;88:104975.
56. Tambe S, Jain D, Rawat R, Mali S, Pagano MA, Brunati AM, Amin P. MeltSerts technology (brinzolamide ocular inserts via hot-melt extrusion): QbD-steered development, molecular dynamics, in vitro, ex vivo and in vivo studies. Int J Pharm. 2023;648:123579.
57. Verma P, Rajpurohit R, Yadav KS. Quality by design steered approach for co-encapsulation of timolol maleate and dorzolamide hydrochloride in injectable liposomes. Int J Pharm. 2024;664:124566.
58. Vikash B, Shashi, Pandey NK, Kumar B, Wadhwa S, Goutam U, Alam A, Al-Otaibi F, Chaubey P, Mustafa G, Gupta G, Dua K, Singh SK. Formulation and evaluation of ocular self-nanoemulsifying drug delivery system of brimonidine tartrate. J Drug Deliv Sci Technol. 2023;81:104226.
59. Youshia J, Kamel AO, El Shamy A, Mansour S. Gamma sterilization and in vivo evaluation of cationic nanostructured lipid carriers as potential ocular delivery systems for antiglaucoma drugs. Eur J Pharm Sci. 2021;163:105887.

Advances and Opportunities of the Novel Functional Nanocarriers

9

Namish Manchanda, Shivam Pathak, Divyanshu Srivastava, Saurabh Mittal, and Sushama Talegaonkar

Abstract

In the realm of pharmaceutical research, ocular drug delivery presents the most. The physiological hurdles, especially the intricate intraocular tissue layers that hinder drug penetration into the anterior and posterior compartments of the eye, remain the fundamental impediment to the oversight of drug delivery to the eye. The targeted delivery and therapeutic utility of pharmacologically valuable molecules, which are otherwise limited by conventional approaches, could potentially be improved by integrating nanotechnology into drug delivery research. Recent developments in nano-assisted drug delivery systems deployed in ophthalmology have brought noteworthy advances in the delivery of small to large molecules across the eye. Because of their distinct features, such as size and surface modifications, nanocarriers foster controlled drug release and improve permeation across ocular barriers, enabling efficient delivery to specific eye locations. The development of novel nano-based drug delivery systems with an eye toward potential clinical translation has been the focus of relentless research.

N. Manchanda
Centre for Pharmaceutical Nanotechnology, Department of Pharmaceutics, National Institute of Pharmaceutical Education & Research (NIPER), Mohali, Punjab, India

S. Pathak · D. Srivastava · S. Mittal
Center of Pharmaceutics, Amity Institute of Pharmacy, Amity University, Noida, Uttar Pradesh, India

S. Talegaonkar (✉)
Department of Pharmaceutics, School of Pharmaceutical Sciences, Delhi Pharmaceutical Sciences & Research University (DPSRU), Government of NCT of Delhi, New Delhi, Delhi (NCT), India
e-mail: stalegaonkar@dpsru.edu.in

N. K. Mehra (ed.), *Complex Ophthalmic Dosage Forms: Advances in Biomedical Applications and Future Perspectives*, https://doi.org/10.1007/978-981-96-6306-4_9

Such developments have brought about an array of nanovehicles for ocular use, ranging from solid lipid nanoparticles, polymeric nanoparticles, niosomes, liposomes, dendrimers, nanoemulsions, and micelles. This chapter sets forth the latest findings of preclinical research around novel nano-based systems for various ocular uses and offers a glimpse of the nano-based drug products that are being approved by regulatory bodies and being explored in clinical trials for the treatment of ocular diseases.

Keywords

Ocular drug delivery · Bioavailability · Nanocarriers · Permeation · Ophthalmology

9.1 Introduction

The ocular drug delivery field presents unique challenges due to the intricate anatomy and physiology of the eye, which limits the effectiveness of conventional drug delivery systems. The eye's protective barriers, such as the blood-ocular barriers, tear film, and continuous drainage of aqueous humor, significantly hinder the penetration and bioavailability of therapeutic agents [1]. While essential for maintaining ocular homeostasis, these barriers pose significant challenges for treating ocular diseases. Traditional methods, such as eye drops or systemic administration, often fail to deliver sufficient drug concentrations to the targeted ocular tissues, leading to suboptimal therapeutic outcomes [2]. This issue has driven the exploration of more advanced drug delivery systems, with nanotechnology emerging as one of the most promising approaches. Nanotechnology-based ocular drug delivery systems have shown tremendous potential in overcoming these barriers and improving drug delivery to various parts of the eye, including the anterior and posterior segments. Nanoparticles, liposomes, niosomes, nanoemulsions, solid lipid nanoparticles (SLNs), and nanostructured lipid carriers (NLCs) are just some of the nanocarriers being investigated for their ability to enhance drug stability, prolong release, and target specific ocular tissues [3]. These nanocarriers can be engineered to provide sustained release of drugs, improve their solubility, and protect them from enzymatic degradation, resulting in enhanced therapeutic efficacy. The advantages of nanotechnology-based systems are multifaceted. First, they can significantly increase the bioavailability of drugs in the eye by avoiding the natural defense mechanisms, such as tear dilution and rapid elimination from the ocular surface. Second, they enable the encapsulation of both hydrophilic and hydrophobic drugs, thus expanding the range of therapeutic agents that can be delivered effectively. Furthermore, these systems can be tailored to achieve targeted drug delivery, reducing systemic side effects and minimizing the frequency of drug administration, which improves patient compliance—one of the major challenges in ocular therapies [4].

Despite these promising advancements, translating nanotechnology-based ocular drug delivery systems into clinical practice faces several hurdles. Regulatory concerns, manufacturing complexities, and the need for extensive preclinical and clinical evaluations pose significant challenges to the widespread adoption of these technologies. Regulatory bodies, such as the FDA and EMA, require comprehensive data on nanocarriers' safety, efficacy, and long-term biocompatibility [5]. The lack of standardized protocols for evaluating nanomedicine, coupled with the high costs associated with their production, further complicates the pathway to clinical translation [6]. Additionally, the unique physiological environment of the eye introduces specific challenges in terms of nanocarrier design. For example, nanocarriers intended for ocular delivery must be nontoxic, non-immunogenic, and capable of withstanding the dynamic aqueous and vitreous environments of the eye. They must also exhibit prolonged retention times to ensure sufficient drug exposure to the target tissues [7]. However, achieving these design criteria is no simple task, as nanocarriers must balance size, surface charge, and drug loading capacity while ensuring that they do not induce adverse immune responses or toxicity. Another critical challenge is the limited number of nanotechnology-based ocular drug delivery systems that have progressed from preclinical studies to clinical trials. The majority of nanomedicines for ocular delivery remain in the experimental stages. This discrepancy highlights the need for more rigorous translational research and the development of strategies to bridge the gap between laboratory success and clinical applicability [3]. However, the future of nanotechnology in ocular drug delivery is promising. Advances in biomaterials, surface functionalization, and stimuli-responsive drug delivery systems are driving the field toward more sophisticated and effective treatments for a wide range of ocular diseases, including glaucoma, age-related macular degeneration, diabetic retinopathy, and uveitis. Personalized medicine, in particular, offers exciting possibilities for tailoring nanocarrier systems to individual patient needs, potentially revolutionizing the treatment landscape for ocular conditions.

9.2 Ocular Drug Delivery: Hikes to Climb Over

Conventional ocular drug delivery methods, such as eye drops, ointments, and systemic administration, have been the cornerstone of ophthalmic treatment for decades. These methods are widely used due to their ease of application, patient compliance, and cost-effectiveness. Eye drops, in particular, are the most common route of administration for treating anterior segment diseases like conjunctivitis, dry eye, and glaucoma. They offer noninvasive delivery, allowing patients to self-administer the drug without medical assistance, which contributes to their popularity. Despite their wide usage, conventional ocular drug delivery methods have several limitations that affect therapeutic outcomes. The primary challenge with these methods is the eye's unique protective mechanisms, such as tear production, blinking, and lacrimal drainage, which rapidly eliminate the drug from the ocular surface. This results in poor drug retention and low bioavailability, with studies showing that less than 5% of the applied dose reaches the target tissue. Consequently, frequent dosing is often required to maintain therapeutic levels, which can reduce patient compliance and increase the risk of side effects. Systemic administration of drugs for ocular diseases, while effective for treating posterior segment conditions such as retinal disorders, presents its own set of challenges. Due to the blood-ocular barriers, only a small fraction of the systemically administered drug reaches the eye. High doses are often needed to achieve therapeutic concentrations, which increases the risk of systemic toxicity and adverse effects in other organs. Moreover, systemic delivery is not always specific to the eye, leading to non-targeted impacts and limiting its overall efficacy. Ointments and gels provide better retention times than eye drops, but they have drawbacks. These formulations can cause blurred vision, discomfort, and poor patient acceptance. Additionally, while they improve drug retention on the ocular surface, they still do not effectively address the challenge of delivering drugs to the posterior segment of the eye.

9.3 Emerging Nano-assisted Platforms for Ocular Delivery

Nanotechnology-based ocular drug delivery systems offer a revolutionary approach to overcoming the limitations of conventional methods. Utilizing nanoparticles, liposomes, nanoemulsions, and other nanocarriers enhances drug bioavailability, prolongs release, and allows for targeted delivery to both anterior and posterior segments of the eye. They protect drugs from degradation and provide sustained therapeutic effects, reducing the need for frequent dosing and improving patient compliance. Additionally, nanotechnology enables the delivery of a wider range of therapeutic agents, including hydrophobic drugs, thereby broadening treatment possibilities for various ocular diseases. Now let's discuss each nanotechnology ocular drug delivery system in detail in the subsequent section.

9.3.1 Vesicular Nanosystems

An intriguing method for seeking ideal ophthalmic delivery attributes, certainly an excellent permeability, tailored and regulated drug distribution, and convenience in drug administration, is through the application of vesicular nanosystems. Vesicles are a viable way to meet the demand for ophthalmic drug delivery with drop simplicity. Furthermore, vesicular carriers aid with both controlled ocular administration and sustained activity of the drug at the corneal surface by blocking its enzymatic metabolism at the corneal epithelium. Nevertheless, they will restrict and prolong the drug's effectiveness at its target site. Consequently, it makes sense to envision vesicles as prospective vehicles, and as that implies, they modulate the drug's disposition while affecting its absorption across the ocular membranes [8–10]. Notably, the two main vesicular carriers employed in ophthalmology include liposomes and niosomes.

9.3.1.1 Liposomes

One of the key research objectives lately concerns the design of optical products that may attain a drug's desired concentration at the site of action long enough to bring about a response while transferring both hydrophilic and hydrophobic therapies onto the inner layers of the eye. Accordingly, liposomes are the ideal means to accomplish these goals in ocular therapy. Furthermore, liposomes can be used to deliver ocular medicines with a high molecular weight and limited solubility. Additionally, the positively charged liposomal surface fosters interactions with the corneal mucous membrane that extends corneal penetration and time of contact [9, 11]. For instance, Bigdeli et al. designed cationic liposomes that delivered brimonidine and timolol together to alleviate glaucoma. By optimizing size, charge, and encapsulation efficiency, the liposomal preparation improved drug transport to the eye and had a dual effect of lowering intraocular pressure (IOP). The cationic nature of the liposomes facilitated better adhesion to the negatively charged ocular surface, and in vivo further studies showed superior IOP reduction with this combination compared to individual drug formulations. The study highlighted the effectiveness of using liposomal technology to improve the therapeutic outcomes of combination therapies in ocular diseases [12]. Likewise, Soni and Saini's study aimed to develop cationic liposomes that had been loaded with brimonidine tartrate with the intent to alleviate glaucoma. The liposomes' positively charged surface augmented their adherence to the negatively charged corneal layer, improving their ocular retention time and overall bioavailability. The findings from in vivo studies showed a notable decrease in IOP, indicating that cationic liposomes may be a useful delivery method to aid in the management of glaucoma [13]. To enhance corneal persistence and penetration for the administration of antiglaucoma drugs, Qiu et al. devised dendritic oligoethylenimine-decorated liposomes. Given the dendritic shape, which facilitated contact with the corneal surface, this novel liposomal formulation boosted the drug's diffusion across the cornea and retention on the ocular surface. In contrast to conventional liposomal formulations, in vivo investigations outlined improved absorption and a prolonged drop in IOP. This strategy offers a potential for more effective and durable treatment for glaucoma and marks a substantial development in topical medication delivery methods [14]. In a similar vein, Moiseev et al. developed PEGylated mucoadhesive liposomes decorated with maleimide for enhanced ocular delivery. The maleimide modification allowed the liposomes to form covalent bonds with the mucins in the tear film, leading to prolonged retention on the ocular surface. These liposomes were designed for sustained drug release, improving ocular bioavailability for anterior segment disorders. In vitro and in vivo studies demonstrated improved mucoadhesion and drug delivery efficiency compared to non-PEGylated formulations, offering a promising strategy for long-term ocular drug administration [15]. Another work, described by Qiao et al., put together liposomes incorporating rebamipide for treating dry eye conditions. To improve its retention and bioavailability, liposomes were employed to deliver rebamipide, which is renowned for being able to induce mucin production, to the ocular surface. Unlike traditional eye drops, the liposomal formula helped mitigate the symptoms of dry eyes by improving patients' hydration and mucin production. The study showed that liposomal carriers could improve rebamipide's therapeutic benefits when used to treat dry eye conditions [16].

9.3.1.2 Niosomes

Nanovesicles developed from nonionic surfactants, otherwise referred to as niosomes, are becoming popular nanocarriers in therapies owing to their affordability and resilience in solution and preservation. Niosomes are appealing nanocarriers for the therapeutic management of eye disorders because of their inherent characteristics. They are likewise flexible in organization, readily biodegradable, and perfectly suited for carrying drugs that are hydrophilic or hydrophobic. Additionally, vesicle component optimization or the design of a multifaceted interface greatly reduces nonspecific absorption [17, 18]. Ozdemir and Uner, for example, built a niosomal preparation of loteprednol for extended ocular release that was intended to relieve dry eye conditions. When evaluated for distribution within a rabbit model of dry eye, the niosomes displayed better drug persistence and tolerability in eye tissues. By bringing down demand for recurrent dosage, the prolonged-release attributes strengthened their therapeutic value in controlling dry eye irritation. According to the investigation, niosomes may be an effective way to distribute corticosteroids for the treatment of long-term eye conditions [19]. To maximize drug retention and absorption, a different study by Kandpal et al. put together a novel niosomal in situ gel that combines the benefits of gel platforms and niosomes. By enabling a prolonged release, this strategy sank the demand for frequent doses. Studies conducted both in vitro and in vivo showed enhanced corneal permeability and sustained drug action, which renders it valuable for ailments that call for persistent treatment, including glaucoma or post-surgical inflammation [20].

For the prolonged use of pilocarpine hydrochloride in managing the symptoms of glaucoma, Alotaibi et al. constructed a mucoadhesive biodegradable ocular insert laden with niosomes. The extended-release ocular inserts were fabricated to maintain optimal concentrations in the ocular tissues while promoting compliance among patients. The study, however, implied the promise of noisome-based inserts for delayed drug release, decreased dose frequency, and enhanced clinical efficacy [21]. Similarly, intending to alleviate ocular fungal infections, Yassin et al. reviewed the design of fluconazole-loaded niosomes integrated into contact lenses for persistent drug release while augmenting antifungal efficacy against chronic infections. By upholding drug concentrations in ophthalmic tissues for a long duration of time, the findings concluded that niosomes could increase the efficacy of treatment for antifungal molecules, thus rendering them an intriguing therapy for chronic ocular fungal infections [22]. In another study, brimonidine-loaded niosomes were put together by Tripathi and Yadav and subsequently added onto contact lenses to combat glaucoma by delaying drug release. The focus of the niosome-laden lenses sought to improve drug adherence across the ocular tissues, translating into a substantial drop in IOP. Following the research's findings, this delivery method extends drug release and limits the requirement for recurrent dosing, proposing a viable strategy for the continual therapy of glaucoma [23]. Comparably, as a means to ameliorate glaucoma, Ramadan et al. gazed into the fabrication of niosomal formula featuring timolol maleate, which was tailored for ocular administration in pursuance of particle dimension, encapsulation efficacy, and in vitro release. The results disclosed that the niosomal formulation delivered constant release and greater biological absorption in contrast to conventional products, and in vivo studies backed up a substantial drop in IOP, suggesting a potential substitute to liposomal systems for the management of glaucoma [24].

9.3.1.3 Other Vesicular Systems

Many vesicular systems, including transferosomes, ethosomes, transethosomes, bilosomes, and cubosomes, have recently been studied for ocular drug delivery. Introducing surfactants, the edge activators to the lipid bilayer in typical liposomes augments the adaptability and elasticity of the vesicles and yields transfersomes, which are ultra-flexible deformable nano-liposomes. They are both biodegradable and biocompatible, having exceptional encapsulation and preventing the payload degradation medications from breaking down in the body [25, 26]. In order to investigate more thoroughly, Mazyed et al. designed transgelosomes, an acetazolamide-loaded transferosomal gel to improve drug delivery to the eyes. The transferosomal gel demonstrated a substantial decline in IOP for 24 h, a longer duration of drug release, and a higher ex vivo permeation when compared to a free drug. Consequently, the transferosomes offered a potentially effective method for enhancing acetazolamide corneal distribution [25]. Ethosomes, another deformable flexible vesicle, had been designed to distribute actives to underneath tissue by altering the vesicle's composition. They can improve drug distribution and localization because of their more compact and more flexible form than conventional liposomes, which facilitates local delivery of both hydrophilic and hydrophobic active molecules [27]. For example, Uner et al. built ethosomes containing timolol maleate for the management of glaucoma. According to an ex vivo permeability investigation, ethosomes could effectively transport drugs to the cornea by releasing them gradually across the ocular membrane. The safety evaluation for the irritation testing was conducted with the HET-CAM approach, and the formulation appears nonirritating. An in vivo investigation further concluded that the new drug delivery method accomplished a similar pharmacological response to that of traditional eye drops [28]. It's noteworthy to consider that transethosomes, another kind of nanovesicles carrying ethanol and an edge activator, were lately put out. Transethosomes combine the benefits of transfersomes with ethosomes and offer extra benefits due to the inclusion of these components. Ethanol improves drug penetration through the microscopic epithelial pores' lipid fluidization while the edge activator weakens the phospholipid bilayer, and the vesicle becomes highly flexible and adaptable [29]. To attain better ocular absorption and longer corneal penetration, these ultradeformable vesicles have become superior vesicular systems. Accordingly, Srivastava et al. designed an innovative vesicular system with enhanced flexibility while comparing it with traditional vesicular carriers entitled transniosomes, which feature a blend of transfersomes, liposomes, and niosomes laden with brinzolamide. In contrast to other traditional vesicles, preliminary studies demonstrated their edge by gaining improved intraocular permeability and a noteworthy IOP reduction, reflecting a sustained therapeutic impact and enhanced ocular bioavailability [30]. In a similar vein, a different study by Patil et al. investigated the fabrication of transniosomes that contained brinzolamide for efficient glaucoma treatment. When compared to the commercial preparation or free drug suspension, these vesicles pointed out ultradeformability, extensive permeability, and prolonged drug release with no irritation. The successful delivery of transniosomes was further reinforced by the comprehensive in vivo studies, implying it is a therapeutic nanocarrier system by lower IOP [31]. Bilosomes represent an added, invaluable vesicular system that promotes corneal surface time while improving corneal permeability and flexibility. Closely linked to niosomes, they constitute closed bilayer vesicles of nonionic surfactants that carry bile salts. The nanosize of these vesicles and incorporation of bile salts and surfactants implies many opportunities to assess them through the ocular pathway [32]. Comparably, the unique and thermodynamically resilient isotropic framework, cubosomes consist of two in-line water channel systems and a bicontinuous bent bilayer of lipids in 3D. Their extensive surface space, outstanding thermal strength, modest viscosity, and capacity to hold water-soluble and insoluble molecules make them ideal options for ocular use [33]. Furthermore, Table 9.1 lists an array of vesicular nanocarriers reported in the scientific literature intended for intraocular drug delivery in several ocular disorders.

9.3.2 Lipid Nanoparticles

Lipid nanoparticles (LNPs) that are both biocompatible and biodegradable have become an appealing alternative for existing ophthalmic delivery methods if one wants to get beyond lots of ocular barriers. Over the preceding decade or so, lipid-based nanosystems brought about significant scientific developments and therapeutic upsides of ocular therapy, ranging from a prolonged drug-release pattern, an extended precorneal residence time, targeted site delivery, and a breakthrough in overall ocular bioavailability [68–70]. There has been much discussion of the special benefits of LNP, notably, solid lipid nanoparticles (SLNs) and nanostructured lipid carriers (NLCs)—in the focused ophthalmic delivery of different drug molecules. Solid lipid nanoparticles (SLN) are regarded as the most promising tactics for managing corneal disorders among lipidic nanocarriers. By honing their contact with the mucous membrane, SLNs' biological compatibility and mucoadhesive features foster and ameliorate the payloads' accessibility, mitigate its systemic and local adverse impacts, and reinforce its corneal residence. SLNs further exhibit exceptional in vivo tolerance, great mechanical strength, and protection of labile molecules against degradation. In addition, they incorporate the benefits of distinct colloidal carriers, like polymeric nanoparticles, emulsions, liposomes, or niosomes, while steering clear of their drawbacks [71, 72]. In this regard, Satyanarayana et al. sought to adopt SLN for introducing the glaucoma drug bimatoprost into the eyes to improve the drug's intraocular bioavailability and clinical impact. While compared to conventional formulations, the bimatoprost-loaded SLNs showed significant strides in drug release patterns and ocular retention. The in vivo observations further demonstrated a favorable IOP-lowering effect, underscoring the hope of SLNs to offer a better approach in glaucoma therapy [73]. The other study conducted by Alhakamy et al. focused on putting together SLN to facilitate the ophthalmic

Table 9.1 An overview of a few vesicular nanocarriers that are utilized for the ocular drug delivery

S. no.	Nanovesicle	Active pharmaceutical ingredient	Indication	In vitro and/or in vivo model(s)	Outcome(s)	References
1.	Liposomes	Latanoprost	Glaucoma	hTERT-HCEC, IM-HConEpiC, New Zealand rabbits	Devised liposomes revealed threefold better ocular bioavailability with no in vivo discomfort and remarkable in vitro tolerance	[34]
2.		Nutlin-3a	Proliferative vitreoretinal diseases	ARPE-19, HRPE	Nutlin-3a was effectively enclosed in the liposomal system and displayed a substantial decline in both migration and viability in RPE models	[35]
3.		Bevacizumab, dexamethasone	Wound healing	HCEC	Liposomes proposed a continuous release pattern alongside their biocompatibility	[36]
4.		Moxifloxacin	Bacterial infection (*Pseudomonas aeruginosa*, *Staphylococcus aureus*)	Albino rabbits	Besides maintaining a regulated drug release pattern, liposomes displayed no symptoms of discomfort in the in vivo model	[37]
5.		Dexamethasone	Intraocular inflammation	ARPE-19, C57BL/6 J mice	Designed liposomes have a strong anti-inflammatory effect, given their significant cellular drug delivery and low carrier cytotoxicity	[38]
6.		Resveratrol	Retinal damage	ARPE-19, Zebrafish mice	Liposomes restored the mitochondrial membrane potential, significantly reduced oxidative damage, and preserved cellular integrity	[39]
7.		Rebamipide	Dry eye disease	New Zealand rabbits, Sprague-Dawley rats	With prolonged corneal retention, liposomal eye drops illustrated outstanding stability, no irritation, and an extended release rate	[16]
8.		Besifloxacin	Bacterial infection (*Pseudomonas aeruginosa*)	–	Liposomes made it easier to distribute topical medications by providing a stable, biocompatible, and injectable formulation	[40]
9.		Brinzolamide	Glaucoma	–	Enhanced IOP reduction effect with similar tolerability to the commercially available formulation	[41]
10.		Tetrodotoxin, dexmedetomidine	Ocular surface diseases	HCLE	Extended corneal anesthesia without postponing the healing of corneal wounds	[42]

(continued)

Table 9.1 (continued)

S. no.	Nanovesicle	Active pharmaceutical ingredient	Indication	In vitro and/or in vivo model(s)	Outcome(s)	References
11.	Niosomes	Loteprednol etabonate	Dry eye disease	–	Niosomal preparation highlighted preliminary efficacy by stimulating the healing of rabbit corneal cells	[19]
12.		Epalrestat	Diabetic eyes	*Danio rerio* (zebrafish) embryos	Excellent permeability toward corneal and scleral tissues, favorable biocompatibility, and no risk of ocular distress	[43]
13.		Brinzolamide	Glaucoma	New Zealand white rabbits	Niosomal in situ gel persistently drops intraocular pressure in rabbits lasting 24 h after a single administration	[44]
14.		Latanoprost	Glaucoma	New Zealand albino rabbits	In contrast to commercial products, prolonged drug absorption with significantly reduced intraocular pressure	[45]
15.		Fluconazole hydroxypropyl-β-cyclodextrin	Fungal keratitis, fungal endopthalmitis	Albino rabbits	Niosomes exhibited extended corneal delivery, enhanced antifungal efficacy, and recommended tolerability	[46]
16.		Brimonidine tartrate	Glaucoma	New Zealand albino rabbits	Niosomal preparation displayed prolonged release characteristics together with better ocular bioavailability	[47]
17.		Natamycin	Fungal keratitis	Rabbit	The fabricated niosomal composition indicated profound transcorneal penetration and longer corneal persistence	[48]
18.		Lomefloxacin	Bacterial conjunctivitis	Albino rabbits	The built vesicular preparations did not seem to cause discomfort and outlined highly significant therapeutic benefits for ocular application	[49]

19.	Ethosomes	Amphotericin B, indocyanine green	Fungal keratitis	HCE, Wistar rats	The vesicular system showed enhanced ex vivo penetrations in the excised goat cornea, with in vivo efficacy in rats underlying the potential to be developed into a treatment for fungal keratitis	[50]
20.		Timololmaleate	Glaucoma	Wistar albino rats	The ethosomes were ideal for ocular use and demonstrated a notable reduction in intraocular pressure without causing any discomfort, bleeding, or lysis	[28]
21.		Ketoconazole	Fungal keratitis	New Zealand rabbits	In addition to exhibiting longer drug release and no ocular discomfort, nanovesicles demonstrated more powerful antifungal efficacy against a conventional strain of *Candida albicans* when compared to a drug suspension	[51]
22.	Transferosomes	Curcumin	–	–	Transferosomes demonstrated improved corneal and conjunctival drug uptake with noticeably deeper and more effective drug delivery, not much burst release, good corneal biocompatibility, and an entrapment efficiency of over 99.96%	[52]
23.		Fluconazole	Fungal keratitis	–	For 24 h, fluconazole from the transfersomal formulation maintained its ex vivo ocular penetration while demonstrating good bioadhesion, outstanding antifungal activity, and ocular safety and non-irritability	[53]
24.		Cyclosporine A	–	New Zealand albino rabbit	The transferosomes showed potential for sustaining the release over the period investigated. The corneal permeability and flux from the Span® and Tween® formulations were similar	[54]

Table 9.1 (continued)

S. no.	Nanovesicle	Active pharmaceutical ingredient	Indication	In vitro and/or in vivo model(s)	Outcome(s)	References
25.	Bilosomes	Betaxolol hydrochloride	Glaucoma	New Zealand albino rabbits	When compared to the commercially available product, the bilosomal formulation demonstrated superior and long-lasting IOP-lowering activity, which may reduce the frequency of drug administration and improve patient compliance	[55]
26.		Brinzolamide	Glaucoma	–	With a trans-corneal permeability of 8.78 ± 0.14 cm/h, the bilosomes were shown to be nonirritating to the eyes, suggesting their promising deployment in glaucoma treatment	[56]
27.		Agomelatine	Glaucoma	New Zealand albino rabbits	The developed product outperformed the free medication in the in vivo pharmacodynamics parameters and displayed a spherical shape with more elasticity than its corresponding niosomes	[57]
28.		Acetazolamide	Glaucoma	New Zealand albino rabbits	When compared to free drug, commercially available dorzolamide eye drops, and commercially available oral tablets, the bilosomal preparation demonstrated a stronger and enduring effect on IOP lowering while being safe after ocular instillation	[58]
29.		Natamycin	–	–	The drug's transcorneal flow was raised by 6–9 times, indicating effective bilosome ocular permeation and an intriguing framework for longer-lasting and better ocular pharmacotherapy	[59]
30.		Terconazole	Ocular fungal infections	Albino rabbits	Following topical ocular administration, the bilosomal formulation demonstrated improved drug flux through rabbit cornea along with confirmed safety	[60]

31.	Cubosomes	Luteolin	Glaucoma, ocular inflammation		In anti-inflammatory trials, phytocubosomes demonstrated a sustained drug release profile and enhanced antioxidant activity, translating to 6.46-fold decreases in intraocular pressure and quicker recovery times	[61]
32.		Fenticonazole nitrate	Candidiasis	New Zealand albino rabbits	The ex vivo transcorneal permeation of the cubosomes was 2.69 times more than that of the drug suspension. Additionally, the in vivo results confirmed that cubosomes are effective in improving drug bioavailability, ocular retention, absorption, and penetration	[62]
33.		Sorafenib tosylate	Diabetic retinopathy	Wistar albino rats	The AUC_{0-24} in the retina was approximately 5 times higher in drug-loaded cubosomes than in drug suspension, improving the drug's bioavailability	[63]
34.		Acetazolamide	Glaucoma	New Zealand albino rabbits	When compared to commercial Azopt® eye drops and Cidamex® tablets, the cubosomal product indicated superior therapeutic effectiveness and a considerable reduction in IOP without causing eye irritation	[64]
35.		Brimonidine tartrate	Glaucoma	New Zealand albino rabbits	In comparison to the commercial product (Alphagan®P), the developed formula revealed enhanced corneal permeability and improved ocular bioavailability, thus increasing patient adherence to therapy	[65]
36.		Timolol Maleate	Glaucoma	New Zealand rabbits	Ex vivo and in vivo tests showed that cubosomes outscored commercially available eye drops in terms of boosting corneal penetration by extending the duration of drug retention and amplifying the drug's ability to reduce intraocular pressure yet being nontoxic and biocompatible	[66]
37.		Beclomethasone dipropionate	Uveitis	New Zealand albino rabbits	The developed cubosomes had a higher relative bioavailability, a markedly controlled release, improved ocular permeability, and a high encapsulation efficiency. When compared to the drug suspension, cubosomes showed a significant increase in $AUC_{0\text{-}10h}$ of up to 5.5-fold	[67]

administration of ofloxacin, an antibiotic widely employed for healing corneal infections. This research explored various formulation strategies to maximize the drug's intraocular bioavailability and retention, while the optimized SLNs illustrated a stronger foundation and release properties, intensifying the drug concentration in ocular tissues. In vivo studies, however, highlighted the SLNs' promise as a workable substitute for antibiotic therapy in ophthalmology by confirming their effectiveness in addressing eye infections [74]. To better deliver ciprofloxacin to the eyes, Onugwu et al. formulated SLNs decorated with chitosan and poly(2-ethyl-2-oxazoline). By strengthening mucoadhesion, the chitosan coat facilitated drug accumulation on the corneal surface. In an attempt to combat bacterial conjunctivitis more effectively, the investigation proved that the devised SLN provides a uniform dispersion and better bioavailability in ocular tissues. Conclusions from the study suggested SLNs can function much better as ocular medication delivery systems when encircled with biocompatible polymers [75]. Furthermore, for topical ophthalmic treatment, in an intensive study published by Nair et al., clarithromycin was embedded in SLN to improve the steadiness and effectiveness of the therapy. The SLNs have drastically decreased the bacterial load in ocular infections by enabling a continuous release pattern and augmenting corneal diffusion of clarithromycin. The findings from this research underline the likelihood of SLNs strengthening the impact of antibiotic therapies for ailments concerning the eyes [76].

By resolving the drawbacks of the SLNs, another lipid system, nanostructured lipid carriers (NLCs) enable better ocular medication delivery. Their increased mucoadhesive properties, absorption rate, comfort, and intraocular penetration contribute to greater interfacial contact with the ocular surface, extending the precorneal residence period. Varela-Fernández et al. devised lactoferrin-loaded NLCs with the intent to maximize the diffusion of drugs into the eyes. To improve the bioavailability and longevity of lactoferrin, a significant protein with anti-inflammatory and antioxidant qualities, this investigation concentrated on the preparation and evaluation of NLCs. The optimized NLCs showed increased corneal barrier crossing and longer retention duration in the eye, indicating their potential for successful treatment of a range of ocular diseases [77]. In a distinct investigation, Gautam et al. looked into the development and optimization of a gel product that included the antibiotic lomefloxacin in NLC for intraocular administration. The goal of the study was to maximize drug loading and sustained release characteristics by adjusting the formulation parameters. According to in vitro tests, the NLC gel exhibited prolonged retention on the ocular surface and increased antibacterial activity, which suggests that it could be a viable treatment for bacterial conjunctivitis [78]. Likewise, for topical ocular delivery, Kumar et al. (2022) looked at itraconazole-loaded NLCs with an intent to improve the itraconazole's solubility and bioavailability in ophthalmic use. Therefore, to effectively treat fungal eye infections, the adjusted NLCs showed enhanced penetration across ocular barriers and sustained release characteristics. When compared to conventional distribution methods, in vivo experiments validated the effectiveness of the NLC formulation [79]. Yet further work by Baig et al. aimed at building cationic NLCs for the delivery to the eye of a separate antibiotic in ophthalmic injection, besifloxacin. The NLCs' cationic nature enhanced their mucoadhesive qualities, which improved the drug's ability to stick to the surface of the eye. According to the study, cationic NLCs had superior antibacterial activity when compared to traditional solutions, indicating a great deal of promise for enhancing the management of ocular infections [80]. Moreover, a quick overview of ocular delivery systems employing LNPs is given in Table 9.2, with special attention to SLNs and NLCs, which have been the subject of multiple recent studies published in the literature.

9.3.3 Polymeric Nanoparticles

Polymeric nanoparticles (PNPs) have a few significant features which renders them especially useful for ocular drug administration. These characteristics include their capacity to stick to the eye's mucous layer, which prolongs drug absorption and increases the absorption rate; their controlled and steady release profiles, which lower dosage frequency and enhance patient compliance; and their biodegradability and biocompatibility, which guarantee long-term ocular use safety [95, 96]. Additionally, PNPs preserve encapsulated medications from enzymatic degradation, improve drug penetration across ocular barriers, and decrease systemic damage by localizing their release at the intended site. They are more stable and efficient at penetrating underlying corneal tissues because of their small dimension and the capacity to modify their surface [97–99]. PNPs are readily developed on natural polymers including chitosan (CS), hyaluronic acid (HA), albumin, cellulose, gelatin, and alginate, or by utilizing synthetic polymers like polylactic acid (PLA), poly-lactic-co-glycolic acid (PLGA), polyacrylamide, and polycaprolactone (PCL) (Xiao et al., 2022). Notably, natural polymers are particularly advantageous because of their low toxicity, biodegradability, and biological compatibility. For instance, chitosan lengthens drug persistence on the corneal surface via improving adhesion. Likewise, gelatin and alginate provide biological compatibility and can create hydrogels that gradually govern the drug release of encapsulated therapeutics. Ocular discomfort and low corneal bioavailability are two problems with the traditional use of nonsteroidal anti-inflammatory medication (NSAID) eye drops for managing eye inflammation. In a work by Mohamed et al., they developed CS-NPs that loaded meloxicam (MLX) via electrostatic forces and improved the drug's solubility. According to their research, in rabbit models, the MLX release through these nanoparticles lasted for 72 h and had a suitable flux and penetration through the cornea and sclera. In contrast to traditional MLX eye drops, in vivo tests indicated that the MLX-CS-NPs composition had better therapeutic properties without irritating the eyes. The mucoadhesive qualities of chitosan prolonged the duration of contact and release, while the nanoparticles served as a reservoir, allowing for consistent drug release. Following 3 days of MLX-CS-NPs eye drop therapy, the histological study revealed a normal scleral epithelium with no edema symptoms and a healthy corneal epithelium with no inflammatory conditions, as seen in Fig. 9.1 [100]. It is pertinent to be aware that mucoadhesive nanoparticles may prolong the drug's pre-corneal dwelling duration and availability, which is favorable over traditional

Table 9.2 Applications of several LNPs being utilized in ocular therapy

Sr. no.	Lipidic nanosystem	Active pharmaceutical ingredient	Indication	Cell line/animal model	Outcome(s)	References
1.	SLN	Vildagliptin	Diabetic retinopathy	Albino rabbits	SLN showcases a sustained release profile that regulates blood sugar levels	[81]
2.		Bimatoprost	Glaucoma	–	A prolonged release pattern alongside ocular acceptability was reported with the proposed SLN system	[82]
3.		Clarithromycin	Bacterial endophthalmitis	Albino rabbits	SLN features a substantially greater drug concentration in aqueous humor with no observed discomfort	[76]
4.		Econazole	Fungal keratitis	New Zealand white rabbits	Improved antifungal potency in tandem with a controlled drug release profile	[83]
5.		Isoniazid	Ocular tuberculosis	HCLE, HCJE, rabbits, rats	Regulation of ocular TB in corneal and conjunctival cell lines in an effective, localized, and reliable manner	[84]
6.		Triamcinolone acetonide	Posterior segment ocular diseases	–	Extended drug persistence across the ocular surface with no adverse effect on the corneal epithelia	[85]

7.	NLC	Apigenin	Anterior segment ocular diseases	New Zealand white rabbits	The proposed NLCs exhibited physiological stability, corneal uptake, and prolonged release across the ocular tissues	[86]
8.		Melatonin	Uveal melanoma	HCE-2 cell line, New Zealand white rabbits	Devised nanostructure revealed substantially improved therapeutic efficacy	[87]
9.		Mangiferin	Oxidative stress–related ocular diseases	ARPE-19	By reducing ROS production and GSH depletion brought on by H_2O_2, mangiferin-loaded NLCs demonstrated their antioxidant qualities	[88]
10.		Sorafenib	Corneal neovascularization	HCEC, New Zealand white rabbits	The nanoformulation maintained a 12-h drug level while being safe in HCEC and nonirritating in rabbit eyes	[89]
11.		Lactoferrin	Keratoconus	–	NLCs owned mucoadhesive qualities, a sustained and regulated release pattern, and were persistent throughout	[77]
12.		Itraconazole	Fungal infection (*Aspergillus niger* and *Aspergillus flavus*)	–	The devised NLCs highlight considerable potency when compared to the commercial formulation	[90]
13.		Quercetin	–	HCEC, New Zealand white rabbits	A clear and aesthetically appealing NLC system that promotes transcorneal permeation and precorneal persistence	[91]
14.		Amphotericin B	Fungal infection (*Candida albicans*—wild-type and drug resistant—and *Aspergillus fumigatus*)	ARPE-19	Developed NLCs were sufficiently effective and showed no undesirable effects	[92]
15.		Dexamethasone	Posterior segment ocular diseases	HCE-T	NLC system illustrates improved absorption, high payload capacity, and narrow size distribution	[93]
16.		Dexamethasone	Posterior segment ocular diseases	New Zealand albino rabbits	A substantial leap in ocular accessibility, retention, and absorption with no redness of the conjunctiva	[94]

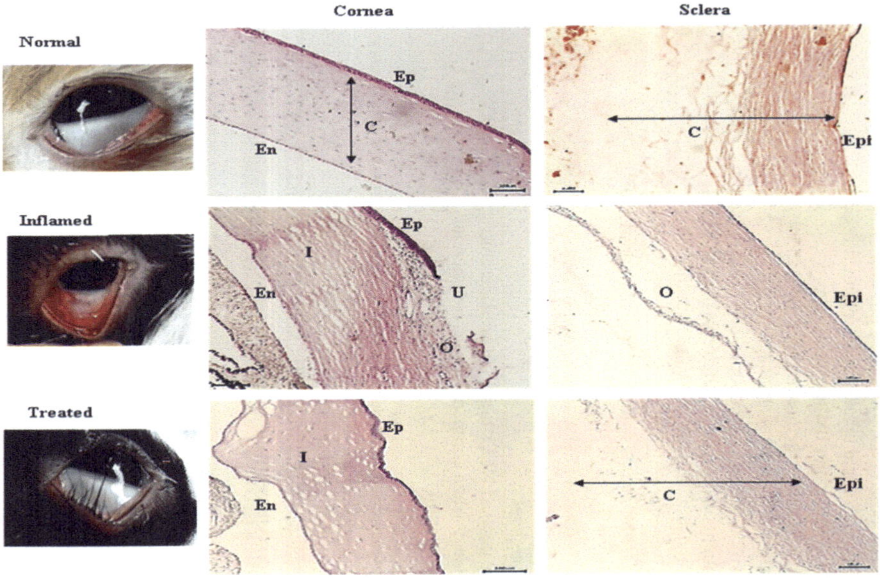

Fig. 9.1 Light microscopy images of rabbit corneal and scleral tissues, both before and after MLX-CS-NPs eye drops treatment. (Reprint from [100]). Images show the epithelium (Ep), endothelium (En), connective tissue (C), episclera (Epi), ulceration (U), edema (O), and inflammatory cell infiltration (I)

preparations that wear off in ocular diseases, including bacterial or fungal infections, because of tear drainage and dose spillage. In a related investigation, Bhosale et al. coated voriconazole-loaded mucoadhesive CS-NPs (VCZ-MA-NPs) with the anionic polymer sodium alginate and subsequently turned them into mucous-penetrating NPs (VCZ-MP-NPs). Under electron microscopic imaging, the NPs, CS-NPs, and modified NPs all had a spherical and uniform shape. For VCZ-MA-NPs, the drug penetration was 58.98% ± 0.54%, while for VCZ-MP-NPs, it was 70.02% ± 0.61%. The outcomes demonstrated that the mucous-penetrating nanoparticles might reliably infiltrate the corneal epithelium, avoiding the ocular fungal covering and mucous barriers, implying the opportunity for treating fungal keratitis. Likewise, the usual treatment for corneal ulcers, whether they are indolent or melting, is the application of corneal healing agents and antibiotic eye drops. Natural remedies, including curcumin, have perks over conventional therapy for ulcerative keratitis given that they have potent anti-inflammatory capabilities and may block proinflammatory cytokines. Recent work by Sohani et al. employed curcumin NPs encased with β-cyclodextrin and HA to formulate an ophthalmic hydrogel that would speed up corneal healing. When compared to alternative therapies, the curcumin NP hydrogel showed noticeably better corneal epithelial healing and restored tissue integrity in the ulcerative keratitis model. Notably, NPs dramatically improved corneal transparency and decreased lesion size [101].

PLA, PCL, and PLGA are examples of synthetic polymers that may yield stable nanoparticles with predictable degradation regimes. Drugs may exhibit a continuous release by varying their tunable degradation rates. These polymers are also perfect for long-term use in delicate tissues like the eye since they break down into nontoxic metabolites like lactic acid and glycolic acid [102]. For example, the commonly used antifungal sertaconazole (STZ) has limited utility in topical preparations due to its struggling water solubility. Eleraky et al. developed Sertaconazole-PEG 2000-laden PLGA NPs to treat ocular keratitis to overcome this issue. For an entire day, the product displayed prolonged absorption. The corneal permeability of PLGA NPs loaded with STZ-PEG2000 was 2.5 times higher than that of PLGA NPs loaded with STZ in the rabbits' corneas. When employing *Candida albicans* to test the NPs' antifungal efficacy, the NP-based preparation proved to be four times more effective than the free drug formulation [103]. A separate investigation by Nguyen et al. focused on how the thickness of the shell affected hollow PLA NPs' ability to distribute drug continuously for the treatment of glaucoma. Pilocarpine-containing PLA NPs with tuned shell thickness (10–100 nm) had no damaging impact on rabbit eyes or human lens-derived BCE cells. In rabbits, moderately thick shells (about 40 nm) were best because they prevented retinal cells and optic nerve degenerative changes in progressive glaucoma, maintained retinal electrophysiology, controlled IOP, and preserved corneal endothelial structure for over 56 days. This technique demonstrates the potential of modifying shell thickness to develop sustained-release medication delivery systems for long-term eye disorders [104]. A novel method for developing a ranibizumab-loaded PLGA/PLA co-polymeric system for the management of AMD was recently discovered. Research on the ocular cell lines D407 and ARPE19 showed a substantial level of cell viability and outstanding trans-well permeability. Furthermore, the formulation demonstrated outstanding biocompatibility by lowering inflammatory cytokines and reactive oxygen species (ROS). These findings suggest that the prepared NPs appeared safe to administer in the eyes and have the potential to improve intraocular AMD treatment [105] Table 9.3 presents a concise summary of various studies investigating PNPs in ocular diseases.

9.3.4 Other Nanostructures

9.3.4.1 Nanoemulsions

Nanoemulsions (NEs) have become more prevalent in the past decade for intraocular drug administration. In light of their prolonged drug release, longer residence time, elevated topical penetrability, lipophilicity, and overall stability, NEs may lower periodic installations and enhance patient compliance. Furthermore, sterilizing NEs is as convenient as with conventional solutions and rarely causes obscured vision, which makes them suited for developing clinically worthwhile ocular drug delivery systems [116, 117]. In a study, Tang et al. came up with organic cation/carnitine transporter (OCTN2)- and Amino acid transporter $B^{0,+}$ ($ATB^{0,+}$)-targeted NEs for bettering drug delivery to the corneal epithelial cells. With an eye on distinct transporters in HCECs, the NEs get over the ocular obstacles, notably lacrimal

Table 9.3 Several PNPs reported in recent research for use in ophthalmology

Sr. no.	Nanocarrier material	Active pharmaceutical ingredient	Indication	Cell types/animal model	Outcome(s)	References
1.	PLGA, PCL	Baricitinib	Uveitis	–	Compared to a free drug, PNPs provided a depot effect and enhanced intraocular penetration without displaying any discomfort	[106]
2.	Thiolated xyloglucan	Moxifloxacin	Bacterial infection	RPE	Enhanced corneal permeation, longer retention, and biological adhesion	[107]
3.	Chitosan, PLGA	Forskolin, rutin	–	–	PNPs confirmed superior persistence, an extended residence period, and a sustained release practice	[108]
4.	Chitosan oligosaccharide, dextran sulfate	Dorzolamide	Glaucoma	–	The proposed PNPs illustrated superior therapeutic potency and intraocular invasion, indicating an effective, reliable substitute	[109]
5.	Eudragit RL	Terconazole	Fungal infection (*Candida* species)	New Zealand albino rabbits	PNPs indicated higher antimycotic action, yet did not provoke any irritation or abnormal changes	[110]
6.	PLGA	Moxifloxacin	Bacterial infection	New Zealand white rabbits	PNPs exhibited improved drug kinetics while avoiding irritation to the eyes	[111]
7.	Chitosan	Clarithromycin	Microbial infection (*Streptococcus aureus*, *Pseudomonas aeruginosa*)	–	PNP significantly raised corneal diffusion, enhanced bactericidal sensitivity, and displayed no evidence of corneal damage	[112]

8.	Montmorillonite, Eudragit RS and RL	Betaxolol	Glaucoma	–	Longer residency duration, outstanding stability, and strong corneal adherence	[113]
9.	Eudragit RL and RS, Tween 80	Gatifloxacin	Microbial infection (*Escherichia coli, Pseudomonas aeruginosa, Staphylococcus aureus*)	Human corneal cells	Significant rigidity, extended residency period, and substantial ocular adherence	[114]
10.	Chitosan, sodium tripolyphosphate	Norfloxacin	Microbial infection (*Staphylococcus aureus*)	–	PNP showed promising outcomes in eradicating extraocular bacterial infections	[115]

dilution and ocular penetration. Implementing stearoyl L-carnitine, the researchers set up NEs that significantly improved drug uptake over conventional formulations and had great entrapment efficiency. In addition, these NEs showed extended-release traits rendering them impeccable for perpetual eye disorders like dry eye syndrome and glaucoma [118]. The most recent research by Jahromy et al. reviewed curcumin-loaded NEs for precise delivery to the posterior region of the eye; the NEs were tailored for particle size and stability, assuring curcumin attained clinical levels in the retina; and results revealed that curcumin-loaded NEs substantially minimized oxidative distress plus inflammation in animal models of age-related macular degeneration, showcasing an intriguing option to combat retinal conditions as an outcome of curcumin's anti-inflammatory and antioxidant effects [119]. In a related work, Guo et al. devised an antioxidant-based NE to better administer the standard glaucoma drug latanoprost. With its anti-oxidative properties, the NE sought to safeguard the cornea from oxidative damage, a key contributor to the development of glaucoma, while simultaneously boosting the uptake of drugs in the corneal tissues. Unlike conventional latanoprost products, the formulation confirmed superior efficacy for significantly reducing IOP. The antioxidants additionally lowered inflammatory distress, which intensified the efficacy of the treatment. The study, thus, integrates corneal protection and IOP management, marking a breakthrough in NE-based treatments for glaucoma [120]. Additionally, a different study by Saquib et al. assessed the usage of dextran-containing NEs to treat dry eye illness. Encapsulated in NE, dextran, a hydrophilic polymer, restored corneal hydration and reduced intraocular irritation. By reducing lacrimal outflow and retaining the fluid on the corneal surface, the Nes might deliver dry eye patients durable symptomatic relief better than traditional therapies [121].

9.3.4.2 Polymeric Micelles

Polymeric micelles (PMs) emerged as one of the most tempting drug delivery systems in modern times for treating disorders of the eye that affect either the anterior compartment (like dry eye syndrome) or posterior (like glaucoma or diabetic retinopathy). Given a highly hydrophilic wrap around a hydrophobic center that self-associates in water, these colloidal molecules produce nanoscale aggregates with a core-shell pattern that arises from a water-friendly and water-resistant block copolymer. Because of their additional hydrophilicity, PMs make lipophilic parts better soluble and provide a clear, aqueous solution, which makes application easier with no forfeiting vision [122–124]. Furthermore, owing to its limited dimensions, which significantly enhance infiltration and durability, it is practicable to target inflammatory areas. Because PMs can increase topical bioavailability, they have attracted a lot of attention for use in clinical settings [125, 126]. Lu et al. recently set up PMs for the management of fungus-related keratitis by embedding butenafine in a TPGS frame aimed at improving its intraocular penetration. These PMs have enticing consistency during preservation and biphasic release layouts. Additionally, these nanomicelles revealed exceptional ophthalmic resilience and profound cellular responsiveness in rabbits. The nanoformulation demonstrated considerably improved in vivo ocular penetration plus notable antifungal capabilities in comparison to the drug suspension [127]. In another pertinent study, Sathe et al. developed durable and clear TPGS-based PMs to cope with questions concerning natamycin's limited ocular bioavailability for the treatment of fungal keratitis. It had demonstrated effectiveness in extending penetration, sustained release, and pre-corneal residency for up to a whole day. In addition, studies carried out in vivo, ex vivo, and in vitro implied its efficacy as well. Furthermore, the rabbit model has been employed to explicate the reliability and acceptability of the developed PM, indicating that the blend is both essentially safe and not bothersome to the eye. Consequently, designed nanomicelles could turn into one of the effective substitute medications for the treatment of fungal keratitis [128].

9.3.4.3 Hydrogels

A remarkably flexible category of biomaterials, hydrogels offer a wide range of possible purposes in ophthalmology. They are currently being researched for potential uses as vitreous alternatives and intravitreal drug delivery systems. They have gained extensive use as soft contact lenses, foldable intraocular lenses, in situ gelling formulations for ophthalmic drug delivery, and ocular adhesives for wound repair. For instance, Ilochonwu et al. developed thermo-sensitive hydrogels to convey an anti-VEGF component and a corticosteroid into the eyes, enabling a controlled drug release when exposed to physiological temperatures. Evaluations conducted both in vitro and in vivo revealed that the drug was effectively retained in the ocular tissues, resulting in enduring therapeutic benefits. This novel method offers a viable way to treat eye conditions that call for long-term administration of anti-inflammatory and anti-VEGF drugs [129]. Similarly, Aslzad et al. established an in situ forming hydrogel containing a blend of chitosan and dialdehyde starch for the transport of betamethasone through the eyes. The purpose of this hydrogel was to provide targeted and prolonged medication release by forming a gel-like structure when it comes into contact with ocular fluids. According to the study, the hybrid hydrogel successfully increased betamethasone's ocular bioavailability while lowering eye tissue inflammation. The findings imply that this hybrid strategy may improve the treatment of several inflammatory disorders of the eyes [130]. Additionally, for use in ophthalmic delivery, Li et al. designed a bioinspired self-assembly supramolecular hydrogel system to build an enduring framework that would make it easier to encapsulate and release therapeutic substances. The investigation showed that the supramolecular hydrogel improved the corneal availability of encapsulated medications by offering a prolonged release profile. This study suggests that bioinspired hydrogels may provide novel approaches to efficiently administering ocular medications [131]. In a distinct work, Varela-Garcia et al. devised bioinspired hydrogels functioning with cytosine to carry transferulic acid, a naturally occurring antioxidant, for application in the eyes. By simulating the natural extracellular matrix, the hydrogels created an environment that was favorable for drug encapsulation and release. According to in vitro research, the functionalized hydrogels boosted transferulic acid's stability and bioavailability, which in turn improved the antioxidant activity of ocular tissues. This study demonstrates the potential of bioinspired hydrogels to deliver therapeutic compounds that can counteract oxidative stress in conditions affecting the eyes [132].

9.3.4.4 Dendrimers

Dendrimers, a new class of polymeric molecules featuring unique tree-like nanostructure drivers that facilitate quick drug distribution and therapeutic targeting, have ophthalmologists' close consideration in recent times. Having their myriad advantages—including their nanosize, numerous surface moieties, quick modification, and, most importantly, the convenience of manufacturing—the setting up of dendrimers out of a range of delivery methods could potentially boost the drug's bioavailability. The wrapped-up product may have a distinct architecture and terminal groups that may provide appropriate pharmacokinetic properties due to tweaking and even deception of the structural foundation of dendrimers during fabrication. These characteristics will further impact how useful it is as a nanocarrier for imaging, drug delivery, and genetic material. Additionally, scientists have engineered close to 100 distinct kinds of dendrimers to decrease the toxicity of medications and increase their lifespan in ocular tissues [133–136]. For instance, Wang et al. constructed dendrimer gel particles and incorporated brimonidine tartrate and timolol maleate to test the effectiveness and safety of each in alleviating glaucoma by evaluating the system's cytocompatibility, biodegradable properties, release kinetics, and ocular penetration. The developed dendrimeric technology provided notable cytocompatibility and enabled zero-order release, while all of them have shown good hemocompatibility. The formulation demonstrated a clear reduction in IOP in the in vivo glaucoma model, in both Brown Norway rats and mice, indicating a positive therapeutic impact for treating glaucoma [137].

9.4 Clinical Portrait of Ocular Nanoformulations

The application of nanotechnology in ocular drug delivery has significantly advanced, with several nanoformulated drugs approved for use and others actively undergoing clinical trials. These formulations address the challenges in traditional ocular therapies, such as limited drug penetration and bioavailability, by using nano-sized carriers that enhance delivery to target ocular tissues. Here, we review prominent nanomedicine-based drugs that have been approved and those in various clinical trial phases, highlighting their therapeutic uses and advancements.

9.4.1 Approved Drug Products

9.4.1.1 Restasis®

Restasis® was the first nanoemulsion formulation approved for treating chronic dry eye. Cyclosporin A in Restasis® is encapsulated within a nanoemulsion, enhancing its penetration across ocular barriers to reduce inflammation and increase tear production in patients with dry eye syndrome. Since its approval, Restasis® has demonstrated effective management of dry eye symptoms, setting a precedent for nanoemulsion-based ocular therapies, such as bacterial infections and post-surgical inflammation [138].

9.4.1.2 Durezol®

Durezol® is another approved nanoemulsion formulation, containing the corticosteroid difluprednate, used to treat ocular inflammation. This nanoemulsion improves the bioavailability of difluprednate, allowing for a more targeted anti-inflammatory effect within the eye. Durezol®'s success has validated the use of nanoemulsions for effective anti-inflammatory ocular treatments, especially post-surgical inflammation management [139].

9.4.1.3 Visudyne®

Approved by the FDA in 2000, Visudyne® is a liposomal formulation that delivers verteporfin to treat age-related macular degeneration (AMD) and choroidal neovascularization. Verteporfin is a photosensitizer activated by light during photodynamic therapy, targeting abnormal blood vessels that cause vision impairment in AMD. Visudyne®'s formulation as a liposome enhances its delivery to target tissues while minimizing systemic exposure and potential side effectsq[140].

9.4.1.4 Macugen®

Macugen®, an anti-VEGF therapy, was approved in 2004 as an intravitreal injection. Its nano-aptamer structure selectively binds to VEGF, reducing abnormal blood vessel growth in the retina associated with wet AMD. This formulation's nano-based approach enables effective inhibition of VEGF while limiting damage to surrounding retinal cells, offering a safer and more targeted option for AMD treatment [140].

9.4.1.5 Ozurdex®

Ozurdex® is a biodegradable dexamethasone implant approved for managing macular edema and noninfectious uveitis. The dexamethasone in Ozurdex® is delivered through a slow-release system, allowing sustained drug delivery over several months, thereby reducing the need for frequent injections and enhancing patient compliance [141].

9.4.2 Drug Products Under Clinical Trials

9.4.2.1 TLC399 (ProDex)

TLC399, a lipid-based nanoformulation containing dexamethasone sodium phosphate, is currently in phase II clinical trials for treating macular edema caused by retinal vascular occlusion. The nano-lipid structure of TLC399 enhances the delivery and retention of dexamethasone within the retina, allowing for prolonged therapeutic effects and potentially decreasing the frequency of treatments necessary for managing macular edema [142].

9.4.2.2 POLAT-001

This innovative nano-liposomal formulation containing latanoprost has completed phase II clinical trials for the treatment of primary open-angle glaucoma and ocular hypertension. POLAT-001's liposomal coating helps latanoprost penetrate ocular tissues more efficiently, enabling improved intraocular pressure control, which is essential in managing glaucoma progression. If successful, POLAT-001 could represent a less invasive and more effective approach for glaucoma patients [142].

9.4.2.3 SYSTANE®

SYSTANE® is a propylene glycol-based nanoemulsion eye drop designed to alleviate symptoms of dry eye disease. Having completed phase IV trials in 2020, SYSTANE® provides extended moisture retention on the ocular surface, offering relief for dry eye sufferers. Its nanoemulsion formulation improves the contact time on the ocular surface, enhancing lubrication and tear stability compared to conventional dry eye treatments [143].

The increasing list of FDA-approved and clinically tested nanoformulations reflects a growing acceptance of nanomedicine within ophthalmology. These advancements underscore the potential for nanocarriers to overcome ocular drug delivery challenges, offering more effective, sustained, and patient-friendly treatments. However, further studies are essential to refine these formulations, optimize therapeutic outcomes, and ensure safety across diverse patient populations.

9.5 Ocular Nanomedicine: Roadblocks in Their Development

The field of ophthalmic nanomedicine has shown remarkable promise in enhancing drug delivery and treatment outcomes for ocular diseases. However, translating these innovations from research to clinical application presents significant challenges that span across biological, technical, and regulatory domains. Understanding and addressing these barriers are essential to making nanomedicine-based therapies viable for widespread clinical use.

9.5.1 Ocular Barriers

The eye's unique anatomy poses one of the most significant challenges to effective nanomedicine delivery. The presence of multiple physiological barriers, such as the tear film, corneal epithelium, and blood-retinal barriers, restricts drug penetration, limiting therapeutic efficacy. Due to these protective mechanisms, only a small fraction of topically applied drugs reach deeper ocular tissues, leading to limited bioavailability [144]. Traditional drug delivery methods, such as eye drops, often fail to maintain therapeutic drug levels due to these barriers. Nanocarrier systems, such as liposomes, nanoparticles, and micelles, have been developed to improve drug penetration, yet their effectiveness remains limited by these intrinsic barriers [96].

9.5.2 Stability and Controlled Release

Stability issues are also critical, as many nanomedicine formulations are susceptible to degradation or aggregation, affecting their shelf life and bioactivity. Additionally, achieving controlled and sustained drug release is challenging. Ocular diseases often require long-term treatment, but frequent drug administration can lead to compliance issues and potential toxicity. Nanocarriers designed for extended drug release could address these issues, yet they require meticulous engineering to balance release rates without compromising stability [145].

9.5.3 Immunogenicity and Biocompatibility

Biocompatibility and immunogenicity of nanocarriers present additional concerns. The eye is highly sensitive to foreign materials, and immune responses can be triggered by nanocarriers, leading to inflammation or toxicity. Developing biocompatible materials that minimize adverse immune reactions is crucial for patient safety. Even minor immunogenic responses can result in inflammation, potentially exacerbating the disease and reducing therapeutic efficacy. Research on materials such as biodegradable polymers and bioinspired materials is ongoing to mitigate these risks, though current options may still induce mild immune responses [146].

9.5.4 Patient Compliance and Administration Challenges

The invasiveness of certain drug delivery methods, such as intravitreal injections, poses further challenges to patient compliance. Although nanoparticles can improve drug delivery efficiency, these methods may still be painful or uncomfortable for patients [96]. Developing minimally invasive or alternative administration routes, such as topical or subconjunctival injections, is a current focus. However, ensuring that nanomedicines delivered through these routes can achieve effective drug concentrations remains a major obstacle [145].

9.5.5 Scale-Up and Manufacturing Complexity

Manufacturing nanomedicine at scale while maintaining uniformity and quality is another significant barrier. Nanoparticles and nanocarriers require precise engineering and consistency, which is challenging to achieve in large-scale production. Issues such as particle size distribution, surface modifications, and encapsulation efficiency must be meticulously controlled, increasing production costs and complexity. Scale-up processes must also comply with stringent regulatory standards, requiring advanced manufacturing practices that are often cost-prohibitive [144].

9.5.6 Regulatory Considerations

Nanomedicines face a complex regulatory landscape, as current guidelines are not fully adapted to assess their unique properties and risks. Regulatory bodies need to consider factors like particle size, surface characteristics, and long-term safety profiles when evaluating these therapies [145]. Standardized protocols for assessing the safety and efficacy of nanomedicines are still under development, which delays approval and commercialization. Ethical concerns, such as the long-term effects of nanomaterials on ocular tissues and patient safety, also play a role in regulatory decisions, requiring further research to establish clear safety profiles [146].

In a nutshell, while ophthalmic nanomedicine holds immense potential for improving ocular disease treatment, the challenges of biological barriers, formulation stability, biocompatibility, patient compliance, manufacturing scalability, and regulatory complexities present substantial hurdles. Continued advancements in nano-formulation technology, interdisciplinary collaboration, and regulatory adaptation are essential to overcoming these challenges and realizing the full therapeutic potential of nanomedicine in ophthalmology.

9.6 Future Perspectives

The future of nanotechnology in ocular drug delivery looks incredibly promising, with ongoing research pushing the boundaries of what is possible. One of the key areas of future development lies in the personalization of nanocarrier systems. By tailoring nanocarriers to individual patient needs, such as genetic makeup or specific disease conditions, the next generation of therapies could offer even more precise and effective treatment. Additionally, advances in biomaterials and surface functionalization will allow for the creation of more sophisticated nanocarriers that are not only more biocompatible but also capable of responding to stimuli such as pH, temperature, or enzymes, enabling controlled and targeted drug release. Moreover, the integration of diagnostic and therapeutic capabilities into a single platform, often referred to as theranostics, could revolutionize ocular disease management by providing real-time monitoring and treatment. Another exciting avenue is the exploration of gene therapy and CRISPR-based technologies in combination with nanotechnology to correct genetic disorders causing ocular diseases. As the regulatory landscape evolves and manufacturing processes become more streamlined, the transition of these advanced systems from bench to bedside is expected to accelerate, offering hope for more effective and accessible treatments for ocular conditions worldwide.

9.7 Conclusion

The advancements in nanotechnology-based ocular drug delivery systems represent a significant leap forward in addressing the limitations of conventional ocular therapies. Traditional methods, such as eye drops and systemic administration, often suffer from poor bioavailability due to the eye's natural barriers, which impede the effective delivery of drugs to the target tissues. Nanotechnology-based systems, including nanoparticles, liposomes, and solid lipid nanoparticles (SLNs), have shown great promise in overcoming these challenges. These systems enhance drug stability, prolong release, and improve drug targeting, making them more effective for both anterior and posterior segment diseases. Additionally, these nanocarriers can encapsulate a broader range of therapeutic agents, including hydrophilic and hydrophobic drugs, improving the scope of treatment for conditions like glaucoma, diabetic retinopathy, and age-related macular degeneration. However, the clinical translation of these technologies remains a significant challenge, with regulatory concerns, manufacturing complexities, and the need for extensive preclinical and clinical evaluations slowing down progress. Despite these hurdles, the potential benefits of nanotechnology in ocular drug delivery are immense, offering solutions that could lead to improved therapeutic outcomes, patient compliance, and overall treatment efficacy in the future.

References

1. Ahmed S, Amin MM, Sayed S. Ocular drug delivery: a comprehensive review. AAPS PharmSciTech. 2023;24:66. https://doi.org/10.1208/s12249-023-02516-9.
2. Chen Y, Ye Z, Chen H, Li Z. Breaking barriers: nanomedicine-based drug delivery for cataract treatment. Int J Nanomedicine. 2024;19:4021–40. https://doi.org/10.2147/IJN.S463679.
3. Li S, Chen L, Fu Y. Nanotechnology-based ocular drug delivery systems: recent advances and future prospects. J. Nanobiotechnol. 2023;21:232. https://doi.org/10.1186/s12951-023-01992-2.
4. Wu Y, Li X, Fu X, Huang X, Zhang S, Zhao N, Ma X, Saiding Q, Yang M, Tao W, Zhou X, Huang J. Innovative nanotechnology in drug delivery systems for advanced treatment of posterior segment ocular diseases. Adv Sci. 2024;11:2403399. https://doi.org/10.1002/advs.202403399.
5. Giri BR, Jakka D, Sandoval MA, Kulkarni VR, Bao Q. Advancements in ocular therapy: a review of emerging drug delivery approaches and pharmaceutical technologies. Pharmaceutics. 2024;16:1325.
6. Joyce P, Allen CJ, Alonso MJ, Ashford M, Bradbury MS, Germain M, Kavallaris M, Langer R, Lammers T, Peracchia MT. A translational framework to DELIVER nanomedicines to the clinic. Nat Nanotechnol. 2024;1–15:1597.
7. Chellathurai MS, Chung LY, Hilles AR, Sofian ZM, Singha S, Ghosal K, Mahmood S. Pharmaceutical chitosan hydrogels: a review on its design and applications. Int J Biol Macromol. 2024;135775:135775.
8. Batur E, Özdemir S, Durgun ME, Özsoy Y. Vesicular drug delivery systems: promising approaches in ocular drug delivery. Pharmaceuticals. 2024;17:511. https://doi.org/10.3390/ph17040511.
9. López-Cano JJ, González-Cela-Casamayor MA, Andrés-Guerrero V, Herrero-Vanrell R, Molina-Martínez IT. Liposomes as vehicles for topical ophthalmic drug delivery and ocu-

lar surface protection. Expert Opin Drug Deliv. 2021;18:819–47. https://doi.org/10.1080/17425247.2021.1872542.
10. Shen H-H, Chan EC, Lee JH, Bee Y-S, Lin T-W, Dusting GJ, Liu G-S. Nanocarriers for Treatment of Ocular Neovascularization in the Back of the Eye: New Vehicles for Ophthalmic Drug Delivery. Nanomedicine. 2015;10:2093–107. https://doi.org/10.2217/nnm.15.47.
11. Gholizadeh S, Wang Z, Chen X, Dana R, Annabi N. Advanced nanodelivery platforms for topical ophthalmic drug delivery. Drug Discov Today. 2021;26:1437–49. https://doi.org/10.1016/j.drudis.2021.02.027.
12. Bigdeli A, Makhmalzadeh BS, Feghhi M, SoleimaniBiatiani E. Cationic liposomes as promising vehicles for timolol/brimonidine combination ocular delivery in glaucoma: formulation development and in vitro/in vivo evaluation. Drug Deliv Transl Res. 2023;13:1035–47. https://doi.org/10.1007/s13346-022-01266-8.
13. Soni PK, Saini TR. Formulation design and optimization of cationic-charged liposomes of brimonidine tartrate for effective ocular drug delivery by design of experiment (DoE) approach. Drug Dev Ind Pharm. 2021;47:1847–66. https://doi.org/10.1080/03639045.2022.2070198.
14. Qiu L, Zhang J, Liu S, Li B, Wang J, Tang J, Pu X, Huang Z, Liao X, Yin G. Dendritic oligoethylenimine decorated liposome with augmented corneal retention and permeation for efficient topical delivery of antiglaucoma drugs. Nano Lett. 2023;23:11193–202. https://doi.org/10.1021/acs.nanolett.3c03691.
15. Moiseev RV, Kaldybekov DB, Filippov SK, Radulescu A, Khutoryanskiy VV. Maleimide-decorated PEGylated mucoadhesive liposomes for ocular drug delivery. Langmuir. 2022;38:13870–9. https://doi.org/10.1021/acs.langmuir.2c02086.
16. Qiao H, Xu Z, Sun M, Fu S, Zhao F, Wang D, He Z, Zhai Y, Sun J. Rebamipide liposome as an effective ocular delivery system for the management of dry eye disease. J Drug Deliv Sci Technol. 2022;75:103654.
17. Durak S, Esmaeili Rad M, Alp Yetisgin A, Eda Sutova H, Kutlu O, Cetinel S, Zarrabi A. Niosomal drug delivery systems for ocular disease – recent advances and future prospects. Nano. 2020;10:1191. https://doi.org/10.3390/nano10061191.
18. Verma A, Tiwari A, Saraf S, Panda PK, Jain A, Jain SK. Emerging potential of niosomes in ocular delivery. Expert Opin Drug Deliv. 2021;18:55–71. https://doi.org/10.1080/17425247.2020.1822322.
19. Ozdemir S, Uner B. Prolonged release niosomes for ocular delivery of loteprednol: ocular distribution assessment on dry eye disease induced rabbit model. AAPS PharmSciTech. 2024;25:119. https://doi.org/10.1208/s12249-024-02838-2.
20. Kandpal N, Dhuliya R, Padiyar N, Singh A, Khaudiyal S, Ale Y, Jakhmola V, Nainwal N. Innovative niosomal in-situ gel: elevating ocular drug delivery synergies. Sci: J. Appl. Pharm; 2024.
21. Alotaibi TA, Iyire A, Assaf S, Dahmash EZ. Development and characterization of niosomes loaded mucoadhesive biodegradable ocular inserts for extended release of pilocarpine HCl. Fut J Pharm Sci. 2024;10:22. https://doi.org/10.1186/s43094-024-00598-1.
22. Yassin GE, Amer MA, Mannaa IM, Khalifa MKA. Fluconazole-niosome-laden contact lens: a promising therapeutic approach for prolonged ocular delivery and enhanced antifungal activity. J Pharm Innov. 2024;19:45. https://doi.org/10.1007/s12247-024-09850-w.
23. Tripathi S, Yadav KS. Development of brimonidine niosomes laden contact lenses for extended release and promising delivery system in glaucoma treatment. DARU J Pharm Sci. 2023;32:161–75. https://doi.org/10.1007/s40199-023-00500-z.
24. Ramadan AA, Eladawy SA, El-Enin ASMA, Hussein ZM. Development and investigation of timolol maleate niosomal formulations for the treatment of glaucoma. J Pharm Investig. 2020;50:59–70. https://doi.org/10.1007/s40005-019-00427-1.
25. Mazyed EA, Abdelaziz AE. Fabrication of transgelosomes for enhancing the ocular delivery of acetazolamide: statistical optimization, in vitro characterization, and in vivo study. Pharmaceutics. 2020;12:465. https://doi.org/10.3390/pharmaceutics12050465.

26. Simrah A, Hafeez SA, Usmani MP. Izhar, transfersome, an ultra-deformable lipid-based drug nanocarrier: an updated review with therapeutic applications. Naunyn-Schmiedebergs Arch Pharmacol. 2024;397:639–73. https://doi.org/10.1007/s00210-023-02670-8.
27. Mbah CC, Builders PF, Attama AA. Nanovesicular carriers as alternative drug delivery systems: ethosomes in focus. Expert Opin Drug Deliv. 2014;11:45–59. https://doi.org/10.1517/17425247.2013.860130.
28. Uner B, Ozdemir S, Nur Pilevne S, A.R. Cenk Celebi, Timolol-loaded ethosomes for ophthalmic delivery: reduction of high intraocular pressure in vivo. Int J Pharm. 2023;640:123021. https://doi.org/10.1016/j.ijpharm.2023.123021.
29. Gupta PN, Mishra V, Rawat A, Dubey P, Mahor S, Jain S, Chatterji DP, Vyas SP. Non-invasive vaccine delivery in transfersomes, niosomes and liposomes: a comparative study. Int J Pharm. 2005;293:73–82. https://doi.org/10.1016/j.ijpharm.2004.12.022.
30. Srivastava V, Patil RK, Mehra NK. A one-platform comparison study of brinzolamide-loaded liposomes, niosomes, transfersomes, and transniosomes for better management of glaucoma. Int J Pharm. 2024;666:124741. https://doi.org/10.1016/j.ijpharm.2024.124741.
31. Patil RK, Srivastava V, Bhawale R, Tryphena KP, Khatri DK, Doijad N, Mehra NK. Revamping the corneal permeability and antiglaucoma therapeutic potential of brinzolamide using transniosomes: optimization, in vitro and preclinical evaluation. Nanomedicine. 2024;19:947–64. https://doi.org/10.2217/nnm-2023-0280.
32. Aburahma MH. Bile salts-containing vesicles: promising pharmaceutical carriers for oral delivery of poorly water-soluble drugs and peptide/protein-based therapeutics or vaccines. Drug Deliv. 2016;23:1847–67. https://doi.org/10.3109/10717544.2014.976892.
33. Nath AG, Dubey P, Kumar A, Vaiphei KK, Rosenholm JM, Bansal KK, Gulbake A. Recent advances in the use of cubosomes as drug carriers with special emphasis on topical applications. J Lipids. 2024;2024:2683466. https://doi.org/10.1155/2024/2683466.
34. Brugnera M, Vicario-de-la-Torre M, González-Cela Casamayor MA, López-Cano JJ, Bravo-Osuna I, Huete-Toral F, González Rubio ML, Carracedo G, Molina-Martínez IT, Andrés-Guerrero V, Herrero-Vanrell R. Enhancing the hypotensive effect of latanoprost by combining synthetic phosphatidylcholine liposomes with hyaluronic acid and osmoprotective agents. Drug Deliv Transl Res. 2024;14:2804–22. https://doi.org/10.1007/s13346-024-01584-z.
35. Esposito E, Pozza E, Contado C, Pula W, Bortolini O, Ragno D, Toldo S, Casciano F, Bondi A, Zauli E, Secchiero P, Zauli G, Melloni E. Microfluidic fabricated liposomes for nutlin-3a ocular delivery as potential candidate for proliferative vitreoretinal diseases treatment. Int J Nanomedicine. 2024;19:3513–36. https://doi.org/10.2147/IJN.S452134.
36. Farooq U, O'Reilly NJ, Ahmed Z, Gasco P, et al. Design of liposomal nanocarriers with a potential for combined dexamethasone and bevacizumab delivery to the eye. Int J Pharm. 2024;654:123958. https://doi.org/10.1016/j.ijpharm.2024.123958.
37. Duman G, Yıldır İ, Macit M, Genç E, Sümer E, Kale S, Deniz İ. Development and evaluation of 3D-printed ocular insert containing liposomal moxifloxacin. J Drug Deliv Sci Technol. 2024;92:105353. https://doi.org/10.1016/j.jddst.2024.105353.
38. Al-Amin MD, Mastrotto F, Subrizi A, Sen M, Turunen T, Arango-Gonzalez B, Ueffing M, Malfanti A, Urtti A, Salmaso S, Caliceti P. Tailoring surface properties of liposomes for dexamethasone intraocular administration. J Control Release. 2023;354:323–36. https://doi.org/10.1016/j.jconrel.2023.01.027.
39. Gu H, Chen P, Liu X, Lian Y, Xi J, Li J, Song J, Li X. Trimethylated chitosan-coated flexible liposomes with resveratrol for topical drug delivery to reduce blue-light-induced retinal damage. Int J Biol Macromol. 2023;252:126480. https://doi.org/10.1016/j.ijbiomac.2023.126480.
40. Dos Santos GA, Ferreira-Nunes R, Dalmolin LF, Dos Santos Ré AC, Anjos JLV, Mendanha SA, Aires CP, Lopez RFV, Cunha-Filho M, Gelfuso GM, Gratieri T. Besifloxacin liposomes with positively charged additives for an improved topical ocular delivery. Sci Rep. 2020;10:19285. https://doi.org/10.1038/s41598-020-76381-y.
41. Wang F, Bao X, Fang A, Li H, Zhou Y, Liu Y, Jiang C, Wu J, Song X. Nanoliposome-encapsulated brinzolamide-hydropropyl-β-cyclodextrin inclusion complex: a potential

therapeutic ocular drug-delivery system. Front Pharmacol. 2018;9 https://doi.org/10.3389/fphar.2018.00091.
42. Zhan C, Santamaria CM, Wang W, McAlvin JB, Kohane DS. Long-acting liposomal corneal anesthetics. Biomaterials. 2018;181:372. https://doi.org/10.1016/j.biomaterials.2018.07.054.
43. Kattar A, Quelle-Regaldie A, Sánchez L, Concheiro A, Alvarez-Lorenzo C. Formulation and characterization of epalrestat-loaded polysorbate 60 cationic niosomes for ocular delivery. Pharmaceutics. 2023;15:1247. https://doi.org/10.3390/pharmaceutics15041247.
44. Gupta P, Yadav KS. Formulation and evaluation of brinzolamide encapsulated niosomal *in-situ* gel for sustained reduction of IOP in rabbits. J Drug Deliv Sci Technol. 2022;67:103004. https://doi.org/10.1016/j.jddst.2021.103004.
45. Fathalla D, Fouad EA, Soliman GM. Latanoprost niosomes as a sustained release ocular delivery system for the management of glaucoma. Drug Dev Ind Pharm. 2020;46:806.
46. Elmotasem H, Awad GE. A stepwise optimization strategy to formulate in situ gelling formulations comprising fluconazole-hydroxypropyl-beta-cyclodextrin complex loaded niosomal vesicles and Eudragit nanoparticles for enhanced antifungal activity and prolonged ocular delivery. Asian J Pharm Sci. 2019;15:617. https://doi.org/10.1016/j.ajps.2019.09.003.
47. Eldeeb AE, Salah S, Ghorab M. Proniosomal gel-derived niosomes: an approach to sustain and improve the ocular delivery of brimonidine tartrate; formulation, in-vitro characterization, and in-vivo pharmacodynamic study. Drug Deliv. 2019;26:509. https://doi.org/10.1080/10717544.2019.1609622.
48. Paradkar MU, Parmar M. Formulation development and evaluation of Natamycin niosomal in-situ gel for ophthalmic drug delivery. J Drug Deliv Sci Technol. 2017;39:113–22. https://doi.org/10.1016/j.jddst.2017.03.005.
49. Khalil RM, Abdelbary GA, Basha M, Awad GEA, El-Hashemy HA. Enhancement of lomefloxacin Hcl ocular efficacy via niosomal encapsulation: in vitro characterization and in vivo evaluation. J Liposome Res. 2017;27:312–23. https://doi.org/10.1080/08982104.2016.1191022.
50. Sushma MV, Sankaranarayanan SA, Bantal V, Pemmaraju DB, Rengan AK. Ethosomal nanoformulations for combinational photothermal therapy of fungal keratitis. Adv Ther. 2023;6:2200331. https://doi.org/10.1002/adtp.202200331.
51. Ahmed TA, Alzahrani MM, Sirwi A, Alhakamy NA. Study the antifungal and ocular permeation of ketoconazole from ophthalmic formulations containing trans-ethosomes nanoparticles. Pharmaceutics. 2021;13:151. https://doi.org/10.3390/pharmaceutics13020151.
52. Barbalho GN, Brugger S, Raab C, Lechner J-S, Gratieri T, Keck CM, Rupenthal ID, Agarwal P. Development of transferosomes for topical ocular drug delivery of curcumin. Eur J Pharm Biopharm. 2024;205:114535. https://doi.org/10.1016/j.ejpb.2024.114535.
53. Das B, Nayak AK, Mallick S. Fluconazole-loaded Hyaluronic acid-modified transfersomal hydrogels containing D-panthenol for ocular delivery in fungal keratitis management. Curr Drug Deliv. 2024;22 https://doi.org/10.2174/0115672018342369241018050810.
54. Uwaezuoke O, Du Toit LC, Kumar P, Ally N, Choonara YE. Linoleic acid-based transferosomes for topical ocular delivery of cyclosporine A. Pharmaceutics. 2022;14:1695. https://doi.org/10.3390/pharmaceutics14081695.
55. Sakr MG, El-Zahaby SA, Al-Mahallawi AM, Ghorab DM. Fabrication of betaxolol hydrochloride-loaded highly permeable ocular bilosomes (HPOBs) to combat glaucoma: *In vitro, ex vivo&in vivo* characterizations. J Drug Deliv Sci Technol. 2023;82:104363. https://doi.org/10.1016/j.jddst.2023.104363.
56. Nair VS, Srivastava V, Bhavana V, Yadav R, Rajana N, Singh SB, Mehra NK. Exploring penetration ability of carbonic anhydrase inhibitor–loaded ultradeformable bilosome for effective ocular application. AAPS PharmSciTech. 2023;24:157. https://doi.org/10.1208/s12249-023-02617-5.
57. Nemr AA, El-Mahrouk GM, Badie HA. Hyaluronic acid-enriched bilosomes: an approach to enhance ocular delivery of agomelatine via D-optimal design: formulation, in vitro characterization, and in vivo pharmacodynamic evaluation in rabbits. Drug Deliv. 2022;29:2343–56. https://doi.org/10.1080/10717544.2022.2100513.

58. Mohsen AM, Salama A, Kassem AA. Development of acetazolamide loaded bilosomes for improved ocular delivery: preparation, characterization and *in vivo* evaluation. J Drug Deliv Sci Technol. 2020;59:101910. https://doi.org/10.1016/j.jddst.2020.101910.
59. Janga KY, Tatke A, Balguri SP, Lamichanne SP, Ibrahim MM, Maria DN, Jablonski MM, Majumdar S. Ion-sensitive in situ hydrogels of natamycin bilosomes for enhanced and prolonged ocular pharmacotherapy: in vitro permeability, cytotoxicity and in vivo evaluation. Artif Cells Nanomed Biotechnol. 2018;46:1039–50. https://doi.org/10.1080/21691401.2018.1443117.
60. Abdelbary AA, Abd-Elsalam WH, Al-mahallawi AM. Fabrication of novel ultradeformable bilosomes for enhanced ocular delivery of terconazole: *In vitro* characterization, *ex vivo* permeation and *in vivo* safety assessment. Int J Pharm. 2016;513:688–96. https://doi.org/10.1016/j.ijpharm.2016.10.006.
61. Omran S, Elnaggar YSR, Abdallah OY. Controlled release, chitosan-tethered luteolin phytocubosomes; formulation optimization to in-vivo antiglaucoma and anti-inflammatory ocular evaluation. Int J Biol Macromol. 2024;254:127930. https://doi.org/10.1016/j.ijbiomac.2023.127930.
62. Nemr AA, Ahmed S, Adel IM. Limonene-enriched ultra-structural cubosomes to augment ocular delivery of a poorly water soluble anti-fungal drug: fabrication, characterization, statistical optimization, in vivo corneal uptake and histopathological evaluation in rabbits. J Drug Deliv Sci Technol. 2024;98:105886. https://doi.org/10.1016/j.jddst.2024.105886.
63. Madhusudhan S, Gupta NV, Rahamathulla M, Chidambaram SB, Osmani RAM, Ghazwani M, Ahmed MM, Farhana SA, Sarhan MY, Tousif AH. Subconjunctival delivery of sorafenib-tosylate-loaded cubosomes for facilitated diabetic retinopathy treatment: formulation development, evaluation, pharmacokinetic and pharmacodynamic (PKPD) studies. Pharmaceutics. 2023;15:2419. https://doi.org/10.3390/pharmaceutics15102419.
64. Teba HE, Khalil IA, El Sorogy HM. Novel cubosome based system for ocular delivery of acetazolamide. Drug Deliv. 2021;28:2177–86. https://doi.org/10.1080/10717544.2021.1989090.
65. Eldeeb AE, Salah S, Ghorab M. Formulation and evaluation of cubosomes drug delivery system for treatment of glaucoma: *Ex-vivo* permeation and *in-vivo* pharmacodynamic study. J Drug Deliv Sci Technol. 2019;52:236–47. https://doi.org/10.1016/j.jddst.2019.04.036
66. Huang J, Peng T, Li Y, Zhan Z, Zeng Y, Huang Y, Pan X, Wu C-Y, Wu C. Ocular cubosome drug delivery system for timolol maleate: preparation, characterization, cytotoxicity, ex vivo, and in vivo evaluation. AAPS PharmSciTech. 2017;18:2919–26. https://doi.org/10.1208/s12249-017-0763-8.
67. Gaballa SA, El Garhy OH, Moharram H, Abdelkader H. Preparation and evaluation of cubosomes/cubosomal gels for ocular delivery of beclomethasone dipropionate for management of uveitis. Pharm Res. 2020;37:198. https://doi.org/10.1007/s11095-020-02857-1.
68. Manchanda N, Vishkarma H, Goyal M, Shah S, Famta P, Talegaonkar S, Srivastava S. Surface functionalized lipid nanoparticles in promoting therapeutic outcomes: an insight view of the dynamic drug delivery system. Curr Drug Targets. 2024;25:278. https://doi.org/10.2174/0113894501285598240216065627.
69. Battaglia L, Serpe L, Foglietta F, Muntoni E, Gallarate M, Del Pozo Rodriguez A, Solinis MA. Application of lipid nanoparticles to ocular drug delivery. Expert Opin Drug Deliv. 2016;13:1743–57. https://doi.org/10.1080/17425247.2016.1201059.
70. Jacob S, Nair AB, Shah J, Gupta S, Boddu SHS, Sreeharsha N, Joseph A, Shinu P, Morsy MA. Lipid nanoparticles as a promising drug delivery carrier for topical ocular therapy – an overview on recent advances. Pharm. 2022;14:533.
71. Gugleva V, Andonova V. Recent progress of solid lipid nanoparticles and nanostructured lipid carriers as ocular drug delivery platforms. Pharmaceuticals. 2023;16:474.
72. Sánchez-López E, Espina M, Doktorovova S, Souto EB, García ML. Lipid nanoparticles (SLN, NLC): overcoming the anatomical and physiological barriers of the eye – Part II – ocular drug-loaded lipid nanoparticles. Eur J Pharm Biopharm. 2017;110:58–69. https://doi.org/10.1016/j.ejpb.2016.10.013.

73. Satyanarayana SD, Abu Lila AS, Moin A, Moglad EH, Khafagy E-S, Alotaibi HF, Obaidullah AJ, Charyulu RN. Ocular delivery of bimatoprost-loaded solid lipid nanoparticles for effective management of glaucoma. Pharmaceuticals. 2023;16:1001.
74. Alhakamy NA, Hosny KM, Aldryhim AY, Rizg WY, Eshmawi BA, Bukhary HA, Murshid SS, Khallaf RA. Development and optimization of ofloxacin as solid lipid nanoparticles for enhancement of its ocular activity. J Drug Deliv Sci Technol. 2022;72:103373.
75. Onugwu AL, Attama AA, Nnamani PO, Onugwu SO, Onuigbo EB, Khutoryanskiy VV. Development and optimization of solid lipid nanoparticles coated with chitosan and poly (2-ethyl-2-oxazoline) for ocular drug delivery of ciprofloxacin. J Drug Deliv Sci Technol. 2022;74:103527.
76. Nair AB, Shah J, Al-Dhubiab BE, Jacob S, Patel SS, Venugopala KN, Morsy MA, Gupta S, Attimarad M, Sreeharsha N. Clarithromycin solid lipid nanoparticles for topical ocular therapy: optimization, evaluation, and in vivo studies. Pharmaceutics. 2021;13:523.
77. Varela-Fernández R, García-Otero X, Díaz-Tomé V, Regueiro U, López-López M, González-Barcia M, Lema MI, Otero-Espinar FJ. Lactoferrin-loaded nanostructured lipid carriers (NLCs) as a new formulation for optimized ocular drug delivery. Eur J Pharm Biopharm. 2022;172:144–56.
78. Gautam D, Chaurasia H, Singh R. Design and optimization of lomefloxacin loaded NLC gel for ophthalmic drug delivery. Int J Health Sci. 2022:7022–44.
79. Kumar M, Tiwari A, Asdaq SMB, Nair AB, Bhatt S, Shinu P, Al Mouslem AK, Jacob S, Alamri AS, Alsanie WF. Itraconazole loaded nano-structured lipid carrier for topical ocular delivery: optimization and evaluation. Saudi J Biol Sci. 2022;29:1–10.
80. Baig MS, Owida H, Njoroge W, Yang Y. Development and evaluation of cationic nanostructured lipid carriers for ophthalmic drug delivery of besifloxacin. J Drug Deliv Sci Technol. 2020;55:101496.
81. Elsayed MMA, Elsayed A, Fouad MA, Mohamed MS, Lee S, Mahmoud RA, Sabry SA, Ghoneim MM, Hassan AHE, Elkarim RAA, Belal A, El-Shenawy AA. Development and optimization of vildagliptin solid lipid nanoparticles loaded ocuserts for controlled ocular delivery: a promising approach towards treating diabetic retinopathy. Int J Pharm. 2024;X 7:100232. https://doi.org/10.1016/j.ijpx.2024.100232.
82. Satyanarayana SD, Abu Lila AS, Moin A, Moglad EH, Khafagy E-S, Alotaibi HF, Obaidullah AJ, Charyulu RN. Ocular delivery of bimatoprost-loaded solid lipid nanoparticles for effective management of glaucoma. Pharm Basel Switz. 2023;16:1001. https://doi.org/10.3390/ph16071001.
83. Liang Z, Zhang Z, Yang J, Lu P, Zhou T, Li J, Zhang J. Assessment to the antifungal effects in vitro and the ocular pharmacokinetics of solid-lipid nanoparticle in rabbits. Int J Nanomedicine. 2021;16:7847. https://doi.org/10.2147/IJN.S340068.
84. Singh M, Guzman-Aranguez A, Hussain A, Srinivas CS, Kaur IP. Solid lipid nanoparticles for ocular delivery of isoniazid: evaluation, proof of concept and in vivo safety & kinetics. Nanomedicine. 2019;14:465–91. https://doi.org/10.2217/nnm-2018-0278.
85. Tatke A, Dudhipala N, Janga KY, Balguri SP, Avula B, Jablonski MM, Majumdar S. in situ gel of triamcinolone acetonide-loaded solid lipid nanoparticles for improved topical ocular delivery: tear kinetics and ocular disposition studies. Nano. 2019;9:33. https://doi.org/10.3390/nano9010033.
86. Bonilla-Vidal L, Espina M, García ML, Baldomà L, Badia J, González JA, Delgado LM, Gliszczyńska A, Souto EB, Sánchez-López E. Novel nanostructured lipid carriers loading Apigenin for anterior segment ocular pathologies. Int J Pharm. 2024;658:124222. https://doi.org/10.1016/j.ijpharm.2024.124222.
87. Bonilla-Vidal L, Espina M, García ML, Cimino C, Carbone C, Baldomà L, Badia J, Gliszczyńska A, Souto EB, Sánchez-López E. Melatonin loaded nanostructured lipid carriers for the treatment of uveal melanoma. J Drug Deliv Sci Technol. 2024;100:106057. https://doi.org/10.1016/j.jddst.2024.106057.

88. Santonocito D, Barbagallo I, Distefano A, Sferrazzo G, Vivero-Lopez M, Sarpietro MG, Puglia C. Nanostructured lipid carriers aimed to the ocular delivery of mangiferin: in vitro evidence. Pharmaceutics. 2023;15:951. https://doi.org/10.3390/pharmaceutics15030951.
89. Luo Q, Yang J, Xu H, Shi J, Liang Z, Zhang R, Lu P, Pu G, Zhao N, Zhang J. Sorafenib-loaded nanostructured lipid carriers for topical ocular therapy of corneal neovascularization: development, in-vitro and in vivo study. Drug Deliv. 2022;29:837–55. https://doi.org/10.1080/10717544.2022.2048134.
90. Kumar M, Tiwari A, Asdaq SMB, Nair AB, Bhatt S, Shinu P, Al Mouslem AK, Jacob S, Alamri AS, Alsanie WF, Alhomrani M, Tiwari V, Devi S, Pathania A, Sreeharsha N. Itraconazole loaded nano-structured lipid carrier for topical ocular delivery: optimization and evaluation, Saudi. J Biol Sci. 2022;29:1–10. https://doi.org/10.1016/j.sjbs.2021.11.006.
91. Yu Y, Xu S, Yu S, Li J, Tan G, Li S, Pan W. A hybrid genipin-cross-linked hydrogel/nano-structured lipid carrier for ocular drug delivery: cellular, ex vivo, and in vivo evaluation. ACS Biomater Sci Eng. 2020;6:1543–52. https://doi.org/10.1021/acsbiomaterials.9b01800
92. Lakhani P, Patil A, Wu K-W, Sweeney C, Tripathi S, Avula B, Taskar P, Khan S, Majumdar S. Optimization, stabilization, and characterization of amphotericin B loaded nanostructured lipid carriers for ocular drug delivery. Int J Pharm. 2019;572:118771. https://doi.org/10.1016/j.ijpharm.2019.118771.
93. Kiss EL, Berkó S, Gácsi A, Kovács A, Katona G, Soós J, Csányi E, Gróf I, Harazin A, Deli MA, Budai-Szűcs M. Design and optimization of nanostructured lipid carrier containing dexamethasone for ophthalmic use. Pharmaceutics. 2019;11:679. https://doi.org/10.3390/pharmaceutics11120679.
94. Mo Z, Ban J, Zhang Y, Du Y, Wen Y, Huang X, Xie Q, Shen L, Zhang S, Deng H, Hou D, Chen Y, Lu Z. nanostructured lipid carriers-based thermosensitive eye drops for enhanced, sustained delivery of dexamethasone. Nanomedicine. 2018;13 1239–53. https://doi.org/10.2217/nnm-2017-0318.
95. Liu LC, Chen YH, Lu DW. Overview of recent advances in nano-based ocular drug delivery. Int J Mol Sci. 2023;24:15352. https://doi.org/10.3390/IJMS242015352.
96. Zhang J, Jiao J, Niu M, Gao X, Zhang G, Yu H, Yang X, Liu L. Ten years of knowledge of nano-carrier based drug delivery systems in ophthalmology: current evidence, challenges, and future prospective. Int J Nanomedicine. 2021;16:6497–530. https://doi.org/10.2147/IJN.S329831.
97. Bachu RD, Chowdhury P, Al-Saedi ZHF, Karla PK, Boddu SHS. Ocular drug delivery barriers – role of nanocarriers in the treatment of anterior segment ocular diseases. Pharmaceutics. 2018;10:28.
98. Boddu SHS. Polymeric nanoparticles for ophthalmic drug delivery: an update on research and patenting activity. Recent Pat Nanomedicine. 2012;2:96–112. https://doi.org/10.2174/1877912311202020096.
99. Zielinska A, Carreiró F, Oliveira AM, Neves A, Pires B, Venkatesh DN, Durazzo A, Lucarini M, Eder P, Silva AM, Santini A, Souto EB. Polymeric nanoparticles: production, characterization, toxicology and ecotoxicology. Molecules. 2020;25:3731. https://doi.org/10.3390/MOLECULES25163731.
100. Mohamed HB, Shafie MAA, Mekkawy AI. Chitosan nanoparticles for meloxicam ocular delivery: development, in vitro characterization, and in vivo evaluation in a rabbit eye model. Pharm. 2022;14:893.
101. Sohani Z, Jamshidi S, Koohi MK, Malakootikhah J, Abarkar M, Golchin D, Roshani S, Naghdi H, et al. Novel ophthalmic hyaluronic acid-hydrogel with curcumin nanoparticles for enhanced healing of ulcerative keratitis in rabbit model. Sci Rep. 2024;14:23046. https://doi.org/10.1038/S41598-024-74195-W.
102. Elmowafy M, Shalaby K, Elkomy MH, Alsaidan OA, Gomaa HAM, Abdelgawad MA, Mostafa EM. Polymeric nanoparticles for delivery of natural bioactive agents: recent advances and challenges. Polymers. 2023;15:1123.

103. Eleraky NE, Attia MA, Safwat MA. Sertaconazole-PLGA nanoparticles for management of ocular keratitis. J Drug Deliv Sci Technol. 2024;95:105539. https://doi.org/10.1016/J.JDDST.2024.105539.
104. Nguyen DD, Luo LJ, Lai JY. Effects of shell thickness of hollow poly(lactic acid) nanoparticles on sustained drug delivery for pharmacological treatment of glaucoma. Acta Biomater. 2020;111:302–15. https://doi.org/10.1016/J.ACTBIO.2020.04.055.
105. Xu FM, et al. Novel fabrication of anti-VEGF drug ranibizumab loaded PLGA/PLA co-polymeric nanomicelles for long-acting intraocular delivery in the treatment of age-related macular degeneration therapy. Regen Ther. 2024;26:620. https://doi.org/10.1016/J.RETH.2024.06.019.
106. Beirampour N, Bustos-Salgado P, Garrós N, Mohammadi-Meyabadi R, Domènech Ò, Suñer-Carbó J, Rodríguez-Lagunas MJ, Kapravelou G, Montes MJ, Calpena A, Mallandrich M. Formulation of polymeric nanoparticles loading baricitinib as a topical approach in ocular application. Pharmaceutics. 2024;16:1092. https://doi.org/10.3390/pharmaceutics16081092.
107. Sufian MA, Abbas G, Rasul A, Irfan M, Khan HU. Moxifloxacin-loaded nanoparticles of thiolated xyloglucan for ocular drug delivery: permeation, mucoadhesion and pharmacokinetic evaluation. Int J Biol Macromol. 2024;270:132522. https://doi.org/10.1016/j.ijbiomac.2024.132522.
108. Dahiya P, Zafar A, Ahmad FJ, Khalid M, Ali A. Development of Forskolin and rutin-loaded polymeric nanoparticles for enhancement of topical ocular delivery: optimization, in-vitro, ex-vivo, and toxicity evaluation. J Drug Deliv Sci Technol. 2023;82:104292. https://doi.org/10.1016/j.jddst.2023.104292.
109. Pardeshi SR, More MP, Pardeshi CV, Chaudhari PJ, Gholap AD, Patil A, Patil PB, Naik JB. Novel crosslinked nanoparticles of chitosan oligosaccharide and dextran sulfate for ocular administration of dorzolamide against glaucoma. J Drug Deliv Sci Technol. 2023;86:104719. https://doi.org/10.1016/j.jddst.2023.104719.
110. Mohsen AM. Cationic polymeric nanoparticles for improved ocular delivery and antimycotic activity of terconazole. J Pharm Sci. 2022;111:458–68. https://doi.org/10.1016/j.xphs.2021.09.019.
111. Khan FU, Nasir F, Iqbal Z, Neau S, Khan I, Hassan M, Iqbal M, Ullah A, Khan SI, Sakhi M. Improved ocular bioavailability of moxifloxacin HCl using PLGA nanoparticles: fabrication, characterization, in-vitro and in-vivo evaluation. Iran J Pharm Res. 2021;20:592. https://doi.org/10.22037/ijpr.2021.114478.15054.
112. Bin-Jumah M, Gilani SJ, Jahangir MA, Zafar A, Alshehri S, Yasir M, Kala C, Taleuzzaman M, Imam SS. Clarithromycin-loaded ocular chitosan nanoparticle: formulation, optimization, characterization, ocular irritation, and antimicrobial activity. Int J Nanomedicine. 2020;15:7861–75. https://doi.org/10.2147/IJN.S269004.
113. Zhao Y, Li J, Han X, Tao Q, Liu S, Jiang G, Zhu G, Yang F, Lv Z, Chen Y, Ping Q, Li W, Hou D. Dual controlled release effect of montmorillonite loaded polymer nanoparticles for ophthalmic drug delivery. Appl Clay Sci. 2019;180:105167. https://doi.org/10.1016/j.clay.2019.105167.
114. Duxfield L, Sultana R, Wang R, Englebretsen V, Deo S, Swift S, Rupenthal I, Al-Kassas R. Development of gatifloxacin-loaded cationic polymeric nanoparticles for ocular drug delivery. Pharm Dev Technol. 2016;21:172–9. https://doi.org/10.3109/10837450.2015.1091839.
115. Upadhayay P, Kumar M, Pathak K. Norfloxacin loaded pH triggered nanoparticulate in-situ gel for extraocular bacterial infections: optimization, ocular irritancy and corneal toxicity. Iran J Pharm Res IJPR. 2016;15:3.
116. Choradiya BR, Patil SB. A comprehensive review on nanoemulsion as an ophthalmic drug delivery system. J Mol Liq. 2021;339:116751. https://doi.org/10.1016/j.molliq.2021.116751.
117. Singh M, Bharadwaj S, Lee KE, Kang SG. Therapeutic nanoemulsions in ophthalmic drug administration: concept in formulations and characterization techniques for ocular drug delivery. J Control Release. 2020;328:895–916. https://doi.org/10.1016/j.jconrel.2020.10.025.

118. Tang B, Wang Q, Zhang G, Zhang A, Zhu L, Zhao R, Gu H, Meng J, Zhang J, Fang G. OCTN2- and ATB0,+-targeted nanoemulsions for improving ocular drug delivery. J Nanobiotechnol. 2024;22:130. https://doi.org/10.1186/s12951-024-02402-x.
119. Jahromy MH, Qomi M, Fazelipour S, Sami N, Faali F, Karimi M, Moghadam FA. Evaluation of curcumin-based ophthalmic nano-emulsion on atropine-induced dry eye in mice. Heliyon. 2024;10:e29009.
120. Guo X, Zhang J, Liu X, Lu Y, Shi Y, Li X, Wang S, Huang J, Liu H, Zhou H. Antioxidant nanoemulsion loaded with latanoprost enables highly effective glaucoma treatment. J Ophthalmol Clin Res. 2023;361:534–46.
121. Saquib MN. A long-acting ocular lubricant eye drop based on dextran hydrogel nanoparticles. Master's Thesis, University of Waterloo; 2019. https://uwspace.uwaterloo.ca/handle/10012/14505. Accessed 10 Oct 2024.
122. Binkhathlan Z, Ali R, Alomrani AH, Abul Kalam M, Alshamsan A, Lavasanifar A. Role of polymeric micelles in ocular drug delivery: an overview of decades of research. Mol Pharm. 2023;20:5359–82. https://doi.org/10.1021/acs.molpharmaceut.3c00598.
123. Luhar M, Viradiya R, Panjabi S, Patel G. Nanotechnology in ocular drug delivery: the potential of polymeric micelles as a drug delivery vehicle. J Ocul Pharmacol Ther. 2024;4_:54. https://doi.org/10.1089/jop.2024.0060.
124. Vaishya RD, Khurana V, Patel S, Mitra AK. Controlled ocular drug delivery with nanomicelles. Wiley Interdiscip Rev Nanomed Nanobiotechnol. 2014;6:422–37. https://doi.org/10.1002/wnan.1272.
125. Mandal A, Bisht R, Rupenthal ID, Mitra AK. Polymeric micelles for ocular drug delivery: From structural frameworks to recent preclinical studies. J Control Release. 2017;248:96–116. https://doi.org/10.1016/j.jconrel.2017.01.012.
126. Özsoy Y, Güngör S, Kahraman E, Durgun ME. Chapter 4 – polymeric micelles as a novel carrier for ocular drug delivery. In: Grumezescu AM, editor. Nanoarchitectonics biomed. William Andrew Publishing; 2019. p. 85–117. https://doi.org/10.1016/B978-0-12-816200-2.00005-0.
127. Lu P, Liang Z, Zhang Z, Yang J, Song F, Zhou T, Li J, Zhang J. Novel nanomicelle butenafine formulation for ocular drug delivery against fungal keratitis: *In Vitro* and *In Vivo* study. Eur J Pharm Sci. 2024;192:106629. https://doi.org/10.1016/j.ejps.2023.106629.
128. Sathe P, Kailasam V, Nagarjuna V, Sharma H, Velpandian T, Garg P, Nirmal J. Nanomicelles empower natamycin in treating fungal keratitis: an in vitro, ex vivo and in vivo study. Int J Pharm. 2024;656:124118. https://doi.org/10.1016/j.ijpharm.2024.124118.
129. Ilochonwu BC, Van Der Lugt SA, Annala A, Di Marco G, Sampon T, Siepmann J, Siepmann F, Hennink WE, Vermonden T. Thermo-responsive Diels-Alder stabilized hydrogels for ocular drug delivery of a corticosteroid and an anti-VEGF fab fragment. J Control Release. 2023;361:334–49.
130. Aslzad S, Savadi P, Abdolahinia ED, Omidi Y, Fathi M, Barar J. Chitosan/dialdehyde starch hybrid in situ forming hydrogel for ocular delivery of betamethasone. Mater Today Commun. 2022;33:104873.
131. Li X, Liu H, Yu A, Lin D, Bao Z, Wang Y, Li X. Bioinspired self-assembly supramolecular hydrogel for ocular drug delivery. Chin Chem Lett. 2021;32:3936–9.
132. Varela-Garcia A, Concheiro A, Alvarez-Lorenzo C. Cytosine-functionalized bioinspired hydrogels for ocular delivery of antioxidant transferulic acid. Biomater Sci. 2020;8:1171–80.
133. Kalomiraki M, Thermos K, Chaniotakis NA. Dendrimers as tunable vectors of drug delivery systems and biomedical and ocular applications. Int J Nanomedicine. 2016;11:1–12. https://doi.org/10.2147/IJN.S93069.
134. Kambhampati SP, Kannan RM. Dendrimer nanoparticles for ocular drug delivery. J Ocul Pharmacol Ther. 2013;29:151–65. https://doi.org/10.1089/jop.2012.0232.
135. Sztandera K, Rodríguez-García JL, Ceña V. In vivo applications of dendrimers: a step toward the future of nanoparticle-mediated therapeutics. Pharmaceutics. 2024;16:439. https://doi.org/10.3390/pharmaceutics16040439.

136. Yavuz B, Bozdağ Pehlivan S, Ünlü N. Dendrimeric systems and their applications in ocular drug delivery. Sci World J. 2013;2013:732340. https://doi.org/10.1155/2013/732340.
137. Wang J, Li B, Huang D, Norat P, Grannonico M, Cooper RC, Gui Q, Chow WN, Liu X, Yang H. Nano-in-nano dendrimer gel particles for efficient topical delivery of antiglaucoma drugs into the eye. Chem Eng J Lausanne Switz. 2021;1996(425):130498. https://doi.org/10.1016/j.cej.2021.130498.
138. Patil S, Sawale G, Ghuge S, Sathaye S. Quintessence of currently approved and upcoming treatments for dry eye disease. Graefes Arch Clin Exp Ophthalmol. 2024;263:269. https://doi.org/10.1007/s00417-024-06587-7.
139. Reimondez-Troitiño S, Csaba N, Alonso MJ, de la Fuente M. Nanotherapies for the treatment of ocular diseases. Eur J Pharm Biopharm. 2015;95:279–93. https://doi.org/10.1016/j.ejpb.2015.02.019.
140. Pooja D, Kadari A, Kulhari H, Sistla R. Chapter 13 – lipid-based nanomedicines: current clinical status and future perspectives. In: Grumezescu AM, editor. Lipid nanocarriers drug target. William Andrew Publishing; 2018. p. 509–28. https://doi.org/10.1016/B978-0-12-813687-4.00013-X.
141. Costello MA, Liu J, Wang Y, Qin B, Xu X, Li Q, Lynd NA, Zhang F. Reverse engineering the Ozurdex dexamethasone intravitreal implant. Int J Pharm. 2023;634:122625.
142. Grumezescu AM. Design of nanostructures for versatile therapeutic applications. William Andrew; 2018.
143. Alcon Research, Evaluation of the clinical efficacy and tolerability of SYSTANE complete in adult patients with dry eye disease following topical ocular use for 4 weeks: a multicenter trial, clinicaltrials.gov, 2019. https://clinicaltrials.gov/study/NCT03492541. Accessed 27 Oct 2024.
144. Srinivasarao DA, Lohiya G, Katti DS. Fundamentals, challenges, and nanomedicine-based solutions for ocular diseases. WIREs Nanomed Nanobiotechnol. 2019;11:e1548. https://doi.org/10.1002/wnan.1548.
145. Meng T, Kulkarni V, Simmers R, Brar V, Xu Q. Therapeutic implications of nanomedicine for ocular drug delivery. Drug Discov Today. 2019;24:1524–38.
146. Iqbal H, Razzaq A, Zhou D, Lou J, Xiao R, Lin F, Liang Y. Nanomedicine in glaucoma treatment; current challenges and future perspectives. Mater Today Bio. 2024:101229.

Safety and Toxicities of Nanocarriers in Ophthalmic Drug Delivery System

10

Sunil Kumar Kadiri, V. T. Iswariya, Prashant Tiwari, and V. V. Rajesham

Abstract

The incorporation of nanocarriers into drug delivery systems for eye conditions is a potential development in targeted treatment and long-term medication release. The safety and toxicity issues surrounding the use of nanocarriers in ophthalmology are thoroughly examined in this chapter. By improving medication solubility, bioavailability, and targeted transport to the ocular tissues, nanocarriers such as liposomes, dendrimers, nanoparticles, and micelles have the potential to overcome the drawbacks of traditional ophthalmic therapies. Nevertheless, their distinct physicochemical characteristics give rise to serious questions regarding toxicity and biocompatibility, requiring in-depth analyses. With the aim to evaluate the toxicity characteristics of these nanocarriers, this chapter focuses on in vitro and in vivo investigations. To assess cellular reactions to nanocarriers, in vitro investigations include genotoxicity tests, oxidative stress indicators, and cytotoxicity assays. Acute and chronic toxicity, organ-specific toxicity, and pharmacokinetic behavior in animal models are all covered by

S. K. Kadiri
College of Pharmaceutical Sciences, Dayananda Sagar University, Bengaluru, Karnataka, India

V. T. Iswariya
Department of Pharmaceutics, CMR College of Pharmacy, Hyderabad, India

P. Tiwari (✉)
College of Pharmaceutical Sciences, Dayananda Sagar University, Bengaluru, Karnataka, India
e-mail: prashanttiwari-sps@dsu.edu.in

V. V. Rajesham
Department of Pharmacology, CMR College of Pharmacy, Hyderabad, India

N. K. Mehra (ed.), *Complex Ophthalmic Dosage Forms: Advances in Biomedical Applications and Future Perspectives*, https://doi.org/10.1007/978-981-96-6306-4_10

in vivo evaluations. Size, surface charge, composition, and dose are some of the elements that affect the safety of nanocarriers. These factors are thoroughly examined to see how they may contribute to any harmful consequences. Other difficulties covered include immunological reactions, the requirement for standardized testing procedures, and the way in which nanocarriers interact with ocular tissues. This study emphasizes how crucial it is to conduct thorough safety evaluations and provide recommendations for the clinical application of nanocarriers in the transport of medications to the eyes. For these cutting-edge drug delivery technologies to be successfully applied in treating a variety of ocular disorders and, eventually, improve patient outcomes, their safety must be guaranteed.

Keywords
Nanocarriers · Genotoxicity · Dendrimers · Cytotoxicity · Micelles · Safety and toxicity

10.1 Introduction

The safety and efficacy of ocular drug delivery depend on the pH of the drug preparation, which should be compatible with the membrane of the eye. The diversity of structures of eye makes it difficult to design medication delivery systems for. The primary issue with the traditional eye drop drug delivery method is that the drug is quickly and completely excreted from the organ, resulting in a significant loss of the medication. Since only a small amount of the medication reaches the site of action and the remainder is drained away, frequent use is required to obtain a therapeutic effect. The posterior segment of the eye is composed of the retina, vitreous humor, and choroid. Implants, intravenous and intravitreal drug delivery devices, and periocular medication administration, which call for a high concentration of the medication, can all be used to treat diseases affecting these regions [1–4].

10.1.1 Challenges for Ocular Drug Administration

Ocular administration of drugs presents significant challenges owing to the eye's distinctive structure and defensive mechanisms. The challenges encompass the following.

10.1.1.1 Suboptimal Medication Bioavailability

Topical formulations, such as eye drops, the predominant mode of medication delivery, may yield inadequate bioavailability. A considerable amount of the medicine is diminished due to tear formation, blinking, or drainage via the nasolacrimal duct, resulting in merely a minor fraction reaching the target area [5–7].

10.1.1.2 Brief Drug Residence Duration

The quick turnover of tears, along with continuous blinking and drainage, reduces the duration medications remain on the ocular surface, requiring frequent re-administration to sustain therapeutic levels [8].

10.1.1.3 Systemic Absorption and Adverse Effects

Medications may be absorbed systemically via the nasolacrimal duct, resulting in possible adverse effects in regions outside the ocular area, including the heart or lungs, especially with the use of strong pharmaceuticals [9–11].

10.1.1.4 Patient Compliance

The necessity for frequent administration, challenges in the accurate application of eye drops, and possible discomfort from formulations may result in suboptimal patient adherence. This is particularly crucial in chronic disorders such as glaucoma, where compliance with medication is vital for averting disease progression [12, 13].

Ocular pharmaceuticals require sterility, complicating their formulation and storage. Certain formulations necessitate preservatives for stability; nonetheless, prolonged usage may induce irritation or allergic responses [14].

These problems underscore the necessity for sophisticated drug delivery technologies, like sustained-release formulations, nanoparticles, or implanted devices, to optimize ocular medication delivery, increase patient outcomes, and mitigate the strain of frequent Administration [15].

10.2 Safety and Toxicity Concerns Associated with Ocular Nanocarriers

Ocular nanocarriers have garnered considerable interest in drug delivery for their promise to enhance therapeutic efficacy by boosting drug bioavailability, extending drug release, and augmenting target specificity within the eye. Nonetheless, guaranteeing the safety and reducing the toxicity of these nanocarriers are essential difficulties that must be resolved prior to their extensive clinical application [16–18].

(a) Composition of Nanocarriers: Lipid-based nanocarriers, such as liposomes and solid lipid nanoparticles, exhibit biocompatibility and biodegradability, rendering them safer for ocular applications. Nevertheless, they may still induce fat buildup and immunological responses if not meticulously prepared. Polymeric nanocarriers (e.g., PLGA, chitosan, and PEG) are extensively utilized for their adjustable qualities; nonetheless, there are concerns over the potential toxicity of polymer degradation products. Metal-based nanoparticles (e.g., gold or silver nanoparticles) offer unique optical qualities but may generate oxidative stress and inflammation, leading to possible cytotoxicity and genotoxicity in ocular tissues. Inorganic nanocarriers, such as silica or cerium oxide nanoparticles, may aggregate and induce chronic toxicity; yet, cerium oxide possesses antioxidant properties that could be advantageous [19–24].

(b) Dimensions and Surface Characteristics.

Nanoparticles smaller than 100 nm typically exhibit enhanced penetration through ocular barriers, including the cornea and conjunctiva; nevertheless, they also possess an elevated chance of inadvertently accessing unwanted regions, such as the retina or optic nerve, resulting in off-target effects. Positively charged nanoparticles exhibit enhanced binding to negatively charged ocular tissues; nonetheless, they are associated with a higher propensity for irritation, inflammation, and toxicity relative to neutral or negatively charged particles [41].

(c) Toxicity Issues.

Certain nanocarriers may induce irritation to the ocular epithelium. Cytotoxicity assays, such as the MTT assay, are frequently performed to assess this. Nanoparticles, particularly those composed of metals, can provoke the generation of reactive oxygen species (ROS), resulting in oxidative harm to ocular tissues, including the cornea and retina. Certain nanocarriers may elicit an immunological response, leading to eye inflammation. The release of pro-inflammatory cytokines can heighten the risk of injury. If nanocarriers reach the retina, they can cause phototoxicity and retinal damage, especially in circumstances where the nanocarrier materials interact with light (e.g., photosensitive nanoparticles) [44].

(d) Biodistribution and Clearance.

Nanoparticles that resist degradation or are ineffectively removed may accumulate in ocular tissues, resulting in persistent toxicity or granuloma development. Nanoparticles delivered through eye drops may be swiftly eliminated by tear fluid, diminishing effectiveness while concurrently decreasing toxicity risk. Extending retention while avoiding heightened toxicity presents a formulation difficulty.

(e) Method of Administration.

Eye drops serve as a prevalent way for delivering ocular nanocarriers. The primary safety issues are irritation and limited absorption, necessitating elevated dosages and heightening the risk of poisoning. Intravitreal injection facilitates direct administration to the posterior segment (retina, vitreous body), although it entails an elevated risk of retinal toxicity, inflammation, and mechanical injury during the procedure. Subconjunctival and periocular routes circumvent certain obstacles to enhance delivery efficacy; yet, they may induce local inflammation and fibrosis.

(f) Regulatory and Long-Term Implications.

Regulatory agencies (FDA, EMA) mandate comprehensive preclinical and clinical investigations on ocular toxicity prior to the approval of nanocarriers for utilization. This includes the assessment of genotoxicity, carcinogenicity, and chronic effects on ocular structures.

Long-term safety data are frequently insufficient, presenting a significant deficiency for chronic ocular disorders (e.g., glaucoma or age-related macular degeneration) that may necessitate extended therapy.

(g) Strategies to Mitigate Toxicity.

Coating nanoparticles with biocompatible compounds, such as PEGylation, can mitigate toxicity by enhancing colloidal stability and reducing immunological response. Ligand-functionalized nanoparticles enhance specificity toward certain ocular tissues, minimizing off-target effects and systemic exposure. Integrating antioxidants into formulations may mitigate oxidative stress and safeguard ocular tissues from damage generated by reactive oxygen species (ROS).

10.2.1 The Negative Aspects of Ophthalmic Drug Delivery Systems

Ophthalmic drug delivery methods, although crucial for addressing numerous ocular disorders, encounter severe drawbacks that hinder their efficacy. A principal problem is inadequate drug retention on the ocular surface caused by fast drainage through tear fluid and blinking, resulting in diminished bioavailability and the requirement for frequent dosing. Numerous drugs are unable to infiltrate deeper ocular tissues, including the retina, due to the corneal barrier and blood-ocular barriers, hence limiting therapeutic choices for disorders impacting the posterior portion of the eye. Moreover, systemic absorption via the nasolacrimal duct may result in unexpected side effects, diminishing the quantity of the drug available in the eye for therapeutic efficacy [24, 66].

Certain formulations, like eye drops, are frequently poorly accepted by patients, resulting in discomfort or irritation [34]. Furthermore, certain sophisticated medication delivery devices, such as implants or injections, are intrusive and carry hazards of infection, inflammation, and patient non-compliance. The expense associated with the development and production of sophisticated ocular drug delivery systems, including nanoparticles and sustained-release formulations, might be substantial, hence restricting their accessibility for wider application. These constraints underscore the necessity for continuous research to enhance the safety, efficacy, and comfort of ocular medication administration.

10.2.2 The Limitations of Delivering Ocular Drugs

- The dosage form cannot be stopped in an emergency.
- Blockage of eyesight.
- Finds it difficult to insert and remove the dosage form.

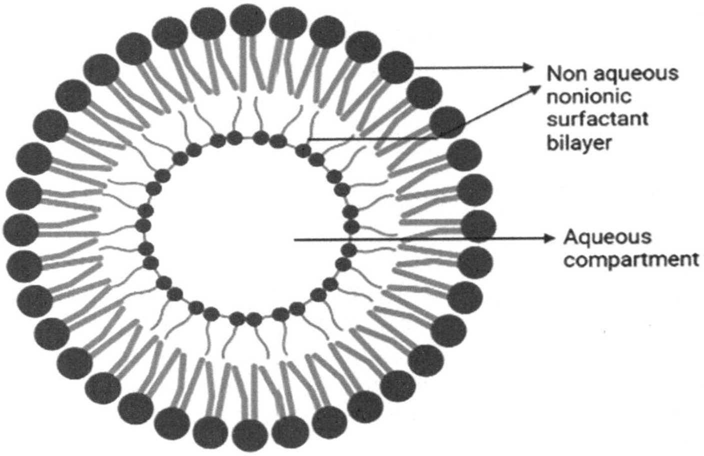

Fig. 10.1 Structure of a niosome. (Self-created)

Occasionally, when you touch your eyes or go to sleep, the medication may be lost.

10.3 Types of Nanocarriers

10.3.1 Niosomes

Niosomes help in improving the safety and reducing the toxicity of nanocarriers by offering better biocompatibility, stability, controlled drug release, and the ability to target specific tissues. These factors collectively enhance the therapeutic efficacy of drugs while minimizing side effects associated with higher doses or non-targeted drug delivery [38, 39] (Fig. 10.1).

10.3.2 Solid Lipid Nanoparticles (SLNs) and Nanostructured Lipid Carriers (NLCs)

Solid lipid nanoparticles (SLNs) and nanostructured lipid carriers (NLCs) are effective in enhancing the safety and reducing the toxicity of nanocarriers due to their biocompatibility, controlled release profiles, low immune activation, and the ability to protect and target drugs more efficiently. These advantages make SLNs and NLCs promising candidates for safer drug delivery systems in various therapeutic areas [40].

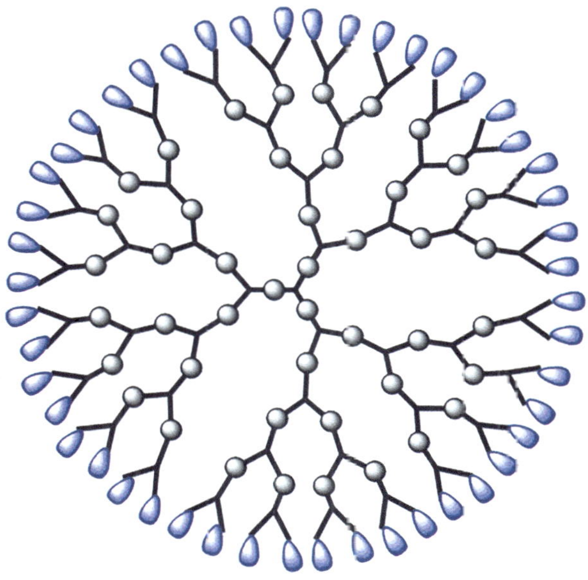

Fig. 10.2 Structure of a dendrimer. (Self-created)

10.3.3 Polymeric Nanoparticles

Polymeric nanoparticles offer a promising strategy to improve the safety and reduce the toxicity of nanocarriers in drug delivery. By utilizing biocompatible and biodegradable materials, providing controlled drug release, targeting specific tissues, and minimizing immune responses and other adverse interactions, polymeric nanoparticles help create safer and more effective drug delivery systems [41].

10.3.4 Dendrimers

Dendrimers are highly effective nanocarriers for improving the safety and reducing the toxicity of drug delivery systems. Their unique structural properties, biocompatibility, and versatility in design allow for precise drug targeting, controlled release, and minimized systemic side effects, making them a promising option in the development of safer and more effective nanocarriers for drug delivery [42] (Fig. 10.2).

10.3.5 Hydrogels

Hydrogels are an excellent platform for improving the safety and minimizing the toxicity of various nanocarriers in drug delivery. Their ability to control drug release, protect nanocarriers, reduce immune responses, and allow for targeted, localized delivery makes them a versatile and effective tool for enhancing the safety of nanomedicine [43] (Fig. 10.3).

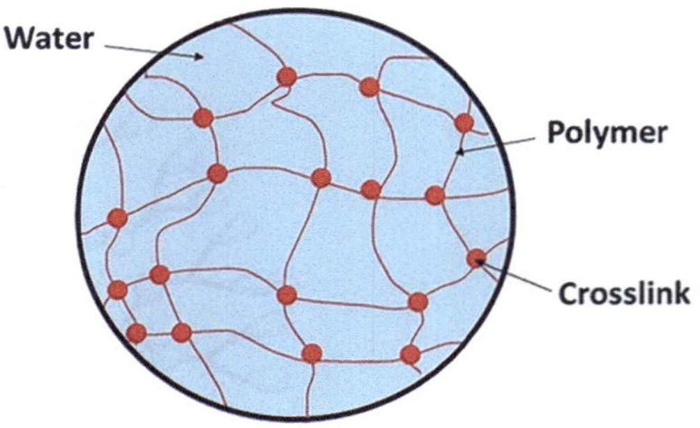

Fig. 10.3 Structure of hydrogel. (Self-created)

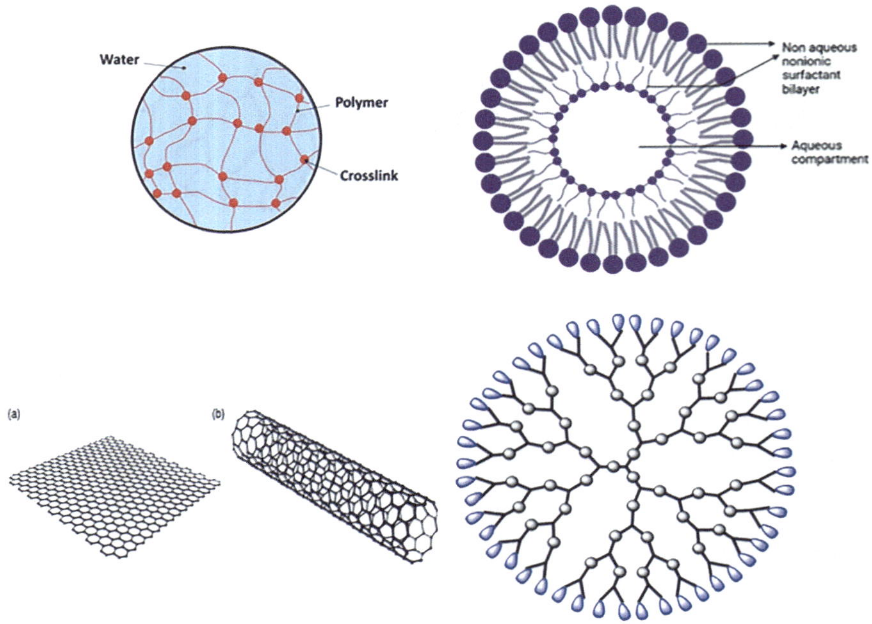

Fig. 10.4 Structure of graphene and carbon nanotubes

10.3.6 Micelles

Description: Self-assembled colloidal structures formed by amphiphilic molecules in aqueous solutions, with a hydrophobic core and hydrophilic shell as shown in Fig. 10.4.

Advantages: Ability to solubilize poorly water-soluble drugs, enhanced permeability, and retention effect.

Applications: Effective in delivering hydrophobic drugs.

10.3.7 Nanoemulsions

Nanoemulsions are highly effective in enhancing the safety and minimizing the toxicity of various nanocarriers. Their ability to solubilize hydrophobic drugs, provide controlled and targeted drug release, reduce systemic toxicity, and prevent aggregation makes them an ideal platform for safe and efficient drug delivery [44].

10.3.8 Carbon Nanotubes and Graphenes

Carbon nanotubes can be highly beneficial in improving the safety and reducing the toxicity of nanocarriers used in drug delivery. Through surface functionalization, they become more biocompatible, reduce unwanted immune responses, and enable targeted drug delivery, all of which contribute to a safer therapeutic profile. By optimizing the design and functionalization of CNTs, their potential for drug delivery systems can be harnessed with minimized risks to the patient's eye [45–46] (Fig. 10.4).

10.4 Popular Nanocarriers of Ocular Drug Release

The choice of nanocarrier depends on the drug's properties, the targeted ocular tissue, and the desired release profile. These technologies are advancing the treatment of a variety of ocular conditions, including glaucoma, retinal diseases, infections, and inflammation. The ocular preparations of DDSs offer benefits of targeted and sustained drug release, and they are beneficial for carrying ophthalmic sustained release systems due to their unique molecular structure and biological properties [47–49].

10.4.1 Nanoparticles

Nanoparticles are formed by natural degradable phospholipids that self-assemble and have bi-continuous microstructural properties in the aqueous phase [126, 127]. Due to the size of the bilayer area, cubic liquid crystal nanoparticles have an advantage over liposomes for loading lipid-soluble and amphiphilic drugs. **Nanoparticles** in drug delivery systems offer remarkable benefits but also present potential toxicity concerns that need to be addressed through careful design and surface modification. By considering size, shape, composition, and functionalization, the safety of **nanocarriers** can be optimized to minimize toxicity while maximizing therapeutic benefits.

10.4.2 Microemulsions/Nanoemulsions

There are several uses for other nanotechnology-based carriers in drug delivery studies, including nanoemulsion (NEs) and microemulsion (MEs). Furthermore,

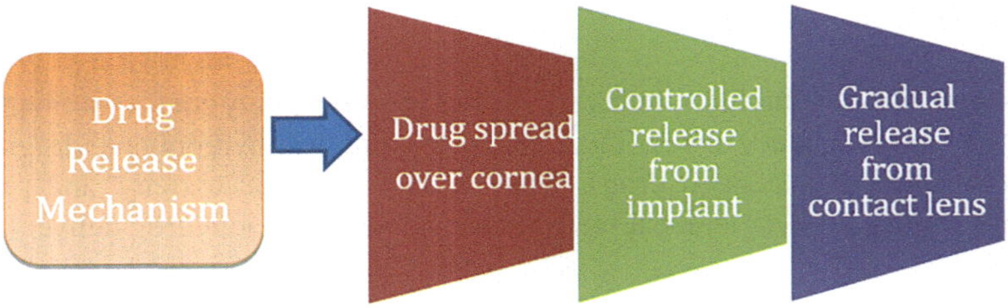

Fig. 10.5 Representation of drug release mechanism. (Self-created)

because of their high surfactant/co-surfactant concentrations, they have the potential to be drug solubilizers that improve penetration. NEs have a high degree of stability, little toxicity and irritation, a sizable capacity for drug loading, and much higher drug bioavailability. Among the most studied and used nanocarriers (NDSs) for local medication delivery in the eyes are NEs. NEs are mostly absorbed through the cornea following local administration to the eye [50–52].

10.4.3 Drug Delivery Systems That Use In Situ Gels and Others [30]

Over the past 10 years, in situ gelling drug delivery devices have drawn a lot of interest from the ophthalmology community. In order to increase the bioavailability and retention duration of medications on the ocular surface, these preparations like exosomes and niosomes were combined [139]. The nanocomposite gel system is thoroughly inspected for quality control. The whole system quality must be taken into account, as well as the gel's stability and the in situ gel's macro-level gelling performance [139]. Given that the nano-gel composite system makes extensive use of auxiliary materials, safety should come first. However, a comparison between MEs and the gel matrix was not performed in this work. To assess the safety of a composite system, it is recommended to compare a single formulation with the matching matrix. This comparison can be used to guide the creation of new composite formulations [53–55] as shown in Fig. 10.5.

10.5 Safety Considerations of Nanocarriers

10.5.1 Biocompatibility

It depicts the capability of a nanocarrier to exhibit its desired purpose without creating any unwanted local or systemic effects in the body [140]. It is a critical parameter for nanocarrier design and development. The factors influencing biocompatibility include:

(a) **Material Composition:** The choice of material—whether organic (like lipids, polymers) or inorganic (such as metals, silica)—greatly influences biocompatibility. Organic nanocarriers are generally preferred due to their biodegradability and lower toxicity profiles.
(b) **Surface Properties:** Surface charge, hydrophobicity, and functionalization can affect protein adsorption and cell interactions, influencing biocompatibility. For example, a hydrophilic and neutrally charged surface is less likely to be distinguished by the immune system.
(c) **Size and Shape:** Smaller nanocarriers can easily navigate through the body and are generally more biocompatible than larger ones [110]. The shape of the nanocarriers can also influence cellular uptake and distribution.
(d) **Degradation Products:** The biocompatibility of nanocarriers also depends on their degradation products. Non-toxic, easily metabolizable, and excretable degradation products are essential for safe applications.

10.5.2 Immunogenicity

Immunogenicity refers to the ability of a substance to provoke an immune response. Nanocarriers may interact with the immune system in various ways:

(a) **Recognition by the Immune System:** Surface properties of nanocarriers determine their interaction with immune cells. Modifications, such as PEGylation (attaching polyethylene glycol chains), can help in reducing recognition and clearance by the immune system.
(b) **Activation of Immune Cells:** Some nanocarriers can activate immune cells, leading to the production of inflammatory cytokines and other mediators. This can be both beneficial (in vaccine delivery) and harmful (in drug delivery).
(c) **Immune Clearance:** Strategies to avoid reticuloendothelial system (RES) clearance include size optimization and surface modification [56–58].

10.5.3 Inflammation and Irritation Potential

Inflammation and irritation are common concerns with nanocarriers, particularly with repeated or prolonged use.

(a) **Pro-inflammatory Responses:** Nanocarriers can induce inflammation through direct activation of immune cells or through indirect mechanisms [59].
(b) **Irritation:** Local irritation at the site of administration can occur due to the physical properties of the nanocarriers or their interaction with biological tissues. Ensuring the nanocarriers are non-irritant involves careful selection of materials and design considerations [60].

(c) **Toxicity Assessment:** Comprehensive in vitro and in vivo studies are required to assess the potential for inflammation and irritation. This includes histological examination, cytokine profiling, and other biochemical markers [61].

10.6 Toxicological Profiles of Nanocarriers

The toxicological profiles of nanocarriers are essential in understanding their safety and potential risks when used for drug delivery or other biomedical applications [126, 127]. Here's a detailed overview that includes preclinical studies, in vitro and in vivo toxicology assessments [62].

10.6.1 Overview of Nanocarriers

They offer advantages like targeted delivery, controlled release, and improved bioavailability. However, their small size and unique properties also raise concerns about toxicity and biocompatibility [63].

10.6.2 Preclinical Studies

Preclinical studies for nanocarriers involve comprehensive evaluations before clinical trials. These studies aim to predict how nanocarriers will interact with biological systems and identify potential toxic effects (Table 10.1).

10.6.2.1 Key Preclinical Focus Areas

Assessment of size, shape, surface charge, and composition, which can influence toxicity. The importance of physicochemical characteristics, including size, shape, surface charge, and composition, in the toxicological profile of nanoparticles is highlighted by the results of Pandey et al. (2020). Designing safer nanomaterials, particularly for biomedical and environmental applications, requires an understanding of these characteristics. Before using nanoparticles in consumer goods, medical procedures, or other areas where human and environmental exposure is possible, a

Table 10.1 Key preclinical focus areas

S. no.	Preclinical focus areas	Key features
1	Physicochemical characterization	Assessment of size, shape, and surface charge
2	Dose-response relationship	Relationship between nanocarrier dosage and biological response
3	ADME profile	Studying the absorption, distribution, Metabolism, and excretion of nanocarriers in animal models

thorough evaluation of the particles is necessary due to the variety in toxicity based on these characteristics [64–70].

10.6.2.2 Dose-Response Relationship

Establishing the relationship between nanocarrier dosage and biological response. The importance of dose-response relationships in the creation of treatments based on nanocarriers is demonstrated by this case study. All of these results highlight how important it is to carefully optimize dosage in order to strike a balance between safety and effectiveness. To fully realize the potential of nanocarrier systems, future developments in nanomedicine must place a high priority on precise control of dosage settings [71].

10.6.2.3 ADME Profile

Studying the absorption, distribution, metabolism, and excretion of nanocarriers in animal models study highlights the absorption, biodistribution, metabolism, and excretion patterns of nanocarriers, offering important insights into their ADME profiles. Designing safer and more efficient nanomedicines requires an understanding of these factors. In order to improve therapeutic results, this case study emphasizes how crucial it is to incorporate ADME studies into the early phases of nanocarrier development [72–74].

10.7 In Vitro Toxicology Assessments

In vitro assessments provide preliminary data on the potential toxicity of nanocarriers using cell-based models.

10.7.1 Key In Vitro Tests

10.7.1.1 Cytotoxicity Assays

Evaluate the impact of nanocarriers on cell viability using assays like MTT (tetrazolium reduction assay), LDH (lactate dehydrogenase release assay), and apoptosis assays [75].

10.7.1.2 Genotoxicity Tests

Assess the potential of nanocarriers to cause DNA damage or mutations. The genotoxicity tests include Ames test, in vitro micronucleus assay, mouse lymphoma assay, and comet assay [76–79].

- **Ames test**: Screens for **mutagenic potential** using bacterial strains.
- **In vitro micronucleus assay**: Detects **chromosomal damage** in eukaryotic cells.
- **Mouse lymphoma assay**: Detects **mutagenic effects** and **clastogenicity** in mouse lymphoma cells.

- **Comet assay**: Detects **DNA damage** at the single-cell level, including strand breaks.

Each of these tests can help identify the **genotoxic risks** posed by nanoparticles, ensuring that they are safe for human use before clinical trials and commercialization.

10.7.1.3 Immunotoxicity

Evaluate the potential of nanocarriers to provoke an immune response or cause immunosuppression.

10.7.1.4 Cell Uptake and Localization

Analyze how nanocarriers are internalized by cells and their intracellular localization.

10.8 In Vivo Toxicology Assessments

In vivo studies provide more comprehensive safety profiles by considering the interactions of nanocarriers within whole organisms.

10.8.1 Key In Vivo Tests

In vivo toxicity testing is crucial for determining the **safety** and **toxicity** profiles of nanocarriers, as it provides insights into how these particles behave in a living organism. The **acute**, **subchronic**, and **chronic toxicity** studies, along with **biodistribution**, **histopathological analysis**, **immunotoxicity**, and **genotoxicity** assessments, form the basis for evaluating the **biocompatibility** of nanocarriers. By carefully assessing these parameters, researchers can identify potential risks associated with nanoparticle use, optimize their design, and ensure that they are safe for clinical applications.

10.8.2 Acute and Chronic Toxicity

Determine the immediate and long-term effects of nanocarriers following single or repeated doses.

10.8.3 Organ Toxicity and Histopathology

Examine potential damage to vital organs such as the liver, kidneys, heart, and lungs through histological analysis.

10.8.4 Biodistribution Studies

Investigate how nanocarriers are distributed and accumulated in different tissues and organs over time.

10.8.5 Pharmacokinetics and Pharmacodynamics

Study the behavior of nanocarriers in the body, including their distribution, metabolism, and elimination. Pharmacodynamic study includes measuring physiological or biochemical changes induced by the drug and to explore the relationship between drug dose and its effect to establish the optimal therapeutic window [85–93].

10.8.6 Carcinogenicity and Reproductive Toxicity

Assess the potential of nanocarriers to cause cancer or affect reproductive health [94–97].

10.9 Challenges and Considerations [142]

10.9.1 Size and Surface Modifications

The small size and possible modifications on the nanocarriers' surface can drastically alter their interaction with biological systems [98, 99].

10.9.2 Aggregation and Solubility

Nanocarriers might aggregate or have solubility issues, affecting their toxicity profile.

10.9.3 Animal Models

Differences in species-specific responses require careful selection of animal models and extrapolation of data to humans.

10.9.3.1 Therapeutic Considerations

Nepafenac-loaded ophthalmic nanocarriers for treatment of uveitis concluded that these nanocarriers not only enhance drug effectiveness but also present an innovative approach in management of ocular inflammation [100–118].

10.10 Safety and Toxicity Concerns for Nanocarriers

The primary concerns for ophthalmic formulations' clinical approval are safety and toxicity. Preclinical studies have examined the ocular toxicity of engineered nanoformulations [53, 97].

10.10.1 Biocompatibility and Immunotoxicity

Immune system activation: Nanocarriers may trigger immune responses, leading to inflammation, hypersensitivity, or allergic reactions. This can cause long-term issues if the body treats the nanocarrier as a foreign invader.

Inflammation and cytokine release: In vivo exposure to certain nanocarriers can stimulate the release of pro-inflammatory cytokines, leading to chronic inflammation.

10.10.2 Cellular Toxicity

Cell membrane disruption: Some nanoparticles may interact with and damage cell membranes, leading to cell lysis or death. Oxidative stress: Nanoparticles can induce the production of reactive oxygen species (ROS), which can lead to oxidative stress and damage cellular components, including lipids, proteins, and DNA. Endocytosis and cytotoxicity: Nanocarriers are typically taken up by cells via endocytosis. However, if nanoparticles accumulate in excessive amounts within cells, they may disrupt normal cellular processes or cause toxicity [119–133].

10.10.3 Organ Toxicity

Accumulation in organs: Nanoparticles can accumulate in organs such as the liver, spleen, kidneys, and lungs, potentially leading to organ-specific toxicity.

Bioaccumulation: Persistent nanoparticles may bioaccumulate in tissues, leading to long-term toxic effects, including tissue damage, fibrosis, or dysfunction [134–142].

10.10.4 Genotoxicity

DNA damage: Certain nanocarriers, especially those containing metals or metal oxides, may cause genetic mutations, chromosomal damage, or DNA strand breaks, increasing the risk of cancer or heritable genetic disorders.

10.10.5 Hemotoxicity

Blood coagulation: Nanocarriers, particularly those that interact with blood components, may induce blood clotting or hemolysis (rupture of red blood cells), leading to adverse effects on circulation and organ perfusion.

10.10.6 Reproductive and Developmental Toxicity

Fertility issues: Some nanocarriers may adversely affect reproductive organs, impair fertility, or influence fetal development. Developmental toxicity: Nanoparticles can cross the placental barrier, potentially causing harm to the developing fetus.

10.10.7 Environmental Toxicity

Ecosystem impact: The potential release of nanoparticles into the environment during manufacturing, usage, or disposal could harm aquatic life, soil microorganisms, and other environmental components.

10.10.8 Long-Term Safety

Chronic toxicity: Long-term exposure to nanocarriers could result in cumulative toxicity, where effects may not be immediately apparent but emerge over time, necessitating long-term safety studies. Due to anatomical constraints, drug distribution to the ocular surface presents significant challenges for formulation scientists. Conventional ocular treatment has limitations, as previously mentioned. On the other hand, serious or chronic eye illnesses necessitate regular medication delivery [118]. Nanocarriers are made to help carry drugs to ocular tissues in a targeted and regulated manner, overcoming the shortcomings of the present ocular therapy options.

With sub-micron-sized particles that range in size from 10 to 1000 nm, nanocarriers are unique particulate systems that can offer a variety of medication delivery options. The conjugation and retention of nanocarriers at a particular spot are determined by their surface charge.

Zeta potential, or this surface charge, is what keeps nanodispersion stable. Particle agglomeration is prevented when two particles with the same high zeta potential and charge reject one another. Because the surface of the cornea and conjunctiva have a negative charge, cationic nanocarriers in ocular administration may be drawn to them via electrostatic interactions. This results in topical medication administration to the anterior portion of the eye and makes it easier for cationic nanocarriers to stay attached to negatively charged ocular tissue.

On the other hand, because of their particle size and surface charge, anionic nanocarriers can diffuse into the retina, while cationic nanocarriers cluster or agglomerate when administered intravitreal (IVT). Submicron-sized liposomes have recently demonstrated promise as a topical medication delivery method for the treatment of disorders that affect the posterior segment when used as eye drops. Drugs can be delivered to the intended locations using nanocarriers, which may be able to get past ocular obstacles.

Using drug-loaded devices or inserts in the conjunctival cul-de-sac or even directly on the cornea is another tactic. These ocular implants can serve as controlled-release medication reservoirs, regardless of whether they are made to be biodegradable or detachable. Ocular drug delivery has also been employed using contact lenses, and the incorporation of so-called container molecules, which can hold the molecules in their holes, can improve the lenses' ability to load drugs. Ocular implants, which are made to either live completely inside the eye or pass through the ocular membrane, are one means of getting beyond all exterior obstacles. These can be prepared to give treatment for persistent eye problems either temporarily or over a period of years [40–44].

10.11 Conclusion

Using nanocarriers in ocular drug delivery systems is a viable way to improve patient compliance, therapeutic effectiveness, and targeting. This study emphasizes how crucial it is to assess these nanocarriers' toxicity and safety profiles in order to guarantee their feasibility for clinical usage. Because of their tiny size and special surface characteristics, which might have unanticipated toxicological consequences, nanocarriers pose particular safety issues despite their potential benefits. Thorough preclinical analyses are thus required, with an emphasis on the physicochemical characteristics of nanocarriers that may affect how they interact with ocular tissues. Studies conducted in vivo and in vitro are essential for determining any harmful effects on ocular tissues as well as for figuring out how nanocarriers distribute, clean up, and remain safe over time in the ocular environment. In order to obtain regulatory approval and guarantee patient safety, the incorporation of nanocarriers into ocular medication delivery requires strict adherence to regulatory standards, guaranteeing their biocompatibility and non-toxicity. It is also necessary to take ethical issues into account, especially those pertaining to possible long-term impacts and the use of animal models in preclinical research. In the future, it will be crucial to create new nanocarriers with enhanced biocompatibility and tailored delivery capabilities. New materials with improved safety and efficacy for ophthalmic application may be discovered as a result of advances in material science and nanotechnology. Further research should concentrate on developing standardized techniques for toxicity evaluation and improving preclinical models.

References

1. Abud MB, Louzada RN, Isaac DLC, Souza LG, Dos Reis RG, Lima EM, de Ávila MP. In vivo and in vitro toxicity evaluation of liposome-encapsulated sirolimus. Int J Retin Vitreous. 2019;5(1):35. https://doi.org/10.1186/s40942-019-0186-7.
2. Aengenheister L, Favaro RR, Morales-Prieto DM, Furer LA, Gruber M, Wadsack C, Markert UR, Buerki-Thurnherr T. Research on nanoparticles in human perfused placenta: state of the art and perspectives. Placenta. 2021;104:199–207. https://doi.org/10.1016/j.placenta.2020.12.014.
3. Agrahari V. A comprehensive insight on ocular pharmacokinetics. Drug Deliv Transl Res. 2016;
4. Ahmad I. Optimization by design of etoposide loaded solid lipid nanoparticles for ocular delivery: characterization, pharmacokinetic and deposition study. Mater Sci Eng C Mater Biol Appl. 2019;
5. Alvarez-Trabado J. Designing lipid nanoparticles for topical ocular drug delivery. Int J Pharm. 2017;
6. Amankwa CE, Gondi SR, Dibas A, Weston C, Funk A, Nguyen T, Nguyen KT, Ellis DZ, Acharya S. Novel thiol containing hybrid antioxidant-nitric oxide donor small molecules for treatment of glaucoma. Antioxidants (Basel). 2021;10(4):575. https://doi.org/10.3390/antiox10040575.
7. Ameh T, Sayes CM. The potential exposure and hazards of copper nanoparticles: a review. Environ Toxicol Pharmacol. 2019;71(103220):103220. https://doi.org/10.1016/j.etap.2019.103220.
8. Apaolaza PS, Busch M, Asin-Prieto E, Peynshaert K, Rathod R, Remaut K, Dünker N, Göpferich A. Hyaluronic acid coating of gold nanoparticles for intraocular drug delivery: evaluation of the surface properties and effect on their distribution. Exp Eye Res. 2020;198:108151. https://doi.org/10.1016/j.exer.2020.108151.
9. Ashammakhi N, Darabi MA, Çelebi-Saltik B, Tutar R, Hartel MC, Lee J, Hussein S, Goudie MJ, Cornelius MB, Dokmeci MR, Khademhosseini A. Microphysiological systems: next generation systems for assessing toxicity and therapeutic effects of nanomaterials. Small Methods. 2020;4(1) https://doi.org/10.1002/smtd.201900589.
10. Awwad S, Mohamed Ahmed AHA, Sharma G, Heng JS, Khaw PT, Brocchini S, Lockwood A. Principles of pharmacology in the eye: principles of pharmacology in the eye. Br J Pharmacol. 2017;174(23):4205–23. https://doi.org/10.1111/bph.14024.
11. Baudouin C, Kolko M, Melik-Parsadaniantz S, Messmer EM. Inflammation in glaucoma: from the back to the front of the eye, and beyond. Prog Retin Eye Res. 2021;83:100916. https://doi.org/10.1016/j.preteyeres.2020.100916.
12. Bayda S, Adeel M, Tuccinardi T, Cordani M, Rizzolio F. The history of nanoscience and nanotechnology: from chemical-physical applications to nanomedicine. Molecules (Basel). 2019;25(1):112. https://doi.org/10.3390/molecules25010112.
13. Begines B, Ortiz T, Pérez-Aranda M, Martínez G, Merinero M, Argüelles-Arias F, Alcudia A. Polymeric nanoparticles for drug delivery: recent developments and future prospects. Nanomaterials (Basel). 2020;10(7):1403. https://doi.org/10.3390/nano10071403.
14. Bhatt P, Fnu G, Bhatia D, Shahid A, Sutariya V. Nanodelivery of resveratrol-loaded PLGA nanoparticles for age-related macular degeneration. AAPS PharmSciTech. 2020;21(8):291. https://doi.org/10.1208/s12249-020-01836-4.
15. Bhattacharjee A, Das PJ, Adhikari P, Marbaniang D, Pal P, Ray S, Mazumder B. Novel drug delivery systems for ocular therapy: with special reference to liposomal ocular delivery. Eur J Ophthalmol. 2019;29(1):113–26. https://doi.org/10.1177/1120672118769776.
16. Cabrera FJ, Wang DC, Reddy K, Acharya G, Shin CS. Challenges and opportunities for drug delivery to the posterior of the eye. Drug Discov Today. 2019;24(8):1679–84. https://doi.org/10.1016/j.drudis.2019.05.035.

17. Ch S, Mishra P, Bhatt H, Ghosh B, Roy S, Biswas S. Hydroxypropyl methacrylamide-based copolymeric nanoparticles loaded with moxifloxacin as a mucoadhesive, cornea-penetrating nanomedicine eye drop with enhanced therapeutic benefits in bacterial keratitis. Colloids Surf B Biointerfaces. 2021;208:112113. https://doi.org/10.1016/j.colsurfb.2021.112113.
18. Chan C-M, Hsiao C-Y, Li H-J, Fang J-Y, Chang D-C, Hung C-F. The inhibitory effects of gold nanoparticles on VEGF-A-induced cell migration in choroid-retina endothelial cells. Int J Mol Sci. 2019a;21(1):109. https://doi.org/10.3390/ijms21010109.
19. Chan TCY, Chow SSW, Wan KHN, Yuen HKL. Update on the association between dry eye disease and meibomian gland dysfunction. Xianggang Yi Xue Za Zhi [Hong Kong Med J]. 2019b;25(1):38–47. https://doi.org/10.12809/hkmj187331.
20. Chen X, Zhu S, Hu X, Sun D, Yang J, Yang C, Wu W, Li Y, Gu X, Li M, Liu B, Ge L, Gu Z, Xu H. Toxicity and mechanism of mesoporous silica nanoparticles in eyes. Nanoscale. 2020;12(25):13637–53. https://doi.org/10.1039/d0nr03208e.
21. Chrishtop VV, Prilepskii AY, Nikonorova VG, Mironov VA. Nanosafety vs. nanotoxicology: adequate animal models for testing in vivo toxicity of nanoparticles. Toxicology. 2021;462:152952. https://doi.org/10.1016/j.tox.2021.152952.
22. Corn KC. Lipids in the tumor microenvironment: from cancer progression to treatment. Prog Lipid Res. 2020;
23. Dave RS. Ocular drug delivery to the anterior segment using nanocarriers: a mucoadhesive/mucopenetrative perspective. J Control Release. 2021;
24. del Amo EM. Topical ophthalmic administration: can a drug instilled onto the ocular surface exert an effect at the back of the eye? Front Drug Deliv. 2022;2:954771.
25. Deval G, Boland S, Fournier T, Ferecatu I. On placental toxicology studies and cerium dioxide nanoparticles. Int J Mol Sci. 2021;22(22):12266. https://doi.org/10.3390/ijms222212266.
26. Dhahir RK, Al-Nima AM, Al-Bazzaz FY. Nanoemulsions as ophthalmic drug delivery systems. Turk J Pharm Sci. 2021;18(5):652–64. https://doi.org/10.4274/tjps.galenos.2020.59319.
27. Dugershaw BB, Aengenheister L, Hansen SSK, Hougaard KS, Buerki-Thurnherr T. Recent insights on indirect mechanisms in developmental toxicity of nanomaterials. Part Fibre Toxicol. 2020;17(1):31. https://doi.org/10.1186/s12989-020-00359-x.
28. Dukovski BJ. Functional ibuprofen-loaded cationic nanoemulsion: development and optimization for dry eye disease treatment. Int J Pharm. 2020;
29. Durgun ME, Güngör S, Özsoy Y. Micelles: promising ocular drug carriers for anterior and posterior segment diseases. J Ocul Pharmacol Ther. 2020;36(6):323–41. https://doi.org/10.1089/jop.2019.0109.
30. Eldin SM, Naim A. Safety and toxicity of nanoparticles in ophthalmic drug delivery: a review. J Nanomed. 2021; https://doi.org/10.2217/nnm-2021-0025.
31. Feghhi M, Sharif Makhmalzadeh B, Farrahi F, Akmali M, Hasanvand N. Anti-microbial effect and in vivo ocular delivery of ciprofloxacin-loaded liposome through rabbit's eye. Curr Eye Res. 2020;45(10):1245–51. https://doi.org/10.1080/02713683.2020.1728777.
32. Gan L. Self-assembled liquid crystalline nanoparticles as a novel ophthalmic delivery system for dexamethasone: improving preocular retention and ocular bioavailability. Int J Pharm. 2010;
33. Garcês A. Formulations based on solid lipid nanoparticles (SLN) and nanostructured lipid carriers (NLC) for cutaneous use: a review. Eur J Pharm Sci. 2018;
34. Gaudana R. Recent perspectives in ocular drug delivery. Pharm Res. 2009;
35. Gharpure S, Akash A, Ankamwar B. A review on antimicrobial properties of metal nanoparticles. J Nanosci Nanotechnol. 2020;20(6):3303–39. https://doi.org/10.1166/jnn.2020.17677.
36. Gorantla S. Nanocarriers for ocular drug delivery: current status and translational opportunity. RSC Adv. 2020;
37. Gote V, Sikder S, Sicotte J, Pal D. Ocular drug delivery: present innovations and future challenges. J Pharmacol Exp Ther. 2019;370(3):602–24. https://doi.org/10.1124/jpet.119.256933.
38. Hanafy BI, Cave GWV, Barnett Y, Pierscionek B. Treatment of human lens epithelium with high levels of nanoceria leads to reactive oxygen species mediated apoptosis. Molecules (Basel). 2020;25(3):441. https://doi.org/10.3390/molecules25030441.

39. Hanafy BI, Cave GWV, Barnett Y, Pierscionek BK. Nanoceria prevents glucose-induced protein glycation in eye lens cells. Nanomaterials (Basel). 2021;11(6):1473. https://doi.org/10.3390/nano11061473.
40. Hassan B. Recent advances on polysaccharides, lipids and protein based edible films and coatings: a review. Int J Biol Macromol. 2018;
41. Haunberger A, Goepferich A. Nanoparticles targeting retinal and choroidal capillaries in vivo. Methods Mol Biol. 2019;1834:391–404. https://doi.org/10.1007/978-1-4939-8669-9_25.
42. He M, Li X, Yu L, Deng S, Gu N, Li L, Jia J, Li B. Double-sided Nano-ZnO: superior antibacterial properties and induced hepatotoxicity in zebrafish embryos. Toxics. 2022;10(3):144. https://doi.org/10.3390/toxics10030144.
43. Jackson TL, Mandava N, Quiroz-Mercado H, Benage M, Garcia-Aguirre G, Morales-Canton V, Wilbur L, Olson J. Intravitreal quantum dots for retinitis pigmentosa: a first-in-human safety study. Nanomedicine (Lond). 2021;16(8):617–26. https://doi.org/10.2217/nnm-2020-0471.
44. Jain S. Recent advances in lipid-based vesicles and particulate carriers for topical and transdermal application. J Pharm Sci. 2017;
45. Javed R, Ain NU, Gul A, Arslan Ahmad M, Guo W, Ao Q, Tian S. Diverse biotechnological applications of multifunctional titanium dioxide nanoparticles: an up-to-date review. IET Nanobiotechnol. 2022;16(5):171–89. https://doi.org/10.1049/nbt2.12085.
46. Jogpal V, Sanduja M, Dutt R, Garg V, Tinku. Advancement of nanomedicines in chronic inflammatory disorders. Inflammopharmacology. 2022;30(2):355–68. https://doi.org/10.1007/s10787-022-00927-x.
47. Jumelle C, Gholizadeh S, Annabi N, Dana R. Advances and limitations of drug delivery systems formulated as eye drops. J Control Release. 2020;321:1–22. https://doi.org/10.1016/j.jconrel.2020.01.057.
48. Kakkar S. Lipid-polyethylene glycol based nano-ocular formulation of ketoconazole. Int J Pharm. 2015;
49. Kamaleddin MA. Nano-ophthalmology: applications and considerations. Nanomedicine. 2017a;13(4):1459–72. https://doi.org/10.1016/j.nano.2017.02.007.
50. Kambhampati SP, Bhutto IA, Wu T, Ho K, McLeod DS, Lutty GA, Kannan RM. Systemic dendrimer nanotherapies for targeted suppression of choroidal inflammation and neovascularization in age-related macular degeneration. J Control Release. 2021;335:527–40. https://doi.org/10.1016/j.jconrel.2021.05.035.
51. Khiev D. Emerging nano-formulations and nanomedicines applications for ocular drug delivery. Nano. 2021;
52. Khorrami-Nejad M. The impact of visual impairment on quality of life. Med Hypothesis Discov Innov Ophthalmol. 2016;
53. Kim YC, Chiang B, Wu X, Prausnitz MR. Ocular delivery of macromolecules. J Control Release. 2014;190:172–81. https://doi.org/10.1016/j.jconrel.2014.06.043.
54. Kim M, Park J-H, Jeong H, Hong J, Park CY. Effects of nonporous silica nanoparticles on human trabecular meshwork cells. J Glaucoma. 2021;30(2):195–202. https://doi.org/10.1097/IJG.0000000000001709.
55. Laradji A, Karakocak BB, Kolesnikov AV, Kefalov VJ, Ravi N. Hyaluronic acid-based gold nanoparticles for the topical delivery of therapeutics to the retina and the retinal pigment epithelium. Polymers. 2021;13(19):3324. https://doi.org/10.3390/polym13193324.
56. Lee H, Park K. In vitro cytotoxicity of zinc oxide nanoparticles in cultured Statens Seruminstitut Rabbit Cornea cells. Toxicol Res. 2019;35(3):287–94. https://doi.org/10.5487/TR.2019.35.3.287.
57. Lee J-H, Shin H, Shaker MR, Kim HJ, Park S-H, Kim JH, Lee N, Kang M, Cho S, Kwak TH, Kim JW, Song M-R, Kwon S-H, Han DW, Lee S, Choi S-Y, Rhyu IJ, Kim H, Geum D, et al. Production of human spinal-cord organoids recapitulating neural-tube morphogenesis. Nat Biomed Eng. 2022;6(4):435–48. https://doi.org/10.1038/s41551-022-00868-4.
58. Leonardi A. Influence of different surfactants on the technological properties and in vivo ocular tolerability of lipid nanoparticles. Int J Pharm. 2014;

59. Leonardi A. Cationic solid lipid nanoparticles enhance ocular hypotensive effect of melatonin in rabbit. Int J Pharm. 2015;
60. Li M, Yang T, Gao L, Xu H. An inadvertent issue of human retina exposure to endocrine disrupting chemicals: a safety assessment. Chemosphere. 2021a;264(Pt 1):128484. https://doi.org/10.1016/j.chemosphere.2020.128484.
61. Li Q, Weng J, Wong SN, Thomas Lee WY, Chow SF. Nanoparticulate drug delivery to the retina. Mol Pharm. 2021b;18(2):506–21. https://doi.org/10.1021/acs.molpharmaceut.0c00224.
62. Li M, Gong J, Gao L, Zou T, Kang J, Xu H. Advanced human developmental toxicity and teratogenicity assessment using human organoid models. Ecotoxicol Environ Saf. 2022a;235:113429. https://doi.org/10.1016/j.ecoenv.2022.113429.
63. Li M, Zeng Y, Ge L, Gong J, Weng C, Yang C, Yang J, Fang Y, Li Q, Zou T, Xu H. Evaluation of the influences of low dose polybrominated diphenyl ethers exposure on human early retinal development. Environ Int. 2022b;163:107187. https://doi.org/10.1016/j.envint.2022.107187.
64. Li X, Kang B, Eom Y, Zhong J, Lee HK, Kim HM, Song JS. Comparison of cytotoxicity effects induced by four different types of nanoparticles in human corneal and conjunctival epithelial cells. Sci Rep. 2022c;12(1):155. https://doi.org/10.1038/s41598-021-04199-3.
65. Liu Y-C, Wilkins M, Kim T, Malyugin B, Mehta JS. Cataracts. Lancet. 2017;390(10094):600–12. https://doi.org/10.1016/S0140-6736(17)30544-5.
66. Liu Y-C, Lin MT-Y, Ng AHC, Wong TT, Mehta JS. Nanotechnology for the treatment of allergic conjunctival diseases. Pharmaceuticals (Basel). 2020;13(11):351. https://doi.org/10.3390/ph13110351.
67. López-Cano JJ, González-Cela-Casamayor MA, Andrés-Guerrero V, Herrero-Vanrell R, Molina-Martínez IT. Liposomes as vehicles for topical ophthalmic drug delivery and ocular surface protection. Expert Opin Drug Deliv. 2021;18(7):819–47. https://doi.org/10.1080/17425247.2021.1872542.
68. Lynch C, Kondiah PPD, Choonara YE, du Toit LC, Ally N, Pillay V. Advances in biodegradable nano-sized polymer-based ocular drug delivery. Polymers. 2019;11(8):1371. https://doi.org/10.3390/polym11081371.
69. Ma DJ, Lim M-S, Park UC, Park J-B, Ji SY, Yu HG. Magnetic iron oxide nanoparticle labeling of photoreceptor precursors for magnetic resonance imaging. Tissue Eng Part C Methods. 2019;25(9):532–42. https://doi.org/10.1089/ten.TEC.2019.0136.
70. Madni A, Rahem MA, Tahir N, Sarfraz M, Jabar A, Rehman M, Kashif PM, Badshah SF, Khan KU, Santos HA. Non-invasive strategies for targeting the posterior segment of eye. Int J Pharm. 2017;530(1–2):326–45. https://doi.org/10.1016/j.ijpharm.2017.07.065.
71. Mahaling B, Baruah N, Ahamad N, Maisha N, Lavik E, Katti DS. A non-invasive nanoparticle-based sustained dual-drug delivery system as an eyedrop for endophthalmitis. Int J Pharm. 2021;606:120900. https://doi.org/10.1016/j.ijpharm.2021.120900.
72. Mandal A, et al. Polymeric micelles for ocular drug delivery: from structural frameworks to recent preclinical studies. J Control Release. 2017a;
73. Mandal A, Cholkar K, Khurana V, Shah A, Agrahari V, Bisht R, Pal D, Mitra AK. Topical formulation of self-assembled antiviral prodrug nanomicelles for targeted retinal delivery. Mol Pharm. 2017b;14:2056–69.
74. Masse F, Desjardins P, Ouellette M, Couture C, Omar MM, Pernet V, Guérin S, Boisselier E. Synthesis of ultrastable gold nanoparticles as a new drug delivery system. Molecules (Basel). 2019a;24(16):2929. https://doi.org/10.3390/molecules24162929.
75. Masse F, Ouellette M, Lamoureux G, Boisselier E. Gold nanoparticles in ophthalmology. Med Res Rev. 2019b;39(1):302–27. https://doi.org/10.1002/med.21509.
76. Mehnert W. Solid lipid nanoparticles: production, characterization and applications. Adv Drug Deliv Rev. 2001;
77. Mohammadi M, Elahimehr Z, Mahboobian MM. Acyclovir-loaded nanoemulsions: preparation, characterization and irritancy studies for ophthalmic delivery. Curr Eye Res. 2021;46(11):1646–52. https://doi.org/10.1080/02713683.2021.1929328.

78. Mohanty A, Uthaman S, Park I-K. Utilization of polymer-lipid hybrid nanoparticles for targeted anti-cancer therapy. Molecules (Basel). 2020;25(19):4377. https://doi.org/10.3390/molecules25194377.
79. Molokhia SA. Anterior eye segment drug delivery systems: current treatments and future challenges. J Ocul Pharmacol Ther; 2013.
80. Mukherjee A, Waters AK, Kalyan P, Achrol AS, Kesari S, Yenugonda VM. Lipid-polymer hybrid nanoparticles as a next-generation drug delivery platform: state of the art, emerging technologies, and perspectives. Int J Nanomedicine. 2019;14:1937–52. https://doi.org/10.2147/IJN.S198353.
81. Naikoo G, Al-Mashali F, Arshad F, Al-Maashani N, Hassan IU, Al-Baraami Z, Faruck LH, Qurashi A, Ahmed W, Asiri AM, Aljabali AA, Bakshi HA, Tambuwala MM. An overview of copper nanoparticles: synthesis, characterisation and anticancer activity. Curr Pharm Des. 2021;27(43):4416–32. https://doi.org/10.2174/1381612827666210804100303.
82. Negri V, Pacheco-Torres J, Calle D, López-Larrubia P. Carbon nanotubes in biomedicine. Top Curr Chem J. 2020;378(1):15. https://doi.org/10.1007/s41061-019-0278-8.
83. Nguyen VP, Li Y, Qian W, Liu B, Tian C, Zhang W, Huang Z, Ponduri A, Tarnowski M, Wang X, Paulus YM. Contrast agent enhanced multimodal photoacoustic microscopy and optical coherence tomography for imaging of rabbit choroidal and retinal vessels in vivo. Sci Rep. 2019;9(1):5945. https://doi.org/10.1038/s41598-019-42324-5.
84. Paiva MRB, Andrade GF, Dourado LFN, Castro BFM, Fialho SL, Sousa EMB, Silva-Cunha A. Surface functionalized mesoporous silica nanoparticles for intravitreal application of tacrolimus. J Biomater Appl. 2021;35(8):1019–33. https://doi.org/10.1177/0885328220977605.
85. Pal D, et al. Nanotechnology for targeted delivery in cancer therapeutics. Int J Pharm Sci Rev Res. 2010;
86. Pandit J, Sultana Y, Aqil M. Chitosan coated nanoparticles for efficient delivery of bevacizumab in the posterior ocular tissues via subconjunctival administration. Carbohydr Polym. 2021;267:118217. https://doi.org/10.1016/j.carbpol.2021.118217.
87. Patel A. Ocular drug delivery systems: an overview. World J Pharmacol. 2013;
88. Peng C, Kuang L, Zhao J, Ross AE, Wang Z, Ciolino JB. Bibliometric and visualized analysis of ocular drug delivery from 2001 to 2020. J Control Release. 2022;345:625–45. https://doi.org/10.1016/j.jconrel.2022.03.031.
89. Pflugfelder SC, Stern ME. Biological functions of tear film. Exp Eye Res. 2020;197:108115. https://doi.org/10.1016/j.exer.2020.108115.
90. Pohanka M. Copper and copper nanoparticles toxicity and their impact on basic functions in the body. Bratisl Lek Listy. 2019;120(6):397–409. https://doi.org/10.4149/BLL_2019_065.
91. Prasad M, Kumar R, Buragohain L, Kumari A, Ghosh M. Organoid technology: a reliable developmental biology tool for organ-specific nanotoxicity evaluation. Front Cell Dev Biol. 2021;9:696668. https://doi.org/10.3389/fcell.2021.696668.
92. Qamar Z, Qizilbash FF, Iqubal MK, Ali A, Narang JK, Ali J, Baboota S. Nano-based drug delivery system: recent strategies for the treatment of ocular disease and future perspective. Recent Pat Drug Deliv Formul. 2019;13(4):246–54. https://doi.org/10.2174/1872211314666191224115211.
93. Quan J-H, Gao FF, Ismail HAHA, Yuk J-M, Cha G-H, Chu J-Q, Lee Y-H. Silver nanoparticle-induced apoptosis in ARPE-19 cells is inhibited by Toxoplasma gondii pre-infection through suppression of NOX4-dependent ROS generation. Int J Nanomedicine. 2020;15:3695–716. https://doi.org/10.2147/IJN.S244785.
94. Rebibo L. The dose-dependent effect of a stabilized cannabidiol nanoemulsion on ocular surface inflammation and intraocular pressure. Int J Pharm. 2022;
95. Reimondez-Troitiño S. Nanotherapies for the treatment of ocular diseases. Eur J Pharm Biopharm. 2015;
96. Ren X, Zhang Q, Yang J, Zhang X, Zhang X, Zhang Y, Huang L, Xu C, Ge Y, Chen H, Liu X, Wang Y. Dopamine imaging in living cells and retina by surface-enhanced Raman scattering based on functionalized gold nanoparticles. Anal Chem. 2021;93(31):10841–9. https://doi.org/10.1021/acs.analchem.1c01108.

97. Rodrigues FSC, Campos A, Martins J, Ambrósio AF, Campos EJ. Emerging trends in nanomedicine for improving ocular drug delivery: light-responsive nanoparticles, mesoporous silica nanoparticles, and contact lenses. ACS Biomater Sci Eng. 2020;6(12):6587–97. https://doi.org/10.1021/acsbiomaterials.0c01347.
98. Sala M. Lipid nanocarriers as skin drug delivery systems: properties, mechanisms of skin interactions and medical applications. Int J Pharm. 2018;
99. Salas-Ambrosio PJ, Bernad-Bernad MJ, Linares-Alba MA, García-Santisteban R, Tonix-Aburto LA, Ornelas-Lobato GJ, Gracia-Mora I, Rivera-Huerta M, Sánchez-Bartez F, Rico-Morales H, García-Sánchez GA. Toxicity evaluation of a novel rapamycin liposomal formulation after subconjunctival and intravitreal injection. J Ocul Pharmacol Ther. 2021;37(5):261–76. https://doi.org/10.1089/jop.2020.0108.
100. Sánchez-López E. PEGylated PLGA nanospheres optimized by design of experiments for ocular administration of dexibuprofen-in vitro, ex vivo and in vivo characterization. Colloids Surf B Biointerfaces. 2016;
101. Santana CP, Mansur AAP, Mansur HS, da Silva Cunha A Jr. Bevacizumab-conjugated quantum dots: in vitro antiangiogenic potential and biosafety in rat retina. J Ocul Pharmacol Ther. 2020;36(6):467–83. https://doi.org/10.1089/jop.2019.0108.
102. Sarwat S, Stapleton F, Willcox M, Roy M. Quantum dots in ophthalmology: a literature review. Curr Eye Res. 2019;44(10):1037–46. https://doi.org/10.1080/02713683.2019.1660793.
103. Scheive M, Yazdani S, Hajrasouliha AR. The utility and risks of therapeutic nanotechnology in the retina. Ther Adv Ophthalmol. 2021;13:25158414211003381. https://doi.org/10.1177/25158414211003381.
104. Schnichels S, Hurst J, de Vries JW, Ullah S, Gruszka A, Kwak M, Löscher M, Dammeier S, Bartz-Schmidt K-U, Spitzer MS, Herrmann A. Self-assembled DNA nanoparticles loaded with travoprost for glaucoma-treatment. Nanomedicine. 2020;29:102260. https://doi.org/10.1016/j.nano.2020.102260.
105. Sharma N, Bagga B, Singhal D, Nagpal R, Kate A, Saluja G, Maharana PK. Fungal keratitis: a review of clinical presentations, treatment strategies and outcomes. Ocul Surf. 2022;24:22–30. https://doi.org/10.1016/j.jtos.2021.12.001.
106. Shayani Rad M, Sabeti Z, Mohajeri SA, Fazly Bazzaz BS. Preparation, characterization, and evaluation of zinc oxide nanoparticles suspension as an antimicrobial media for daily use soft contact lenses. Curr Eye Res. 2020;45(8):931–9. https://doi.org/10.1080/02713683.2019.1705492.
107. Shi H, Ding J, Chen C, Yao Q, Zhang W, Fu Y, Wang X, Ruan J. Antimicrobial action of biocompatible silver microspheres and their role in the potential treatment of fungal keratitis. ACS Biomater Sci Eng. 2021;7(11):5090–8. https://doi.org/10.1021/acsbiomaterials.1c00815.
108. Singh M, Bharadwaj S, Lee KE, Kang SG. Therapeutic nanoemulsions in ophthalmic drug administration: concept in formulations and characterization techniques for ocular drug delivery. J Control Release. 2020a;328:895–916. https://doi.org/10.1016/j.jconrel.2020.10.025.
109. Singh R, Batoki JC, Ali M, Bonilha VL, Anand-Apte B. Inhibition of choroidal neovascularization by systemic delivery of gold nanoparticles. Nanomedicine. 2020b;28:102205. https://doi.org/10.1016/j.nano.2020.102205.
110. Singh TA, Sharma A, Tejwan N, Ghosh N, Das J, Sil PC. A state of the art review on the synthesis, antibacterial, antioxidant, antidiabetic and tissue regeneration activities of zinc oxide nanoparticles. Adv Colloid Interf Sci. 2021;295(102495):102495. https://doi.org/10.1016/j.cis.2021.102495.
111. Soares F, Domingues DC, Viana SC, Tebaldi DB. Polymer-hybrid nanoparticles: current advances in biomedical applications. Biomed Pharmacother. 2020;131 https://doi.org/10.1016/j.biopha.2020.110695.
112. Song HB, Wi J-S, Jo DH, Kim JH, Lee S-W, Lee TG, Kim JH. Intraocular application of gold nanodisks optically tuned for optical coherence tomography: inhibitory effect on retinal neovascularization without unbearable toxicity. Nanomedicine. 2017;13(6):1901–11. https://doi.org/10.1016/j.nano.2017.03.016.

113. Srinivasarao DA, Lohiya G, Katti DS. Fundamentals, challenges, and nanomedicine-based solutions for ocular diseases. Wiley Interdiscip Rev Nanomed Nanobiotechnol. 2019;11(4):e1548. https://doi.org/10.1002/wnan.1548.
114. Sun J-G, Jiang Q, Zhang X-P, Shan K, Liu B-H, Zhao C, Yan B. Mesoporous silica nanoparticles as a delivery system for improving antiangiogenic therapy. Int J Nanomedicine. 2019;14:1489–501. https://doi.org/10.2147/IJN.S195504.
115. Suri R, Neupane YR, Mehra N, Jain GK, Kohli K. Sirolimus loaded polyol modified liposomes for the treatment of posterior segment eye diseases. Med Hypotheses. 2020;136:109518. https://doi.org/10.1016/j.mehy.2019.109518.
116. Suri R, Neupane YR, Mehra N, Nematullah M, Khan F, Alam O, Iqubal A, Jain GK, Kohli K. Sirolimus loaded chitosan functionalized poly (lactic-co-glycolic acid) (PLGA) nanoparticles for potential treatment of age-related macular degeneration. Int J Biol Macromol. 2021;191:548–59. https://doi.org/10.1016/j.ijbiomac.2021.09.069.
117. Tayel S, et al. Promising ion-sensitive in situ ocular nanoemulsion gels of terbinafine hydrochloride: design, in vitro characterization and in vivo estimation of the ocular irritation and drug pharmacokinetics in the aqueous humor of rabbits. Int J Pharm. 2013;
118. Thomas A, John P. Nanoparticles in ophthalmic drug delivery: safety and toxicity review. Curr Pharm Des. 2022;28(10):1258–71.
119. Tisi A, Flati V, Delle Monache S, Lozzi L, Passacantando M, Maccarone R. Nanoceria particles are an eligible candidate to prevent age-related macular degeneration by inhibiting retinal pigment epithelium cell death and autophagy alterations. Cells (Basel). 2020;9(7):1617. https://doi.org/10.3390/cells9071617.
120. Tisi A, Feligioni M, Passacantando M, Ciancaglini M, Maccarone R. The impact of oxidative stress on blood-retinal barrier physiology in age-related macular degeneration. Cells (Basel). 2021;10(1):64. https://doi.org/10.3390/cells10010064.
121. Ung L, Pattamatta U, Carnt N, Wilkinson-Berka JL, Liew G, White AJR. Oxidative stress and reactive oxygen species: a review of their role in ocular disease. Clin Sci (Lond.). 2017;131(24):2865–83. https://doi.org/10.1042/CS20171246.
122. Vaneev A. Nanotechnology for topical drug delivery to the anterior segment of the eye. Int J Mol Sci. 2021;
123. Vaneev A, Tikhomirova V, Chesnokova N, Popova E, Beznos O, Kost O, Klyachko N. Nanotechnology for topical drug delivery to the anterior segment of the eye. Int J Mol Sci. 2021;22(22):12368. https://doi.org/10.3390/ijms222212368.
124. Wang R. A review of nanocarrier-mediated drug delivery systems for posterior segment eye disease: challenges analysis and recent advances. J Drug Target. 2021;
125. Wang Z, Wang Z. Nanoparticles induced embryo-fetal toxicity. Toxicol Ind Health. 2020;36(3):181–213. https://doi.org/10.1177/0748233720918689.
126. Wang R, Gao Y, Liu A, Zhai G. A review of nanocarrier-mediated drug delivery systems for posterior segment eye disease: challenges analysis and recent advances. J Drug Target. 2021a;29(7):687–702. https://doi.org/10.1080/1061186X.2021.1878366.
127. Wang T, Fan Q, Hong J, Chen Z, Zhou X, Zhang J, Dai Y, Jiang H, Gu Z, Cheng Y, Li Y. Therapeutic nanoparticles from grape seed for modulating oxidative stress. Small. 2021b;17(45):e2102485. https://doi.org/10.1002/smll.202102485.
128. Wen Y. Cross-linked thermosensitive nanohydrogels for ocular drug delivery with a prolonged residence time and enhanced bioavailability. Mater Sci Eng C Mater Biol Appl. 2021;
129. Widjaja LK. Hyaluronic acid-based nanocomposite hydrogels for ocular drug delivery applications. J Biomed Mater Res A. 2014;
130. Wilson SE. Bowman's layer in the cornea- structure and function and regeneration. Exp Eye Res. 2020;195:108033. https://doi.org/10.1016/j.exer.2020.108033.
131. Wong CW, Wong E, Metselaar JM, Storm G, Wong TT. Liposomal drug delivery system for anti-inflammatory treatment after cataract surgery: a phase I/II clinical trial. Drug Deliv Transl Res. 2022;12(1):7–14. https://doi.org/10.1007/s13346-021-00912-x.
132. Xu Q. Nanotechnology approaches for ocular drug delivery. Middle East Afr J Ophthalmol. 2013;

133. Yadav KS, Rajpurohit R, Sharma S. Glaucoma: current treatment and impact of advanced drug delivery systems. Life Sci. 2019;221:362–76. https://doi.org/10.1016/j.lfs.2019.02.029.
134. Yu F, Zheng M, Zhang AY, Han Z. A cerium oxide loaded glycol chitosan nano-system for the treatment of dry eye disease. J Control Release. 2019;315:40–54. https://doi.org/10.1016/j.jconrel.2019.10.039.
135. Zare H, Ahmadi S, Ghasemi A, Ghanbari M, Rabiee N, Bagherzadeh M, Karimi M, Webster TJ, Hamblin MR, Mostafavi E. Carbon nanotubes: smart drug/gene delivery carriers. Int J Nanomedicine. 2021;16:1681–706. https://doi.org/10.2147/IJN.S299448.
136. Zeng Y, Li M, Zou T, Chen X, Li Q, Li Y, Ge L, Chen S, Xu H. The impact of particulate matter (PM2.5) on human retinal development in hESC-derived retinal organoids. Front Cell Dev Biol. 2021;9:607341. https://doi.org/10.3389/fcell.2021.607341.
137. Zhang Y. Biomacromolecules as carriers in drug delivery and tissue engineering. Acta Pharm Sin B. 2018;
138. Zhang J. Cationic nanoemulsions with prolonged retention time as promising carriers for ophthalmic delivery of tacrolimus. Eur J Pharm Sci. 2020;
139. Zhang F, Chen H, Lan J, Song K, Wu X. Preparation and in vitro/in vivo evaluations of novel ocular micelle formulations of hesperetin with glycyrrhizin as a nanocarrier. Exp Eye Res. 2021a;202:108313. https://doi.org/10.1016/j.exer.2020.108313.
140. Zhao G, Sun H, Zhang T, Liu J-X. Copper induce zebrafish retinal developmental defects via triggering stresses and apoptosis. Cell Commun Signal. 2020;18(1):45. https://doi.org/10.1186/s12964-020-00548-3.
141. Zhao N, Gui X, Fang Q, Zhang R, Zhu W, Zhang H, Li Q, Zhou Y, Zhao J, Cui X, Gao G, Tang H, Shen N, Chen T, Song H, Shen W. Graphene quantum dots rescue angiogenic retinopathy via blocking STAT3/Periostin/ERK signaling. J Nanobiotechnol. 2022;20(1):174. https://doi.org/10.1186/s12951-022-01362-4.
142. Zhou Y, Li L, Li S, Li S, Zhao M, Zhou Q, Gong X, Yang J, Chang J. Autoregenerative redox nanoparticles as an antioxidant and glycation inhibitor for palliation of diabetic cataracts. Nanoscale. 2019;11(27):13126–38. https://doi.org/10.1039/c9nr02350j.

Validation in Sterile Ophthalmic Formulation

11

Naureen Afrose, Rideb Chakraborty, Pratibha Bhowmick, and Mithun Bhowmick

Abstract

Ophthalmic preparations that are intended for sterile topical application in the eyes require a rigorously controlled quality standard. Quality assurance requires validation, and since these products are susceptible to contamination, constant validation becomes important for maintaining quality and standard. This chapter gives a comprehensive description of the valve, process valve, and cleaning validation in sterile ophthalmic formulation. Process validation focuses on the desire to provide sufficient evidence that a process capable of producing a product that meets the standards set will indeed deliver such a product. In the context of sterile ophthalmic formulations, process validation means media fill simulation for assessing aseptic processing techniques, sterility for ensuring microbial-free product, and environmental control to check manufacturing environment freedom. Through these validation studies, the manufacturers can get confirmation of their processes' stability and performance. Equipment validation means confirming the suitability of equipment that is used during the manufacturing of sterile ophthalmic products to perform its intended operations. That is design qualification, installation qualification, operational qualification, and performance qualification. Design qualification ensures that an item of equipment has been designed to provide requisite performance and function satisfactorily. Installation qualification ensures that an equipment is installed properly and harmonizes with the other manufacturing procedures. Operational qualification decides the competence of the equipment to perform at predetermined level. Performance qualification shows that the equipment continuously generates quality-related products. A particular emphasis is made to the specifics of ophthalmic devices and the discussion discovers the difficulties that appear during

N. Afrose · R. Chakraborty · P. Bhowmick · M. Bhowmick (✉)
Bengal College of Pharmaceutical Sciences and Research, Durgapur, West Bengal, India

N. K. Mehra (ed.), *Complex Ophthalmic Dosage Forms: Advances in Biomedical Applications and Future Perspectives*, https://doi.org/10.1007/978-981-96-6305-4_11

the use of sterile ophthalmic products due to the sensitivity of the ocular surface and the risk of microbial invasion. This chapter thus discusses various validation methods that are carried out in order to ensure aseptic processing. They include media fill simulations, sterility test, and environmental monitoring.

Keywords

Ophthalmic preparations · Sterile · Validation · Validation methods · Quality assurance

Abbreviations

CPPs	Critical Process Parameters
DQ	Design Qualification
EMA	European Medicines Agency
FDA	Food and Drug Administration
IQ	Installation Qualification
OQ	Operational Qualification
PAT	Process Analytical Technology
PQ	Performance Qualification
QbD	Quality by Design
URS	User Requirement Specifications
USFDA	US Food and Drug Administration

11.1 Introduction

Sterile ophthalmic formulations are specific types of pharmaceutical preparations intended to be applied directly into the eye. These formulations are to be used for the management of a wide range of ocular disorders such as microbial and fungal infections, inflammation, uveitis, age-related macular degeneration, cataract, dry eye, and glaucoma [1–3]. Since the ocular surface is a sensitive area that can be easily infected, the type and cleanliness of the samples obtained must also be highly regarded.

Ocular surface can be described as a sophisticated and sensitive localization that is part of our body. It is always in touch with the exterior environment and therefore prone to bacterial invasion. The introduction of even the smallest contamination into ophthalmic formulations puts the health of the patient at high risk due to possible ocular infections such as conjunctivitis, keratitis, and endophthalmitis [4, 5]. These infections may lead to substantial loss of vision, discomfort and, in the worst-case scenario, establish a permanent vision loss condition. To address these risks some of the regulatory authorities in the region such as the US Food and Drug Administration (USFDA) in the United States and the European Medicines Agency (EMA) in

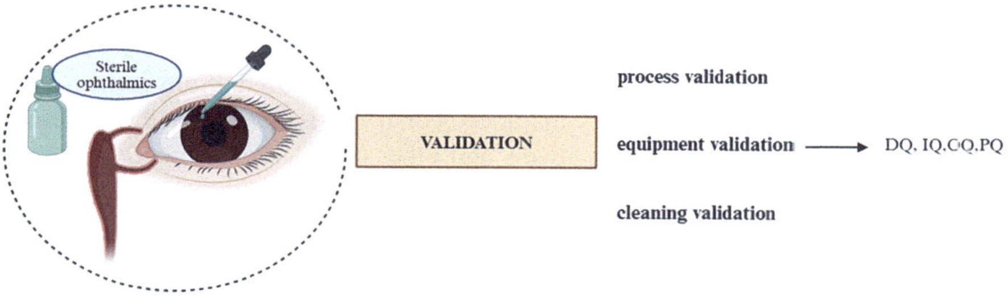

Fig. 11.1 Validation approaches for sterile ophthalmic

Europe have set high-quality standards for sterile ophthalmic formulations [6–8]. These standards are meant to protect consumers and guarantee effectiveness and quality of the products. It is important to abide by these standards in order to safeguard the health of the masses and product trust.

The process of validation is of utmost importance to quality assurance and consistency of sterile ophthalmic formulations [9–12]. It is a process that is well structured to show that it is possible to produce products in manufacturing that conform to some set standard. Through adherence to proper validation measures, drug producers shall be in a position to control for risk, correct for failures, and meet the standards set down by applicable law. The concept of validation is multidimensional comprising process validation, equipment validation, and cleaning validation. In process validation, the subject proves that for any given manufacturing process, the production of the final products is always up to standard [13]. This includes Media fill simulations to evaluate the mode of aseptic processing, sterility testing to establish the lack of microbial presence, and media fill for assessment of the manufacturing environment. Equipment validation to check that the equipment being used in the product manufacturing process meets the requirements of an ideal manufacturing tool and functions optimally. This includes design qualification of equipment, installation and operational qualification, and performance qualification [14–18]. Cleaning validation is designed to show that the cleaning operations are able to remove residues and other elements from the equipment as well as facilities [8, 19–21]. Manufacturing of sterile ophthalmic formulation may therefore require very stringent validation principles in order to enhance quality and meet regulatory standards and safety features that are approved by the various regulatory authorities across the world (Fig. 11.1).

11.2 Importance of Sterile Ophthalmic Formulations

Sterile ophthalmic preparations are essential in the management of various ocular diseases. We know that the eye is a very tender organ of the human body and is easily prone to infections. Even a small number of microorganisms will produce serious contamination such as conjunctivitis, keratitis, or endophthalmitis [22–25].

These infections are uncomfortable, may harm vision, and in severe cases, could ultimately result in blindness. The ophthalmic formulations come into direct contact with the eye by getting over the area of the eye that is composed of the cornea, conjunctiva, and eyelids [26, 27]. They include specialized tissues that are very sensitive and can in fact hold microorganisms and other debris. Consequently, there is need to ensure that formulations are sterile to reduce the chances of infections. Aseptic practices allow the active ingredients in the formulation to perform their therapeutic role free from interference from microbial activity [28–30]. This may lower the effectiveness of the medication to work, effectively increasing the chances of the treatment not to work. Sterile ophthalmic formulations aim at avoiding any instance of ocular infection by duly reducing the rate of infection. It is especially important with immune-compromised patients as well as with patients who are undergoing eye surgery. There are well set legal requirements regarding the production and quality of ophthalmic preparations [15, 31, 32]. Compliance with the mentioned standards such as sterility assurance is crucial for product safety and effectiveness. Finally, it can be inferred that only sterility of ophthalmic formulations is a way to provide a safe and efficient treatment for many of the eye-related problems.

11.3 Regulatory Requirements and Quality Standards

For ophthalmic preparations that are sterile, there is usually a very high regulatory, compounding, and quality standard that is put in place to protect the formulation's safety and efficacy. These regulations are enforced by regulatory agencies all over the world including the Food and Drug Administration (FDA) in the United States and the European Medicines Agency (EMA) in Europe [33–36].

The key regulatory requirements and quality standards for the sterile ophthalmic formulations include but are not limited to points formulated as follows: sterility assurance—the product must not contain any microbial load. It is done through aseptic processes combined with sterilization and sterility control measures. Purity and potency—it has to be the right concentration of active therapeutic substances and no admixtures. This is facilitated in the form of quality control tests where various analytical tests such as High-Performance Liquid Chromatography (HPLC), Gas Chromatography (GC), and particle contamination control the product that must have a certain particle size and particle count to reduce the chances of ocular irritation and inflammation [37, 38]. This is done through filtration and other particulate matter controlling procedures. Packaging and labeling—the packaging and labeling should be comprehensible, complete, and unambiguous. It must contain the requisite data that impact the performance of the drug, including the name of the product, substances contained in it, dosage, date of expiry, and storage information. Stability is where the functionalities of the active ingredient must remain intact over the shelf life of the product.

Stability testing is therefore done with the aim of determining the best storage condition for the product [39–43]. Clinical trials and where activities such as the intensive clinical trials are conducted to show the reliability of the product. These trials are meant to measure the effect of the product on people as a treatment intervention for the disease and other emergent effects of the product. Therefore, strict federal regulatory specifications and quality benchmarks should be followed by pharmaceutical companies to produce safe, effective, sterile ophthalmic formulations.

11.4 Role of Validation in Ensuring Product Quality

As previously mentioned, validation is a critical component for zero compromise in the quality and standard of sterile ophthalmic formulations. They include a set of procedures for establishing that the manufacturing process is delivering products that meet prespecified quality standards. The regular validation protocols help the pharmaceutical manufacturer to avoid risks, reduce the chance of product failure, and adhere to the regulations set.

Process validation and equipment validation are important aspects of validation that also embrace cleaning validation [44–46]. Process validation in a manufacturing environment requires confirmation that a certain manufacturing process is producing reliable products consistently. This includes media fill simulation to evaluate the aseptic processing method, sterility test to check on the absence of microorganisms, and environmental test to check the level of cleanliness in the manufacturing area. Equipment validation guarantees equipment utilized in the production process meet the required specifications to perform precisely as intended for their use [45, 47–50]. This includes design qualification, installation qualification, operational qualification, and performance qualification. Cleaning validation shows that cleaning processes provide necessary proof that residues and contaminants can be effectively eliminated from equipment and facilities in need of cleaning [51–53]. The crucial and complex nature of validation for, and manufacture of, sterile ophthalmic formulations highlights the importance of strict validation and quality assurance practices in pharmaceutical production to meet regulatory and patient needs.

11.5 Process Validation

Process validation is an important tool that can assist in the ability to guarantee the sterility and stability of ophthalmic formulations [54–56]. It is a formal process of establishing a belief that a manufacturing process delivers end products that conform to certain standard standards. These are the media fill simulations, the sterility testing, and the environmental monitoring.

Media fill simulations are used in evaluating the aseptic processing that may be used during production of manufactured products [57–59]. The above simulations require an inoculum to be placed in various kinds of containers containing sterile growth media and left under specific conditions. In the absence of microbial growth, there is evidence that the aseptic processing techniques will not allow microbial proliferation.

Media filling and sterility testing are done to check whether the product is free of microbial contamination or not [58]. This includes performing vigorous and sterile collection of samples from the product and then culturing the samples on different media. The media is then cultured under certain conditions, and evidence from lack of microbial growth means that the product is sterile [57].

Environmental monitoring is utilized in evaluating cleanliness and microbiological load in the manufacturing environment. This encompasses factors like analyzing environmental factors such as the level of air-borne particles and surfaces and personnel's contamination levels of their hands. This way the environmental conditions with regard to the safety of the product from microbial contamination can be closely monitored and thus reduced to the minimum.

11.5.1 Media Fill Simulations

For sterile ophthalmic formulations, media fill simulations within the process are a crucial part of the general process validation [57, 60–63]. They are applied during manufacturing to assure the firm that aseptic processes are used in production and only sterile products are made.

In media fill simulation, vials or bottles are filled with media that's sterile and is a nutrient broth which may be soybean casein digest medium. The filling process involves conditions that replicate the manufacturing process in order to avoid contamination. Following filling, the containers are sealed and then stored at directed temperature and humidity conditions to help favor the growth of microbes across the surface of the container. Media that remain negative after the incubation period is an indication that the applied aseptic processing practices such as personnel decontamination, gowning, environmental control, and equipment sterilization do not allow "take" or promote microbial contaminants [61, 62]. Any contamination that is noticed in the media shows that the process is non-aseptic, and corrective measures must be taken with subsequent revalidation.

By subjecting their products to media fill simulations, the manufacturers of pharmaceutical products can realize any areas that have a weakness in their aseptic processing methods hence being in a position to produce valid sterile ophthalmic formulations every time.

Thus, the primary goal of media fill simulations is to test aseptic processing techniques and to define possible areas of contamination. The specific objectives of a media fill simulation include many points: aseptic technique examination where there is evaluation of skills and techniques used by people administering aseptic procedures; determining potential sources of contamination by locating where contamination may exist, such as a piece of equipment, environment implicated, or operator technique; cleanliness techniques by verifying that cleaning and sterilization techniques are capable of eradicating suspected contaminants; verifying process performance by ensuring that the process is able to deliver sterilized goods regardless of the prevailing circumstances.

Ideally, a media fill simulation should be very realistic in order to give accurate results. Other factors to take into consideration comprise: product volume and type of container wherein the simulation should utilize the same container type and product volume as the actual product destined for the production line critical aseptic processing steps where all critical steps into aseptic processing that involve filling, stoppering, and capping should be conducted on the simulation [62, 64–70]; environmental conditions where the simulation should be conducted in ideal environment with ability to control the temperature, humidity, and air flow; people engagement where the simulation should engage all people who are employed in the actual production process. The process shown in Table 11.1 is followed for the conduction of Media Fill.

For evaluation of results, the following is done. Any form of microbial growth, which can be identified, denotes that the aseptic process of the particular item failed. Documents and observations should reveal specifics of aseptic technique violation or lack of it. Conduct a test to evaluate the state of the environment where manufacturing is done based on the results of environmental monitoring. Verify any other equipment and material that could possibly provide contamination. Pharmaceutical manufacturers can use values for media fill simulations provided here to assess possible challenges in the aseptic management process and can make corrections for steady production of sterile ophthalmic formulations.

11.5.2 Sterility Testing

Several methods can be used for sterility testing, as shown in Table 11.2.

Sampling and testing procedures as well as the interpretation of results can be done by following the methods shown in Table 11.3.

Table 11.1 Media fill conduction process flow

Process flow no.	Process flow specification
Step 1	The preparation of the media fill area involves washing and sanitizing all surfaces found within this area. Gown and glove personnel issues addressed in compliance with aseptic processing. Washing all utensils and equipment that we are going to use in the simulation [69]
Step 2	The stages involve acting of filling bottles, stoppering and capping by emphasizing aseptic measures to reduce contact with the environment. Recording all activities as well as what is observed during the process [70]
Step 3	Storing the filled containers at an appropriate temperature (temperature condition: 30–35 °C, duration: 14 days) [55, 71, 72]
Step 4	Intraclass correlation coefficient (ICC) can visually inspect the containers for any tendency of contamination in terms of turbidity or particulate matter [73]

Table 11.2 Methods for sterility testing

Name of method	General methodology
Direct inoculation	A sample of the product is directly streaked onto an appropriate growth medium, for example, soybean-casein digest medium, or thioglycollate medium [74, 75]
Membrane filtration	Sample is filtered, using a sterile membrane filter, and the filter is then placed onto the surface of a solid culture media [74]
Direct transfer	A part of the product is then directly taken to a suitable culture media using a sterile transfer equipment [74, 75]

Table 11.3 Sampling and testing procedures and interpretation of results of sterility testing

Sampling and testing procedures	
Procedure	**Description**
Aseptic sampling	Samples are taken through aseptic means to prevent contamination from getting into the filled containers in-charge [68, 70]
Inoculation	The samples are then streaked into suitable culture media, such as thioglycollate media for the anaerobic bacteria and soybean casein digest media for aerobic bacteria/ fungi
Incubation	The inoculated media are incubated at certain temperatures; for example, 30–35 °C for aerobic bacterial and 20–25 °C for fungal counts are incubated for a period of 14 days [55, 72]
Inspection	Microbial growth and turbidity and color change in the growth media incubated are observed
Interpretation of results	
Result type	**Interpretation criteria**
Absence of growth	If there is no microbial growth in any of the culture media, then the product is said to be sterile
Presence of growth	The assessment of microbial growth suggests that the aseptic process was violated. The specific area having contamination needs to be established to take proper measures in this regard

Pharmaceutical manufacturers can achieve valid results on the sterility of ophthalmic formulations through the performance of efficient sterility testing within media fill simulations of the respective production.

11.5.3 Environmental Monitoring

Media fills have become an important tool in measuring bacterial growth on environmental surfaces and therefore requires monitoring of the environment [76–78]. They are used to determine the extent of contamination by microbes and the degree of cleaning in the manufacturing environment. Environmental factors also have an effect on contamination through microbial growth can be controlled by the manufacturers in order to reduce such incidences.

Evaluations are made on several parameters in media fill simulation to determine the environmental conditions. The service quality focuses on particulate matter in the air as well as microbial density. This is usually done with air samplers which deposit particles of air on a growth medium. Environmental monitoring includes counting the bioburden on walls, floor, equipment, etc., for which contact plates or swabs are used. These are employed in the removal of microorganisms from surfaces and in the transfer of the removed microorganisms to culture media. Microbial sampling of the hands of personnel is taken using hand wipes or glove finger samples [79].

Several forms of sampling are employed in measuring environmental conditions. Air monitoring includes active air sampling whereby air samplers pull a fixed amount of air through a filter media or directly onto a culture medium. Techniques of air sampling include passive air sampling whereby microorganisms are allowed to diffuse and settle on a culture plate [80]. Surface sampling: in this, contact plates or swabs are used to pick up microorganisms from the surface. Personnel hand hygiene monitoring: here, hand swabs or glove fingertips are sampled to test the levels of microbial presence on the hands of personnel.

Alert and action levels are set in order to demarcate the maximum allowable mean number of microorganisms. When the monitoring data is above the alert value, required measures might need to be taken in order to understand the source of the contamination and avoid similar problems. If the results go beyond the action level, then more intensive measures will be required to correct the violations which may include cleaning, disinfection, or revalidation of the affected area.

Environmental monitoring, therefore, can play an important role in preventing microbial contamination in the manufacturing process area through efficient and proper control of the physical environment and implementing proper sterilizing methods to meet the Pharmaceutical code of sterile products.

11.6 Equipment Validation

Equipment validation is one other important factor in relation to the quality of the sterile ophthalmic formulation [52, 56, 81]. It entails a set pattern of ascertaining that the machinery employed in production is of right quality and standard so as to serve its intended purpose and without failure. These include design qualification, installation qualification, operational qualification, and performance qualification.

Design qualification proof that the equipment being designed to meet established standards and requirements of its function. Installation qualification confirms that the equipment in question is properly installed as a part of the total production process [56]. Operational qualification shows that the equipment operates to a required level and meets requirements in its normal mode of use. Performance qualification makes certain that the specific equipment is capable of delivering products of the required quality each time it is used.

Thorough equipment validation can help pharmaceutical manufacturers reduce the possibility of equipment-related failure and assure consistent production of excellent sterile ophthalmic formulations.

11.6.1 Design Qualification (DQ)

Design Qualification or DQ is a documented confirmation that the design for equipment meets the user requirement with respect to an item or activity for which the equipment will be used [82–85]. This includes checking design documents, including user requirement specifications (URS), design specifications, and engineering drawings to ensure that the designed equipment conforms to acceptable quality standards [86–88].

Purpose and objectives are as follows:

- Suitability: Endeavor to ensure that the equipment design will satisfactorily complement the requirements of the manufacturing process involved.
- Potential risks: Consider possible drawbacks to the design of the equipment and control or avoid them.
- Regulatory requirements: Meet all established legal requirements for designing and validating equipment.
- A foundation for subsequent validation stages: It should serve and be the basis for Installation Qualification (IQ), Operational Qualification (OQ), and Performance Qualification (PQ) [55, 80, 83, 89].

DQ documentation typically includes the following:

- User Requirement Specifications (URS)
- Deviation & Change Control Records
- Design Specifications
- Engineering Drawings
- Vendor Documentation
- Design Review Records
- Deviation and Change Control Records
- DQ Protocol and Report

11.6.2 Installation Qualification (IQ)

Installation Qualification (IQ) is an important part of the validation process begins with, as it confirms that all equipment has been installed appropriately, and is ready for further operational qualification [83, 89–91]. This includes a confirmation process, which checks compliance with the design intents and procedures of the manufacturer. The installation process involves the preparation of the site, installation of equipment, connection to utilities, setting up, and the final documentation check.

For confirmation of installation, a physical examination is carried out to evaluate the exterior material appearance of the equipment as well as confirm correct positioning. Dimensional verification reassures design specifications in terms of the size of the equipment and the clearances required. Procedures in connection with utilities availability and functionality compliance with the laid down standards are also checked. Last but not least all documentation related to installation is reviewed and kept as a record for the same.

11.6.3 Operational Qualification (OQ)

A rigorous process of equipment qualification is used during operations, and operational Qualification (OQ) describes the equipment's performance based on its operation within a number of parameters [84, 92]. This is a process of checking the machines or equipment for their proper working at different operating conditions.

During OQ, system attributes like temperature, pressure, flow rate, and cycle time are reviewed, to ensure that the equipment being tested stabilizes at the desired values. Functional tests are carried out in order to ensure that the equipment can perform simple actions, while performance test is done for the purpose of checking the equipment's ability to perform certain actions up to the set level of acceptance. Also discussed is checking the correct functioning of alarm and safety interlock systems with a view to protecting the equipment and persons. In effect, what OQ allows is for an organization to have confidence that its equipment is both reliable and consistent.

11.6.4 Performance Qualification (PQ)

Performance Qualification (PQ) is one of the most important stages of the validation procedure that confirms the equipment's ability to consistently deliver products of the right quality [82, 84, 92]. This means checking whether the equipment performs in the way required for actual or replica production scenarios. In this regard, the present study has established that for continuous performance verification, calibration of critical equipment parameters is necessary from time to time. Preventive maintenance is mostly done to ensure that the equipment is maintained in their best operating condition most of the time. The analysis of trends of the discovered data can help predict equipment failure, and so maintain it in advance. Re-qualification is required when the equipment has been either overhauled or modified in such a way that its quality is questionable [91]. Through these specific listed steps as well as complete validation activities, it is possible for pharmaceutical manufacturers involving equipment to be fit for its use in preparation of sterile ophthalmic formulations.

11.7 Cleaning Validation

Cleaning validation is one of the many essential components for preparing sterile ophthalmic formulations of quality and homogeneity [84]. This refers to an orderly way of proving the efficiency of cleaning activities as a way of eliminating residues and used in the effort to prevent cross-infection as well as making sure that subsequent batches are sterile. Cleaning practices therefore depend on the selection of the right cleaning solvents. These agents must possess the ability to get rid of residues while not causing further harm to the equipment, or leaving behind toxic traces. Material compatibility, type of residue to be left behind, and damaging effects of the cleaning agents on the environment are some of the choices made. Cleaning materials that can be used in the production of pharmaceuticals may include detergents, solvents, and disinfectants. Cleaning cycles incorporate a series of procedures and techniques in order to deal with contaminants [89, 92]. A cycle may consist of pre-wash, wash, intermediate rinse, sanitation, and post-rash. The period for completing every step depends on contamination level as well as on equipment material and kind of formulations.

Cleaning validation studies therefore show that cleaning processes are effective. Such studies normally consist of residue analysis which may include the use of HPLC or GC-MS to test the surfaces of equipment for any residues [73, 92]. Microbiological monitoring checks the microbial count before and after cleaning or using any disinfecting product and therefore determines the efficiency of the cleaning process. Water quality assessment enables cleaning water to be safe without any element in it.

Cleaning validation requires several sampling and analysis techniques. Swab sampling means that swabs are taken from equipment surfaces after which the swabs are analyzed microbiologically or chemically [93, 94]. Product water sampling looks at water used in the final rinse of a piece of equipment to determine the presence or absence of product residue. Lastly methods such as ATP bioluminescence or surface scanning measure residual surface cleanliness.

The criteria that may be used in the acceptance of the cleaning validation studies are set down usually in consideration of the regulatory demands and the trends set in the industry. These criteria may include the maximum or minimum acceptance levels of active ingredients and excipients, trace substances, cleaners, and microorganism counts.

Cleaning validation studies should be periodically repeated, particularly when some changes in the process, the equipment, or the cleaning agent have been made. Basically, revalidation could also be essential in case of inadequate cleaning confirmation or product contamination. It has been found that through proper consequences of cleaning and specific cleaning validation studies, that is, continuous monitoring and controlling of cleaning, can significantly contribute to the poor cleanliness and sterility of the manufacturing environment and the quality of ophthalmic formulations that are produced.

11.8 Challenges and Considerations

The processes of developing and manufacturing sterile ophthalmic formulations can be a challenge because of the sensitive nature of the site on the ocular surface and high regulatory demands.

Table 11.4 Key regulatory expectations

Criterion	Expectory processes
Sterility assurance	Rigorous sterilization processes and aseptic techniques are essential to guarantee the product's freedom from microbial contamination
Purity and potency	Maintaining the correct concentration and identity of active ingredients is crucial for product efficacy
Particle contamination control	Minimizing the presence of particulate matter in the formulation is essential to prevent eye irritation and infection
Packaging and labeling	Accurate and informative labeling is crucial for patient safety and proper product use
Stability	Maintaining product stability over its shelf life ensures consistent efficacy and safety
Clinical trials	Rigorous clinical trials are necessary to demonstrate the safety and efficacy of the product

Ocular Surface Sensitivity where the ocular surface is also very sensitive to any foreign structure, organism, or particle in this case bacteria, viruses, and other infectious particles may cause ocular surface inflammation and vision impairment. This sensitivity requires the formulation to be controlled strictly for its pH level and tonicity of the particles as well as any particulate matter content. What about the variability, even if it is less, that can result in discomfort to the patient hence reduced therapeutic value. Also, the formulations that can cause harm to the delicate ocular tissues must not contain irritants like preservatives.

Microbial contamination is always a big issue in sterile ophthalmic formulations. Pathogens can be found in food during processing, packing, or storage [93]. To reduce this risk proper aseptic processing techniques such as media fill simulation and environmental monitoring must be strictly followed. Additionally, the formulation may stabilize with preservatives to help counteract the microbial load but added preservatives need to be chosen carefully since they are known to cause ocular surface irritation.

Regulatory agencies worldwide, such as the FDA and EMA, impose stringent regulations on the development, manufacturing, and quality control of sterile ophthalmic formulations. These regulations ensure product safety, efficacy, and quality, which are discussed in Table 11.4.

Adherence to these regulations requires meticulous attention to detail, robust quality control systems, and comprehensive validation studies. The evolving regulatory landscape and increasing consumer expectations for innovative and effective treatments further complicate the development and manufacturing of sterile ophthalmic formulations. Pharmaceutical companies must continuously adapt to these challenges and invest in research and development to bring safe and effective products to market.

11.9 Emerging Validation Approaches

The pharmaceutical industry is constantly evolving, and with it, the approaches to validation. As technology advances and regulatory expectations become more stringent, emerging validation strategies are being implemented to ensure the quality and safety of sterile ophthalmic formulations. These approaches are designed to improve efficiency, reduce costs, and enhance the overall quality of the manufacturing process.

Risk-based validation is a strategic approach that focuses on identifying and mitigating risks associated with the manufacturing process. By prioritizing high-risk areas, manufacturers can allocate resources effectively and reduce the overall validation effort. Key steps involved in risk-based validation include:

- *Risk Assessment:* A thorough risk assessment is conducted to identify potential hazards and assess their severity and likelihood of occurrence.
- *Risk Control:* Appropriate control measures are implemented to mitigate identified risks. These may include additional validation activities, process changes, or enhanced monitoring.
- *Validation Planning:* A validation plan is developed based on the risk assessment, outlining the specific validation activities required.
- *Validation Execution:* The validation activities are executed according to the plan, including media fill simulations, sterility testing, and environmental monitoring.
- *Validation Reporting:* Detailed validation reports are prepared, documenting the results of the validation activities and any deviations or issues identified.

By adopting a risk-based approach, manufacturers can tailor their validation efforts to the specific needs of their products and processes, resulting in more efficient and effective validation. Quality by Design (QbD) is a scientific and risk-based approach to pharmaceutical development and manufacturing [95]. It involves designing products and processes with quality built in from the beginning. QbD principles can be applied to validation by:

- *Identifying Critical Quality Attributes (CQAs):* CQAs are the quality attributes that are critical to the product's safety and efficacy [95, 96].
- *Identifying Critical Process Parameters (CPPs):* CPPs are the process parameters that have a significant impact on CQAs [96].
- *Designing Robust Processes:* Developing robust processes that are less susceptible to variations.
- *Validating the Design Space:* Defining the design space, which is the range of operating conditions within which the process is expected to consistently produce quality products.

By applying QbD principles to validation, manufacturers can achieve greater control over the manufacturing process and improve product quality.

Continuous manufacturing is an emerging trend in the pharmaceutical industry that involves a continuous flow of materials through the manufacturing process, eliminating batch-based production. This approach can offer several benefits, including reduced cycle times, increased efficiency, and improved product quality. Validation of continuous manufacturing processes requires a different approach than traditional batch processes. Key considerations include:

- *Process Analytical Technology (PAT):* Real-time monitoring of critical process parameters to ensure consistent product quality [97]
- *Design Space:* Defining the design space for the continuous process and validating its performance within that space
- *Cleaning Validation:* Ensuring effective cleaning of equipment and piping to prevent cross-contamination
- *Change Control:* Implementing robust change control procedures to manage changes to the continuous manufacturing process

Advanced analytical techniques, such as mass spectrometry, nuclear magnetic resonance spectroscopy, and near-infrared spectroscopy, can be used to enhance validation activities. These techniques can provide more detailed information about the composition and quality of products and raw materials. By using these techniques, manufacturers can improve the accuracy and reliability of their validation data. Emerging validation approaches are transforming the pharmaceutical industry, enabling the development and manufacturing of high-quality sterile ophthalmic formulations. By embracing these new approaches, manufacturers can improve product quality, reduce costs, and meet the increasing regulatory demands. As technology continues to advance, it is likely that even more innovative validation strategies will emerge in the future.

11.10 Conclusion

The sterile ophthalmic formulations play a significant role in addressing diverse eye-related complications. For their safety and effectiveness, then, validation processes must be tightened. These are a set of processes such as process validation, equipment validation, cleaning valve as well as installation validation, which are key when it comes to showing that the manufacturing processes of a product of that particular company are in a position to deliver product value within a given quality.

Process validation procedures are used to cover media fill simulations that check aseptic processing, sterility testing that checks for possible microbial growth, and environmental monitoring of the manufacturing area. Equipment validation simply makes sure that all equipment used in the manufacturing process are right for the intended application and meets standards before it is used on the production line. These include design qualification, installation qualification, operational qualification, and performance qualification. Cleaning validation proves that cleaning is capable of removing residues and objectionable microorganisms from equipment and facility surfaces.

The field of validation is dynamic, according to the advancement in technology and ever-changing expectations of the regulatory bodies. Current trends like risk-based validation, Quality by Design (QbD), continuous manufacturing, substantial analytical methods, and digitalization are what the future of validation looks like. While risk-based validation targets hazard identification and risk control measures, QbD targets creating products and processes with inherent quality. Continuous manufacturing holds certain potential, but validation approaches are different to batch processes. These measures can improve validation work by extending the amount of information about product and material quality provided by analytical techniques. By use of digitalization and automation, we are able to achieve high efficiency and in most cases the human factor is eliminated.

There are growing demands placed on the regulatory agencies for validation focus and a series of new and more rigorous regulations are being set. This means that the manufacturers are required to become familiar with the latest regulatory requirements and have to be ready to make changes to companies' validation plans to reflect them. The reason is that both quality and consistency of the sterile ophthalmic formulations must be effectively maintained. Through conduct and reinvention of validation routines, embracing efficient change control procedures, staff and other relevant personnel training and awareness, utilization of data analysis and trendy tracking, and acceleration of regulatory parties' solid relations, pharmaceutical manufacturers can ensure quality, safety, and efficiency of sterile ophthalmic formulations.

Acknowledgments The authors are grateful to the Bengal College of Pharmaceutical Sciences and Research, Durgapur, West Bengal-713212, for providing the required environment for the work.

References

1. Bisen AC, Dubey A, Agrawal S, Biswas A, Rawat KS, Srivastava S, Bhatta RS. Recent updates on ocular disease management with ophthalmic ointments. Ther Deliv. 2024;10:1–8.
2. Agarwal P, Craig JP, Rupenthal ID. Formulation considerations for the management of dry eye disease. Pharmaceutics. 2021;13(2):207.
3. Sarkar T, Singh M, Jana BK, Mazumder B. Current formulation strategies to design novel carriers for targeted drug delivery and management of infectious keratitis: a comprehensive review on the present state of the art. Lett Drug Des Discov. 2024;21(13):2567–89.
4. Yilmaz OF, Sarmis A, Mutlu MA, Ersoy EE, Askarova U, Oguz H. Bacterial contamination of multi-use antibiotic steroid eye ointments and drops. Graefes Arch Clin Exp Ophthalmol. 2023;261(6):1691–700.
5. Mazet R, Yaméogo JB, Wouessidjewe D, Choisnard L, Gèze A. Recent advances in the design of topical ophthalmic delivery systems in the treatment of ocular surface inflammation and their biopharmaceutical evaluation. Pharmaceutics. 2020;12(6):570.
6. Selmin F, Musazzi UM, Magri G, Rocco P, Cilurzo F, Minghetti P. Regulatory aspects and quality controls of polymer-based parenteral long-acting drug products: the challenge of approving copies. Drug Discov Today. 2020;25(2):321–9.

7. Isles M. Patient safety issues associated with the use of compounded medicines as alternatives to approved pharmaceutical products in Europe and how best practice can improve outcomes. Int J Risk Saf Med. 2020;31(3):133–44.
8. Hasegawa A, Gulmezian-Sefer M, Cheng Y, Srikumar R. Microbiological considerations for ophthalmic products: sterility, endotoxin limits, and preservatives. In: Ophthalmic product development: from bench to bedside. Cham: Springer; 2022. p. 199–227.
9. Mohite P, Gholap A, Pardeshi S, Puri A, Rajput T. Quality by design in pharmaceutical development: current advances and future prospects. Softw Progr Tools Pharm Res. 2024;25:68–107.
10. Ortolani F, Scilimati N, Gialletti R, Menchetti L, Nannarone S. Development and preliminary validation of a pain scale for ophthalmic pain in horses: The Equine Ophthalmic Pain Scale (EOPS). Vet J. 2021;278:105774.
11. Grzybowski A, Brona P. Approval and certification of ophthalmic AI devices in the European Union. Ophthalmol Therapy. 2023;12(2):633–8.
12. D'Oria F, Bordinone MA, Rizzo T, Puzo P, Favale RA, Guerriero S, Alessio G. Validation of a new system for triage of ophthalmic emergencies: the alphabetical triage score for ophthalmology (ATSO). Int Ophthalmol. 2020;40:2291–6.
13. Goldsack JC, Coravos A, Bakker JP, Bent B, Dowling AV, Fitzer-Attas C, Godfrey A, Godino JG, Gujar N, Izmailova E, Manta C. Verification, analytical validation, and clinical validation (V3): the foundation of determining fit-for-purpose for biometric monitoring technologies (BioMeTs). NPJ Digit Med. 2020;3(1):55.
14. Kawalkar R, Dubey HK, Lokhande SP. A review for advancements in standardization for additive manufacturing. Mater Today Proc. 2022;50:1983–90.
15. World Health Organization. Quality assurance of pharmaceuticals: a compendium of guidelines and related materials. Volume 2. Good manufacturing practices and inspection. Geneva: World Health Organization; 2024.
16. Bhusari V, Potdar M, Nandalwar P, Warbhe P, Ghante M. Developing a design qualification protocol as standard operating procedure for a pharmaceutical plant facility meeting cGMP requirement for tablets and capsules manufacturing department. In: Modern aspects of pharmaceutical quality assurance: developing & proposing application models, SOPs, practical audit systems for Pharma Industry. Singapore: Springer Nature; 2024. p. 229–326.
17. DeSantis P. Biotech facility design for validation. In: Process validation in manufacturing of biopharmaceuticals. Boca Raton: CRC Press; 2023. p. 213–54.
18. Bannerjee S, Santra S, Sen KK. Introduction to pharmaceutical validation. In: Dosage forms, formulation developments and regulations. London: Academic Press; 2024. p. 153–79.
19. Zhou W, Ye C, Huang X, Zhang P, Zheng S, Qin L, Chen Y. Efficacy of cleaning methods for ophthalmic microscopic instruments: a comparison study. AORN J. 2020;112(2):112–21.
20. Viinamäki J. Validation of a method for analysing rinsing waters from production of ophthalmic solutions; 2022.
21. Thapar S, Chawla A, Gupta G. Method development and validation for ophthalmic formulations containing antibiotics: a comprehensive review. Indian J Pharm Sci. 2023;85(3) https://doi.org/10.36468/pharmaceutical-sciences.1123.
22. Sahay P, Goel S, Nagpal R, Maharana PK, Sinha R, Agarwal T, Sharma N, Titiyal JS. Infectious keratitis caused by rare and emerging micro-organisms. Curr Eye Res. 2020;45(7):761–73
23. Ting DS, Ho CS, Deshmukh R, Said DG, Dua HS. Infectious keratitis: an update on epidemiology, causative microorganisms, risk factors, and antimicrobial resistance. Eye. 2021;35(4):1084–101.
24. Tan CL, Sheorey H, Allen PJ, Dawkins RC. Endophthalmitis: microbiology and organism identification using current and emerging techniques. Ocul Immunol Inflamm. 2023;31(2):393–401.
25. Azari AA, Arabi A. Conjunctivitis: a systematic review. J Ophthalmic Vis Res. 2020;15(3):372.
26. Wroblewska KB, Jadach B, Muszalska-Kolos I. Progress in drug formulation design and delivery of medicinal substances used in ophthalmology. Int J Pharm. 2021;607:121012.
27. Singh M, Bharadwaj S, Lee KE, Kang SG. Therapeutic nanoemulsions in ophthalmic drug administration: concept in formulations and characterization techniques for ocular drug delivery. J Control Release. 2020;328:895–916.

28. Figus M, Agnifili L, Lanzini M, Brescia L, Sartini F, Mastropasqua L, Posarelli C. Topical preservative-free ophthalmic treatments: an unmet clinical need. Expert Opin Drug Deliv. 2021;18(6):655–72.
29. Murty R. Aseptic packaging and labeling. In: Aseptic pharmaceutical manufacturing II. Boca Raton: CRC Press; 2020. p. 337–67.
30. Gukasyan HJ, Graham R. Liquid ophthalmic drug products: physicochemical properties, formulations, and manufacturing considerations. In: Ophthalmic product development: from bench to bedside. Cham: Springer; 2022. p. 249–315.
31. Gorantla S, Rapalli VK, Waghule T, Singh PP, Dubey SK, Saha RN, Singhvi G. Nanocarriers for ocular drug delivery: current status and translational opportunity. RSC Adv. 2020;10(46):27835–55.
32. Pistollato F, Madia F, Corvi R, Munn S, Grignard E, Paini A, Worth A, Bal-Price A, Prieto P, Casati S, Berggren E. Current EU regulatory requirements for the assessment of chemicals and cosmetic products: challenges and opportunities for introducing new approach methodologies. Arch Toxicol. 2021;95:1867–97.
33. Ahmed R. Ensuring quality medicine: a comprehensive overview of EMA and DGDA's history, structure, and functions. https://doi.org/10.20944/preprints202408.0031.v1.
34. Ayalew K, Ning YM, Foringer MJ, Leibenhaut S, Sellers JW, Yu B, Kronstein PD, Higgerson A, Mihaescu C, Rodriguez M, Williams L. Comparison of good clinical practice inspection processes for marketing applications between the United States Food and Drug Administration and the European Medicines Agency. Ther Innov Regul Sci. 2023;57(1):79–85.
35. Mofid S, Bolislis WR, Kühler TC. Real-world data in the postapproval setting as applied by the EMA and the US FDA. Clin Ther. 2022;44(2):306–22.
36. Makwana RG, Desai KV, Kikani V, Vaja MD. Regulatory advances and prospects of variation filing for the registered parenteral products in USA and Europe. Int J Drug Regul Aff. 2021;9(2):52–65.
37. Thangaraju P, Varthya SB. ISO 10993: biological evaluation of medical devices. In: Medical device guidelines and regulations handbook. Cham: Springer; 2022. p. 163–87.
38. Rico F, Mazabel A, Egurrola G, Pulido J, Barrios N, Marquez R, García J. Meta-analysis and analytical methods in cosmetics formulation: a review. Cosmetics. 2023;11(1):1.
39. Barrieu M, Chennell P, Yessaad M, Bouattour Y, Wasiak M, Jouannet M, Le Basle Y, Sautou V. Physicochemical stability of a novel tacrolimus ophthalmic formulation for the treatment of ophthalmic inflammatory diseases. Pharmaceutics. 2022;14(1):118.
40. Ghiglioni DG, Martino PA, Bruschi G, Vitali D, Osnaghi S, Corti MG, Beretta G. Stability and safety traits of novel cyclosporine a and tacrolimus ophthalmic galenic formulations involved in vernal keratoconjunctivitis treatment by a high-resolution mass spectrometry approach. Pharmaceutics. 2020;12(4):378.
41. Sri-In J, Sisan W, Kingkhangphloo P, Jutasompakorn P, Chandranipapongse W, Chatsiricharoenkul S, Buranakan O, Pornseth A, Surachatkumtonekul T. Stability and sterility of extemporaneously prepared 0.01% atropine ophthalmic solution in artificial tears and balanced salt solution. Siriraj Med J. 2022;74(2):91–9.
42. Gawin-Mikołajewicz A, Nartowski KP, Dyba AJ, Gołkowska AM, Malec K, Karolewicz B. Ophthalmic nanoemulsions: from composition to technological processes and quality control. Mol Pharm. 2021;18(10):3719–40.
43. Berton B, Chennell P, Yessaad M, Bouattour Y, Jouannet M, Wasiak M, Sautou V. Stability of ophthalmic atropine solutions for child myopia control. Pharmaceutics. 2020;12(8):781.
44. Jameel F, Alexeenko A, Bhambhani A, Sacha G, Zhu T, Tchessalov S, Kumar L, Sharma P, Moussa E, Iyer L, Fang R. Recommended best practices for lyophilization validation—2021 part I: process design and modeling. AAPS PharmSciTech. 2021;22(7):221.
45. Karthick C, Kathiresan K. Pharmaceutical process validation: a review. J Drug Deliv Ther. 2022;12(1-S):164–70.
46. Rowan NJ, Kremer T, McDonnell G. A review of Spaulding's classification system for effective cleaning, disinfection and sterilization of reusable medical devices: viewed through a modern-day lens that will inform and enable future sustainability. Sci Total Environ. 2023;878:162976.

47. Tambon F, Laberge G, An L, Nikanjam A, Mindom PS, Pequignot Y, Khomh F, Antoniol G, Merlo E, Laviolette F. How to certify machine learning based safety-critical systems? A systematic literature review. Autom Softw Eng. 2022;29(2):38.
48. Rué A. Design and Automation: Vacuum transport system for the eye drop medicine filling process; 2023.
49. Ng LL. Development and validation of analytical procedures. In: Analytical testing for the pharmaceutical GMP laboratory, vol. 9. Hoboken: Wiley; 2022. p. 143–67.
50. Amaral C, Paiva M, Rodrigues AR, Veiga F, Bell V. Global regulatory challenges for medical devices: impact on innovation and market access. Appl Sci. 2024;14(20):9304.
51. Hart KM, Stapleton F, Carnt N, Arundel L, Lian KY. Optometry Australia's infection control guidelines 2020. Clin Exp Optom. 2021;104(3):267–84.
52. Hout SA. Sterile processing of pharmaceutical products: engineering practice, validation, and compliance in regulated environments. Hoboken: Wiley; 2021.
53. Link T. Guidelines in practice: instrument cleaning. AORN J. 2021;114(3):241–51.
54. Ergin AD, Uner B. Quality by design for parenteral formulations. In: InIntroduction to quality by design (QbD) from theory to practice. Singapore: Springer Nature; 2024. p. 217–42.
55. Leanpolchareanchai J, Tangteerakoon P, Supapsophon P, Sukavatcharin S, Simaroj P, Suksiriworapong J. Extemporaneous preparation of 20 mg/mL ganciclovir in artificial tears in comparison with sterile water for ophthalmic administration: formulation and stability study. Pharmaceutics. 2023;15(1):208.
56. Ivanov IS, Akhmedova DA, Koroleva YA, Shatalov DO. Sterility of ophthalmic solutions as a factor in the evolution of primary packaging for eye drops: a literature review. Sci Pharm. 2024;3(1):40–50.
57. Baseman H. Aseptic processing. In: Assurance of sterility for sensitive combination products and materials. London: Academic Press; 2020. p. 39–77.
58. Boom F, Beaney A. Aseptic handling. In: Practical pharmaceutics: an international guideline for the preparation, care and use of medicinal products. Cham: Springer; 2023. p. 749–65.
59. Anderson NM, Benyathiar P, Mishra DK. Aseptic processing and packaging. In: Food safety engineering. Cham: Springer; 2020. p. 661–92.
60. Adrianto MF, Annuryanti F, Wilson CG, Sheshala R, Thakur RR. In vitro dissolution testing models of ocular implants for posterior segment drug delivery. Drug Deliv Transl Res. 2022;12(6):1355–75.
61. Srivastava V, Chary PS, Rajana N, Pardhi ER, Singh V, Khatri D, Singh SB, Mehra NK. Complex ophthalmic formulation technologies: advancement and future perspectives. J Drug Deliv Sci Technol. 2023;82:104317.
62. Wu Q, Sawaya A. Understanding the current acceptance criteria of media fill-contamination rates in aseptic manufacturing processes. https://doi.org/10.13140/RG.2.2.33179.59685.
63. Ergin AD, Uner B. Quality by design for parenteral formulations. In: Introduction to quality by design (QbD) from theory to practice. Springer Nature: Singapore; 2024. p. 217–42.
64. Nagarsenkar MS, Dhawan VV. Parenteral preparations. In: Remington. London: Academic Press; 2021. p. 577–603.
65. Maulvi FA, Ranch KM, Desai AR, Desai DT, Shukla MR. Ophthalmic preparations. In: Remington. London: Academic Press; 2021. p. 565–75.
66. Singha S, Mehtab D. Sterilization of pharmaceutical dosage forms. In: Drug delivery aspects: Volume 4: expectations and realities of multifunctional drug delivery systems, vol. 27. London: Elsevier; 2020. p. 169.
67. Jacob S, Kather FS, Boddu SH, Shah J, Nair AB. Innovations in nanoemulsion technology: enhancing drug delivery for oral, parenteral, and ophthalmic applications. Pharmaceutics. 2024;16(10):1333.
68. Bagul US, Nazirkar MV, Mane AK, Khot SV, Tagalpallewar AA, Kokare CR. Fabrication of architectonic nanosponges for intraocular delivery of brinzolamide: an insight into QbD driven optimization, in vitro characterization, and pharmacodynamics. Int J Pharm. 2024;650:123746.

69. Chawnani D, Ranch K, Patel C, Jani H, Jacob S, Al-Tabakha MM, Boddu SH. Design and optimization of acetazolamide nanoparticle-laden contact lens using statistical experimental design for controlled ocular drug delivery. J Biomater Sci Polym Ed. 2024;9:1–25.
70. Sun C, Tian L, Wei Y, Chen P, Wu X, Jie Y. Novel bisdemethoxycurcumin@phytomicelle ophthalmic solution: In vitro formulation appraisal and in vivo prompting rapid corneal wound healing evaluations. Exp Eye Res. 2023;234:109608.
71. Yoo H, Na YG, Jin M, Won JH, Song B, Yun TS, Hwang YR, Lee HK, Cho CW. Considerations related to the in-use stability of sterile pharmaceutical dosage forms. J Pharm Investig. 2024;54(1):1–2.
72. Dhara AK, Bandyopadhyay PK, Nayak AK. Pharmaceutical sterilization. In: Dosage forms, formulation developments and regulations. London: Academic Press; 2024. p. 543–75.
73. Hollander ME. Evaluating future biopharmaceutical inspection needs, infrastructure capability gaps, and technology development strategies. Doctoral dissertation, Massachusetts Institute of Technology; 2012.
74. Louis M, Leblanc L, Julian FA. Introduction to culture media in pharmaceutical microbiology for non-sterile products. In: Pharmaceutical microbiological quality assurance and control: practical guide for non-sterile manufacturing, vol. 5; 2019. p. 79–104. https://doi.org/10.1002/9781119356196.ch4.
75. English DJ. Microbial monitoring of a manufacturing facility. In: Cosmetic microbiology. Boca Raton: CRC Press; 2020. p. 143–55.
76. Ferone M, Gowen A, Fanning S, Scannell AG. Microbial detection and identification methods: bench top assays to omics approaches. Compr Rev Food Sci Food Saf. 2020;19(6):3106–29.
77. Mota JD, Boué G, Prévost H, Maillet A, Jaffres E, Maignien T, Arnich N, Sanaa M, Federighi M. Environmental monitoring program to support food microbiological safety and quality in food industries: a scoping review of the research and guidelines. Food Control. 2021;130:108283.
78. Flemming HC, van Hullebusch ED, Little BJ, Neu TR, Nielsen PH, Seviour T, Stoodley P, Wingender J, Wuertz S. Microbial extracellular polymeric substances in the environment, technology and medicine. Nat Rev Microbiol. 2024;27:1–9.
79. King MF, López-García M, Atedoghu KP, Zhang N, Wilson AM, Weterings M, Hiwar W, Dancer SJ, Noakes CJ, Fletcher LA. Bacterial transfer to fingertips during sequential surface contacts with and without gloves. Indoor Air. 2020;30(5):993–1004.
80. Ilori MO, Obayori OS, Salam LB, editors. Aeromicrobiology. London: Elsevier; 2023.
81. Kropp M, Harmening N, Bascuas T, Johnen S, De Clerck E, Fernández V, Ronchetti M, Cadossi R, Zanini C, Scherman D, Ivics Z. GMP-grade manufacturing and quality control of a non-virally engineered advanced therapy medicinal product for personalized treatment of age-related macular degeneration. Biomedicines. 2022;10(11):2777.
82. Shukla SS, Pandey RK, Gidwani B, Kalyani G. Objectives of qualification. In: Pharmaceutical calibration, validation and qualification: a comprehensive approach. Springer Nature: Singapore; 2023. p. 49–61.
83. Ghosh G, Bodroth RP, Dutta B. Impact of process validation and equipment qualification in production of bio-therapeutics. J Adv Biotechnol Exp Ther. 2022;5(1):163.
84. Nygård A. Commissioning and qualification of process equipment for the manufacturing of advanced therapy medicinal products; 2024.
85. DeSantis P. Facility design for validation. In: Handbook of validation in pharmaceutical processes. 4th ed. Boca Raton: CRC Press; 2021. p. 5–31.
86. Niewiadomski L. Incorporating model-based user requirements in product development. Doctoral dissertation; 2021.
87. Shahizan S. A critical genre analysis of the user requirement specification report in the information technology industry. Doctoral dissertation, University of Malaya (Malaysia)); 2020.
88. DeSantis P, Ostrove S. Control systems validation. In: Handbook of validation in pharmaceutical processes. 4th ed. Boca Raton: CRC Press; 2021. p. 873–82.

89. Borkenstein C, Haibl MA, Harer J. Integrated qualification of manufacturing systems for medical devices. In: Medical devices and in vitro diagnostics: requirements in Europe. Cham: Springer; 2023. p. 1–56.
90. Vega-Mercado H. Validation, verification and qualification considerations. In: Handbook of cell and gene therapy. Boca Raton: CRC Press; 2023. p. 135–51.
91. Gübitz B. How to implement a risk-based and life-cycle approach for commissioning and qualification. In: Medical devices and in vitro diagnostics: requirements in Europe. Cham: Springer; 2023. p. 413–47.
92. Poudyal B. Predictive analysis of installation and operational qualification issues vs. process severity events. Doctoral dissertation, Massachusetts Institute of Technology; 2020.
93. Jones SL, Ricke SC, Keith Roper D, Gibson KE. Swabbing the surface: critical factors in environmental monitoring and a path towards standardization and improvement. Crit Rev Food Sci Nutr. 2020;60(2):225–43.
94. Zasada AA, Zacharczuk K, Woźnica K, Główka M, Ziółkowski R, Malinowska E. The influence of a swab type on the results of point-of-care tests. AMB Express. 2020;10:1–6.
95. Lee SH, Kim JK, Jee JP, Jang DJ, Park YJ, Kim JE. Quality by design (QbD) application for the pharmaceutical development process. J Pharm Investig. 2022;52(6):649–82.
96. Wright JF. Quality control testing, characterization and critical quality attributes of adeno-associated virus vectors used for human gene therapy. Biotechnol J. 2021;16(1):2000022.
97. Kim EJ, Kim JH, Kim MS, Jeong SH, Choi DH. Process analytical technology tools for monitoring pharmaceutical unit operations: a control strategy for continuous process verification. Pharmaceutics. 2021;13(6):919.

Packaging, Container, and Closure of Ophthalmic Products

12

J. Josephine Leno Jenita, Seema S. Rathore, D. Manjula, and Wilson Barnabas

Abstract

Packaging, container, and closure systems are vital in maintaining ophthalmic products' safety, efficacy, and quality. These components act as protective barriers against environmental factors such as light, moisture, temperature, and contamination, which can significantly impact the stability and effectiveness of ophthalmic formulations. The unique sensitivity of these formulations necessitates careful packaging design to maintain sterility, ensure precise dosing, and enhance patient safety and compliance. This comprehensive overview examines various aspects of ophthalmic product packaging, including materials selection, container and closure systems, design considerations, quality control measures, and regulatory compliance. Common packaging materials such as glass, plastics, polymers, and aluminum are discussed, along with their advantages and limitations for ophthalmic use. The importance of ergonomic design, ease of use, and compatibility with ophthalmic formulations is emphasized to improve patient experience and treatment adherence. Quality control measures and testing protocols, including visual inspection, extractable and leachable testing, and stability studies, are outlined to ensure the integrity and efficacy of ophthalmic packaging. This chapter also highlights the critical role of regulatory compliance in maintaining consistent quality and safety of ophthalmic products, particularly adherence to FDA and EMA guidelines and Good Manufacturing Practices. Emerging trends and innovations in ophthalmic packaging are explored, including smart packaging technologies for enhanced traceability and patient adherence, as well as sustainable and eco-friendly packaging solutions to address

J. J. L. Jenita (✉) · S. S. Rathore · D. Manjula · W. Barnabas
Department of Pharmaceutics, College of Pharmaceutical Sciences, Dayananda Sagar University, Bengaluru, India
e-mail: josephine-sps@dsu.edu.in

N. K. Mehra (ed.), *Complex Ophthalmic Dosage Forms: Advances in Biomedical Applications and Future Perspectives*, https://doi.org/10.1007/978-981-96-6306-4_12

environmental concerns. The multidisciplinary nature of packaging development is emphasized, underscoring the need for collaboration between formulators, packaging engineers, regulatory experts, and healthcare professionals. In conclusion, the abstract emphasizes the integral role of packaging, container, and closure systems in ophthalmic product development and their significant impact on product quality, safety, and patient outcomes. Ongoing research and innovation in this field are crucial for addressing emerging challenges and improving overall eye health outcomes globally.

Keywords

Ophthalmic packaging · Sterility maintenance · Regulatory compliance · Quality control · Smart packaging

Abbreviations

AAO	American Academy of Ophthalmology
cGMP	Current Good Manufacturing Practice
COC	Cyclic Olefin Copolymer
EMA	European Medicines Agency
FDA	Food and Drug Administration
GMP	Good Manufacturing Practices
LDPE	Low-Density Polyethylene
NFC	Near Field Communication
PET	Polyethylene Terephthalate
PE	Polyethylene
PP	Polypropylene
QA	Quality Assurance
QC	Quality Control
TRS	Technical Report Series
WHO	World Health Organization

Learning objectives on packaging, container, and closure systems for ophthalmic products:

- Understand the critical role of packaging in preserving ophthalmic product safety, efficacy, and quality by protecting against environmental factors and maintaining sterility.
- Identify and compare different packaging materials and container systems used for ophthalmic products, including their advantages and limitations.
- Analyze key design considerations for ophthalmic packaging, focusing on ergonomics, ease of use, and compatibility with product formulations.

- Outline quality control measures and regulatory requirements (FDA, EMA, GMP) essential for ensuring the integrity and compliance of ophthalmic product packaging.
- Evaluate emerging trends and innovations in ophthalmic packaging, including smart technologies and sustainable solutions.
- Assess the impact of packaging design on patient experience, treatment adherence, and overall therapeutic outcomes in ophthalmic care.

12.1 Introduction

The World Health Organization (WHO) estimates that 314 million people worldwide have vision impairment, with 45 million of them being blind. Approximately 82% of all visually impaired people are 50 years of age or older, making up 19% of the global population. Age-related vision impairment and disorders associated with the eye are becoming more common as the world's population rises and the demographics change to include a greater percentage of older individuals, even in developing nations [21, 47].

The safety, efficacy, and quality of ophthalmic formulations depend on the proper packaging, container, and closure systems. Ophthalmic products, such as eye drops, ointments, and gels, are administered directly to the eye; hence, they require specialized packaging design, proper containers, and closures as they play a crucial role due to their distinct properties and specific needs [75, 131]. The packaging, containers, and closures must ensure that the product is maintained sterile from the time of manufacture to the last dose that is administered to prevent any contamination, which may lead to serious adverse effects like eye infections and irritations [101, 153]. In other words, the stability and efficacy of ophthalmic products should be maintained throughout their shelf life as they are sensitive to light, air, and moisture. Therefore, the packaging, containers, and closures must provide sufficient barriers to such fundamentals, thereby preserving the potency of ophthalmic medications and preventing any untoward reactions to ensure patient safety [30, 130]. Packaging, containers, and closures must be designed in such a way that it is easy to use by patients who have limited skills and are visually challenged [89, 123]. Accurate dosing is one of the prime factors to be considered in ophthalmic products as it requires precise dosing to ensure the exact quantity of medication or solution or gel, as it is directly instilled into the eye, thereby minimizing the possible risk that may occur due to under- or overdosing. In addition to this, the design of packaging should give more importance to patient safety and must be tamper-evident, which means the packaging must be designed to reveal any interference with the contents. In addition, the closures should be designed to prevent any illicit access or accidental administration by anyone, especially geriatrics and pediatrics [19]. A container closure system refers to the ensemble of packaging components that collaborate to secure and contain the dosage form. This includes both primary and secondary packaging elements, provided the latter are intended to offer additional protection (Fig. 12.1, [69]).

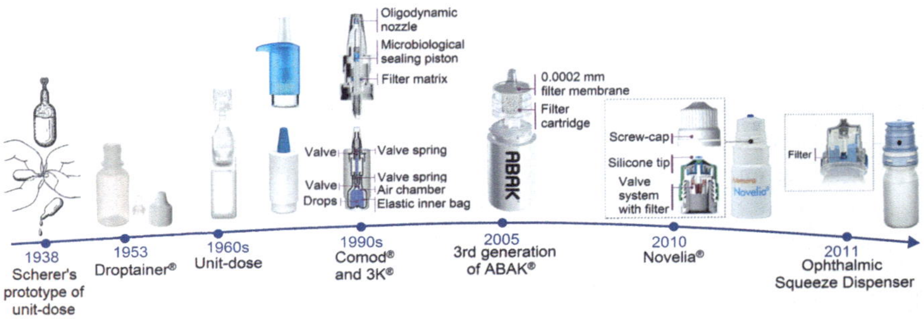

Fig. 12.1 The evolution of primary packaging for eye drops: a historical perspective. (Reproduced with copyright permission from Ref. [69])

Enhanced patient compliance and adherence to the treatment using ophthalmic formulations can be achieved by designing user-friendly packaging that increases the compatibility and ease of usage by the patients, thereby allowing the patients to administer the ophthalmic formulations with accuracy and convenience. Emphasizing and prioritizing the design and materials of these components ensures that pharmaceutical manufacturers significantly improve the safety, efficacy, and quality of ophthalmic formulations, thereby benefiting patient requirements [126, 136, 137]. The key aspect in the formulation and development of ophthalmic products mainly relies on the packaging, containers, and closures as they ensure the sterility, safety, and suitability of the product.

A packaging component refers to any part of a container closure system, including containers (such as ampules, vials, bottles), container liners (like tube liners), closures (such as screw caps, stoppers), closure liners, stopper overseals, container inner seals, administration ports (on large-volume parenteral), overwraps, administration accessories, and container labels. A primary packaging component is in direct contact with the dosage form, whereas a secondary packaging component is not. A container closure system comprises both primary and secondary components, with the latter providing additional protection to the drug product [29]. Essentially, a packaging system is the same as a container closure system. Every proposed packaging system must be proven suitable for its intended use, offering adequate protection to maintain the product's stability and effectiveness throughout its shelf life. It must be compatible with the dosage form, ensuring that the materials do not negatively interact with the drug, potentially compromising its effectiveness or safety. Additionally, the packaging components should be made from materials considered safe for contact with the dosage form and the route of administration, preventing any harmful substances from leaching into the product. This thorough evaluation ensures the drug product's safety, efficacy, and quality throughout its lifecycle [21, 47].

Since ophthalmic products must meet stringent regulatory requirements to guarantee their quality, safety, and efficacy, regulatory compliance has become another critical aspect in packaging, containers, and closures of ophthalmic products [20,

101]. To meet the quality assurance and safety standards that are mandated by regulatory authorities, packaging design requirements must comply with guidelines such as Good Manufacturing Practices (GMP).

Finally, packaging and labeling are crucial for brand differentiation and product recognition. Thoughtfully designed packaging boosts brand visibility, effectively conveys product information, and distinctly differentiates ophthalmic products in the market.

12.2 Requirements for Packaging of Ophthalmic Products

Due to the unique characteristics and delicate nature of the eye, it is essential to ensure the safety, efficacy, and ease of administration of ophthalmic products and use specialized packaging, containers, and closures [77]. To maintain sterility, it is mandatory to prevent contamination and tampering by using tamper-evident seals and user-friendly administration tips to confirm accurate dosing and reduce waste of the ophthalmic product [42]. Table 12.1 emphasizes some key requirements of ophthalmic products that have an impact on packaging, containers, and closure design.

By adhering to the above guidelines, pharmaceutical manufacturers can ensure ophthalmic products that are safe, effective, and easy to administer, which in turn enhances patient compliance. Achieving these targets mainly relies on the design of packaging, containers, and closures, emphasizing innovative and creative approaches in product development.

12.3 Types of Packaging Materials Used in Ophthalmic Products

The selection of packaging materials for ophthalmic products is very critical in maintaining their quality, safety, efficacy, and user-friendliness. Plastic bottles such as low-density polyethylene (LDPE) are more flexible and convenient to squeeze, making them very ideal for eye drops [108, 126]. Among the various glasses as packaging materials, Type I borosilicate glass is superior in chemical resistance (to both acidic and alkaline substances), and its durability is preferred in most of the pharmaceutical industries [72, 104]. Aluminum tubes effectively safeguard ophthalmic ointments from light and air, thereby preserving their potency. Plastic tubes produce flexibility in the ease of use for ophthalmic gels and creams [13]. Vials, either made of glass or plastic, ensure the sterility of ophthalmic products for both single- or multi-dose uses. For single-dose units, blister packs offer an excellent sterile environment and thereby prevent contamination. Lastly, to provide additional protection and provide important usage information, cartons and labels play a pivotal role. Based on the specific needs of the products, the efficacy, and safety for patients, each material for packaging, closures, and containers is chosen [56, 119, 129, 140]. Overall, the selection of the packaging, closures, and containers mainly

Table 12.1 Key requirements of packaging, containers, and closures for ophthalmic products

Considerations	Description	References
Barrier properties	To ensure the stability and efficacy of ophthalmic products, the packaging, containers, and closures must serve as shields to protect them from light, heat, moisture, and air	[22, 86, 87]
Sterility	Ensuring sterility is a vital factor in preventing infections and irritations; therefore, packaging systems must guarantee sterility throughout the shelf life of the ophthalmic products	[55, 56, 97]
Particle generation	The packaging materials must minimize the risk of particle generation throughout the storage, handling, and administration, as the eye is susceptible to particulate matter	[8, 81]
Dosage accuracy	Accurate dosing is very important for effective treatment, so the packaging systems need to ensure consistent and precise dosing to avoid under- or overdosing [67]	[23, 61]
Ease of use	The design of packaging, containers, and closures must be intuitive and user-friendly, especially for self-administration, and should also focus on supporting people with visual and motor impairments	[69, 141]
Compatibility	The packaging, containers, and closures must be compatible with the ophthalmic formulations to maintain the efficacy of the product and prevent degradation	[15]
Regulatory requirements	To ensure safety, quality, and efficacy, the packaging design must comply with regulatory standards such as GMP and pharmacopeial guidelines	[101]
Brand differentiation	A thoughtfully designed packaging boosts brand recognition, efficiently communicates product features, and thereby distinguishes products from competitors in the market	[72]

relies on the compatibility with the formulation, barrier properties, and regulatory requirements, thereby ensuring the integrity and efficacy of the ophthalmic products.

12.4 Overview of the Different Types of Packaging Materials Commonly Used for Ophthalmic Products

12.4.1 Glass as Packaging Materials

Due to their nonreactive and exceptional barrier characteristics, glass containers are widely used in the packaging of ophthalmic formulations [70, 83]. A strong chemical resistance is provided by glass, thereby proving no interaction between product formulation and packaging materials [32, 55]. It shields the ophthalmic formulations from the external environment such as air, light, and moisture, safeguarding its stability and efficacy [32, 90, 130]. Finally, glass containers are one of the favored options for packaging ophthalmic products as they are eco-friendly and recyclable.

12.4.2 Plastic as Packaging Materials

Ophthalmic formulations are usually packaged in plastic containers such as polypropylene (PP), polyethylene terephthalate (PET), and polyethylene (PE), as they offer the advantage of being lightweight, flexible, and cost-effective [38, 94]. Plastic containers are usually found in various ranges of forms such as tubes, bottles, and vials, thereby providing flexibility in packaging design, and also based on the requirement of the product for sensitivity to light, the plastic containers may be produced as transparent or opaque [105, 127]. On the other hand, some plastics may have issues such as permeability that require barrier coatings or additives to increase their shielding properties [24, 133]. Due to their adaptable properties like durability, resistance to chemicals, and flexibility, polymers have been tremendously used for packaging ophthalmic formulations. Polymers that are frequently and often used in pharmaceutical packaging include PET, PE, and cyclic olefin copolymer (COC) [43, 102]. To have particular barrier properties, these polymers can be modified to protect the ophthalmic formulations from light, oxygen, and moisture [15, 43].

12.4.3 Aluminum

Aluminum as a packaging material for ophthalmic formulations is usually preferred for packing of ophthalmic ointments or creams as it presents outstanding barrier properties, preventing the product from light, air, and moisture, thereby certifying the stability of ophthalmic formulation during its shelf life [6, 29]. Additionally, aluminum provides several advantages such as less weight, being easily extrudable, and portable, thereby providing easy and accurate dosing for patients [74].

12.4.4 Multilayer Packaging

As the name itself indicates, multilayer, types of packaging usually combine various layers of materials, like polymers, plastics, and barrier coatings, to increase the shielding properties of the packaging material [95, 160]. Multilayer packaging ensures product integrity and shelf life as it offers better light protection, barrier properties, and stability for ophthalmic products [58, 65].

Some of the marketed products available with multilayer packaging are discussed below:

Acuvue Oasys: These contact lenses from Johnson & Johnson are packaged in multilayer blister packs. The packaging helps keep the lenses hydrated and protected until they are ready to be used.

CooperVision Proclear: Packaged in multilayer blister packs, these lenses are designed to provide all-day comfort and maintain their moisture content.

Bausch + Lomb Ultra: These daily disposable lenses come in multilayer packaging that ensures the lenses remain sterile and easy to handle.

Table 12.2 comprehensively summarizes the merits, demerits, and considerations for each type of packaging material regularly or normally used in ophthalmic formulations.

Ultimately, each type of packaging material for ophthalmic formulations presents unique advantages, disadvantages, and considerations that must be carefully assessed based on the product's specific needs, patient safety, regulatory requirements, and environmental sustainability. Balancing these factors will help manufacturers choose the most suitable packaging material for their ophthalmic products to ensure product stability, safety, and patient satisfaction.

12.4.5 Color Codes for Topical Ocular Drugs

The American Academy of Ophthalmology (AAO) supports the standardization of a color-coding scheme for topical ocular drug caps and labeling. The pharmaceutical industry was involved in the development of this color-coding scheme and has generally supported the policy. These recommendations have taken into account available color options, common prescribing patterns, and patient safety considerations, even though it is impossible to reach a 100% consensus among the industry representatives when selecting from a small number of available colors. Color coding in ophthalmic drugs aims to prevent medication errors and help patients distinguish among various medicines [31]. Due to the large number of options available, including those for glaucoma, infections, inflammation, and allergies, it can be difficult to differentiate between different types of topical ocular drugs. Color coding ensures the right drug is given by providing a visual indication that enables prompt identification [17, 18, 84, 146]. Table 12.3 depicts the color code for topical ocular medications (American Academy of Ophthalmology).

12.5 Challenges in the Selection of Packaging Materials for Ophthalmic Formulations

12.5.1 Chemical Compatibility

To maintain the stability and efficacy of ophthalmic formulations, it is crucial to ensure that the packaging material does not react with the product [156].

Loss of potency: Consider an eye drop formulation containing Timolol, a drug utilized for the treatment of glaucoma. When stored in a plastic bottle composed of materials such as polyethylene or polypropylene, the active component may adsorb onto the walls of the container, leading to a decrease in its concentration and efficacy [149].

Degradation of active drug: For instance, consider an eye drop with an active ingredient that degrades upon exposure to certain chemicals. If the plastic container releases small amounts of a stabilizer, it could react with the drug, leading to its breakdown into inactive or harmful compounds [79].

Table 12.2 Comparative assessment of different types of packaging materials used for ophthalmic products

Type of packaging material	Advantages	Disadvantages	Considerations	Examples of marketed product	References
Glass packaging	Inert in nature Exceptional barrier properties concerning moisture, air, and light Recyclable/reusable and eco-friendly Appearance wise it is superior and is of high quality	Highly fragile and prone to breakage, thereby enhancing the threat of glass particles in the product Unlike plastic or polymer, comparatively heavy weight Costly when compared to plastic Fewer options for container designs when compared to plastic packaging	To shield from light, amber glass is to be selected, or else colorless glass is sufficient To increase barrier properties, specialized undercoatings or treatments to be used Use additional protective packaging like cartons and trays to avoid any accidental breaking in transport	Lucentis (Ranibizumab, Roche's): macular edema, diabetic retinopathy Avastin (Bevacizumab, Roche): wet age–related macular degeneration (AMD), diabetic retinopathy Atropine eye drop 1%: Mydriatic	[107, 124, 129]
Plastic packaging	As they are lightweight and flexible, it makes them suitable for portable and squeezable containers Readily available in several sizes and shapes and also economical Compared to glass, it is less subjected to breakage Depending on light sensitivity, available in light opaque or transparent	Probability of affecting stability is more as they are highly permeable to light, air, and oxygen Possibility of interaction or leaching of chemicals into ophthalmic formulations is high As they are not eco-friendly with less recyclability, it leads to environmental issues Compared to glass, it is less premium	Based on the suitability of the formulation, plastic material needs to be selected To improve barrier qualities, use coatings or some additives Either opaque or transparent packaging is selected based on the sensitivity of product to light Extractables and leachables must comply with the regulatory requirements	Visine Eye Drops (1% Polyethylene glycol 400, Johnson & Johnson's): dry eyes and protect against irritation Pataday (Olopatadine HCl, Alcon Lab Inc.): Allergic conjunctivitis Natacyn (5% Natamycin Ophthalmic Suspension, Eyevance Pharmaceuticals): fungal keratitis Voriconazole Ophthalmic Solution: Fungal infections	[33, 37, 43]

(continued)

Table 12.2 (continued)

Type of packaging material	Advantages	Disadvantages	Considerations	Examples of marketed product	References
Aluminum packaging	To guarantee product stability, safety, and patient satisfaction, producers can select the best packaging material for their ophthalmic products by considering these elements	Unlike plastics, they are high cost Opaque, hence visibility is limited Unlike glass or plastics, they are not recyclable	Cost-benefit ratio should be taken into account for premium products Appropriate labeling should be ensured to compensate for the nontransparency Depending on local regulations and requirements, the recyclability should be evaluated	FML® (Fluorometholone ophthalmic ointment) 0.1%: For swelling, redness, itching, and allergic reactions Erythromycin Ophthalmic Ointment USP: Neonatal conjunctivitis and ophthalmia neonatorum Maxitrol® (Neomycin, Polymyxin B, and Dexamethasone): for bacterial eye infection	[65, 74]

Table 12.3 Color codes and descriptions for topical ocular medications

Medication type	Color code	Description
Adrenergic agonist combinations	Light green (Pantone 373C)	Used to treat glaucoma by reducing intraocular pressure through various mechanisms
Adrenergic agonists	Purple (Pantone 2583)	Stimulate adrenergic receptors, decreasing intraocular pressure and used in glaucoma treatment (Brimonidine 0.1%)
Anti-infectives	Tan (Pantone 467)	Used to treat bacterial infections of the eye
Anti-inflammatory, nonsteroidal	Gray (Pantone 4)	Nonsteroidal drugs are used to reduce inflammation and pain in the eye
Anti-inflammatory, steroids	Pink (Pantone 197)	Steroidal drugs are used to reduce inflammation in ocular conditions
Anti-inflammatory, immunomodulators	Olive green (Pantone 5763C)	Used to modulate the immune response and reduce inflammation in the eye (Bimatoprost 0.01%. Travoprost 0.004%)
Beta-blocker combinations	Dark blue (Pantone 281)	Combination drugs that reduce intraocular pressure by decreasing aqueous humor production (Dorzolamide/Timolol 0.5%)
Beta-blockers	Yellow (Pantone Yellow C)	Lower intraocular pressure by reducing aqueous humor production, used in glaucoma treatment (Timolol 0.5%)
Carbonic anhydrase inhibitors	Orange (Pantone 1585 C)	Reduce intraocular pressure by decreasing aqueous humor production (Dorzolamide Brinzolamide 1%)
Cytotoxic	Black (Pantone 6)	Used for the treatment of certain types of eye cancers
Miotics	Dark green (Pantone 348)	Constrict the pupil, increasing the outflow of aqueous humor and lowering intraocular pressure
Mydriatics and cycloplegics	Red (Pantone 1797)	Dilate the pupil for examination or surgery and paralyze the ciliary muscles
Prostaglandin analogs	Turquoise (Pantone 326)	Increase the outflow of aqueous humor to reduce intraocular pressure, used in glaucoma treatment

Excipient degradation: An ophthalmic solution may include a preservative such as benzalkonium chloride to inhibit microbial growth. However, if the preservative is absorbed by container materials like LDPE or PP, its concentration decreases, increasing the risk of microbial contamination and potential infections [27].

Physical changes: An eye drop formulated at a specific pH may experience changes due to interactions with the packaging. For example, if the container material releases acidic or basic substances, it can alter the solution's pH, potentially causing eye irritation or reducing the drug's stability.

12.5.2 Barrier Properties

To preserve the integrity of the product, the packaging material should be selected to shield it from air, light, and moisture [29, 137].

12.5.3 Regulatory Compliance

It is very difficult to comply with the strict regulatory guidelines and requirements for ophthalmic products, especially those related to extractables and leachables [124].

12.5.4 Patient Safety

Patient safety in ophthalmic products hinges on maintaining sterility and minimizing particulate matter. Sterility prevents microbial contamination, crucial for avoiding infections and ensuring the drug's potency. Visual inspections and stringent quality control measures help identify and eliminate particulate matter, safeguarding the product from physical contaminants that could harm patients. Together, these factors ensure the ophthalmic product is safe and effective for patient use [48].

12.5.5 Environmental Impact

The requirement of efficient packaging along with environmental sustainability and recyclability poses a challenge while deciding on a container-closure system for the ophthalmic product [51, 157].

12.5.6 Economy

Another important consideration is achieving a balance between cost-effective solutions and maintaining uncompromised quality and safety [63].

12.6 Container and Closure Systems

One of the vital considerations for assuring the sterility and integrity of pharmaceutical products, specifically, ophthalmic formulations, are containers and closures [103]. Containers and closures are the prime barriers for contaminants such as particulate matter, air dust, and microorganisms, assuring the sterility of the ophthalmic products [26, 91]. Preventing air and moisture from entering the packaging system, along with using closures and containers that protect against light, temperature, and oxidation, helps maintain product stability, prevents microbial growth, and conserves the efficacy of ophthalmic products during their shelf life [89]. Special features such as tamper-evident and child-resistant assure the safety and integrity of the product by limiting accidental access to the product. In addition to the above safety, the containers and closures minimize medication errors by enabling accurate and appropriate dosing. Following regulatory guidelines is mandatory to assure the

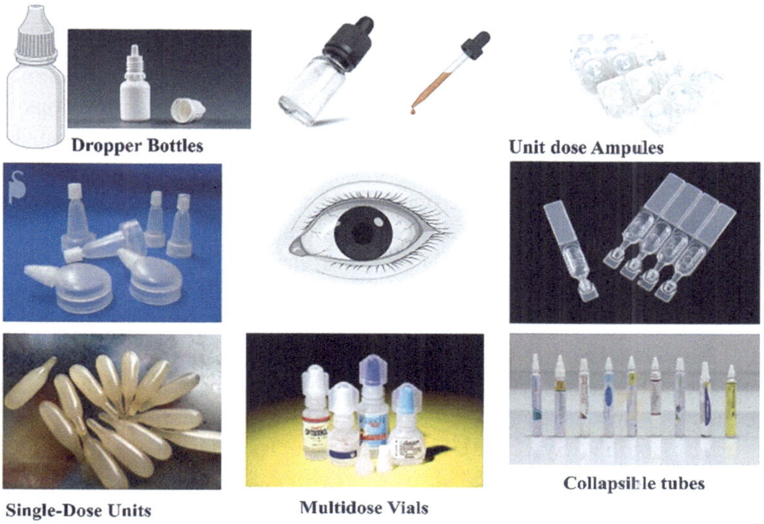

Fig. 12.2 Containers and closure systems for ophthalmic products. (Self-created with biorender.com licensed version)

quality, safety, and efficacy of the product. Generally, well-designed closures and containers increase patient compliance and safety by adhering to sterility, stability, and quality, thereby making them indispensable in pharmaceutical packaging.

For the packaging of pharmaceutical products, including ophthalmic products, a variety of containers and closures are being used [1]. Various types of container and closure systems are used for packaging pharmaceutical products, including ophthalmic products. The following is an outline of various types of container and closure systems that include dropper bottles, vials, and single-dose units (Fig. 12.2).

12.6.1 Dropper Bottles

These are normally used for packaging eye drops. They usually comprise a bottle made of either glass or plastic with a dropper tip and a screw-on or snap-on closure. The accurate and controlled administration of the ophthalmic formulation is done by dropper tip, which aids in precise dosing in patients [23, 126]. The design of the closure system is in such a way that it prevents contamination and thereby maintains sterility by ensuring a tight seal [34, 82, 130]. Dropper bottles are available in variant sizes to fill different volumes of formulation and are designed to be user-friendly and ergonomic for patients' convenience [62, 80], e.g., Roche's Lucentis (Ranibizumab), Bimatoprost Ophthalmic Solution, and Latanoprost Eye Drops IP.

12.6.2 Vials

Vials are usually small, cylindrical containers normally used for ophthalmic solutions, suspensions, and injectable medications. They are available in various shapes and sizes and are made of either glass or plastic [126, 159]. Normally, these vials are featured with a screw-on or crimp-on closure system to seal the container and thereby maintain the sterility of the ophthalmic products. Vials are utilized for the storage and preservation of the sterility of ophthalmic preparations. In contrast, dropper bottles have been designed for the controlled dispersion of liquids in drops, rendering them suitable for administering eye drops. To facilitate aseptic access to the ophthalmic products, the rubber stoppers or septa are used to provide a secured closure [126]. These rubber stoppers are often used for multi-dose formulations, allowing repeated access to the formulation without affecting its stability and sterility [11]. Such closures are commonly employed in medications that require frequent dosing, e.g., Lomb's Systane Ultra Lubricant Eye Drops, Alcon's Pataday (Olopatadine Hydrochloride Ophthalmic Solution and Johnson & Johnson's Visine All Day Comfort for dryness and irritation throughout the day; e.g., for vials include SYSTANE Complete Unit-Dose Preservative-Free Eye Drops 30ct Single-use vials for Dry Eye Relief, Soothe Preservative Free Lubricant Eye Drops, Long-Lasting Dry Eye Relief, Moisturizing & Comforting, Suitable for Sensitive Eyes, Single Use Vials, 30 Count, Refresh Refresh Celluvisc Lubricant Eye Drops Single-Use Containers.

12.6.3 Single-Dose Units

These types of containers are pre-filled with a single dose of medication, which is meant for one-time use. They are normally used for preservative-free eye drops, to ensure precise dosing and prevent contamination [36, 126]. They are typically made of plastic or foil packaging, thereby ensuring a sterile environment for the ophthalmic formulations. For easy access and disposal after single use, they are fabricated as tear-off or twist-off closures. As they are single-dose units, they provide the advantage of a fresh dose of medication for each administration, thereby reducing the risk of contamination and microbial growth and maintaining sterility [68, 89, 145]. They provide the advantage of eliminating the need for preservatives to be added to the formulation and are also convenient for the patients [78, 126].

In deciding whether the containers or closures system for ophthalmic formulations are bottles, vials, or single-dose units, each one of them offers unique benefits for dosing accuracy, sterility, convenience of use, and patient safety [73, 90]. The specific requirements of the product, mode of administration, and the intended use by the patient are some of the factors to be considered in the selection of the most appropriate containers and closures; for example, Allergan's Refresh Optive Advanced Lubricant Eye Drops are typically packed in single-use vials. These vials are designed to maintain the sterility and convenience, with each vial containing a single dose of the preparation.

Table 12.4 Case studies for containers and closure systems for ophthalmic products

Case study	Details
Ophthalmic Product Package Inspection (PTI Packaging and Inspection Systems) [121]	Evaluation of different packaging formats, including soft containers with snap-on or torque screw caps, as well as foil-lidded blister packs. Emphasis on identifying defects such as micro-leaks and flange issues
Ophthalmic Drug Packaging: Safety Risks and Patient Acceptability Issues [26]	Analysis of safety risks and patient acceptability concerns Emphasis on identifying design flaws in container closure systems and the FDA's measures to mitigate these risks. Highlighting the critical importance of maintaining sterility
Ophthalmic Product Development: Key Considerations [113]	Important factors in the development of ophthalmic products involve advanced dispensing systems, navigating regulatory process challenges, conducting in vitro testing, and performing extractable and leachable testing

These preservatives, such as benzalkonium chloride (BAK), polyquaternium-1, sorbic acid, chlorhexidine, sodium perborate, Purite® (stabilized oxychloro complex), and SofZia®, play a crucial role in maintaining the sterility and efficacy of multi-dose ophthalmic products, ensuring patient safety, and extending the shelf life of the medications [109, 115].

Table 12.4 includes the case studies for containers and closure systems for ophthalmic products. These case studies focus on various aspects of ophthalmic product packaging, aiming to address and resolve issues related to packaging effectiveness, safety, and development; Fig. 12.2 includes containers and closure systems for ophthalmic products.

12.7 Mechanisms for Ensuring Tamper-Evident Packaging and Child-Resistant Closures

Ensuring essential safety measures in pharmaceutical packaging, mainly for formulations that pose a risk if improperly accessed, the tamper-evident packaging and child-resistant closure mechanisms can be utilized [53, 92]. To safeguard patients, manufacturers are required to adhere to regulatory norms and regulations while creating and installing safety measures. These measures, when paired with proper labeling and user education, improve product safety, reduce the possibility of misuse or accidental ingestion, and give patients and caregivers peace of mind [14, 147]. Both mechanisms are highlighted in the following.

12.7.1 Tamper-Evident Packaging

Tamper-evident packaging for eye drops is designed to ensure product safety and integrity by providing visible evidence if the packaging has been opened or tampered with. Tamper-evident rings or seals that shatter when the bottle is first opened are typical characteristics that let the user know the product is brand new. The Food

Table 12.5 Overview of tamper-evident packaging

Tamper-evident feature	Description
Heat induction seals	These types of seals form a bond between the packaging and closure that is broken upon opening, leaving a visible seal on the package
Tear bands	These bands are intended to break or tear when the package is opened, signifying that the product has been accessed
Frangible components	Breaks or frangible caps are designed to break easily when tampered with packaging
Visual indicators	Color-changing ink, holographic seals, or other visual signs can indicate that the packaging has been tampered with

and Drug Administration (FDA) has warned that eye drop bottles with loose safety seals could be dangerous if they fall into the eye. In addition to preventing contamination, proper tamper-evident packaging guarantees that patients receive products that are both safe and effective [128, 135]. Some mechanisms for attaining tamper-evident packaging are as follows (Table 12.5).

12.7.2 Child-Resistant Closures

Child-resistant closures for ophthalmic products are designed to prevent young children from accidentally ingesting or accessing medications while allowing adults easy access. These closures, such as "press-turn" and "squeeze-turn" designs, require specific actions to open, reducing the risk of child access. To fulfill child-resistant criteria, they are put through a rigorous testing process that includes adult and kid tests to guarantee their usefulness and efficacy. These requirements are guided by regulatory standards such as UNI EN ISO 8317:2016. The design strikes a balance between adult accessibility and child safety, making the packaging safe for kids while still being used by adults, especially senior citizens. Both patient safety and compliance depend on this equilibrium [96, 144]. Methods used for child-resistant closures are depicted in Table 12.6.

12.8 Key Considerations in the Design of Packaging of Ophthalmic Formulations

Various factors must be considered in designing packaging for ophthalmic products to assure safety, efficacy, and patient compliance. To increase the comfort of the patient, ease of handling, and effective administration of the product, ergonomic design is crucial in ophthalmic product packaging [99, 106]. This design mainly focuses on the comfortable fitting in the hand, secured grip, and ease of operation during use. For patients with less manual dexterity, a nonslip material or textures can ensure a secure grip [64, 91]. Additionally, features such as finger grips, contours, and indentations facilitate easy handling and accurate dispensing of ophthalmic

Table 12.6 Overview of child-resistant closures

Child-resistant closure	Description	Additional details
Push-and-turn closures	To open these closures, one needs to push and turn the closure, which requires skill and strength that is beyond the reach of children [16, 158, 152]	Commonly used for medications and household chemicals to prevent accidental ingestion by children
Squeeze-and-turn closures	These types of closures usually utilize squeezing and turning simultaneously, wherein children find it very difficult to open [162]	Often found on bottles of cleaning products and pharmaceuticals, providing an extra layer of safety
Sliding closures	These closures restrict children's access by requiring a sequence of sliding or twisting movements to unlock [16]	Used in various packaging, including some food containers and personal care products, to ensure safety
Push-down-and-turn closures	These closures involve pushing down and then turning to open, combining two movements that are challenging for young children [139]	Widely used for prescription bottles and certain over-the-counter medications to prevent accidental access

products [117]. Lastly, the caps should be designed with intuitive mechanisms that minimize the effort, prevent spillage or leakage, and make it easy to open.

To facilitate patients with precise dosing and appropriate administration, the ophthalmic products must be user-friendly and intuitive for patients. The administration mechanism should be accurate and allow for a controlled dispensing of the ophthalmic product whether a dropper, or pump, or squeeze bottle is used [35, 83]. For patients with visual impairments, the packaging should be designed in such a way that it is clear and easy to read dosage markings on the bottle which is very important for precise dosing [86, 87]. The ease of administration can be further enhanced by including clear and concise directions for the product administration on the packaging, which includes illustrations or figures for visual guidance, some special instructions, etc. To still further enhance patient compliance, the packaging must be designed in such a way that it is easy to operate with one hand and allows the patients to self-administer the dose efficiently and precisely [25, 61].

Another crucial factor to be considered for ophthalmic product packaging is the compatibility of the ophthalmic formulation. These packaging materials should be inert and compatible with the ophthalmic formulations to prevent any possible interaction or leaching of harmful materials [15, 76, 123]. The packaging should be done very cautiously to safeguard the light-sensitive drugs, which is achieved by using light-resistant materials or use of coatings that prevent the formulations from degradation caused by exposure to ultraviolet light [77, 142] Another factor that needs consideration in maintaining product stability is to prevent water vapor permeation which is achieved by using packaging materials that offer a moisture barrier [40, 128]. Additional stability can be achieved by preventing the entry of oxygen and preserving the integrity of the product by using a closure system that provides an air-tight seal [130].

The design of packaging for ophthalmic formulations can increase patient compliance, product safety, and adherence to treatment by integrating ergonomic design principles, easy-to-use features, and compatibility with the formulations [7, 86, 87]. Next, to meet the specific needs of ophthalmic patients and healthcare providers, the optimized packaging designs should prioritize user-centered design and product compatibility. Thus, this holistic approach of packaging design not only ensures the efficacy and safety of the ophthalmic products but also improves patient satisfaction [154, 161]. A significant difference in the overall effectiveness of ophthalmic treatments, providing patients with a reliable and user-friendly means of administering medications, is achieved by perfectly designing the packaging containers [8, 59]. In conclusion, the selection of the most appropriate container and closure system mainly relies on the specific requirements of the product, the mode of administration, and the intended use by patients (Fig. 12.3).

12.9 Labeling Requirements, Regulatory Considerations, and Patient Safety Aspects

In designing the packaging for ophthalmic products, several issues must be carefully taken into account to ensure patient convenience, product efficacy, and safety. Important factors in this process include labeling guidelines, regulatory requirements, and patient safety considerations. Manufacturers can guarantee regulatory compliance, improve patient safety, and encourage the safe and effective use of pharmaceutical products by taking labeling requirements, regulatory considerations, and patient safety elements into account while designing and developing the packaging for ophthalmic drugs [49, 112, 150]. Adherence to regulatory protocols

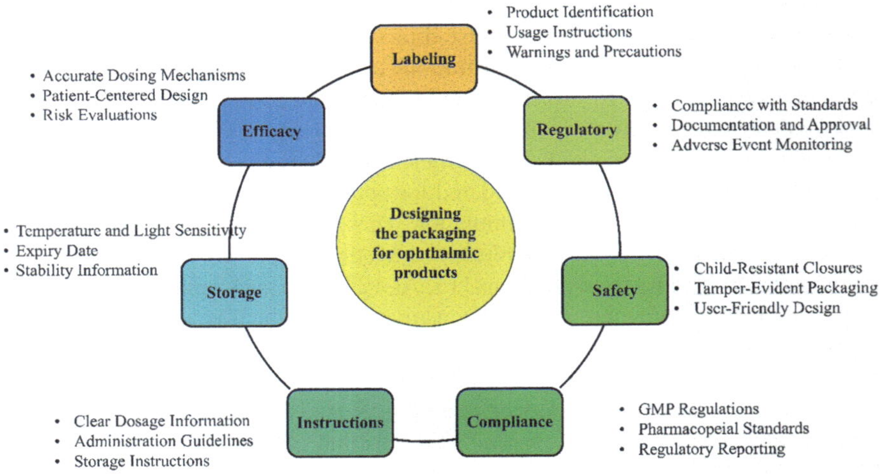

Fig. 12.3 Essential factors in designing packaging for ophthalmic formulations. (Self-created with biorender.com licensed version)

and standards is vital in assuring the security and effectiveness of pharmaceutical products while safeguarding the well-being of patients. Health authorities have established regulatory standards for the labeling of ophthalmic products [153]. The product name, strength, and dosage form must all be identified on the label, along with the active ingredients and their concentrations in the formulation [6, 141]. Patients must also receive precise and concise dosage instructions, along with information on how often to use the product and how to administer it. Moreover, warnings, precautions, and contraindications of its use must be included, and it must also specify storage conditions, such as requirements for temperature and light sensitivity, to maintain product stability and the product's expiry date [29, 111, 151]. To guarantee product safety and efficacy, ophthalmic product packaging must abide by the regulatory requirements of GMP and quality standards. The regulatory considerations include filing, packaging design, and documentation for regulatory review and approval before product launch, adhering to pharmacopeial standards for packaging quality and performance, monitoring and documenting any adverse events related to packaging defects or issues and reporting them to the regulatory authorities, and complying with GMP regulations to ensure manufacturing processes meet quality and safety standards [63, 139]. Pharmaceutical packaging design must prioritize patient safety, particularly for ophthalmic medications where precise administration and high-quality products are essential [60]. Using child-resistant closures to prevent accidental ingestion by children and ensure safe storage, implementing tamper-evident packaging to prevent tampering and ensure product integrity [96, 138, 144]. To achieve appropriate packaging solutions for ophthalmic products, it is imperative to prioritize patient-centered design, engage with regulatory agencies, and conduct extensive risk evaluations [81]. When it comes to giving patients a dependable and convenient way to take their medication, well-designed packaging can have a major impact on the overall efficacy of ophthalmic therapies [55, 157, 162]. As a result, the unique needs of the product, the method of administration, and the planned use by patients all play a role in choosing the best container and closing system.

12.10 Quality Control and Testing

Ophthalmic preparations require rigorous testing procedures and quality control (QC) measures as integral parts of their package development process. By implementing such measures, the effectiveness, safety, and integrity of the closure, container, and packaging systems can be assured. Stability studies are especially important for ophthalmic products because of the delicate nature of the eyes and the requirement for sterility [12, 82]. Any deterioration in the product's quality may pose a major threat to health. Stability studies therefore aid in guaranteeing that the product is safe, efficient, and devoid of contamination till its expiration date [66, 93]. Pharmaceutical companies can guarantee the efficacy and safety of their ophthalmic medications for the duration of their shelf life by carrying out comprehensive analyses (Fig. 12.4).

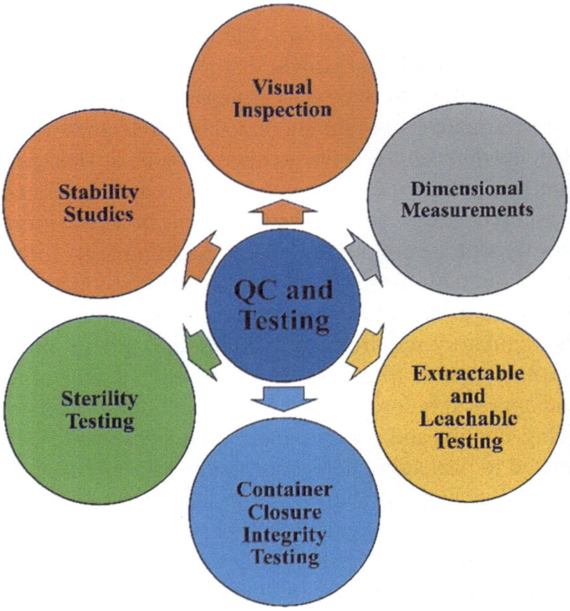

Fig. 12.4 Outline of QC and testing procedure for packaging systems. (Self-created with biorender.com licensed version)

12.10.1 Outline of Quality Control Measures and Testing Protocols for Assessing Packaging Systems Integrity and Efficacy

12.10.1.1 Visual Inspection

An essential QC procedure for assessing the external characteristics of packaging materials, containers, and closures is visual inspection. This procedure entails inspecting the product for flaws, harm, leaks, or possible impurities that could compromise its integrity [93, 148]. Color consistency, surface polish, clarity, and the general appearance of the package components are important visual inspection factors.

12.10.1.2 Dimensional Measurements

It guarantees that the packaging components' physical dimensions and specifications adhere to the specifications of the design. This includes confirming the packing system's overall dimensional accuracy, closure alignment, and correct fit. Accurate measurements of package component dimensions were made using precision tools including gauges, calibers, and micrometers [56].

12.10.1.3 Extractable and Leachable Testing

This test is performed to assess the possibility of material migration from packaging materials into the product formulation. It determines the leachable and extractable compounds from the container-closure system to evaluate the safety and compatibility of packaging materials [24, 116]. By the protocols, samples are taken from the packaging materials and analyzed for the presence of leachable or extractables that may compromise the safety or quality of the final product [64, 76, 129].

12.10.1.4 Container Closure Integrity Testing

To make sure packaging systems offer a reliable barrier against microbiological penetration, contamination, or product leakage, this testing evaluates the seal integrity of those systems. The integrity of the container closing system is assessed using techniques such as high voltage leak detection, vacuum decay testing, and dye ingress testing [37, 140].

12.10.1.5 Sterility Testing

In addition to confirming that the product is sterile, sterility testing is carried out It evaluates how well the packaging system keeps the product sterile throughout its shelf life. This is especially important to protect patients from infection and to maintain their safety [134]. To detect microbial growth, protocols entail inoculating samples from the packaging system onto culture media and allowing them to incubate under the right circumstances and observe the presence/absence of microbial growth [69, 89, 126].

12.10.1.6 Stability Studies

Stability studies assess how well packaging systems function in different environmental settings, including exposure to light, humidity, and temperature. These studies evaluate how storage conditions affect the packing system's effectiveness and integrity over time [57, 143]. For ophthalmic products, the shelf life, compatibility, and effectiveness of the packaging, container, and closure systems are determined using both accelerated and real-time stability studies [13, 22].

Pharmaceutical companies can guarantee the dependability, effectiveness, and security of ophthalmic product packaging, containers, and closure systems by putting in place thorough QC procedures and testing guidelines [10]. The maintenance of pharmaceutical packaging systems' quality and integrity is contingent upon strict adherence to regulatory regulations and comprehensive testing, which ultimately facilitates the safe and efficient administration of pharmaceuticals to patients [132] (Fig. 12.4).

12.11 Regulatory Compliance

The packaging, container, and closure of ophthalmic items are governed by requirements set forth by the FDA in the United States and the European Medicines Agency (EMA) in Europe, among other regulatory norms and standards [83]. The quality, safety, and effectiveness of pharmaceutical packaging are greatly impacted by regulatory standards and regulations, particularly regarding the packaging, containers, and closing systems used for ophthalmic drugs [46, 52]. It is imperative to adhere to the regulations established by reputable regulatory bodies, such as the FDA and EMA, to secure market acceptance and guarantee patient safety [135].

12.12 FDA Guidelines

12.12.1 FDA Guidelines for Ophthalmic Product Packaging

The FDA has established extensive regulations and guidelines to guarantee the security and integrity of pharmaceutical containers, closure systems, and packaging (FDA 2023, [98, 125]). The following are some important rules concerning the packaging of ophthalmic products.

12.12.2 Current Good Manufacturing Practice (cGMP)

Packaging operations must comply with cGMP regulations to guarantee the quality, safety, and efficacy of pharmaceutical products [97, 100]. This includes stringent controls and procedures to maintain high standards throughout the manufacturing processes [45, 147].

12.12.3 Container Closure Systems for Packaging Human Drugs and Biologics

The FDA provides detailed guidelines for testing container closure systems to ensure they maintain product integrity and sterility. This is a crucial step preferred for preventing contamination and ensuring the safety of the ophthalmic preparations [24, 92, 103].

12.12.4 Guidance for Industry: Container and Closure System Integrity Testing Instead of Sterility Testing

As an alternative to sterility testing, this guideline describes procedures for evaluating the integrity of the container closure system. To make sure that the packaging system offers a reliable barrier against contaminants, it focuses on evaluating the seal quality and overall integrity of those systems [25, 28].

12.12.5 Labeling Requirements

The guidelines issued by the FDA outline certain standards for pharmaceutical product labeling. The product name, active components, dosage, cautions, storage requirements, and expiry date must all be listed on the labels [60, 63]. This guarantees that all the information required for safe and efficient use is available to the patients and healthcare professionals.

Pharmaceutical manufacturers must make sure that the quality, safety, and efficacy of the preparations are maintained throughout their shelf life by following these FDA recommendations when packaging them.

12.13 EMA Guidelines

The EMA in the European Union uses stringent regulatory guidelines to guarantee the safety and quality of products.

12.13.1 Annex 9 Guidelines on Packaging for Ophthalmic Products

Guidelines for packaging pharmaceutical items, including ophthalmic medications, are provided in Annex 9 of WHO Technical Report Series (TRS) No. 902. Packaging has several uses, such as informational, presentational, protective, and confinement purposes. Additionally, it must stop counterfeiting and tampering. The specifications and compatibility of the product should guide the selection of materials (such as glass, plastics, and metals) and closures (such as rubber closures and caps). To ensure that packaging satisfies safety and quality standards, it must adhere to regulatory rules established by agencies like the FDA and EMA [50]. To maintain product integrity and patient safety, the recommendations mentioned in Annex 9 are followed. It outlines pharmaceutical packaging specifications, including design, materials, labeling, and regulatory compliance [9].

12.13.2 Guideline on the Evaluation of Packing of Ophthalmic Products for Human Use

This guideline offers a systematic method for evaluating the packaging of ophthalmic drugs, assuring compliance with critical safety, quality, and efficacy standards. It emphasizes the essential function of packaging in maintaining the sterility and integrity of ophthalmic products, which are especially vulnerable due to their direct application to the eye. Essential factors encompass the selection of materials that are compatible with the product and resistant to moisture, light, and other environmental elements that may compromise the medication. The recommendation emphasizes the significance of tamper-evident features to avert contamination and guarantee customer safety. Additionally, the evaluation method entails examining the packaging's efficacy in preventing the entrance of particulate matter, which may jeopardize patient safety. This guideline functions as a thorough reference for producers and regulators, guaranteeing that the packaging of ophthalmic products adequately safeguards patients' health while preserving the therapeutic advantages of the items [4, 5, 54].

12.13.3 Guidance on GMP for Pharmaceutical Products

The required quality assurance (QA) and QC procedures for pharmaceutical manufacturing processes, including packaging processes, are outlined in the EMA's GMP rules [71, 120]. These regulations aid in upholding strict requirements for product safety and quality throughout the production process [63, 108]. Key aspects include maintaining sterility throughout the product's shelf life, preventing microbial contamination, and visually inspecting for particulate matter. Manufacturers must thoroughly document all manufacturing processes, quality control measures, and compliance with cGMP requirements (Fig. 12.5).

By following FDA and EMA recommendations, pharmaceutical companies may maintain regulatory compliance, guarantee product quality and safety, and demonstrate that their ophthalmic products adhere to stringent packaging, container, and closure system standards [45]. Adhering to these standards is imperative to secure market approval and maintain customer confidence in the efficacy and safety of pharmaceutical products [2, 3, 39].

12.14 Future Trends and Innovations

Several new developments and trends are being seen in the field of ophthalmic product packaging to improve the sustainability, efficacy, and safety of the products. Aiming to improve patient outcomes, lessen environmental impact, and improve overall product quality, these innovations include the use of sustainable packaging solutions and the integration of smart packaging technologies (Fig. 12.6) [76, 131].

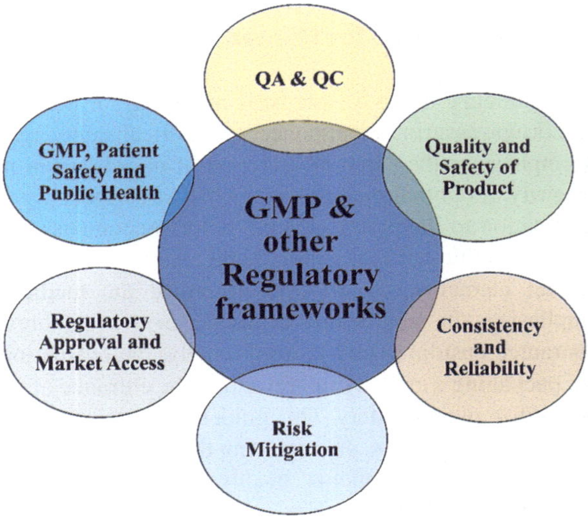

Fig. 12.5 Overview of GMP and regulatory frameworks. (Self-created with biorender.com licensed version)

Fig. 12.6 Smart technologies for packaging of ophthalmic preparations. (Self-created with biorender.com licensed version)

Further sections discuss new developments and trends in the design of ophthalmic product packaging, containers, and closures in detail.

12.14.1 Smart Packaging Technologies

The pharmaceutical industry, particularly the packaging of ophthalmic products, is transforming to smart packaging technology. These technologies integrate elements like bar codes, radio frequency identification, near field communication (NFC), and sensors to track consumption, maintain product quality, and give patients and healthcare professionals access to real-time information. These technologies increase patient adherence, treatment outcomes, and medication management by enabling improved traceability, authentication, and dosage monitoring for ophthalmic products [65, 131]. These technologies can also offer useful data on product storage conditions, usage patterns, and expiration dates to improve QC and regulatory compliance.

12.14.1.1 COMOD®

The COMOD® system is an innovative, preservative-free eye drop dispenser developed by URSAPHARM. The name COMOD stands for Continuous Mono Dose, which describes its integrated airless application system. This technique maintains the liquid's cleanliness and airtight conditions, eliminating the need for preservatives that might cause irritation or trigger allergies. The COMOD® technology ensures constant and dependable application by enabling precise, drop-by-drop doses.

12.14.1.2 ABAK® System

The ABAK® system by Thea is another advanced preservative-free eye drop dispenser. It uses a high-tech bottle with a filter that prevents microbial contamination, allowing the contents to remain sterile for up to 3 months after opening. The ABAK® system is designed to be user-friendly and environmentally friendly, reducing packaging waste compared to single-dose containers.

12.14.1.3 Novelia®

The Novelia® system by Nemera is a multidose preservative-free eye dropper that uses a nonreturn valve and a silicone membrane to prevent bacterial contamination. This system ensures that only one calibrated drop is delivered each time, improving patient adherence and comfort. Novelia® is suitable for various ophthalmic formulations and has been approved for both prescription and over-the-counter products.

12.14.1.4 Ophthalmic Squeeze Dispenser

The ophthalmic squeeze dispenser is a general term for various multidose droppers that rely on mechanical measures to prevent microbial contamination. These dispensers are designed to be easy to use and safe, providing preservative-free formulations for long-term treatments such as chronic dry eye and glaucoma. They are widely used in both over-the-counter and prescription products.

Temperature sensors for tracking cold chain storage, signs for product expiration, and connectivity capabilities for dose reminders are examples of smart packaging technology such as: Allergan Refresh Tears Eye Drops, Roche's Lucentis for retinal damage in diabetes, and Systane Ultra for dry eye syndrome [44].

Allergan Refresh Tears Eye Drops

Allergan Refresh Tears Eye Drops feature PureFlow™ Technology, which includes a one-way valve and a fine silicon membrane. This design prevents contamination by ensuring that no liquid can enter the bottle after dispensing, allowing the drops to be preservative-free.

Roche's Lucentis for Retinal Damage in Diabetes

Roche's Lucentis employs sustainable packaging solutions. The packaging is designed to be eco-friendly, using recyclable materials and reducing waste. This aligns with Roche's commitment to sustainability and environmental stewardship.

Systane Ultra for Dry Eye Syndrome

Systane Ultra also uses PureFlow™ technology with a one-way valve and a silicon membrane to prevent contamination. This allows the eye drops to be preservative-free, ensuring safety and comfort for users.

12.14.2 Sustainable Packaging Solutions (Eco-friendly Design)

A developing trend in the pharmaceutical industry is the use of sustainable packaging solutions, which are motivated by consumer demand for eco-friendly products, regulatory regulations, and concern for the environment. In the packaging of ophthalmic products, sustainable packaging options seek to minimize waste and carbon footprints and encourage environmental care [118, 143].

Compostable materials, recyclable packaging, bio-based materials, and less packaging waste are some examples of sustainable packaging options. To increase the sustainability of packaging systems, manufacturers are looking into substitutes, including plant-based polymers, recycled content, and biodegradable materials [85, 110, 155].

During a product's lifecycle, sustainable package design incorporates eco-friendly techniques, lightweight, minimalistic design, and efficient material and energy use. Environmental factors are increasingly influencing packaging, container design, and closure systems. Designers emphasize materials and production techniques that support the circular economy, encourage recyclability, and have the least negative environmental effects [38, 114].

Reusable containers, recyclable closures, bio-based polymers, and biodegradable packaging materials are some examples of innovative eco-friendly packaging designs [122]. To strike a balance between consumer convenience, environmental responsibility, and product preservation, designers are looking into sustainable solutions.

To produce package systems that support sustainability objectives, eco-friendly design concepts also include cutting down on packaging waste, optimizing packaging dimensions, and putting responsible sourcing procedures into practice.

The pharmaceutical industry is leading innovation in ophthalmic product packaging to suit changing consumer expectations, regulatory needs, and sustainability goals. It does this by integrating smart packaging technologies, eco-friendly design principles, and sustainable packaging solutions [41]. In designing and developing packaging solutions for ophthalmic products, these new trends demonstrate a dedication to promoting environmental responsibility and improving patient safety and product quality.

12.15 Conclusion

Systems for closure, packaging, and containers are essential for guaranteeing the security and effectiveness of ophthalmic goods. Throughout the product's lifecycle, these elements function as barriers to protect the product's integrity, stop contamination, and uphold sterility. Ophthalmic products are protected from external elements like light, moisture, and air that could deteriorate the product's active ingredients by packaging, container, and closure methods. A product's stability and effectiveness are preserved with the aid of well-designed packaging, guaranteeing its continued safety for use. Packaging is essential in preventing ophthalmic product contamination, which could cause infections or unfavorable patient reactions. Systems for containers and closures offer a tight seal that shields the product from impurities and guarantees its safety and purity. Sterility is an essential component of product safety for ophthalmic products since they come into contact with delicate eye parts. Systems for closing containers must keep them sterile, preventing the entry of microorganisms and guaranteeing that the substance is free of dangerous pathogens that could infect the eyes [122]. Systems for containers and closures help ensure that ophthalmic medications are dosed accurately. Accurate dosage delivery systems and well-thought-out closures assist patients in delivering the recommended quantity, reducing the possibility of over- or under-dosing, and guaranteeing the best possible therapeutic results. Ophthalmic products must adhere to regulatory criteria concerning packaging, containers, and closure mechanisms to guarantee product quality, safety, and effectiveness. Following the rules established by organizations like the FDA and EMA shows that a manufacturer is dedicated to upholding customer safety and high standards of quality. Patient safety must always come first when packaging pharmaceuticals. To keep patients safe and guarantee that the substance is administered safely and effectively, packaging, containers, and sealing mechanisms must adhere to strict safety regulations. Good packaging can help ophthalmic items last longer on the shelf by shielding them from outside influences that might hasten their deterioration. An effective container design helps a product remain stable throughout time, increasing its usefulness and efficacy. In conclusion, preserving the safety, efficacy, and integrity of ophthalmic products depends greatly on the packaging, container, and closing methods. Pharmaceutical producers may provide safe and high-quality ophthalmic products that satisfy the demands of patients and healthcare providers by putting regulatory compliance, sterility assurance, dosage accuracy, product protection, and contamination prevention first in packaging design.

12.16 Summary

Packaging, container, and closure systems are essential for ensuring the safety, efficacy, and quality of ophthalmic products. These systems protect against environmental factors like light, moisture, temperature, and contamination, which can affect product stability and effectiveness. The sensitivity of ophthalmic formulations requires careful packaging design to maintain sterility, ensure accurate dosing, and enhance patient safety and compliance. Common materials for ophthalmic packaging include glass, polyethylene (PE), polypropylene (PP), polyethylene terephthalate (PET), polymers, aluminum, and multilayer packaging. Each material has unique advantages and considerations, chosen based on chemical compatibility, barrier properties, and regulatory compliance. Container and closure systems like dropper bottles, vials, and single-dose units help with precise product administration while maintaining integrity. Safety features such as tamper-evident and child-resistant closures prevent unauthorized access and accidental misuse. Packaging design must prioritize ergonomics, ease of use, and compatibility with the formulation to improve patient experience and treatment adherence. Features like easy-to-grip containers, clear dosage markings, and intuitive dispensing mechanisms are crucial for proper use and maximizing therapeutic outcomes. Quality control measures, including visual inspection, dimensional measurements, extractable and leachable testing, container closure integrity testing, sterility testing, and stability studies, are essential to maintain packaging system integrity and efficacy throughout the product's shelf life. Regulatory compliance with guidelines from the FDA, EMA, and adherence to Good Manufacturing Practices (GMP) ensures that packaging meets high-quality and safety standards. These regulations cover material selection, manufacturing processes, labeling requirements, and quality control procedures. GMP compliance ensures not only regulatory approval but also consistent product quality, risk mitigation, and patient safety. Future trends in ophthalmic packaging include smart technologies like sensors, RFID, and QR codes to enhance traceability, monitor product integrity, and improve patient adherence. There is also a growing emphasis on sustainable packaging solutions, driven by environmental concerns and regulatory pressures. This includes bio-based materials, recyclable packaging, and eco-friendly design practices to reduce waste and carbon footprint. Ophthalmic packaging requires a multidisciplinary approach involving formulators, packaging engineers, regulatory experts, and healthcare professionals. Considerations such as product stability, dosage accuracy, sterility maintenance, and user experience must be balanced to develop optimal packaging solutions. Ongoing research and development in packaging technologies are crucial for addressing challenges and improving patient outcomes, including advancements in materials science, novel closure systems, and innovative dispensing mechanisms. In summary, ophthalmic closure, container, and package technologies have a big influence on the efficacy, safety, and quality of products. Manufacturers can provide high-quality, safe ophthalmic products by giving priority to product protection, contamination prevention, sterility assurance, dosage accuracy, and regulatory compliance. Continued advancements in packaging, coupled with a commitment to sustainability and patient-centric design, will shape the future of ophthalmic product development and improve global eye health outcomes.

Acknowledgments The authors would like to express their gratitude to the Department of Pharmaceutics, College of Pharmaceutical Sciences, Dayananda Sagar University, Bengaluru, for providing facilities during the manuscript writing. Furthermore, the authors acknowledge biorender.com for the fabrication of some figures.

Conflict of Interest The authors declare no competing financial interest.

Data Availability Not applicable.

References

1. Agarwal NB, Karwa M. Chapter 12 – pharmaceutical regulations in India. In: Pharmaceutical medicine and translational clinical research. Elsevier; 2018. p. 215–31.
2. Akers MJ. Sterile product packaging and delivery systems. Int J Pharm Compd. 2015a;19(6):491–500.
3. Akers MJ. Quality assurance and quality control, part 2. Int J Pharm Compd. 2015b;19(3):215–21.
4. Alghamdi EAS, Al Qahtani AY, Sinjab MM, Alyahya KM. Extemporaneous ophthalmic preparations. Springer; 2020a.
5. Alghamdi EAS, Al Qahtani AY, Sinjab MM, Alyahya KM. Guidelines of The American Society of Health-System Pharmacists (ASHP) on pharmacy-prepared ophthalmic products. In: Extemporaneous ophthalmic preparations; 2020b. p. 21–3. https://doi.org/10.1007/978-3-030-27492-4_8.
6. Allen LV Jr. Dosage form design and development. Clin Ther. 2008;30(11):2102–11. https://doi.org/10.1016/j.clinthera.2008.11.015.
7. Allen LV Jr. Sterile basics of compounding: repackaging, part 1. Int J Pharm Compd. 2024;28(4):298–303.
8. Amarji B, Kulkarni A, Deb PK, Deepika, Maheshwari R, Tekade RK. Chapter 15 – package development of pharmaceutical products: aspects of packaging materials used for pharmaceutical products. In: Dosage form design parameters, Advances in pharmaceutical product development and research. II. Elsevier; 2018. p. 521–52.
9. Annex 9 Guidelines on packaging for pharmaceutical products. 2002. WHO Technical Report Series, No. 902:137–142.
10. Baertschi SW, Clapham D, Foti C, Kleinman MH, Kristensen S, Reed RA, et al. Implications of in-use photostability: Proposed guidance for photostability testing and labeling to support the administration of photosensitive pharmaceutical products, part 2: topical drug product. J Pharm Sci. 2015;104(9):2688–701.
11. Bakhrushina EO, Anurova MN, Demina NB, Lapik IV, Turaeva AR, Krasnuk II. Ophthalmic drug delivery systems (review). Drug Dev Registration. 2021;10(1):57–66. https://doi.org/10.33380/2305-2066-2021-10-1-57-66.
12. Banker GS, Rhodes CS. Modern pharmaceutics. 4th ed. Marcel Dekker; 1990. p. 587–600.
13. Banker GS, Siepmann J, Rhodes C. Design and evaluation of ophthalmic pharmaceutical products. In: Modern pharmaceutics. CRC Press; 2002. p. 647–738.
14. Baranowski P, Karolewicz B, Gajda M, Pluta J. Ophthalmic drug dosage forms: characterization and research methods. Sci World J. 2014:861904. https://doi.org/10.1155/2014/861904.
15. Barnes AR. Compatibility of a commercially available low-density polyethylene eye-drop container with antimicrobial preservatives and potassium ascorbate. J Clin Pharm Ther. 1995;20(6):341–4. https://doi.org/10.1111/j.1365-2710.1995.tb00708.x.
16. Beard CJ. An investigation into child-resistant closures. Arh Hig Rada Toksikol. 1970;21(3):255–63.
17. Benjamin DM. Reducing medication errors and increasing patient safety: case studies in clinical pharmacology. J Clin Pharmacol. 2003;43(7):768–83. https://doi.org/10.1177/0091270003254794.
18. Berman A. Reducing medication errors through naming, labeling, and packaging. J Med Syst. 2004;28(1):9–29. https://doi.org/10.1023/b:joms.0000021518.60670.10.
19. Berry Global. Improving patient experience through ophthalmic packaging. 2020. https://www.berryglobal.com/en/events/webinars/improving-ophthalmic-packaging.

20. Bhirud D, Agrawal G, Shah H, Patel A, Palkar MB, Bhattacharya S, Prajapati BG. Nitrosamine impurities in pharmaceuticals: an empirical review of their detection, mechanisms, and regulatory approaches. Curr Top Med Chem. 2024;24(6):503–22. https://doi.org/10.2174/0115680266278636240125113509.
21. Birkhoff M, Marx D. New devices for dispensing ophthalmic treatments may be the key to managing the life-cycle of established products with low investments in filling technology. IPI. 2010;2:44–6.
22. Bokser AD, O'Donnell PB. Stability of pharmaceutical products. Remington; 2013. p. 37.
23. Brandt JD. Human factors and ophthalmic drug packaging: time for a global standard. Ophthalmology. 2015;122(12):2368–70. https://doi.org/10.1016/j.ophtha.2015.08.035.
24. Broschard TH, Glowienke S, Bruen US, Nagao LM, Teasdale A, Stults CL, et al. Assessing safety of extractables from materials and leachables in pharmaceuticals and biologics – current challenges and approaches. Regul Toxicol Pharmacol. 2016;81:201–11. https://doi.org/10.1016/j.yrtph.2016.08.011.
25. Brown H, Mahler HC, Mellman J, Nieto A, Wagner D, Schaar M, et al. Container closure integrity testing-practical aspects and approaches in the pharmaceutical industry. PDA J Pharm Sci Technol. 2017;71(2):147–62. https://doi.org/10.5731/pdajpst.2016.006999.
26. Cai B, Raw A, Uppoor R, Bernard KA, Chambers WA, Sharadendu A, et al. Ophthalmic drug packaging: safety risks and patient acceptability issues. Pharm Technol. 2016;40(10):44–51.
27. Chaudhary S. Evaluation of antimicrobial preservative efficacy test of Benzalkonium chloride as IPC. Indian J Pharm Sci. 2023;86(5):1628–41.
28. Cheryl L, Stults M, Harper D, Zane D. Challenges associated with biological safety assessments for drug-device combination products. PDA J Pharm Sci Technol. 2024;78(1):100–24. https://doi.org/10.5731/pdajpst.2022.012822.
29. Chillon A, Pace A, Zuccato D. Introducing the Alba® primary packaging platform. Part 1: particle release evaluation. PDA J Pharm Sci Technol. 2018;72(4):382–92. https://doi.org/10.5731/pdajpst.2018.008623.
30. Chowhan M, Lang JC, Missel P. Ophthalmic preparations. Remington; 2012. p. 541.
31. Color codes for topical ocular medications – American Academy of Ophthalmology. n.d.. https://www.aao.org/about/policies/color-codes-topical-ocular-medications
32. Colsoul ML, Hecq JD, Soumoy L, Charles O, Goderniaux N, Bihin B, Jamart J, Galanti L. Long-term stability of an infusion containing paracetamol, alizapride, ketorolac and tramadol in glass bottles at 5±3°C. Eur J Hosp Pharm. 2020;27(e1):e74–8. https://doi.org/10.1136/ejhpharm-2019-001966.
33. Cuadros-Rodríguez L, Lazúen-Muros M, Ruiz-Samblás C, Navas-Iglesias N. Leachables from plastic materials in contact with drugs. State of the art and review of current analytical approaches. Int J Pharm. 2020;583:119332. https://doi.org/10.1016/j.ijpharm.2020.119332.
34. Dave P, Villarreal G Jr, Friedman DS, Kahook MY, Ramulu PY. Ability of bottle cap color to facilitate accurate patient-physician communication regarding medication identity in patients with glaucoma. Ophthalmology. 2015;122(12):2373–9. https://doi.org/10.1016/j.ophtha.2015.06.013.
35. De La Fuente J, Bix L. A tool for designing and evaluating packaging for healthcare products. J Patient Compl. 2011;1(1):48–52.
36. Degenhard M, Birkhoff M. Ophthalmic squeeze dispenser–eliminating the need for additives in multidose preservative-free eye care formulations. Drug Dev Deliv. 2017;17(7):40–4.
37. Dillingh J, Smith J. Containers. In: Practical pharmaceutics: an international guideline for the preparation, care and use of medicinal products. Springer; 2015. p. 501–36.
38. Dowaliby M. Geriatric ophthalmic dispensing. Am J Optom Physiol Optic. 1975;52(6):422–7. https://doi.org/10.1097/00006324-197506000-00008.
39. Dunkelberg H, Wedekind S. Preliminary results for a new final package test to assess the quality of sterile package systems. Infect Control Hosp Epidemiol. 2004;25(1):26–9. https://doi.org/10.1086/502287.

40. Eissenberger K, Ballesteros A, De Bisschop R, Bugnicourt E, Cinelli P, Defoin M, et al. Approaches in sustainable, biobased multilayer packaging solutions. Polymers (Basel). 2023;15(5):1184. https://doi.org/10.3390/polym15051184.
41. Erdner J, Armengol A. Addressing challenges of sterile ophthalmic product filling. 2023. https://scientificproducts.com/wp-content/uploads/sites/58/2023/08/SP-Hull-SP-i-Dositecno-Tech-Note-Addressing-Challenges-of-Sterile-Ophthalmics-Filling.pdf.
42. Eye Product Packaging. The improvements in ophthalmic packaging. 2021. https://www.lfofamerica.com/contract-packaging/the-improvements-in-ophthalmic-packaging.
43. Falk YZ, Runnsjo A, Pettigrew A, Scherer D, Engblom J, Kocherbitov V. Interactions of perfluorohexyloctane with polyethylene and polypropylene pharmaceutical packaging materials. J Pharm Sci. 2020;109(7):2180–8. https://doi.org/10.1016/j.xphs.2020.03.026.
44. Farhana M. Brand elements lead to brand equity: differentiate or die. Inform Manage Bus Rev. 2012;4(4):223–33. https://doi.org/10.22610/imbr.v4i4.983.
45. FDA. Pharmaceutical CGMPs for the 21st century – a risk-based approach. Final Report; FDA, September 2004. fda.gov/media/77391/download
46. FDA. Focus Area: Advanced Manufacturing. FDA.gov/science-research, updated September 6, 2022. Accessed 18 Sept 2024.4067969.
47. FDA, C. C. Container closure systems for packaging human drugs and biologics. Guidance for industry. 1999. https://www.fda.gov/regulatory-information/search-fda-guidance-documents/container-closure-systems-packaging-human-drugs-and-biologics.
48. Feghhi M, Mahmoudabadi AZ, Mehdinejad M. Evaluation of fungal and bacterial contaminations of patient-used ocular drops. Med Mycol. 2008;46(1):17–21. https://doi.org/10.1080/13693780701487979.
49. Fontaine AL. Current requirements and emerging trends for labelling as a tool for communicating pharmacovigilance findings. Drug-Saf. 2004;27:579–89. https://doi.org/10.2165/00002018-200427080-00009.
50. Food and Drug Administration, HHS. Ophthalmic drug products for over-the-counter human use; final monograph; technical amendment. Final rule; technical amendment. Fed Regist. 2003;68(33):7919–21.
51. Forcinio H. Evaluating package integrity. Pharm Technol. 2015;39(4) http://www.pharmtech.com/evaluating-package-integrity
52. Fortinguerra F, Clavenna A, Bonati M. Ocular medicines in children: the regulatory situation related to clinical research. BMC Pediatr. 2012;12:8. https://doi.org/10.1186/1471-2431-12-8.
53. Frey LT, Tilburg WC. Child-resistant packaging for E-liquid: a review of US State legislation. Am J Public Health. 2016;106(2):266–8. https://doi.org/10.2105/AJPH.2015.302957.
54. Garrido A, Lepailleur A, Mignani SM, Dallemagne P, Rochais C. hERG toxicity assessment: useful guidelines for drug design. Eur J Med Chem. 2020;195:112290. https://doi.org/10.1016/j.ejmech.2020.112290.
55. Gaynes BI, Singa RM, Schaab G, Sorokin Y. Impact of administration angle on the cost of artificial tear solutions: does bottle positioning minimize wastage? J Ocul Pharmacol Ther. 2007;23(2):196–201. https://doi.org/10.1089/jop.2006.0122.
56. Gibson M. 12 Ophthalmic dosage forms. In: Pharmaceutical preformulation and formulation: a practical guide from candidate drug selection to commercial dosage form. Routledge; 2016. p. 431.
57. Gopal AD, Rathi VK, Teng CC, Del Priore L, Ross JS. Incremental revisions across the life span of ophthalmic devices after initial food and drug administration premarket approval, 1979-2015. Ophthalmology. 2017;124(8):1237–46. https://doi.org/10.1016/j.ophtha.2017.03.040.
58. Gorbet MB, Tanti NC, Jones L, Sheardown H. Corneal epithelial cell biocompatibility to silicone hydrogel and conventional hydrogel contact lens packaging solutions. Mol Vis. 2010;16:272–82.
59. Gore A, Pujara C. Topical ophthalmic dosage form development: key components and critical quality attributes. In: Ophthalmic product development, AAPS advances in the

pharmaceutical sciences series, vol. 37. Cham: Springer; 2021. p. 153–80. https://doi.org/10.1007/978-3-030-76367-1_7.
60. Gu T, Wang G, van den Oord EJC, Goldman E, Yang C, Xie N, Yao L, Wang CY, Jablonski M, Ray K, Liu F, Pan W, Flores G, Aleya L, Meng X, Jiao Y, Li M, Wang Y, Gu W. A perspective on evaluating life stage differences in drug dosages for drug labeling and instructions. AAPS J. 2024;26(5):95. https://doi.org/10.1208/s12248-024-00964-0.
61. Haack D, Koeberle M. Pharmaceutical packaging–how advances in pharmaceutical packaging are better meeting patients' needs. Drug Dev Deliv. 2017;17(6):20–4.
62. Hallaji NAE, Trope GE. Generic glaucoma drop bottle cap colours. Can J Ophthalmol. 2016;51(5):392–3. https://doi.org/10.1016/j.jcjo.2016.04.027.
63. Hogg P, Vere R, Elen D, Lomas R, Chandrasekar A, Rooney P. Validation of a closed system for dispensing serum eye drops. BMJ Open Ophthalmol. 2022;7(Suppl 2):A12. https://doi.org/10.1136/bmjophth-2022-EEBA.28.
64. Houston CT, Rodrigues AD, Smith BB, Wang T, Richardson M. Principles for management of extractables and leachables in ophthalmic drug products. PDA J Pharm Sci Technol. 2022a;76(3):278–94. https://doi.org/10.5731/pdajpst.2022.012744.
65. Ibrahim ID, Hamam Y, Sadiku ER, Ndambuki JM, Kupolati WK, Jamiru T, Eze AA, Snyman J. Need for sustainable packaging: an overview. Polymers (Basel). 2022;14(20):4430. https://doi.org/10.3390/polym14204430.
66. Indian Pharmacopoeia. 2007. Volume-1, The Indian Pharmacopoeia Commission, Central Indian Pharmacopoeia Laboratory, pp 363–371.
67. Iskandar K, Marchin L, Kodjikian L, Rocher M, Roques C. Highlighting the microbial contamination of the dropper tip and cap of in-use eye drops, the associated contributory factors, and the risk of infection: a past-30-years literature review. Pharmaceutics. 2022;14(10):2176.
68. Ivanov IS, Bakhrushina EO, Turaeva AR, Shatalov DO, Aydakova AV, Akhmedova DA, Anatolevich Kedik S. Approaches to the search of the optimum packaging of eye drops. Int J App Pharm. 2022;14(5):1–7. https://doi.org/10.22159/ijap.2022v14i5.45402.
69. Ivanov IS, Akhmedova DA, Koroleva YA, Shatalov DO. Sterility of ophthalmic solutions as a factor in the evolution of primary packaging for eye drops: a literature review. Sci Pharm. 2024;3(1):40–50.
70. Iwatsuka K, Inada K, Ueoka H, Otsuka T, Maeda M, Yamaguchi M, Yasueda S. Effect of textured eye drop bottles on the photostability of pranoprofen 0.1% ophthalmic solution. Drug Dev Ind Pharm. 2015;41(9):1526–31. https://doi.org/10.3109/03639045.2014.971028. Epub 2015 Jul 21
71. Jenke D. Evaluation of the chemical compatibility of plastic contact materials and pharmaceutical products; safety considerations related to extractables and leachables. J Pharm Sci. 2007;96(10):2566–81. https://doi.org/10.1002/jps.20984.
72. Jenke D. Materials in manufacturing and packaging systems as sources of elemental impurities in packaged drug products: an updated literature review. PDA J Pharm Sci Technol. 2020;74(3):324–47. https://doi.org/10.5731/pdajpst.2019.010033.
73. Jenke D, Carlson T. A compilation of safety impact information for extractables associated with materials used in pharmaceutical packaging, delivery, administration, and manufacturing systems. PDA J Pharm Sci Technol. 2014;68(5):407–55. https://doi.org/10.5731/pdajpst.2014.00995.
74. Jenke D, Castner J, Egert T, Feinberg T, Hendricker A, Houston C, et al. Extractables characterization for five materials of construction representative of packaging systems used for parenteral and ophthalmic drug products. PDA J Pharm Sci Technol. 2013;67(5):448–511. https://doi.org/10.5731/pdajpst.2013.00933.
75. Jenke DR, Stults CL, Paskiet DM, Ball DJ, Nagao LM. Materials in manufacturing and packaging systems as sources of elemental impurities in packaged drug products: a literature review. PDA J Pharm Sci Technol. 2015;69(1):1–48. https://doi.org/10.5731/pdajpst.2015.01005.
76. Jenke D, Egert T, Hendricker A, Castner J, Feinberg T, Houston C, et al. Simulated leaching (migration) study for a model container-closure system applicable to parenteral and ophthal-

mic drug products. PDA J Pharm Sci Technol. 2017;71(2):68–87. https://doi.org/10.5731/pdajpst.2016.007229.
77. Jerkins WA, Osborn KJ. Packaging drugs and pharmaceutical products edition. Technomic; 1993. p. 1372–13899.
78. Jezek J, Darton NJ, Derham BK, Royle N, Simpson I. Biopharmaceutical formulations for pre-filled delivery devices. Expert Opin Drug Deliv. 2013;10(6):811–28. https://doi.org/10.1517/17425247.2013.780023.
79. Jumelle C, Gholizadeh S, Annabi N, Dana R. Advances and limitations of drug delivery systems formulated as eye drops. J Control Release. 2020;321:1–22. https://doi.org/10.1016/j.jconrel.2020.01.057.
80. Kairuz TE, Gargiulo D, Bunt C, Garg S. Quality, safety and efficacy in the 'off-label' use of medicines. Curr Drug Saf. 2007;2(1):89–95. https://doi.org/10.2174/157488607779315471.
81. Kartop RA, Güleli M, Faruzlu FA, Caliskan C. Evaluation of a novel LC-MS/MS based analytical method for the risk assessment of nitrosamines in pharmaceutical products and packaging materials. J Pharm Sci. 2024;113(6):1597–606. https://doi.org/10.1016/j.xphs.2024.01.011.
82. Kastango ES, Douglass K. Quality assurance for sterile products. Int J Pharm Compd. 2001;5(4):246–53.
83. KBV Research. Global ophthalmic packaging market by dose (multidose and single dose), by material (plastic, glass and other materials), by type (prescription and OTC), by regional outlook, industry analysis report and forecast, 2021–2027. 2022. https://www.kbvresearch.com/ophthalmic-packaging-market. Last accessed on 14 Apr 2022.
84. Kenagy JW, Stein GC. Naming, labeling, and packaging of pharmaceuticals. Am J Health Syst Pharm. 2001;58(21):2033–41. https://doi.org/10.1093/ajhp/58.21.2033.
85. Keyes A, Saffron CM, Manjure S, Narayan R. Biobased compostable plastics end-of-life: environmental assessment including carbon footprint and microplastic impacts. Polymers (Basel). 2024;16(21):3073. https://doi.org/10.3390/polym16213073.
86. Kim YJ, Maeng J, Irazoqui PP. Eyeglasses-powered, contact lens-like platform with high power transfer efficiency. Biomed Microdevices. 2015a;17(4):75. https://doi.org/10.1007/s10544-015-9979-0.
87. Kim FJ, da Silva RD, Gustafson D, et al. Current issues in patient safety in surgery: a review. Patient Saf Surg. 2015b;9:26. https://doi.org/10.1186/s13037-015-0067-4.
88. Kohn LT, Corrigan JM, Donaldson MS. To err is human: building a safer health system. Washington: National Academies Press (US); 2000. Setting Performance Standards and Expectations for Patient Safety. https://www.ncbi.nlm.nih.gov/books/NBK225181/
89. Koziy T, Pochtarev S. The influence of antimicrobial packaging on ensuring the quality and safety of food products in the conditions of war. Univ Econ Bull. 2022;53:67–72. https://doi.org/10.31470/2306-546X-2022-53-67-72.
90. Kuhn F, Mester V, Morris R, Dalma J. Serious eye injuries caused by bottles containing carbonated drinks. Br J Ophthalmol. 2004;88(1):69–71. https://doi.org/10.1136/bjo.88.1.69.
91. Kulkarni S, Agrawal A, Sharma SB, Jain S. Creative Innovations in Pharmaceutical Packaging. Indian J Pharm Pharmacol. 2015;2(4):230–5.
92. Kweder SL. Drugs and biologics in pregnancy and breastfeeding: FDA in the 21st century. Birth Defects Res A Clin Mol Teratol. 2008;82(9):605–9. https://doi.org/10.1002/bdra.20500.
93. Lachman Leon, Lieberman HA, Kanig JL. The theory and practice of industrial pharmacy. Third Indian Edition 1990. Dadar Bombay: Varghese; 1990. p. 711–32.
94. Lantos PR. Plastics in medical applications. J Biomater Appl. 1998;2(3):358–71. https://doi.org/10.1177/088532828700200305.
95. Lee JS, Park MA, Yoon CS, Na JH, Han J. Characterization and preservation performance of multilayer film with insect repellent and antimicrobial activities for sliced wheat bread packaging. J Food Sci. 2019;84(11):3194–203. https://doi.org/10.1111/1750-3841.14823.
96. Lembersky RB, Nichols MH, King WD. Effectiveness of child-resistant packaging on toxin procurement in young poisoning victims. Vet Hum Toxicol. 1996;38(5):380–3.

97. Lezotre P-L. Introduction. In: International cooperation, convergence and harmonization of pharmaceutical regulations; 2014. p. 1–5. https://doi.org/10.1016/b978-0-12-800053-3.00001-x.
98. Lippman RE. The FDA role in contact lens development and safety. Cornea. 1990;9(Suppl 1):S64–7.; discussion S68. https://doi.org/10.1097/00003226-199010001-00026.
99. Liu K, O Donovan JF. Pharmacy packaging and inserts. In: StatPearls [Internet]. Treasure Island: Stat Pearls Publishing; 2023.
100. Liu XM, Harmon PS, Maziarz EP, Rah MJ, Merchea MM. Comparative studies of Hyaluronan in marketed ophthalmic products. Optom Vis Sci. 2014;91(1):32–8. https://doi.org/10.1097/OPX.0000000000000100.
101. Lodha S, Patel H, Joshi S, Kalyankar G, Mishra A. Chapter 6 – regulatory requirements of regulated market. In: Regulatory affairs in the pharmaceutical industry; 2022. p. 113–61.
102. Masmoudi F, Fenouillot F, Mehri A, Jaziri M, Ammar E. Characterization and quality assessment of recycled post-consumption poly(ethylene terephthalate) (PET). Environ Sci Pollut Res Int. 2018;25(23):23307–14. https://doi.org/10.1007/s11356-018-2390-7.
103. Mathaes R, Mahler HC, Roggo Y, Huwyler J, Eder J, Fritsch K, et al. Influence of different container closure systems and capping process parameters on product quality and Container Closure Integrity (CCI) in GMP drug product manufacturing. PDA J Pharm Sci Technol. 2016;70(2):109–19. https://doi.org/10.5731/pdajpst.2015.005918.
104. Matthews BR. Drug substance and drug product – container/closure. In: International pharmaceutical product registration. CRC Press; 2016. p. 185–94.
105. McKeen LW. 3-Plastics used in medical devices. In: Handbook of polymer applications in medicine and medical devices. Plastics Design Library; 2014. p. 21–53. https://doi.org/10.1016/B978-0-323-22805-3.00003-7.
106. Melchore JA, Berdovich D. Considerations for design and use of container challenge sets for qualification and validation of visible particulate inspection. PDA J Pharm Sci Technol. 2012;66(3):273–84. https://doi.org/10.5731/pdajpst.2012.00862.
107. Mellinas C, Valdés A, Ramos M, Burgos N, Garrigos MDC, Jiménez A. Active edible films: current state and future trends. J Appl Polym Sci. 2016;133
108. Melo GB, Cruz NFSD, Emerson GG, Rezende FA, Meyer CH, Uchiyama S, et al. Critical analysis of techniques and materials used in devices, syringes, and needles used for intravitreal injections. Prog Retin Eye Res. 2021;80:100862. https://doi.org/10.1016/j.preteyeres.2020.100862.
109. Messmer EM. Konservierungsmittel in der Ophthalmologie [Preservatives in ophthalmology]. Der Ophthalmologe Zeitschrift der Deutschen Ophthalmologischen Gesellschaft. 2012;109(11):1064–70. https://doi.org/10.1007/s00347-012-2638-4.
110. Mo Y, Huang X, Hu C. Recent advances in the preparation and application of bio-based polyurethanes. Polymers (Basel). 2024;16(15):2155. https://doi.org/10.3390/polym16152155.
111. Muto T, Machida S. Vitreous reflux frequency and intraocular pressure after first-time intravitreal aflibercept injections: comparison of 30- and 32-gauge needles. Clin Ophthalmol. 2020;14:625–34. https://doi.org/10.2147/OPTH.S243370.
112. Nanjwade BK, Sonaje DB, Manvi FV. Preparation and evaluation of eye-drops for the treatment of Bacterial Conjunctivitis. IJPI's J Pharm Cosmetol. 2011;1(2):43–9.
113. Next Breath, LLC. Ophthalmic product development: key considerations recommendations to reduce risks, mitigate failure, and drive timelines in a competitive market. 2013. In White Paper. https://www.oindpnews.com/wp-content/uploads/2013/06/Next-Breath-Opthlamic-Product-Development.pdf?form=MG0AV3
114. Nikam SM, Davange MD, Patil DM. A review on advancements in the packaging of medicines. Int J in Pharm Sci. 2023;1(12):416–32. https://doi.org/10.5281/zenodo.10389787.
115. Noecker R. Effects of common ophthalmic preservatives on ocular health. Adv Therapy. 2001;18:205–15. https://doi.org/10.1007/BF02853166.
116. Norwood DL, Nagao LM, Stults CL. Perspectives on the PQRI extractables and leachables safety thresholds and best practices recommendations for inhalation drug products. PDA J Pharm Sci Technol. 2013;67(5):413–29. https://doi.org/10.5731/pdajpst.2013.00934.

117. Notenboom K, Leufkens HG, Vromans H, Bouvy ML. Learning from patients: Identifying design features of medicines that cause medication use problems. Int J Pharm. 2017;517(1–2):128–34. https://doi.org/10.1016/j.ijpharm.2016.12.004.
118. Novack GD. Confusion in product packaging. Ocul Surf. 2014;12(1):77–81. https://doi.org/10.1016/j.jtos.2013.10.005.
119. Okeke CC, Hunt DG, Mohr N, Medwick T, Sheinin EB, Williams RL. Packaging systems: compendial requirements. In: Encyclopedia of pharmaceutical science and technology, six volume set (print). CRC Press; 2013. p. 2409–30.
120. Omotola EO, Supriyanto G. Occurrence, detection and ecotoxicity of microplastics in selected environments-a systematic appraisal. Heliyon. 2024;10(14):e32095. https://doi.org/10.1016/j.heliyon.2024.e32095.
121. Ophthalmic Product Package Inspection. PTI packaging and inspection systems. n.d.. https://www.pharmaceuticalonline.com/doc/ophthalmic-product-package-inspection-case-0001?form=MG0AV3.
122. Ordabayeva N, Chandon P. In the eye of the beholder: visual biases in package and portion size perceptions. Appetite. 2016;103:450–7. https://doi.org/10.1016/j.appet.2015.10.014.
123. Parenky AC, Wadhwa S, Chen HH, Bhalla AS, Graham KS, Shameem M. Container closure and delivery considerations for intravitreal drug administration. AAPS Pharm Sci Tech. 2021;22(3):100. https://doi.org/10.1208/s12249-021-01949-4.
124. Paskiet D, Jenke D, Ball D, Houston C, Norwood DL, Markovic I. The Product Quality Research Institute (PQRI) leachables and extractables working group initiatives for Parenteral and Ophthalmic Drug Product (PODP). PDA J Pharm Sci Technol. 2013;67(5):430–47. https://doi.org/10.5731/pdajpst.2013.00936.
125. Quality Considerations for Topical Ophthalmic Drug Products Guidance for Industry. U.S. Department of Health and Human Services Food and Drug Administration Center for Drug Evaluation and Research (CDER). 2023. https://www.fda.gov/media/172937/download.
126. Rahman MQ, Tejwani D, Wilson JA, Butcher I, Ramaesh K. Microbial contamination of preservative free eye drops in multiple application containers. Br J Ophthalmol. 2006;90(2):139–41. https://doi.org/10.1136/bjo.2005.078386.
127. Raina H, Jindal A. Packaging of non-injectable liquid pharmaceuticals: a review. J Appl Pharm Sci. 2017;7(2):248–57.
128. Rashed NM, Memon SA, Turki SMA, Shalaby TA, El-Mogy MM. An analysis of conventional and modern packaging approaches for cut flowers: a review article. Front Plant Sci. 2024;15:1371100. https://doi.org/10.3389/fpls.2024.1371100.
129. Reape D. Evaluation and identification of leachables from the container closure system in ophthalmic product (Doctoral dissertation, Dublin City University). 2014.
130. Roche M, Lannoy D, Bourdon F, Danel C, Labalette P, Berneron C, et al. Stability of frozen 1% voriconazole eye-drops in both glass and innovative containers. Eur J Pharm Sci. 2020;141:105102. https://doi.org/10.1016/j.ejps.2019.105102.
131. Sah SK, Kaity S. Chapter 22 – pharmaceutical packaging: recent trends and challenges. In: Dosage forms, formulation developments and regulations recent and future trends in pharmaceutics, vol. 1; 2024. p. 663–84.
132. Samanta KK, Basak S, Chattopadhyay SK. Potentials of fibrous and nonfibrous materials in biodegradable packaging. In: Environmental footprints of packaging. Singapore: Springer; 2016. p. 75–113.
133. Sanchez C, Hortal M, Aliaga C, Devis A, Cloquell-Ballester VA. Recyclability assessment of nano-reinforced plastic packaging. Waste Manag. 2014;34(12):2647–55. https://doi.org/10.1016/j.wasman.2014.08.006.
134. Sandle T. Sterile ophthalmic preparations and contamination control. J GXP Compliance. 2014;18:1–5.
135. Shekhani R, Steinacher L, Swen JJ, Ingelman-Sundberg M. Evaluation of current regulation and guidelines of pharmacogenomic drug labels: opportunities for improvements. Clin Pharmacol Ther. 2020;107(5):1240–55. https://doi.org/10.1002/cpt.1720.

136. Shire DB, Gingerich MD, Wong PI, Skvarla M, Cogan SF, Chen J, Wang W, Rizzo JF. Microfabrication of components for a high-density sub-retinal visual prosthesis. Micromachines (Basel). 2020;11(10):944. https://doi.org/10.3390/mi11100944.
137. Singh G, Gollapalli R, Blinder A, Patel M. Identification of leachable impurities in an ophthalmic drug product originating from a polymer additive Irganox 1010 using mass spectroscopy. J Pharm Biomed Anal. 2018;152:197–203. https://doi.org/10.1016/j.jpba.2018.01.053.
138. Sklubalova Z, Zatloukal Z. Systematic study of factors affecting eye drop size and dosing variability. Pharmazie. 2005;60(12):917–21.
139. Sklubalova Z, Zatloukal Z. Classification of plastic eye dropper tips using Harkins and Brown's factor. Pharmazie. 2007;62(10):750–5.
140. Smith J, Larmene-Beld K. Containers. In: Practical pharmaceutics: an international guideline for the preparation, care and use of medicinal products. Cham: Springer; 2023. p. 169–206.
141. Spaniol K, Koerschgen L, Sander O, Koegler G, Geerling G. Comparison of application systems for autologous serum eye drops. Curr Eye Res. 2014;39(6):571–9. https://doi.org/10.3109/02713683.2013.855237.
142. Sudol E, Kozikowska E. Mechanical properties of polyurethane adhesive bonds in a mineral wool-based external thermal insulation composite system for timber frame buildings. Materials (Basel). 2021;14(10):2527. https://doi.org/10.3390/ma14102527, PMID 3 FDA, Pharmaceutical CGMPs for the 21st Century – a risk-based approach. Final report; FDA, September 2004. fda.gov/media/77391/download
143. Tan JCG, Webb RG, Marks DC. Serum growth factor stability in different eye drop packaging systems during storage. Transfus Apher Sci. 2020;59(1):102608. https://doi.org/10.1016/j.transci.2019.06.032.
144. Tenenbein M. Child-resistant closure: yesterday, today and tomorrow. Inj Prev. 2018;24(1):2–4. https://doi.org/10.1136/injuryprev-2017-042577.
145. Teping C, Wiedemann B. The COMOD system. A preservative-free multidose container for eye drops. Klin Monatsbl Augenheilkd. 1994;205(4):210–7. https://doi.org/10.1055/s-2008-1045518.
146. Traynor K. FDA works toward product naming, packaging guidance. Am J Health Syst Pharm. 2010;67(16):1308–10. https://doi.org/10.2146/news100057.
147. U.S. Food and Drug Administration (FDA). Guidance for industry: container and closure system integrity testing in lieu of sterility testing as a component of the stability protocol for sterile products. 2008. https://www.gmp-compliance.org/files/guidemgr/UCM146076.pdf.
148. Uddin M, Abdullah M, Kabir M, Jinnat S, Sonia Z, Yesmin B, Amran M. Quality control tests for ophthalmic pharmaceuticals: pharmacopoeial standards and specifications. J Adv Med Pharm Sci. 2017;14(2):1–17.
149. Urtti A, Rouhiainen H, Kaila T, Saano V. Controlled ocular timolol delivery: systemic absorption and intraocular pressure effects in humans. Pharm Res. 1994;11(9):1278–82. https://doi.org/10.1023/a:1018938310628.
150. US Food and Drug Administration. Requirements on content and format of labelling for human prescription drugs and biologics; requirements for prescription drug product labels; proposed rule 65. Fed Regist. 2000;247:81082–131.
151. Van Santvliet L, Ludwig A. Determinants of eye drop size. Surv Ophthalmol. 2004;49(2):197–213. https://doi.org/10.1016/j.survophthal.2003.12.009.
152. Veronin M. Packaging and labeling of pharmaceutical products obtained from the internet. J Med Internet Res. 2011;13(1):e22. https://doi.org/10.2196/jmir.1441.
153. Vijayakumar A, Vairavasundaram S, Koilraj JAS, Rajappa M, Kotecha K, Kulkarni A. Real-time visual intelligence for defect detection in pharmaceutical packaging. Sci Rep. 2024;14(1):18811. https://doi.org/10.1038/s41598-024-69701-z.
154. Vrabic BU. Hybrid materials of bio-based aerogels for sustainable packaging solutions. Gels. 2023;10(1):27. https://doi.org/10.3390/gels10010027.
155. Wagner JR. Multilayer flexible packaging. 2nd ed. Norwich: William Andrew; 2016.

156. Wang X, Huang L, Zhang C, Deng Y, Xie P, Liu L, et al. Research advances in chemical modifications of starch for hydrophobicity and its applications: A review. Carbohydr Polym. 2020;240:116292. https://doi.org/10.1016/j.carbpol.2020.116292.
157. Ward J, Buckle P, John CP. Designing packaging to support the safe use of medicines at home. Appl Ergon. 2010;41(5):682–94. https://doi.org/10.1016/j.apergo.2009.12.005.
158. Wiseman HM, Guest K, Murray VS, Volans GN. Accidental poisoning in childhood: a multicentre survey. 2. The role of packaging in accidents involving medications. Hum Toxicol. 1987;6(4):303–14. https://doi.org/10.1177/096032718700600407.
159. Xu L, Wang X, Wu M. Topical medication instillation techniques for glaucoma. Cochrane Database Syst Rev. 2017;2(2):CD010520. https://doi.org/10.1002/14651858.CD010520.pub2.
160. Zaki M, Pardo J, Carracedo G. A review of international medical device regulations: contact lenses and lens care solutions. Cont Lens Anterior Eye. 2019;42(2):136–46. https://doi.org/10.1016/j.clae.2018.11.001.
161. Zhang L, Zhang AN, He SM, Zheng GQ, Zeng FR, Wang YZ, Liu BW, Zhao HB. Biomimetic nanoporous transparent universal fire-resistant coatings. ACS Appl Mater Interfaces. 2024;16(15):19519–28. https://doi.org/10.1021/acsami.4c00570.
162. Saurabh M. Nikam, Madhuri D. Davange, Dhananjay M Patil. A Review on Advancements in The Packaging of Medicines. Int J in Pharm Sci. 2023;1(12):416–432. https://doi.org/10.5281/zenodo.10389787.

Regulatory Perspectives in Ophthalmic Products

13

Vivek Prabhakar Rajkule
and Chandrakantsing Vijaysing Pardeshi

Abstract

The eye is a highly delicate organ of the human body having various barriers that protect the eye from foreign particles. Ophthalmic drug products are preparations that deliver a drug either as drops, ointments, or injections on the eye, into the eye, or onto the conjunctiva and are used to treat various eye conditions. Strict regulatory considerations need to be followed by the industries involved in the manufacturing of ophthalmic drug products to bring a quality product with assured patient safety and therapeutic efficacy. The present chapter highlights the key regulatory aspects relevant to ophthalmic drug products, either as simple drug preparations or as drug-device combination products. In the end, a brief success story of the recent VIUTY™ eye drop is given for readers' discretion

Keywords

Eye · Ocular · Ophthalmic · Regulatory · Extractables · Leachables · Food and drug administration

13.1 Introduction

Fabrication of ocular products intended for administration directly into the eye or onto the eye surface for the management of a variety of ocular disease conditions is currently an area of choice for researchers with a focus on novel drug delivery

V. P. Rajkule · C. V. Pardeshi (✉)
Industrial Pharmacy Laboratory, Department of Pharmaceutics, R. C. Patel Institute of Pharmaceutical Education and Research, Shirpur, India
e-mail: chandrakant.pardeshi@rcpatelpharmacy.co.in

N. K. Mehra (ed.), *Complex Ophthalmic Dosage Forms: Advances in Biomedical Applications and Future Perspectives*, https://doi.org/10.1007/978-981-96-6306-4_13

systems [1]. Ocular products are generally sterile formulations containing one or more active pharmaceutical ingredients (API), applied either topically (as ointments), instilled into subconjunctival region (as solution or suspension), or given parenterally (intravitreal or intracameral). The pH, tonicity, and buffer capacity need to be carefully controlled in the ophthalmic products. In general, anesthetics, antioxidants, antivirals, miotics, and lubricants may be incorporated into the ocular products [2].

When considering regulatory perspectives on ophthalmic products, the key focus is on ensuring the safety and efficacy of eye drops, ointments, or other ocular medications by strictly regulating factors like sterility, particulate matter, pH balance, tonicity, preservatives, and packaging. Appropriate attention must be paid to the unique physiological traits of the eye and potential for serious complications, if not properly formulated and administered; this often involves stringent quality control testing and specific considerations for drug-device combination products [3].

The fabrication of ophthalmic products requires stringent compliance with the sterility requirements to safeguard the patients' safety. Sterility is much essential requirement for the manufacturing of these ophthalmic products, as the presence of contamination may lead to ocular damage. It is essential for industries to follow the current Good Manufacturing Practice (cGMP) guidelines so that the sterility of the ophthalmic formulation is ensured. Control of foreign particulate matters, testing preservative effectiveness, sterility tests, and leaching tests are a few of the special considerations that need to be controlled while formulating the ophthalmic formulations [4, 5].

The regulatory landscape for ophthalmic products is complex and vital for ensuring the product's quality, patient's safety, and therapeutic efficacy. The US Food and Drug Administration (FDA) primarily oversees the regulations relevant to the ophthalmic products throughout the ophthalmic product lifecycle, from development to post-marketing surveillance.

The present chapter focuses on the key regulatory frameworks and processes governing the approval of ophthalmic products, including drug-device combination products. The chapter is entailed with the details of the microbiological and particulate matters considerations, testing of extractables and leachables, drug release testing, and stability testing as a part of regulations on the ophthalmic drug products or drug-device combination products for ophthalmic use.

13.2 New Drug Development Process: An Introduction

Pharmaceutical industries engaged in the development and manufacturing of ophthalmic products need to seek prior approval from the USFDA or other suitable regulatory bodies before investigating new drugs into human subjects or bringing them into the commercial market. The development process starts with the application from the pharmaceutical industries to the FDA seeking approval of ophthalmic drugs, with the primary focus being the demonstration of the safety and efficacy of the ophthalmic products being developed. Application and review during the product approval is a multistage process involving both the manufacturer and the FDA, with inputs obtained from the experts in clinical studies design, biostatistics, microbiology, and pharmacology or toxicology [6]. Several drugs undergo stringent screening through a variety of assays and tests during the preclinical investigations in the drug discovery stage of the drug development process. A lead molecule is picked for further therapeutic and toxicity investigations in suitable preclinical models. Further, investigations of the safety and efficacy of the lead candidate are carried out at the clinical level using different phases of the studies, as per the defined regulations. Clinical trials are categorized into four phases, viz. phase I to phase IV, with each phase enduring from a few months to several years. Upon successful completion of each phase of the trial, necessary approvals are obtained from the appropriate regulatory agency for the advancement to the next phase. An investigational new drug (IND) is approved for marketing following a satisfactory completion and approval of phase I to phase III. The phase IV investigations are usually carried out after a drug is approved, for post-marketing surveillance, by the regulatory body. Recently, phase 0 (human micro-dosing) studies, also known as *exploratory IND studies*, have been initiated to quickly detect safety concerns [7]. Usually sponsors (who develop new drugs) submit the IND application to the FDA to seek approval to initiate the phase I clinical trials. IND application describes the data of nonclinical studies and a detailed protocol for the clinical studies. If the FDA review committee feels confident about the safety of the proposed drugs in clinical settings, the sponsors can begin their clinical trials. Once the sponsor completes all the studies essential to support the approval of their newly developed drug for marketing, they submit either NDA (new drug application for new small molecules) or BLA (biological license application for new biological molecules). The indication, dose regiment, and potential side effects of the drug are mentioned on the package insert and are referred to as *labeling*. Labeling also contains data from genotoxicity, reproductive toxicity, and carcinogenicity studies [7]. An expert team at the FDA, CDER (Center for Drug Evaluation and Research), and CBER (Center for Biologics Evaluation and Research) evaluates both the clinical and nonclinical data presented by the sponsor for review.

Table 13.1 illustrates the different phases of clinical trials with their objectives and sample size.

FDA's guidelines also provide sponsors an opportunity to meet FDA personnel during the entire drug development process [8]. The meetings of the FDA are categorized as Type A, Type B, and Type C. The sponsor must submit a formal request for a meeting with the FDA. If the sponsor is granted with meeting protocol, the sponsor has to present the IND enabling plan (nonclinical study data supporting the safety of projected API in clinical studies) during the pre-IND meeting while during the end-of-phase II meeting, the sponsor has to present data on the already completed nonclinical studies and a plan to support phase III clinical trials. Finally, during the pre-NDA/BLA meeting, the sponsor has to confirm to the FDA that there is no gap in their nonclinical programs [7].

Table 13.1 Different phases of clinical trials

Phase	Objectives	Dose	Approximate sample size
0	Half-life and bioavailability of drugs	Subtherapeutic	~10 healthy subjects
I	Assessing the safety and likely therapeutic dose of drug on healthy human volunteers	Often subtherapeutic, ascending doses	15–20 healthy subjects
II	Testing the drugs on patients to examine the safety and efficacy	Therapeutic doses	Up to 300 patients
III	Testing the drugs on patients to assess the safety, efficacy, and effectiveness	Therapeutic doses	>300 patients
IV	Drug monitoring following approval, post-marketing surveillance	Therapeutic doses	Anyone seeking treatment from doctor

13.3 Regulation of Ophthalmic Products: Ophthalmic Drugs and Devices

The FDA holds rights for the regulation of all classes of drugs and drug-device combination products (which are used to deliver ophthalmic drugs) in the United States. Figure 13.1 illustrates the centers, offices, and divisions of the FDA which are involved in regulating ophthalmic products.

The FDA's guidance for Industries[1] discusses the quality considerations relevant to topical ophthalmic drug products (ointment, cream, gel, and liquid formulations, viz. solution, suspension, and emulsion) which are explicitly administered topically into or around the eye [9].

These guidelines specifically discuss the following considerations:

- Microbiological considerations
- Evaluation of particulate matters, extractables and leachables, and impurities and degradation products
- Drug release and/or dissolution tests, in vitro
- Recommendations on container-closure systems (CCSs) considering their design, delivery, and dispensing characteristics
- Recommendations on stability studies

[1] *Consistent with the current good manufacturing practice (cGMP) requirements outlined in section 501(a)(2)(B) of the Federal Food, Drug, and Cosmetic Act (FD&C Act) and 21 CFR parts 210 and 211 for all drug products, part 601 for biological products, and part 4 for combination products.*

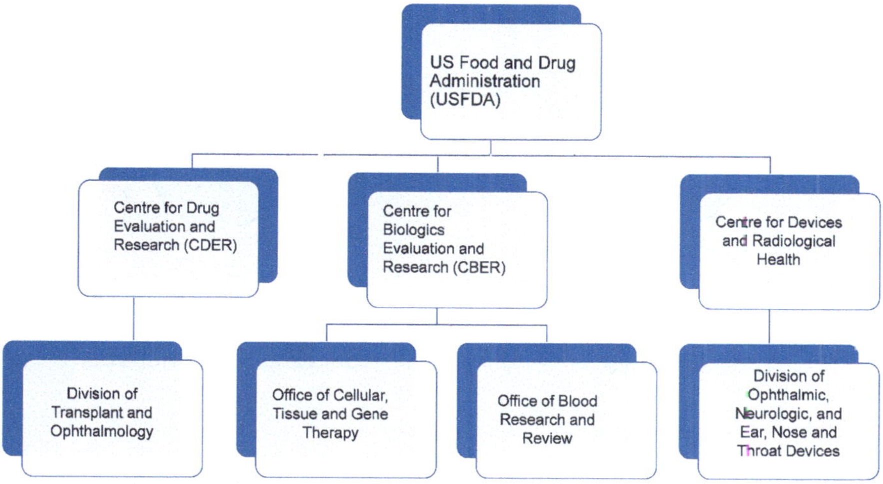

Fig. 13.1 The centers, offices, and divisions of the FDA, which are involved in the regulation of ophthalmic products (Modified from ref. [7])

13.4 Microbiological Considerations

13.4.1 Ophthalmic Product Sterility

A major critical quality attribute (CQA) with prime importance to ophthalmic products is the product's sterility. The formulation considerations of ophthalmic preparations include a mandatory requirement of stringent observance of the sterility of ophthalmic products to make sure the patients' safety during ophthalmic therapy. Maintaining sterility is a mandatory requirement for the manufacturers of ophthalmic formulations, as contaminated eye products may damage the eyesight. The current Good Manufacturing Practice (cGMP) guidelines must be complied with by the ophthalmic product producers to confirm the product sterility. If ophthalmic preparations are not appropriately sterilized, they may pose severe consequences when used clinically. Ophthalmic formulations are directly applied to the eye tissues, including the conjunctiva or eyelid, which are susceptible to infections due to mucous membranes. Foreign particles from the ophthalmic formulations can lead to various infections, viz. conjunctivitis, as shown in Fig. 13.2. Long-term use can lead to loss of sight and other critical complications [10].

Key Regulatory Considerations for Ophthalmic Product Sterility

- **Strict Requirements of the Ophthalmic Product Sterilization**

All the ophthalmic preparations must be appropriately sterilized using well-validated techniques, viz. autoclaving, membrane filtration, or radiation

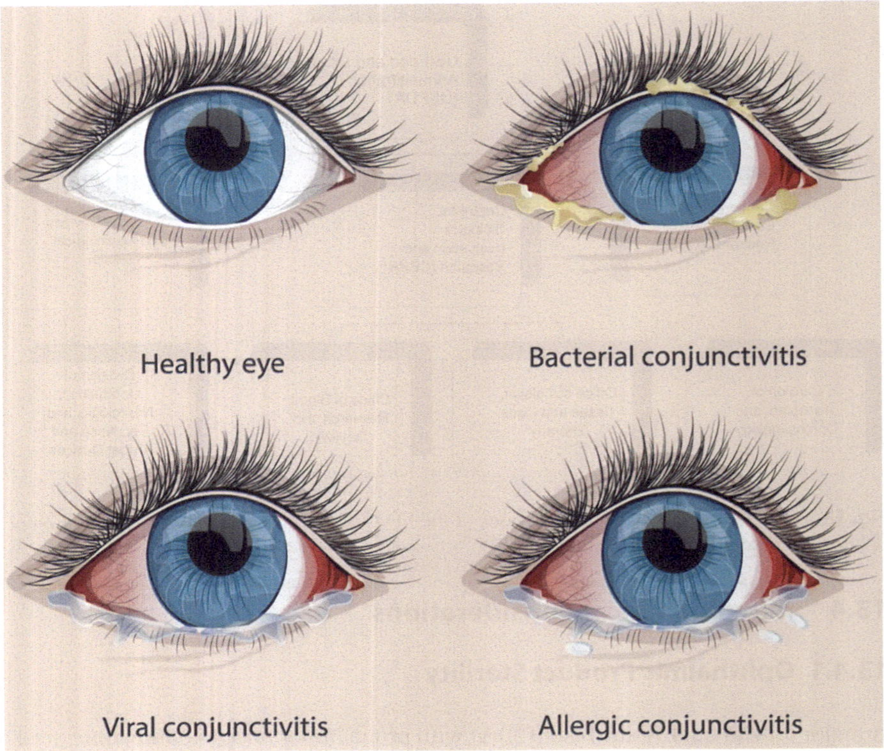

Fig. 13.2 Comparison of healthy eyes and conjunctivitis following sterility concerns. (Reproduced with copyright permission from ref. [10])

sterilization. The selection of appropriate sterilization technique depends on the product stability while ensuring complete microbial eradication.

- **Aseptic Manufacturing Practices**

The facility of ophthalmic product manufacturing must comply with the stringent aseptic guidelines to prevent product contamination during manufacturing. The major aspects should involve the proper training of the personnel, the design of the aseptic cleanroom area, and periodic monitoring of the environmental parameters.

- **Sterility Testing**

A regular examination of the finished ophthalmic products for sterility is highly mandatory to validate the absence of viable microorganisms. This should be normally performed by utilizing well-validated methods, viz. membrane filtration, incubation, etc.

- **Packaging and Container Design**

The containers used for ophthalmic products must be designed in a way to maintain sterility. It should prove the tamper-proof sealing and proper closure systems to avoid product contamination while the product is used by the end user.

- **Documentation and Quality Control**

A full documentation of the data concerning the manufacturing of the ophthalmic preparations, records of the sterilization, data related to environmental monitoring, and sterility testing is essential and should be done appropriately for compliance with the regulatory guidelines outlined by the FDA (in the United States) and other such equivalent agencies globally.

Failure to comply with these requirements will cause affected products to be deemed adulterated under section 501(a)(2)(B) of the Federal Food, Drug, and Cosmetic Act (FD & C Act). For recommendations on how to meet cGMP requirements for product sterility, see guidance for industry *Sterile Drug Products Produced by Aseptic Processing—Current Good Manufacturing Practice* (September 2004).

13.4.2 Multidose Ophthalmic Drug Products

The ocular products for multidose drug administration should be appropriately designed and controlled to avoid infective contamination during the product shelf life, while the product is in use by the patient. In addition, these should be supported by the stability data. In case of single-dose ophthalmic products, the CCSs are designed in such a way that prevents the hazards associated with in-use contamination and the growth of microorganisms between the doses. In the case of multidose ophthalmic products, the in-use contamination and growth of microorganisms can occur whenever the multidose CCSs are opened (multiple times) over the course of their shelf life. Liquid ophthalmic formulations intended for multiple dosing should comprise appropriate preservative(s) to preserve the product during multiple-dose instillations and should also minimize the risk of possible damage occurring due to incidental product contamination while the product is being used by the patient. While designing multidose ophthalmic products, careful selection of preservatives is a very crucial stage that inhibits the growth of harmful microorganisms without causing ocular irritation or other such ocular complications [9].

The selection of an appropriate preservative or a combination of preservatives is a very crucial phase in the manufacturing of multidose ophthalmic products. Preservatives protect the ophthalmic drug product from potential microbial ingress following multiple uses of the product by the patients. Such microbial ingress could occur, for example:

(a) If the multidose drug product is contaminated by the surrounding air (possibly containing microorganisms) following administration

(b) If the tip of a dropper is contaminated by a nonsterile surface (i.e., the fluid path is contaminated)
(c) If a contaminated drop of liquid formulation returns to the product reservoir

Regardless of whether a multidose drug product possesses inherent antimicrobial activity or contains one or more added preservatives, manufacturers should implement a well-designed and rigorous antimicrobial effectiveness testing program that covers the product's shelf life.

Silver sulfate or silver-containing substances are strictly prohibited by the FDA for use as preservatives in multidose ophthalmic products owing to important safety issues, which may lead to argyria (a condition characterized by the permanent discoloration of the eyes) and granular deposits of silver in the conjunctiva and cornea [9, 11].

Some manufacturers of ophthalmic products pursued to use of a preservative-free formulation for a multidose liquid drug product integrated with an appropriately designed CCSs system. The objective being elimination of the potential in-use microbial contamination. These formulations and associated presentations should provide guaranteed protection for each of the unit doses instilled into the eye to prevent ocular damage arising from exposure to incidental contaminations during the multiple uses of the product. There are several ways when such products might be unsuccessful in inhibiting microbial contamination. An ophthalmic drug product is susceptible to microbial attack and may harm the patient severely if it does not contain sufficient concentration of appropriate preservatives and is exposed to in-use contamination. CCSs should be able to avoid contamination due to foreseeable external factors during storage and while the product is in-use, which can cause deterioration [9, 11].

Key Regulatory Considerations for Ophthalmic Preservatives [12]

- **Efficacy Against a Wide Range of Microbes**

The preservatives used in multidose ophthalmic products must be able to effectively inhibit the growth of a variety of microbes, viz. viruses, bacteria, fungi, etc., that can contaminate the product during in-use and can pose potential damage to the eyes following topical instillation.

- **Ocular Safety**

The preservatives used in multidose ophthalmic products should not pose any cytotoxic effects on the corneal and conjunctival surface upon ocular instillation. Also, they should elicit minimal ocular irritation and inflammatory effects.

- **Compatibility with Other Ingredients**

The preservatives in the ophthalmic formulations must be compatible, both chemically and physically, with the ocular therapeutics and other formulation components. The instability or precipitation problems must be considered while selecting the preservatives for multidose ophthalmic products.

- **pH Stability**

A suitable mix of preservatives is generally used to maintain their antimicrobial potential over a desired pH range of the multidose ophthalmic solution.

- **Preservative Concentration**

The appropriate concentration of the preservative is very essential with respect to fabrication of stable ophthalmic preparation and should be carefully optimized, during preformulation, to provide adequate protection against a wide range of microbial contamination while minimizing ocular toxicity.

- **Packaging Considerations**

The ideal container-closure combination and the respective material of fabrication should be appropriately selected to minimize the chances of possible contaminations. Also, it should be ensured that the chosen preservative must remain effective through the entire shelf life of the multidose ophthalmic preparation.

The FDA provides guidelines for the development and testing of ophthalmic products, including specific requirements for preservatives used in multidose formulations. Also, European Pharmacopoeia sets standards for the quality and testing of ophthalmic products, including preservatives.

13.5 Particulate Matters

A vigorous visual inspection system and execution of cGMP guidelines is essential to make sure the regulatory compliance of the manufactured ophthalmic preparations. Topical ophthalmic preparations, which are packaged in opaque containers, are appropriately examined, both qualitatively and quantitatively, for the presence of particulate matters using X-ray spectroscopy or destructive testing. This testing also ensures that the particulate matters are present within a range acceptable for clinical administration.

Ophthalmic products and non-compendial ophthalmic drug products which are official in United States Pharmacopoeia (USP) are needed to see the particulate matter requisites in the USP General Chapter <771> titled *Ophthalmic Products—Quality Tests*. Adherence to compendial standards can assist applicants and manufacturers in complying with cGMP regulations (e.g., 21 CFR 211.165(e), 211.167(b), and 211.194(a)(2)) [9].

13.6 Extractable and Leachables

Ophthalmic drug products (specifically ophthalmic suspensions and solutions) have been classified into a high-risk class regarding safety concerns over extractables and leachables (E&L). Despite this categorization, literature or even regulatory bodies lack definite published guidance on the management of E&L in ophthalmic drug products [13, 14].

Extractables are the chemical compounds that can be released or extracted from a container or packaging materials (viz. additives, stabilizer, or impurities added during manufacturing) when it is intentionally exposed to certain extraction conditions. Extractables are generally tested under exaggerated or aggressive conditions to identify and quantify the potential substances that could migrate into the ophthalmic product [14, 15].

Leachables are the chemical compounds that can transfer from a container or packaging materials (viz. additives, stabilizer, or impurities added during manufacturing) into the ophthalmic product during storage or use. Leachables are typically estimated by subjecting the container or packaging material to actual realistic or accelerated conditions that simulate the intended storage or use of the product. These conditions may include factors such as temperature, light exposure, humidity, and actual contact with the ophthalmic product preparation. Leachables are of highly serious concern as they have potential effects on the quality, safety, and efficacy of the ophthalmic product [14, 15] (Fig. 13.3).

The evaluation of the extractables and leachables should judge the primary, secondary, and tertiary packaging elements of the CCSs, including labeling components. Low molecular weight compounds (viz. plasticizers, lubricants, pigments, stabilizers, antioxidants, binding agents) can leach from the CCS components or from labeling components (viz. inks, adhesives, varnishes) of the semipermeable

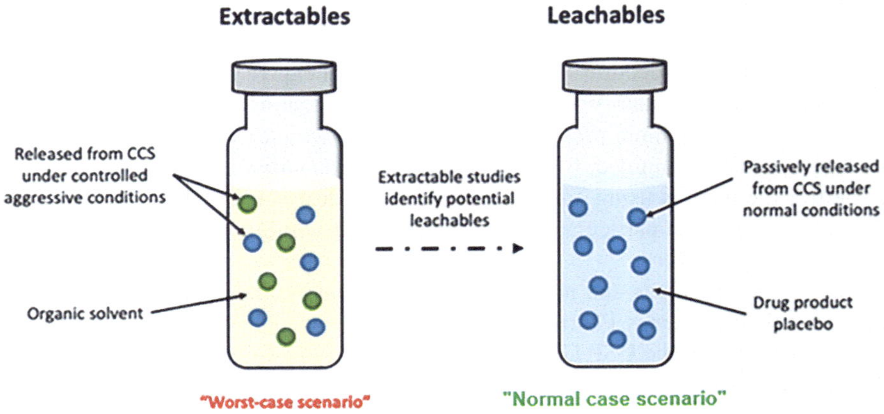

Fig. 13.3 Diagram illustrating the difference between extractables and leachables. (Reproduced with copyright permission from ref. [15])

CCSs into the product. However, it is not a big concern for the ocular products packaged in glass containers (viz. biologicals). The applicants and manufacturers should refer to USP General Chapters <1663> *Assessment of Extractables Associated with Pharmaceutical Packaging/Delivery Systems* and < 1664> *Assessment of Drug Product Leachables Associated with Pharmaceutical Packaging/Delivery Systems* for comprehensive details on testing of extractables and leachables [9].

Product Quality Research Institute Leachables and Extractables Working Group for Parenteral and Ophthalmic Drug Products (PQRI-PODP) was established to extrapolate the PQRI-Orally Inhaled and Nasal Drug Products (OINDP) E&L threshold concepts and best practices to PODP based on the following three-point hypotheses [16]:

1. Threshold concepts established for safety qualification of leachables in OINDP and the existing FDA/EMA guidance are extrapolated to evaluation and safety qualification of packaging systems (such as CCSs) for PODP.
2. Good science procedures elaborated for OINDP pharmaceutical development process are extrapolated to the packaging systems for PODP.
3. The above threshold and good science practice concepts are integrated into a broad process for characterizing the packaging systems with respect to leachables and the associated impression on the safety of PODP.

13.6.1 Extractables Testing

Manufacturers of ophthalmic drug products should document the following information along with their application when they conduct the extractables testing of their fabricated ophthalmic preparation to comply with the cGMP requirements [9, 16]:

- A risk assessment in support of their study approach
- Data obtained from the extractable studies conducted according to the guidance mentioned in USP General Chapter <1663> and should consider the primary, secondary, and tertiary packaging elements
- Information on the use of extraction conditions (viz. media, temperature, time, analytical techniques)
- Information on the use of analytical procedures (viz. gas or liquid chromatography–mass spectrometry), including method validation information.
- An assessment of the resultant extractables profiles

13.6.2 Leachables Testing

Applicants and manufacturers of ophthalmic drug products should conduct the identification and characterization of the possible hazards accompanying leachables from the CCSs or the formulation, as leachables can be obtained from various sources and are dependent on the formulation. It is also expected that the manufacturers should conduct the leachables studies and explain how these hazards can be alleviated.

Manufacturers of ophthalmic drug products should document the following information along with their application when they conduct the leachables testing of their fabricated ophthalmic preparation to comply with the cGMP requirements [9, 16]:

- Data obtained from three preliminary stability samples, as per the guidelines described in USP General Chapter <1664>
- Information on usage of analytical techniques, viz. gas or liquid chromatography–mass spectrometry, including method validation information
- An assessment of the obtained leachables reports
- The acceptance criteria contained in drug product specifications

A separate toxicological risk assessment of leachables may be asked by the regulatory agencies.

13.7 Drug Release/Dissolution Testing for Quality Control of Ophthalmic Products

The rate at which and the extent to which a drug is released from the ophthalmic products are often used as quality criteria to study the impact of formulation- and process-related variables that are controlled with the objective to confirm the reliable product quality [17]. One such methodology that the applicants or manufacturers can think of as a part of their quality control policy for specific ophthalmic products (viz. emulsions, suspensions, semisolids) is the application of drug release/dissolution studies, in vitro. Other methodologies can also be used, viz. use of CQAs, which are relevant to the formulation and/or process variables. It should be justifiable by the applicant how the quality control approach ensures the steady product quality [9].

13.8 CCS Design, Delivery, and Dispensing Characteristics

The recommendations of the FDA or other respective regulatory bodies regarding the design elements and delivery and dispensing characteristics for the CCSs of the ophthalmic drug products should be considered by the applicants and product manufacturers. A CCS holding or containing an ophthalmic drug is considered a combination product when that CCS or its constituent part delivers the drug contained within it. The cGMP requirements under 21 CFR part 4, subpart A are generally utilized for regulating the combination products.

13.8.1 CCS Design [18, 19]

- *Tamper-Proof Packing*

Containers of the ophthalmic drug products containing ophthalmic formulation need to be sterilized during filling and sealing and remain sterile until it is opened. Also, such containers are designed in such a way that they are sealed to prevent the product use until its seal is destructed or broken. Additionally, ophthalmic drug products that are over-the-counter (OTC, drugs that can be bought without a prescription) drugs must be in compliance with the tamper-proof packaging requirements of 21 CFR 211.132. If any CCS contains a non-retaining tamper-proof ring (viz. collar-type or band-type) while sealing the bottle and the cap, then a special precaution is needed so that the ring should not cut off from the container while in use, which may lead to an eye injury. An OTC drug product with a tamper-proof ring should also contain a positive-retention mechanism to avoid the ring detachment while it is being used by the patient [20].

- *Tips*

For CCSs designs where the tip is sealed until it is opened, multistep practices are strongly prevented, as there is a possibility that a patient may touch the tip and contaminate it while trying to uncap it. The FDA suggests using a single-step procedure involving simple directions and twisting the cap without removing it so that the tip remains untouched and product sterility is maintained throughout its use by the patient.

- *Torque Requirements*

The applicants and manufacturers of the ophthalmic drug products should keep in consideration about the specific requirements of torque specifications for the CCSs. This is because some patients may face difficulties while twisting off the cap of CCSs and patients may put in some more effort to open the cap. The FDA recommends low torque requirements so that elderly people can open the caps without any difficulty, but torque should be high enough so that caps remain in place during the fabrication, transport, storage, and handling of the product [9, 21].

- *Color Coding*

Providing a color code to the caps of ophthalmic products can be used as an efficient way of describing the therapeutic category of that product. The FDA recommends using a standardized color-code system as described in the American Academy of Ophthalmology's *Color Codes for Topical Ocular Medications* policy statement [9, 19, 22]. It is also recommended to follow the guidance for industry *Safety Considerations for Container Labels and Carton Labeling Design to Minimize Medication Errors* (May 2022) [23].

13.8.2 Delivery and Dispensing Characteristics

- ***Unit Dose Containers***

It is recommended by the FDA that for all topical ophthalmic drug products containing solutions, suspensions, or emulsions, the maximum fill volume of a unit dose (without preservative) container should not exceed 0.5 mL. As per FDA recommendations, the maximum fill should be not more than 1.0 *g* for unit dose ointment or gel. Also, the unit dose containers should not be able to be recapped.

- ***Multidose Containers***

(a) Drop Size or Volume

The FDA recommends that for all topical ophthalmic drug products, the drop size for multidose containers must be between 20 and 70 microliters. It is also recommended that applicants should accomplish a one-time drop volume or drop weight study to decide on drop size during the drug delivery or dispensing of the generic ophthalmic formulation presented for approval under the ANDA application. The difference in drop size should be within ±10% for the generic product and the reference listed drug (RLD) product. Also, the drop size should lie within the recommended size of 20 to 70 microliters. For any deviations from the RLD, the ANDA applicant should provide a justification to demonstrate that there will be a similar number of delivered doses as the RLD. The ANDA application should contain information on the estimation of drop volume/drop weight and various testing specifications, viz. number of drops in the container and its holding angle during the dose delivery [9, 24].

(b) Dose Uniformity of Suspended Drug Product

All ophthalmic suspension-based drug products should undergo a resuspendability or redispersibility test as per the guidelines mentioned in USP General Chapter <771> *Ophthalmic Products—Quality Tests.*

Additionally, for multiple-dose containers, suitable information on one-time dose-uniformity study (from top, middle, and bottom of the container) should be provided by the manufacturers. This data should be obtained from at least three pilot or exhibit samples to prove the uniform drug distribution. Also, the ophthalmic product is expected to deliver the same labeled dose throughout its shelf life. Alternatively, applicants or manufacturers may also provide the data obtained from the development batches (investigational new drug batches) that signify the preparation ready-to-be-marketed to validate the dose uniformity of the suspended drug product [9, 25].

13.9 Stability Considerations

The manufacturers of ophthalmic drug products must develop an appropriate protocol to assess the stability of the incorporated drug or drug suspension. The results of the stability studies should be used to determine the suitable storage conditions and expiration dates (21 CFR 211.166) of the ophthalmic products. While developing a stability protocol, the manufacturer should consider the following stability recommendations [9, 26].

13.9.1 Container Orientation During Storage

The storage of the ophthalmic drug product container with varied orientations can affect the product stability significantly. Prior to conducting the initial stability studies, the NDA applicants, or manufacturers should conduct a preliminary product development study to examine the product's storage in different container orientations: (i) an upright position and (ii) either an inverted or horizontal position. After completion, the data obtained from this primary development study should be utilized to note and characterize the variations in quality attributes of the finished product and to determine the worst-case orientation of the product. The applicant should utilize this worst-case orientation while performing stability studies using samples that correspond to the commercial manufacturing procedure [9, 27].

The ophthalmic products submitted under a BLA application, in actual, are not dependent on the preliminary developmental studies to determine the storage requirements during stability. On the contrary, these products depend on primary stability studies, often involving process validation samples, to establish storage in a real-time environment. If the interactions between biological ophthalmic drug product and the CCSs (other than sealed ampules) cannot be excluded, the applicants should place the samples for stability testing in an upright position and in either inverted or horizontal position (i.e., all the CCSs surfaces should be in contact with the product) to precisely estimate the effects of all product-contact CCSs elements on the quality of the developed ophthalmic product [9].

The ophthalmic products submitted for approval under an ANDA application require that the applicants should put the samples of the initial stability studies in an upright position and either an inverted or horizontal position. Then, the data from both orientations should be incorporated for regulatory review in the original submission. The manufacturers should then determine the worst-case orientation by comparing the data from both orientations, which should be utilized further to align the use of that orientation for usual stability tests following ANDA approval [9].

Manufacturers must possess a stability studies protocol in written format, which includes the storage conditions for samples taken for testing (21 CFR 211.166(a)(2)) and should generally follow similar principles to determine the worst-case orientation for stability studies [9].

13.9.2 Water Loss

The applicants and manufacturers of ophthalmic products packaged in semipermeable CCSs must perform a water loss test to estimate the moisture transmission properties of the CCSs as well as the protection capabilities of the secondary packaging used, if any. The water loss tests are performed in accordance with the cGMP guidelines. The manufacturers should document and incorporate in their application the information on test conditions and acceptance criteria used (21 CFR 211.194(a)) [9, 28].

13.9.3 Freeze-Thaw Testing for Emulsions and Suspensions

The applicants and manufacturers should accomplish a one-time freeze-thaw testing on their ophthalmic drug products containing drug-loaded emulsion or suspension, to examine the effects of temperature variations (that may be encountered during transportation and handling), which could affect the products' quality as well as performance. This should be done in accordance with the guidance for industry titled *Drug Stability Guidelines* (December 2008) published by the Center for Veterinary Medicine, but the FDA recommends that its thermal cycling study recommendations shall also be applied to the drugs intended for human use [29].

The FDA advises that such a study should entail three cycles for a duration of a minimum of three days and with the temperature cycling between the freezing (−20 °C to 0 °C) and ambient (25 °C to 35 °C) temperatures. The studied samples should be analyzed for various quality attributes and compared with the control drug product at the end of a scheduled number of cycles as well as throughout the entire study period. Any alterations in the test conditions or durations must be properly justified by the applicants [9].

13.9.4 In-Use Stability Studies

The in-use stability tests are utilized to estimate the expiry dates and validate the label claims for appropriate storage conditions that may change after opening, such as a change in temperature or light exposure (see 21 CFR 211.166, 21 CFR 211.137(b)). Manufacturers should include such information on in-use stability tests in their documentation, and applicants should present such data in their applications to the corresponding regulatory body [9].

Under 21 CFR 211.137(h), the OTC drugs are exempted from the expiry date and labeling requirements, as they do not bear any dose limitations and are stable for at least 3 years. Accelerated stability study protocols can correctly establish stability for the purposes of meeting these requirements [9].

13.10 Regulatory Considerations of Complex Ophthalmic Products

The foremost challenge in translating the ophthalmic drug product from the laboratory to the commercial market is through the technology and scaleup [30]. The biggest challenge before ophthalmic researchers and manufacturers is delivering the complex ophthalmic product from the laboratory to the end user (patient) following successful laboratory experimentations, pilot scaling, technology scaling, and finally commercialization. The highly crucial factor that needs to be considered during the scaling-up process of these complex ophthalmic drug products is the confirming robustness and reproducibility. The design, selection, and evaluation of active pharmaceutical ingredients (APIs), polymers, excipients, CQAs, and CMAs of the complex ocular products should be done in accordance with the quality by design (QbD) and quality risk management (QRM) tools and should be based on the appropriate ICH guidelines [31].

A new molecule or delivery system for ocular administration must be followed by a precise methodology for successful clinical translation. Regulatory bodies are highly cautious about the regulations and recommendations on ophthalmic products, as these are to be used on the very delicate organ of the human body, i.e., the eyes. USFDA federal law recommends that the applicants and manufacturers of complex ophthalmic drug products must fulfill an additional requirement of 21 CFR 211.167, which controls foreign, harsh, or abrasive particles in the ophthalmic preparations [32]. As per the 21 CFR 200.50 guidance, ophthalmic products and devices need to be sterile, pyrogen-free, isotonic, and biocompatible. In addition, the complex multi-dose ophthalmic products must contain preservatives to get rid of microbial infection [33].

It is very difficult for regulatory organizations to decide whether an ophthalmic delivery system should be considered as a drug or a device. Contact lenses are most often used as ocular drug delivery systems. If these lenses are intended as a drug delivery platform, they are considered as a drug, while if the same lenses are intended for drug delivery with refractive corrections, they are considered as drug-device combination product. For example: Retisert® (NDA 21737) and Vitrasert® (NDA 20569) were treated as drugs, while Visudyne® (NDA 21–119) was treated as a drug-device combination ophthalmic product because it needs a laser light source to elicit its action [34].

13.11 Combination Product Implications for Ophthalmic Drug Development

The approval procedure for the drug-device combination product is highly complicated as compared to the drug product only, as the combination product needs to meet the regulatory requirements under both the heads simultaneously.

At the initial stage, it is checked whether the drug-device combination product is a sole product with several components, or if several products are packed together in a single package. Sometimes, it is also determined whether several products are packed separately but they are intended for use altogether. Eye drops, a prefilled drug delivery system in a bottle/dropper—fall in the first class of products. Although an IND application is normally adequate for a combination product, detailed information on the whole combination product must be included. Also, the details of the drug in IND and details of the device in IDE (Investigational Device Exemption) applications should be submitted. Eye drops which were earlier considered as drugs but now coming under drug-device combination products must follow the regulatory expectations posed by pharma QMS and cGMP guidelines, for both drug and bottle/dropper systems [35, 36].

13.12 VUITY™: A Success Story

Submission of new ophthalmic drug products in compliance with drug–device combination regulations can lead to the success of advanced ophthalmic therapies for applicants or manufacturers. Recently in October 2021, VUITY™ (an ophthalmic solution pilocarpine HCl) was approved by the FDA to AbbVie Inc. who invented this new once-daily treatment for presbyopia (a condition characterized by blurry near vision), which mostly affects the adult population over the age of 40 years [37].

VUITY™ is an established eyecare therapeutic containing pilocarpine, delivered with the help of proprietary pHast™ technology. This allows VUITY™ to adjust quickly to the physiologic pH of the tear film. VUITY™ improvise the near vision without affecting distance vision by utilizing the eye's own ability to shrink the pupil size. Positive data from the two pivotal phase III clinical trials brought the FDA's approval for VUITY™ [37].

13.13 Conclusion

Developing ophthalmic drug products and getting them approved by regulatory agencies is a highly challenging task for manufacturers. The restrictions posed by the simple ophthalmic products, viz. less bioavailability, less permeability into the deep tissues, and frequent dose requirements, can be overcome either by formulating the complex ophthalmic drug products or by using a drug-device combination product. The development of complex ocular drug products and drug-device combination products holds great promise for the efficient management of ocular diseases, alarming industries, and academia for more fruitful outcomes in this area. In the future, these complex ocular products and devices are anticipated to bring a sound impact for ophthalmologists, provided they must fulfill the requirements and recommendations levied by the various regulatory authorities.

References

1. Weir AB, Wilson SD. Nonclinical regulatory aspects for ophthalmic drugs. In: Assessing ocular toxicology in laboratory animals. Totowa, NJ: Humana Press; 2012. p. 259–94.
2. Baranowski P, Karolewicz B, et al. Ophthalmic drug dosage forms: characterization and research methods. Sci World J. 2014:1–15. https://doi.org/10.1155/2014/861904.
3. Available from https://www.abbviecontractmfg.com/. Accessed on 13 Nov 2025.
4. Raghavan RR. Regulatory considerations for the manufacture and quality controls for sterile products. In: Parenteral medications. 4th ed. CRC Press; 2019. p. 1065–78.
5. Kaufman B, Novack GD. Compliance issues in manufacturing of drugs. Ocul Surf. 2003;1(2):80–5.
6. Code of Federal Regulations Title 21: Part 312, Section 312.21; http://www.accessdata.fda.gov/scripts/cdrh/cfdocs/cfcfr/CFRSearch.cfm?fr=312.21
7. Friedman LM, et al. Introduction to clinical trials. In: Fundamentals of clinical trials. 5th ed. Springer; 2015. p. 1–24.
8. Formal Meetings Between the FDA and Sponsors or Applicants of PDUFA Products. Available online at: https://www.fda.gov/regulatory-information/search-fda-guidance-documents/formal-meetings-between-fda-and-sponsors-or-applicants-pdufa-products. Accessed on 16 Dec 2024.
9. Quality Considerations for Topical Ophthalmic Drug Products (December 2023). Available online from: https://www.fda.gov/regulatory-information/search-fda-guidance-documents/quality-considerations-topical-ophthalmic-drug-products. Accessed on 2 Dec 2024.
10. Available online from: https://www.escopharma.com/solutions/eye-care-ophthalmic-preparations#:~:text=Ophthalmic%20formulations%20also%20require%20that,degeneration%20and%20nutritional%20support%20formulas. Accessed on 10 Dec 2024.
11. Ghelardi E, Celandroni F, Gueye SA, Salvetti S, Campa M, Senesi S. Antimicrobial activity of a new preservative for multiuse ophthalmic solutions. J Ocul Pharmacol Ther. 2013;29(6):586–90.
12. Nentwich MM, Kollmann KH, Meshack J, Ilako DR, Schaller UC. Microbial contamination of multi-use ophthalmic solutions in Kenya. Br J Ophthalmol. 2007;91(10):1265–8.
13. United States Pharmacopoeia, <1664> Assessment of Drug Product Leachables Associated with Pharmaceutical Packaging/Delivery Systems. 2015.
14. Houston CT, Rodrigues AD, Smith BB, Wang T, Richardson M. Principles for management of extractables and leachables in ophthalmic drug products. PDA J Pharm Sci Technol. 2022;76(3):278–94.
15. Available online at: https://pharmasciences.in/what-is-an-extractables-and-leachables-testing-test-in-pharmaceuticals/ (Accessed on January 10, 2025).
16. Paskiet D, Jenke D, Ball D, Houston C, Norwood DL, Markovic I. The product quality research institute (PQRI) leachables and extractables working group initiatives for parenteral and ophthalmic drug product (PODP). PDA J Pharm Sci Technol. 2013;67(5):430–47.
17. Uddin M, Mamun A, Kabir MD, Setu J, Zaman S, Begum Y, Amran M. Quality control tests for ophthalmic pharmaceuticals: pharmacopoeial standards and specifications. J Adv Med Pharm Sci. 2017;14(2):1–17.
18. Parenky AC, Wadhwa S, Chen HH, Bhalla AS, Graham KS, Shameem M. Container closure and delivery considerations for intravitreal drug administration. AAPS PharmSciTech. 2021;22:1–13.
19. Matthews BR. Drug substance and drug product—container/closure. In: International pharmaceutical product registration. CRC Press; 2016. p. 185–94.
20. Cai B, Raw A, Uppoor R, Bernard K, Chambers W, Sharadendu A, et al. Ophthalmic drug packaging: safety risks and patient acceptability issues. Pharm Technol. 2016;40(10):44–51.
21. Moore DB, Hammer JD, Akhtari R, Beck J, Sanders S, Kryscio RJ. Squeeze me if you can: variability in force requirements to extract a drop from common glaucoma bottles. J Glaucoma. 2016;25(9):780–4.

22. Color Codes for Topical Ocular Medications. Available online from: https://www.aao.org/about/policies/color-codes-topical-ocular-medications. Accessed on 16 Jan 2025.
23. Safety Considerations for Container Labels and Carton Labeling Design to Minimize Medication Errors. 2022; Available online from: https://www.fda.gov/regulatory-information/search-fda-guidance-documents/safety-considerations-container-labels-and-carton-labeling-design-minimize-medication-errors. Accessed on 16th Jan 2025
24. Maulvi FA, Ranch KM, Desai AR, Desai DT, Shukla MR. Ophthalmic preparations. In: Remington. Academic Press; 2021. p. 565–75.
25. Ali Y, Lehmussaari K. Industrial perspective in ocular drug delivery. Adv Drug Deliv Rev. 2006;58(11):1258–68.
26. Yoo H, Na YG, Jin M, Won JH, Song B, Yun TS, Hwang YR, Lee HK, Cho CW. Considerations related to the in-use stability of sterile pharmaceutical dosage forms. J Pharm Investig. 2024;54(1):1–12.
27. Bokser AD, O'Donnell PB. Stability of pharmaceutical products. Remington. 2013;37
28. Matthews BR, Wall GM. Stability storage and testing of ophthalmic products for global registration. Drug Dev Ind Pharm. 2000;26(12):1227–37.
29. CVM GFI #5 Drug Stability Guidelines. Available online from: https://www.fda.gov/regulatory-information/search-fda-guidance-documents/cvm-gfi-5-drug-stability-guidelines. Accessed on 17 Jan 2025.
30. Agrahari V, Hiremath P. Challenges associated and approaches for successful translation of nanomedicines into commercial products. Nanomedicine. 2017;12(8):819–23.
31. Rapalli VK, Gorantla S, Waghule T, Mahmood A, Singh PP, Dubey SK, Saha RN, Singhvi G. Nanotherapies for the treatment of age-related macular degeneration (AMD) disease: recent advancements and challenges. Recent Pat Drug Deliv Formul. 2019;13(4):283–90.
32. CFR-Code of Federal Regulations Title 21. n.d.; Available online from: https://www.accessdata.fda.gov/scripts/cdrh/cfdocs/cfcfr/cfrsearch.cfm. Accessed on 18 Jan 2025.
33. Ophthalmic preparations and dispensers., Code of Federal Regulation. n.d.; Available online from: https://www.ecfr.gov/current/title-21/chapter-I/subchapter-C/part-200/subpart-C/section-200.50. Accessed on 18 Jan 2025.
34. V. FDA, Label, Highlights of Prescribing Information, VISUDYNE® (Verteporfin for Injection), 2000, pp. 1–13. www.fda.gov/medwatch. Accessed 18 Jan 2025.
35. Genus v. FDA D.C. Circuit Court Decision and transition from drugs to devices. U.S. Food and Drug Administration. Guidance and Regulatory Information.
36. Losantos, Isabel. "Delivery to the eye: overcoming complex regulatory hurdles for combination products." Cambridge Consultants. Available online from: https://www.cambridgeconsultants.com/overcoming-regulatory-hurdles-for-ophthalmic-combination-products/. Accessed 30 Jan 2025.
37. Eyenovia Announces Reclassification of MydCombi™ as Drug-Device Combination Product by FDA. Eyenovia. Available online from: https://eyenovia.com/eyenovia-announces-reclassification-of-mydcombi-as-drug-device-combination-product-by-fda/. Accessed on 30 Jan 2025.

Polymeric Microsponges for Drug Delivery to Ocular Candidiasis Infections: Preclinical and Clinical Evidence

14

Aayush Singh, Atul Mourya, Dipak Panchal, Roshani Jaiswal, Atreyee Bhattacharyya, Neelesh Kumar Mehra, Snigdha Singh, Ramesh Chandra, and Jitender Madan

Abstract

In this chapter, we explore the latest advancements in microsponge delivery systems (MS-DS) for ocular drug delivery, along with their potential applications in combating ocular fungal infections, particularly candidiasis. The eye provides a unique challenge for drug delivery due to its protective barriers, rapid tear turnover, and limited residence time. Microsponges offer a promising approach by enhancing drug retention, providing controlled release, and improving corneal penetration, thereby ensuring sustained antifungal activity. Given that fungi thrive in moist environments like the conjunctiva and corneal surface, ocular candidiasis remains a serious concern, often leading to severe complications if untreated. Conventional antifungal treatments struggle with low bioavailability and frequent dosing requirements, which microsponge-based formulations can

A. Singh · A. Mourya · D. Panchal · R. Jaiswal · A. Bhattacharyya · S. Singh · J. Madan (✉)
Department of Pharmaceutics, National Institute of Pharmaceutical Education and Research, Hyderabad, Telangana, India

N. K. Mehra
Pharmaceutical Nanotechnology Research Laboratory, Department of Pharmaceutics, National Institute of Pharmaceutical Education and Research, Hyderabad, Telangana, India

R. Chandra
Department of Chemistry, University of Delhi, Delhi, India

Institute of Nanomedical Sciences, University of Delhi, Delhi, India

Dr. B.R Ambedkar Centre for Biomedical Research, University of Delhi, Delhi, India

Maharaja Surajmal Brij University, Bharatpur, Rajasthan, India

N. K. Mehra (ed.), *Complex Ophthalmic Dosage Forms: Advances in Biomedical Applications and Future Perspectives*, https://doi.org/10.1007/978-981-96-6305-4_14

effectively overcome. Additionally, this chapter provides an in-depth review of microsponge fabrication techniques, with a special focus on the quasi-emulsion solvent diffusion method, a widely adopted approach for developing ophthalmic drug carriers. We further discuss the advantages, challenges, and future prospects of microsponge-based ocular drug delivery systems in treating fungal infections and other ophthalmic disorders.

Keywords
Polymeric microsponge · Ocular drug delivery · Quasi-emulsion solvent diffusion · Preclinical evidences · Clinical applications

14.1 Introduction

Fungi are eukaryotic organisms that have their genetic material, DNA, encapsulated in a nucleus surrounded by a nuclear membrane. Additionally, their cell wall is primarily composed of chitin [1]. Over 80% of the healthy sapiens population is prone to prevalent fungal diseases, such as candidiasis, ringworm, and aspergillosis. More than 1.5 million people die each year from fungal infections worldwide, primarily those with weakened immune systems [2]. When *Candida* organisms hematogenously move to the eye and impact organs like the retina and choroid it results in ocular candidiasis. Ocular involvement occurs in 2% to 20% of patients with candidemia, and early detection is made more challenging by the fact that many of these patients are asymptomatic or unable to express their symptoms. Candidemia can lead to endophthalmitis and chorioretinitis, which can cause severe vision loss. Hematogenous seeding of the choroid and retina through tiny capillaries results in endogenous Candida endophthalmitis. Following their proliferation, the organisms produce localized abscess formation as well as focused inflammation. A subsequent spread to the vitreous body could result in retinal detachment and frank endophthalmitis. Sub-acute onset of impaired vision with low-grade apprehension photophobia, and injection are common presenting symptoms [4]. Ocular candidiasis, a fungal eye infection caused by *Candida* species, occurs in a notable proportion of patients with candidemia. A comprehensive meta-analysis encompassing 70 studies found that ocular candidiasis affects approximately 10.7% of individuals diagnosed with candidemia. Within this group, 1.8% develops concordant Candida endophthalmitis, a severe form involving vitreous involvement. Notably, studies from Asian countries reported a higher prevalence of concordant endophthalmitis at 3.6%, compared to 1.4% in European and American countries [5]. Another study reported that ocular candidiasis was present in 12.5% of patients with candidemia, with visual symptoms reported by 28% of those affected [6]. Several commercial formulations are available to combat various fungal infections, such as Miconazole, Terbinafine, and Amphotericin-B. The mechanism of action of several antifungal drugs is illustrated in Fig. 14.1. The last three decades have witnessed a steady and elevated demand in the formulation's universe, offering safe, effective, and stable formulation. Further advancement has been brought by microengineering of the existing delivery systems to provide sustained or controlled release with improved

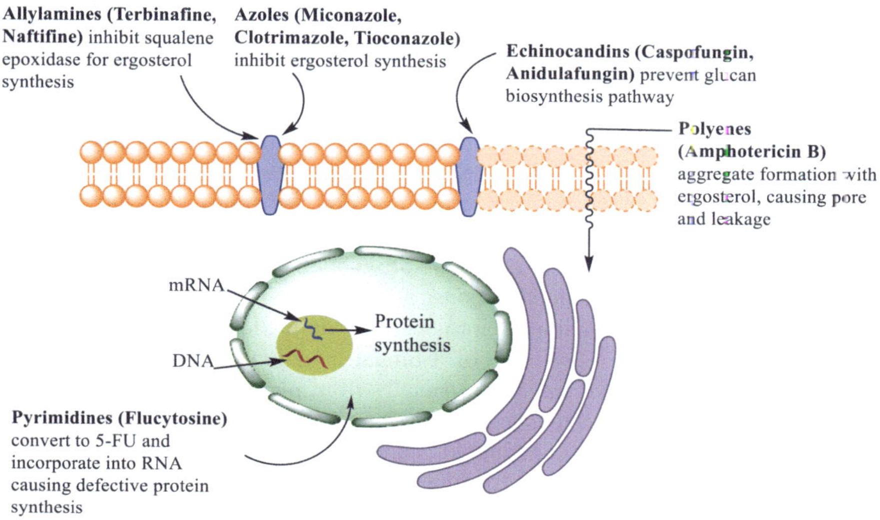

Fig. 14.1 Illustration of the mechanism of several antifungal agents on fungal cell. Created with biorender.com licensed version

physicochemical properties. Among other formulations, the MS-DS meets the aforementioned demands to almost the full extent. Since its inception in 1987 by Won, the system has laid down the basis of multi-functional solid formulation offering micro-engineering for capitalization through several routes. Later the patent for the system was awarded to Advanced Polymer Systems, Inc. (USA).

The MS-DS is identical to a porous microsphere consisting of numerous interwoven semi-rigid skeletons with a multitude of arrays of pores and voids for entrapping drug/active ingredients. It offers the regulation of particle size employed in an MS-DS throughout the whole polymerization process, and drug localization on the skin surface is assured by it within the epidermis, avoiding the bloodstream to a higher degree and reducing its local cutaneous and systemic side effects. The system likewise advises programmable release, and additionally, this idea exhibits several advantages from drug entrapment thanks to improved formulation flexibility, fewer incidents of adverse effects, and stability than its nonporous counterparts [3, 7–9].

14.2 Pathophysiology of Ocular Fungal Infection

Ocular fungal infections occur when fungal pathogens infiltrate the eye structures, leading to inflammation, tissue damage, and even vision loss. The infection commonly causes fungal keratitis (cornea) or fungal endophthalmitis (intraocular compartments) [33]. *Aspergillus, Candida, Cryptococcus, Coccidioides, Fusarium, Penicillium,* and *Pseudallescheria* are the main fungal pathogens that damage the eye. Ocular infections can also be caused by dimorphic fungus, including

Histoplasma capsulatum, *Blastomyces dermatitidis*, *Sporothrix* species, and *Coccidioides* species (*C. immitis* and *C. posadasii*) [10]. *Candida albicans* is the leading cause of fungal endophthalmitis, usually occurring after intraocular surgery or as a complication of systemic candidiasis. *Cryptococcus neoformans* can affect the retina and optic nerve, especially in immunocompromised individuals, leading to chorioretinitis and optic neuropathy. Severe infections like mucormycosis, caused by *Rhizopus* and *Mucor* species, can invade the orbit and sinuses, leading to rapid vision loss and even fatal complications if not treated promptly [9].

The pathophysiology involves a cascade of sequential events comprising fungal adhesion, tissue invasion, immune response activation, and progressive ocular damage, resulting in corneal scarring and intraocular complications. A damaged corneal epithelium is typically the first sign of a fungal infection of the eye. This often occurs as a consequence of trauma, especially from plant matter, which enables fungal spores to make it into the corneal stroma. Ocular operations, immune-suppression, and contact lens contamination are additional risk factors because they can let fungus invade intraocular structures. Upon being introduced, fungi deploy adhesions to cling to the corneal surface or intraocular tissues and start to colonize the region that is affected [11].

After adhesion, fungi release a variety of virulence agents, such as extracellular enzymes that break down intraocular and corneal tissues, such as lipases, proteases, and phospholipases. *Aspergillus* and *Fusarium* are examples of filamentous fungi that produce hyphae that burrow deeply into the corneal stroma and mechanically damage the tissue causing ulceration and gradual corneal thinning. Yeast cells can create biofilms and pseudohyphae in *Candida* infections, which boosts their resistance to antifungal medications and immunological responses [12].

A key factor in the pathophysiology of ocular fungal infections is the host immune response. Toll-like receptors (TLRs) and dectin-1 are examples of pattern recognition receptors (PRRs) that identify fungal components like chitin and β-glucans upon fungal invasion. Inflammatory cytokines such as interleukin-1 (IL-1), interleukin-6 (IL-6), and tumor necrosis factor-alpha (TNF-α) are released as an outcome, luring neutrophils and macrophages to the infection site. Nevertheless, corneal opacification, neovascularization, and further tissue damage might result from excessive inflammatory response [13].

Antifungal resistance and immunological evasion are significantly influenced by fungal biofilms. *Candida* and other yeast-like fungi have a tendency to form biofilms on the surface of the eyes, which makes them immune to the host's defenses and antifungal drugs. Some fungi also alter their surface antigens in order to evade immune recognition. The infection is able to thrive and spread in spite of immune system activity owing to its protracted survival in the ocular environment [14]. Fungal virulence factors, host immune responses, and tissue death interact intricately in ocular fungal infections. To avoid major issues including vision impairment, early detection and treatment are essential (Fig. 14.2).

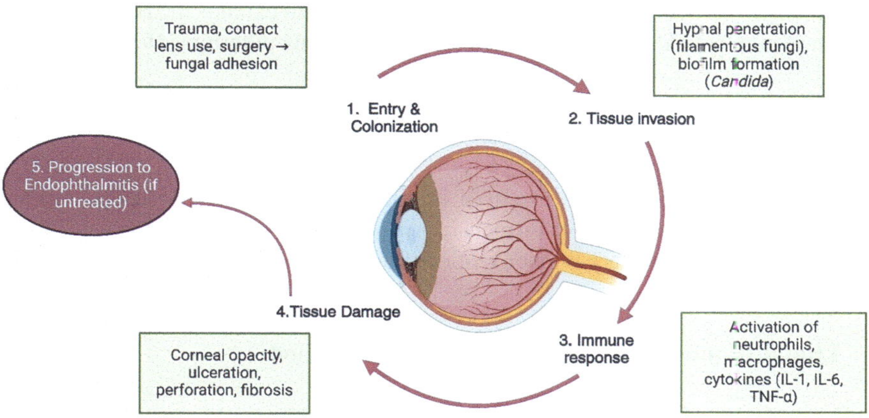

Fig. 14.2 Pathophysiology of ocular fungal infection. Created with biorender.com licensed version

14.3 Microsponge-Based Delivery System (MS-DS)

Microsponges are porous polymeric drug delivery systems made up of microscopic sponge-like perforations that spread over the entire surface ranging in size from 5 to 300 μm and average pore size of 0.25 μm [15]. It is an extensive cross-linked, perforated, micro-polymeric system with particle size ranging from 5 μm to 300 μm and a typical 25 μm sphere has about 2,50,000 holes, with a total porous volume constituting around 1 ml per gram [16]. There are numerous properties of microsponges that put them forward in comparison to other nano- and microparticulate or vesicular systems, and they offer excellent compatibility with a variety of vehicles and substances. The system has excellent soaking properties and can absorb up to sixfold of its mass, owing to its perforated surface offering extra void spaces within its structure [17]. It displays better entrapping efficiency by up to 60%. The system possesses free-flowing characteristics due to its uniform, smaller particle size. Their intrinsic perforations limit the entry of pathogenic invasions (e.g., bacteria, fungi, etc.) owing to their low average particle size (0.25 μm), making it impossible for germs to pass through and preserve them without the need for sterilization or preservative addition, granting them "Self-Sterilizing" potential [17]. In terms of stability, these systems demonstrate reasonable stability at elevated temperatures (up to 130 °C) and across the pH range of 1 to 11, making them chemically, thermally, and biologically stable. Their intrinsic stability indicates that the system can be easily micro-engineered over a range of temperatures and pH [17]. Microsponges can significantly minimize side effects, like discomfort, while maintaining effectiveness. The MS-DS formulation of benzoyl peroxide offers reduced irritation and strong efficacy, which is one such example [18]. The release rate of active ingredient in microcapsules is often unpredictable since they are discharged upon wall rupture. The unpleasant smell uncontrolled active component evaporation and potential active component in compatibility with the vehicle are other drawbacks of topical preparations. On the other hand, the microsponge system extends the active ingredient's retention period on the skin's surface or inside the epidermis [19].

14.3.1 Methods of Preparation

Porous microparticles can be prepared in multiple ways, including conventional methods such as liquid-liquid suspension method and quasi-emulsion diffusion method. However, there are nonconventional or novel methods, which are listed in Table 14.1.

(a) **Liquid-Liquid Suspension Polymerization Method**

Suspension is utilized for the preparation of perforated microspheres. In this approach, immiscible monomeric residues are initially dissolved in an appropriate medium together with active components and dispersed in the aqueous solvent monomer stages that include chemicals such as surfactants (agents that aid the system to remain suspended). Application of heat, radiation, or catalyst triggers the polymerization process [28], and the polymer is formed as a result of these polymerization processes contributing to the formation of a continuous reservoir system and retaining spherical dimensions. Finally, the solvent is eliminated during the polymerization method leaving behind spherical microsponges as given in Fig. 14.3a [29].

(b) **Quasi-emulsion Solvent Diffusion Method**

It is a dual-step procedure through which the sponges can be employed for drugs that are prone to the harsh polymerization process, e.g., heat, radiation, catalyst, etc. [30]. The preparation method utilizes diffusion of solvent using a variety of polymers in varying quantities. The method commences by intermixing two distinct phases, e.g., volatile and aqueous phases, followed by continuous stirring at room temperature leading to the generation of globule-rich emulsion or quasi (emulsion)

Table 14.1 Nonconventional methods for microsponge preparation

Nonconventional methods	Applications	References
Double emulsion solvent evaporation	Silver sulfadiazine loading for wound healing	[20]
	Percutaneous delivery of diclofenac sodium in controlled fashion	[21]
	Doxorubicin loading for pulmonary malignancy	[22]
Oil in oil emulsion solvent diffusion	Nebivolol loading for diabetic wound healing	[23]
	Hydroxyzine hydrochloride for inflamed tissue healing	[24]
Lyophilization	Skin tissue engineering	[25]
	Sustained release of basic fibroblast growth factor	[26]
Electrodynamic atomization	Drug delivery	[27]

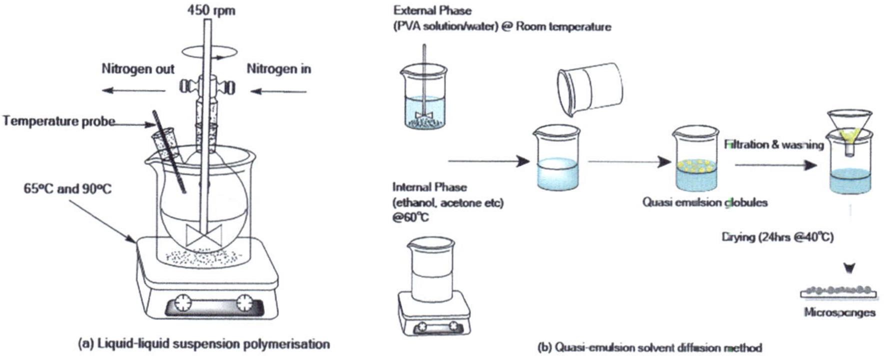

Fig. 14.3 Methods of preparation of microsponges: (**a**) Liquid-liquid suspension polymerization method and (**b**) quasi-solvent evaporation method. Created with biorender.com licensed version

droplet/globules [31]. The volatile phase gradually diffuses out of the emulsion droplets into the aqueous phase, followed by counter diffusion of the aqueous phase into the droplets by which the drug crystallizes inside the volatile phase droplets resulting in the formation of hollow microparticles called microsponges. The emulsion is made by the continuous swirling of the former internal phase at 60 °C and adding into the latter external phase, which is at room temperature [32]. In order to separate microsponges from the obtained mixture, filtration and washing are used, followed by separation and drying, as given in Fig. 14.3b.

14.4 Physical Characteristics

14.4.1 Shape and Size

By regulating the dimensions of the particles during the polymerization process, it is possible to produce free-flowing powders with acceptable aesthetic qualities. Utilizing the laser light diffractometry method or other appropriate technique, the particle size of drug-loaded or free microsponges can be readily analyzed, and the values of d50 for the majority of the formulations can be depicted as a mean size range. Plotting the cumulative percentage of drug release from various micro sponge particles against time allows one to investigate how particle size affects drug release. Particles having a diameter between 10 and 25 μm are suitable for topical formulations since particles beyond 30 μm can aggravate a gritty texture [17]. SEM and conventional light microscopy (LM) are the two primary techniques used to observe microparticles. Both methods can be employed to analyze the external morphology and dimensions of microparticles. LM offers control on the coating settings for double-walled microparticles. The morphological features of multi-walled microparticles are characterized via confocal fluorescence microscopy, and other widely utilized techniques include laser light scattering and multi-size coulter counter [33].

14.4.2 Pore Structure

The volume and diameter of pores play a crucial role in determining the sharpness and duration of the major bioactive works. The movement of bioactive from microsponges into the vehicle in which the material is disseminated is affected by void diameter as well. Mercury intrusion porosimetry can often be employed to examine how pore sizes and release rate of the drug from microsponges are related, and it can also be used to determine the porosity parameters of microsponges, including the intrusion-extrusion isotherms, distribution of pore size, total porous surface area, mean pore diameters, bulk density, and apparent density. With the help of an ultrapycnometer and helium gas, the true density of microparticles can be determined, and the Washburn equation is utilized to calculate the pore diameter [34]. The equation provides pore volume, total open pore volume, porosity, and measurable pore size that should be 0.0018 to 400 μm [34].

$$D = \frac{-4\gamma \cos\theta}{P}$$

where D is pore diameter, θ is the contact angle between pore wall and mercury, γ is mercury's surface tension, and P is applied pressure.

14.4.3 Inner Porosity

Inner porosity in microsponges refers to the network of interconnected voids within their structure, which influences drug loading capacity, release kinetics, and overall performance in pharmaceutical applications. This porosity allows microsponges to absorb and retain active ingredients efficiently, facilitating controlled and sustained drug release. The degree of inner porosity affects parameters such as permeability, entrapment efficiency, and surface area, making it a crucial factor in designing optimized drug delivery systems [35].

By adjusting the drug and polymer concentrations in the emulsion droplet, it is quite easy to control the intrinsic porosity of microsponges. The microspheres with a more significant porosity can be produced at a lower temperature. Internal porosity has a crucial impact on the release profile of the drug. The Higuchi model is the best fit for the drug release profile from these carrier systems, which demonstrates a reliance on internal porosity. Regardless of their surface porosity, particles with similar interior textures displayed nearly the same tortuosity [36], and it can be computed by the formula [37, 38].

$$\text{Porosity}\left(\%\right) = 1 - \frac{V_p}{V_b} \times 100$$

where V_p and V_b are the volume and bulk volume of particles, respectively.

14.4.4 Flow Properties

Changes in particle size, shape, electrostatic charge, and moisture that may result from processing techniques have a considerable impact on most of the flow characteristics. A free-flowing drug candidate may consequently become cohesive during development, mandating the use of a whole new formulation method [39]. For the development of new products, better topographical attributes are necessary since they provide better compression, filling of dies and capsules, and other processes. Due to their spherical shape, microsponges offer excellent flow characteristics, and their microstructure enhances compression [37, 40]. Applying the compressibility index, Hausner's ratio, and angle of repose given by the equation offers a simple indicator of how easily a material may be made to flow.

$$\text{Carr's compressibility index} = \frac{(TD - BD)}{TD} \times 100$$

where *TD* is tapped density, and *BD* is bulk density.

$$\text{Hausner's ratio} = \frac{TD}{BD}$$

$$\text{Angle of repsoe}(\tan\theta) = \frac{2h}{D}$$

where ***h*** is height of powder bed, and ***D*** is diameter of powder bed.

14.4.5 Surface Topography

The superficial topography of the produced microsponges can be investigated by scanning electron microscopy (SEM) after being coated with gold-palladium in argon—rich environment at room temperature. It is also possible to use SEM images of fragmented microsponge particles to show their ultra-porous structure [41].

14.4.6 Compatibility and Crystallinity

Using thin layer chromatography (TLC) and Fourier Transform Infrared (FT-IR) spectroscopy, the compatibility of the bioactive with its excipients can be investigated. In contrast, Powder X-ray Diffraction (PXRD) and Differential Scanning Calorimetry (DSC) can be utilized to evaluate the impact of polymerization on the drug's crystallinity [42]. For DSC, samples of around 5 mg can be precisely weighted on aluminum pans, sealed, and heated at a rate of 15 °C per minute throughout a temperature range of 25–430 °C in a nitrogen-rich atmosphere. The extent of crystallinity can be determined using DSC to detect sharp melting point (indicating its crystalline nature), followed by Hot Stage Microscopy where crystalline and noncrystalline particles can be visually inspected.

14.4.7 Loading Efficiency and Production Yield

It is defined as the ratio of the quantity of drug in the microparticle to the total quantity of drug applied in the formulation of the microparticles. The given equation can be employed to compute the microsponge's loading efficiency (in percentage) [43].

$$\%\text{Loading Efficiency} = \frac{\text{Observed Drug Content}}{\text{Theoretical Drug Content}} \times 100$$

By accurately determining the original mass and final mass of raw materials and microsponges, respectively, it is possible to determine the production yield of the microparticles. Production yield denotes the efficiency of the process involved in the production of the formulation.

$$\text{Production Yield} = \frac{\text{Observed Mass of Microparticles}}{\text{Theoretical Mass}\left(\text{mass of drug} + \text{mass of polymer}\right)}$$

14.5 Factors Affecting the Formulation of Microsponges

There are numerous factors that affect the formulation, and they can be categorized as formulation-related and process-related, which directly or indirectly have an influence on the quality attributes, release profile, and production yield of the designed formulation.

14.5.1 Drug-Polymer Ratio

Drug loading capacity is not significantly influenced by changes in the ratio of drug-polymer; however, manufacturing yield can be altered from a minimal to a maximal ratio. The drug-polymer ratio can have a significant impact on the yield of manufacture, loading effectiveness, and average particle size of microsponges. The production yield was recorded to be relatively low when the drug-polymer ratio was kept at 1:1 (less than 15%). Usually, there is a proportional relationship between the drug-polymer ratio and production yield and vice versa [44]. The reason behind this relationship could be attributed to the decreased diffusional rate of the organic phase (dichloromethane) from the concentrated phase into the aqueous one, offering additional time for droplet generation and thus boosting the number of microparticles produced [45]. Additionally, it was discovered that smaller particle size was achieved by raising the drug-polymer ratio. For instance, the average diameter of microsponges dropped to 310.3 μm from 442.2 μm when the ratio was increased to 13:1 from 3:1.

14.5.2 Type and Amount of Emulsifier

The emulsifier is key to the emulsification process, owing to its potential to diminish the interfacial tension between the organic and aqueous phases. It serves as a droplet stabilizer until the polymer concentration reaches a certain level or until an adequate amount of solvent has been withdrawn from oil droplets to retain the particle shape. Generally, 1% of polyvinyl alcohol (PVA) is employed as an emulsifying agent in the continuous phase during the production of PLGA microparticles. Others include poly(N-vinylpyrrolidone) (PVP) [46], poly (ethylene glycol) (PEG) [47], poloxamer [48], and hydroxypropyl methylcellulose (HPMC) [49]. When compared to PVA at 1%, 0.1% PVA was found to have superior encapsulation potential for Ibuprofen in both organic solvent dichloromethane and ethyl acetate [50]. The difference in interfacial tension between organic and aqueous phases may have an impact on the kinetics of organic phase extraction and, thus, on the overall dimension of microparticles. Both variables, which have a considerable impact on the drug release kinetics, can alter the morphological aspects of the microparticles and the distribution of drugs within them. Owing to the greater probability of phase separation between PLGA and the loaded drug, it may eventually cause a huge burst and minimize the duration of release; thus, it is not advised to use PLGA in microparticles with lower molecular weight drugs.

14.5.3 Effect of External and Internal Phases

The apparent viscosity of the dispersed phase is directly proportional to the microsponge particle size. When the higher viscosity dispersed phase is poured into the continuous phase/external phase, the resultant emulsion globules can scarcely be split into small particles, and larger droplets are observed, leading to a rise in average particle size. Suitable microsponges can be generated only by using a minor quantity of internal phase. The manufacturing yield and drug content seem to be inversely related to the volume of the internal phase, owing to a smaller amount of drug in the internal phase's higher volume [45].

14.5.4 Stirring Speed

Smaller microsponges can be developed as the stirring rate is elevated. The production yield declines as the stirring rate rises, but the drug content increases. It suggests that drug loss can be lowered with a rising stirring rate. As a result, owing to the presence of agitation in the external phase leads polymers to bind to the paddle and reduce the manufacturing yield [45].

14.5.5 Stirring Time

The shape, size, and volume of the generated microparticles are significantly dependent on the stirring time and rate. The stirring time has shown a substantial impact on the shape of the microparticles, encapsulation efficiency, and release profile of Rifampicin-PLGA microparticles. As per the observations obtained by Kong et al., the shape of microparticles is not considerably affected by adjusting the stirring speed. However, if stirring time is prolonged, more fragmented PLGA microparticles are produced with smaller dimensions [51].

Additionally, the impact of particle size on the drug loading and release profile in the MS-DS is larger [52]. The surface morphology of the system is substantially affected by the choice of the solvent, solution, and the maintenance of temperature during formulation. When the polymer precipitates quickly during formulation, there is a corresponding decrease in porosity [53]. As per the porous dimension, the array of perforations forms a continuous pattern exposed to the external surface and allows the drug molecule to diffuse extrinsically at a predetermined rate [54]. Numerous investigations have confirmed the existence of nanoscale pores on the MS-DS particle surface. For instance, by using Atomic Force Microscopy, [55] discovered nanometric holes in Carvedilol-loaded Eudragit RS100 MS-DS. The MS-DS technology was initially developed for topical delivery to lessen the drug's systemic and local adverse effects by delivering a controlled and temporal release of the active ingredients. This new technology is used in several applications for intravenous, ocular, pulmonary, and oral delivery of bioactive agents. MS-DS offers the possibility of protecting the drug within a shielded structure and delivering the drug in a regulated fashion targeting the lower section of the gastrointestinal tract (i.e., colon), where on stimulation, the drug gets released, e.g., by particular enzymes, pH, etc. [56]. Lower pore size (0.07–0.25 μm) causes the particle size to be reduced when the drug molecules entrap within the voids. Drug solubility is improved as particle size shrinks because of the increased surface area [53]. Such MS-DS-based methods may give birth to a new possibility for the spatial and temporal movement of bioactive substances. The entrapment of bioactives in the voids found in the MS-DS increases the rate and extent of solubilization of weakly water-miscible drugs [57].

14.6 Mechanism of Drug Release from MS-DS

The porous micro-entities can absorb a plethora of active substances, including antifungals, anti-infectives, anticancer agents, sunscreens, etc. Until equilibrium is attained, the bioactive is free to stream in and out of the open-structured microsponge particles and into the vehicle.

The porous structure of microsponges, which resembles sponge-like polymeric microspheres, enables them to capture active substances and control their release [58, 59]. These microspheres, which are usually between 5 and 30 μm in size are nowadays renowned as flexible drug delivery techniques, especially for ocular applications [60, 61]. Pressure, temperature, pH, and the solubility of the drug significantly affect the process of drug release from microsponges. These variables allow for regulated drug diffusion via the associated channels in the porous structure of the microsponges [62].

The pharmacokinetics of a drug through ocular route is a complicated process. Following the administration, the drug passes through the blood-aqueous barrier and enters the anterior chamber. It is then transported to the Schlemm's canal and trabecular meshwork, wherein the aqueous humor turnover removes it. Additionally, the medication crosses the blood-aqueous barrier to leave the aqueous humor and reach the systemic circulation. The drug eventually enters the circulation, crosses the blood-retina barrier, and enters the eye's posterior chamber [62]. Drug delivery to the eyes is severely hampered by the rapid clearance driven by tear fluid turnover, which reduces the drug's precorneal residence period. Only a small percentage of the instilled amount makes it through the cornea to the intraocular tissues, resulting in a limited bioavailability of the drug [61]. Microsponges can be added to in situ gels to solve this problem by prolonging the duration of contact between the active component and the ocular surface [58]. This phase shift increases the duration of residency on mucosal surfaces and permits accurate drug administration as shown in Fig. 14.4 [58, 61].

The advantages offered by microsponges for ocular delivery:

- Controlled drug delivery: The sustained release of drug made feasible by microsponges guarantees a consistent therapeutic impact over a longer period of time [60].
- Improved bioavailability: By increasing its contact period with the cornea, microsponges increase the bioavailability of the drug [60].
- Decreased systemic side effects: Because localized drug distribution restricts systemic absorption, adverse effects are less likely to occur [58].

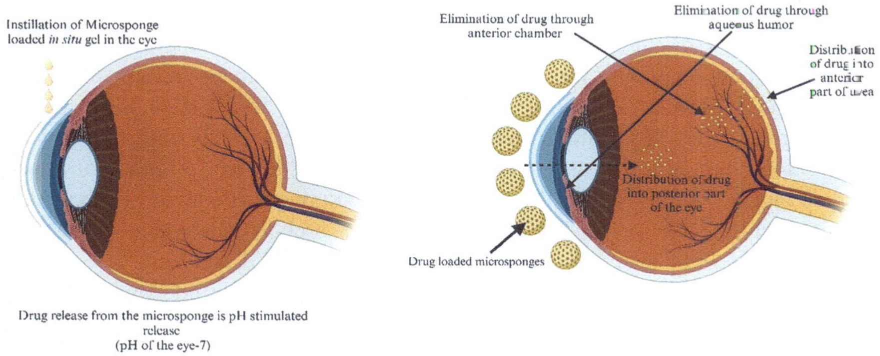

Fig. 14.4 Mechanism of drug release from microsponges for ocular delivery. Created with bio-render.com licensed version

- Enhanced patient compliance: Patients are more likely to adhere to their therapy when there is less need for frequent doses due to sustained release of the drug [61].

14.7 Applications

14.7.1 Ketotifen-Loaded Microsponge-Based Ocular Gel for Ocular Infection

Owing to the limitations posed by the ocular anatomy and biology, such as the relatively impermeable epithelial membrane of the cornea, tear dynamics, nasolacrimal drainage, and poor blood-ocular barrier penetration efficiency, the bioavailability of commercial ocular dosage forms is quite limited. Ketotifen belongs to benzocycloheptathiophene class; owing to its potent antihistaminic and mast cell stabilizing properties, it is employed for the treatment of itchy eyes associated with allergies. J.R. Kumar et al. developed a microsponge gel loaded with Ketotifen and delivered the medication via the ocular route [63]. Due to its porous dimension and plasticity, the porous particulate system serves as a suitable means for the progression of the ocular delivery system, boasting the potential to reduce quick and excessive drug loss through precorneal sites owing to tear production and its drainage. To enable sustained retention of the gel in the eyes, its viscosity was modified to a range between 1299 and 1600 centipoise. Later it was revealed that Ketotifen was released from the gel in 8 hrs, as per the in vitro release study, showing a prolonged drug release pattern and the potential for increasing its bioavailability [63].

14.7.2 Microsponges for Management of Glaucoma

Abd-Elal et al. formulated timolol maleate (TMM)-loaded microsponges to obtain tailored extended release of TMM and to enhance its corneal penetration. TMM-entrapped microsponges were developed through double emulsion (water/oil/oil) quasi-emulsion solvent diffusion method. The particle diameter of TMM-encapsulated microsponges was 6283.33 ± 145.71 nm. The encapsulation efficiency and percent yield were found to be $72.00 \pm 6.08\%$ and $96.55 \pm 4.01\%$, respectively. The drug released from microsponges was $42.12 \pm 3.93\%$ within 6 hours. Furthermore, the corneal penetration was effectively enhanced by TMM-entrapped microsponges. The tailored TMM-entrapped microsponges illustrated in the study exhibit potential to be further examined in future research to boost the in vivo ocular bioavailability of the drug in comparison to aqueous TMM drops [64].

Rantham et al. developed in situ gel loaded with dorazolamide hydrochloride (DHCl) encapsulated in microsponges for management of glaucoma, to enhance its therapeutic adequacy and cut down its adverse effects. The DHCl-entrapped microsponges were developed through quasi emulsion solvent diffusion technique. The particle diameter and PDI of microsponges were found to be 2 μm and 0.77, respectively. The microsponges exhibited higher stability with encapsulation efficiency of 85% and drug loading efficiency of 78%. The in vitro release of DHCl-loaded microsponges incorporated in pluronic F-127 gel was higher than the DHCl loaded in pluronic F-127 gel. By reducing the possibility of systemic adverse effects, above-mentioned novel DHCl microsponges loaded in situ gel formulations may be used as an efficient topical ocular drug delivery method for the treatment of glaucoma [65].

Dhyani et al. formulated atenolol-encapsulated microsponges loaded into in situ gel to provide tailored sustained release of the drug. The atenolol microsponges were prepared by oil-in-oil solvent emulsion diffusion technique. The optimized atenolol microsponges had average particle diameter of 9.21 ± 0.76 μm, which is suitable for ocular delivery. The production yield, encapsulation efficiency, and drug content exhibited were 86.32%, 94.19%, and 93.89%, respectively. The in vitro release of microsponges loaded in situ gel exhibited sustained release of atenolol. The above-mentioned atenolol microsponges loaded in situ gel is a potential candidate to reduce the dose of the drug and to achieve sustained effect of the drug through ocular route of administration [61].

14.7.3 Diclofenac Sodium-Loaded Microsponges for Postoperative Ocular Pain and Inflammations

Ambikar et al. developed diclofenac sodium (DIC)-loaded microsponges for preventing intraoperative miosis after cataract surgery and managing pain in conditions of corneal epithelial anomalies that require surgery. The optimized DIC-loaded microsponges in polymer-to-drug ratio 1:7 exhibited production yield, entrapment efficiency, and drug content as 68.13%, 62.86%, and 80.73%, respectively. The particle diameter was found to be 7.52 ± 2.29 μm which is optimum for ocular delivery. The in vitro release of DIC-loaded microsponges was observed to be 87.94% in 6 hours. The DIC-loaded microsponges were further loaded into in situ gel. The in situ gel formulation offered sustained release of DIC for up to 8 hours, which is appropriate for alleviating ocular pain [66] (Table 14.2).

14.8 Challenges and Future Prospects

In this chapter, we tried to outline the latest developments in the formulation of microsponges, considering their characteristics, benefits, manufacturing techniques, characterization methods, and biofunctionalization, along with biological and theranostics applications. Advanced development for tackling difficulties, particularly in drug delivery, tissue engineering, antimicrobial, and customized release pattern, has been made possible by sponge particulate platforms (micro and nano) throughout the past few decades. The microsponge particles offer a practical and effective method for multidrug protection, customized release pattern, and adaptable treatment regimens. Multidrug formulation has much potential for improving treatment

Table 14.2 Microsponge formulations for ocular delivery

S. no.	Drug	Formulation	Key findings	References
1	Acetazolamide	Microsponge-loaded in situ gel	Developed using the quasi-emulsion solvent diffusion method; achieved high drug entrapment (~82%) and optimal particle size (~10 μm); demonstrated enhanced therapeutic efficacy and nonirritating properties in rabbit eyes	[58]
2	Atenolol	Microsponge-loaded in situ gel	Formulated using ethyl cellulose and Eudragit RS-100 polymers; incorporated into a Poloxamer 407-based in situ gel; exhibited sustained drug release and potential for reduced dosing frequency in ocular hypertension treatment	[61]
3	Dorazolamide hydrochloride	Microsponge-loaded in situ gel	Prepared via the quasi-emulsion solvent diffusion method; resulted in microsponges with a mean particle size of ~2 μm. The in situ gel demonstrated suitable rheological properties and sustained drug release, with nonirritating effects in rabbit eye models	[65]
4	Timolol maleate	Microsponge	Developed timolol maleate (TMM)-loaded microsponges using a double emulsion solvent diffusion method to achieve extended drug release and improved corneal penetration. The microsponges had an average size of 6283.33 nm, with 72% encapsulation efficiency and 96.55% yield. Drug release reached 42.12% within 6 hours, demonstrating sustained delivery. The study suggests that these microsponges could enhance ocular bioavailability compared to traditional TMM eye drops	[64]
5	Ketotifen	Microsponge-loaded gel	Developed a microsponge-based ocular gel containing Ketotifen, designed to minimize drug loss due to tear drainage. The gel's viscosity (1299–1600 cP) was optimized for prolonged eye retention. In vitro studies showed sustained drug release over 8 hours, indicating enhanced bioavailability and effective ocular drug delivery	[63]

(continued)

Table 14.2 (continued)

S. no.	Drug	Formulation	Key findings	References
6	Diclofenac sodium	Microsponge	Formulated diclofenac sodium (DIC)-loaded microsponges to prevent intraoperative miosis and manage ocular pain after surgery. Optimized microsponges (polymer-to-drug ratio 1:7) showed 68.13% yield, 62.86% entrapment efficiency, and 80.73% drug content, with a particle size of 7.52 μm, making them suitable for ocular delivery. In vitro release reached 87.94% in 6 hours, and when incorporated into an in situ gel, drug release was sustained for up to 8 hours, ensuring prolonged pain relief	[66]

effectiveness, reducing adverse effects, and expanding applicability. These developments signal the emergence of a new class of porous materials for use in biomedical applications, and the thorough study will be helpful for exploring and advancing the field of porous particles. However, there are still issues with particle size optimization, pores consistency, particularly flow properties, and in vivo targeted distribution that affect its commoditization in the long run. The distribution of bioactive agents and optimum release pattern depends on the uniformity of both particle and void dimensions. The uniformity and reproducibility of sponge particles are now the prime obstacles to their use in clinical investigations and mass production [67]. But by identifying critical quality parameters (polymer ratio, stirring speed, surfactant concentration, etc.) through response surface optimization [68] and employing design of experiment (DOE)-based approach, one can achieve microsponges with optimized attributes such as particle size, pore structure, and void consistency [69]. In vivo targeted distribution, microsponges are typically administered via non-intravenous methods, which lack appropriate bioavailability for the treatment and aid of deep lesions but the issue can be eradicated by formulating injectable formulations such as powder for reconstitution [70]. However, void inconsistency still remains a significant trammel in their commoditization and effectiveness with desired and consistent quality attributes. Although Hajra et al. found a way to circumvent this quandary by employing in silico modeling where with the assistance of binding free energies, they found that the binding affinity of the mono-polymeric form is weaker than its counter co-polymeric form [71]. The co-polymer complex SLS-Eudragit-HPMC-Naproxen demonstrated the best binding affinity and energy, which was 3.9 kcal/mol. The experimental findings were supported by molecular modeling and simulation studies, which explained Eu-HPMC-SLS as an efficient hybrid system with the maximum binding energy to enhance the encapsulation of Naproxen with an altered release profile [71]. Author laboratory had focussed on in silico docking techniques to accelerate the process of formulation development while reducing the cost as well as the time involved in pre-formulation studies. Here

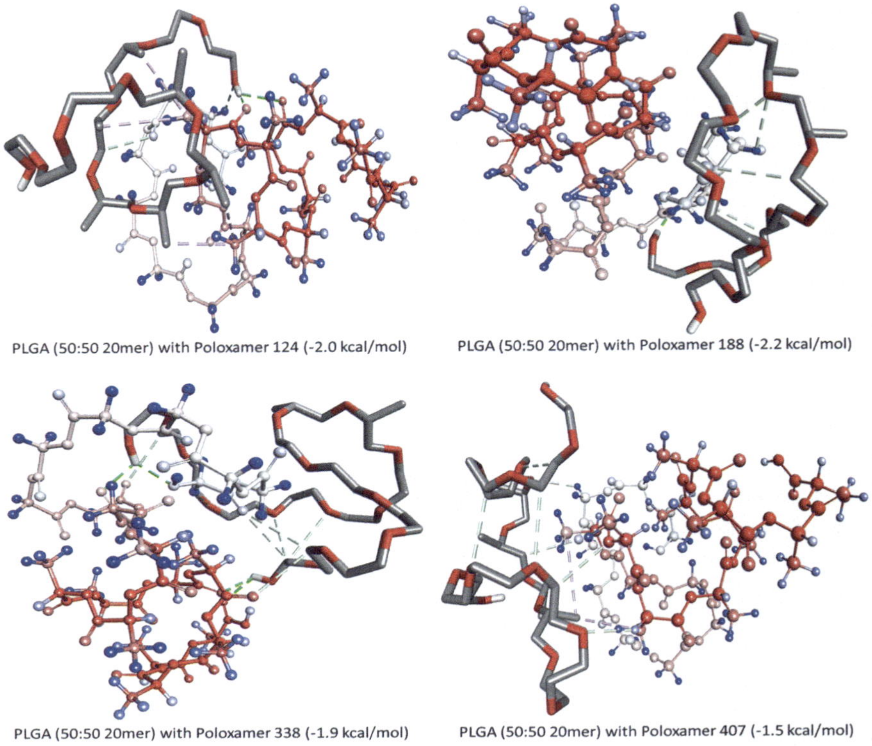

Fig. 14.5 The molecular interaction between PLGA (50:50) 20mer and poloxamer 124, 188, 338, and 407 using in silico molecular docking. (Unpublished data)

we attempted to explore molecular docking studies for screening various poloxamer grades as a stabilizer for PLGA microsphere using AutoDock Vina. The molecular interaction between PLGA (50:50) 20-mer and poloxamer 124, 188, 338, and 407 are displayed in Fig. 14.5. In silico molecular docking study suggests poloxamer 188 as a better stabilizer for PLGA than other poloxamer grades. Interestingly, results from molecular modeling are in agreement with the reported literature by various research groups.

The future prospects of microsponges in ophthalmic applications are promising, offering sustained and targeted drug delivery for various eye disorders. Their porous structure enhances drug retention, bioavailability, and controlled release, making them ideal for treating conditions like glaucoma, dry eye syndrome, and ocular infections [58, 61]. Advancements in stimuli-responsive microsponges, such as pH-sensitive and temperature-controlled systems, could enable on-demand drug release for ailments like conjunctivitis and post-surgical inflammation [62, 66]. Additionally, injectable and implantable microsponge-based systems hold potential for managing chronic retinal diseases, ensuring prolonged therapeutic effects. With further research and clinical validation, microsponges could revolutionize ophthalmic treatments, providing more effective and patient-friendly alternatives to conventional eye drops and injections [17].

Acknowledgments The authors are highly thankful to the Department of Pharmaceuticals, Ministry of Chemicals and Fertilizers, Government of India, New Delhi, India, for providing financial assistance.

References

1. Ranade V, v. Drug delivery systems. 6. Transdermal drug delivery J Clin Pharmacol. 1991;31:401–18. https://doi.org/10.1002/j.1552-4604.1991.tb01895.x.
2. Rayens E, Norris KA. Prevalence and healthcare burden of fungal infections in the United States, 2018. Open forum. Infect Dis. 2022;9:9. https://doi.org/10.1093/OFID/OFAB593.
3. Shah CP, McKey J, Spirn MJ, Maguire J. Ocular candidiasis: a review. Br J Ophthalmol. 2008;92:466–8. https://doi.org/10.1136/BJO.2007.133405.
4. Khalid A, Clough LA, Symons RCA, Mahnken JD, Dong L, Eid AJ. Incidence and clinical predictors of ocular candidiasis in patients with Candida Fungemia. Interdiscip Perspect Infect Dis. 2014;2014:650235. https://doi.org/10.1155/2014/650235.
5. Won R, Alto P, Assignee C. Method for delivering an active ingredient by controlled time release utilizing a novel delivery vehicle which can be prepared by a process utilizing the active ingredient as a porogen 1985.
6. Vyas SP, Suresh P, Khar RK, Roop K. Targeted & controlled drug delivery : novel carrier systems. 2004:594.
7. Florence AT. Targeted and controlled drug delivery: novel carrier systems. Int J Pharm. 2003;267:157. https://doi.org/10.1016/S0378-5173(03)00356-9.
8. van der Meer JWM, van de Veerdonk FL, Joosten LAB, Kullberg BJ, Netea MG. Severe Candida spp. infections: new insights into natural immunity. Int J Antimicrob Agents. 2010;36(Suppl):2. https://doi.org/10.1016/J.IJANTIMICAG.2010.11.013.
9. Thomas PA. Fungal infections of the cornea. Eye. 2003;17(8):852–62. https://doi.org/10.1038/sj.eye.6700557.
10. Ramírez-Soto MC, Bonifaz A. Ocular fungal infections. J Fungi. 2022;8:1078. https://doi.org/10.3390/JOF8101078.
11. Klotz SA, Penn CC, Negvesky GJ, Butrus SI. Fungal and parasitic infections of the eye. Clin Microbiol Rev. 2000;13:662–85. https://doi.org/10.1128/CMR.13.4.662.
12. Kumari A, Tripathi AH, Gautam P, Gahtori R, Pande A, Singh Y, et al. Adhesins in the virulence of opportunistic fungal pathogens of human. Mycology. 2021;12:296–324. https://doi.org/10.1080/21501203.2021.1934176.
13. Patin EC, Thompson A, Orr SJ. Pattern recognition receptors in fungal immunity. Semin Cell Dev Biol. 2019;89:24–33. https://doi.org/10.1016/J.SEMCDB.2018.03.003.
14. Kernien JF, Snarr BD, Sheppard DC, Nett JE. The interface between fungal biofilms and innate immunity. Front Immunol. 2018;8:328714. https://doi.org/10.3389/FIMMU.2017.01968/PDF.
15. Singhvi G, Manchanda P, Hans N, Dubey SK, Gupta G. Microsponge: an emerging drug delivery strategy. Drug Dev Res. 2019;80:200–8. https://doi.org/10.1002/DDR.21492.
16. Al-Sum BA. Kumar Mishra S. NATIONAL SCIENTIFIC COMMITTEE NATIONAL EDITORIAL ADVISORY BOARD: Srivastava D; n.d.
17. Srinatha N, Battu S, Vishwanath BA. Microsponges: a promising frontier for prolonged release-current perspectives and patents. Beni Suef Univ J Basic Appl Sci. 2024;13:1–19. https://doi.org/10.1186/S43088-024-00519-4/TABLES/3.
18. Wester RC, Patel R, Nacht S, Leyden J, Melendres J, Maibach H. Controlled release of benzoyl peroxide from a porous microsphere polymeric system can reduce topical irritancy. J Am Acad Dermatol. 1991;24:720–6. https://doi.org/10.1016/0190-9622(91)70109-F.

19. Jyothi NN, El; Naga Jyothi K, Kumar D, Arshad P, Karthik P, Panneerselvam M. Microsponges: a promising novel drug delivery system. Journal of Drug Delivery and Therapeutics. 2019;9:188–94. https://doi.org/10.22270/JDDT.V9I5-S.3649.
20. Kumar PM, Ghosh A. Development and evaluation of silver sulfadiazine loaded microsponge based gel for partial thickness (second degree) burn wounds. Eur J Pharm Sci. 2017;96:243–54. https://doi.org/10.1016/J.EJPS.2016.09.038.
21. Maiti S, Kaity S, Ray S, Sa B. Development and evaluation of xanthan gum-facilitated ethyl cellulose microsponges for controlled percutaneous delivery of diclofenac sodium. Acta Pharma. 2011;61:257–70. https://doi.org/10.2478/V10007-011-0022-6.
22. Bae SE, Son JS, Park K, Han DK. Fabrication of covered porous PLGA microspheres using hydrogen peroxide for controlled drug delivery and regenerative medicine. J Control Release. 2009;133:37–43. https://doi.org/10.1016/J.JCONREL.2008.09.006.
23. Pandit AP, Patel SA, Bhanushali VP, Kulkarni VS, Kakad VD. Nebivolol-loaded microsponge gel for healing of diabetic wound. AAPS PharmSciTech 2016 18:3 2016;18:846–54. doi:https://doi.org/10.1208/S12249-016-0574-3.
24. Zaki Rizkalla CM, Latif Aziz R, Soliman II. In vitro and in vivo evaluation of hydroxyzine hydrochloride microsponges for topical delivery. AAPS PharmSciTech 2011 12:3 2011;12:989–1001. doi:https://doi.org/10.1208/S12249-011-9663-5.
25. Chen G, Sato T, Ohgushi H, Ushida T, Tateishi T, Tanaka J. Culturing of skin fibroblasts in a thin PLGA–collagen hybrid mesh. Biomaterials. 2005;26:2559–66. https://doi.org/10.1016/J.BIOMATERIALS.2004.07.034.
26. Kanematsu A, Marui A, Yamamoto S, Ozeki M, Hirano Y, Yamamoto M, et al. Type I collagen can function as a reservoir of basic fibroblast growth factor. J Control Release. 2004;99:281–92. https://doi.org/10.1016/J.JCONREL.2004.07.008.
27. Pancholi K, Ahras N, Stride E, Edirisinghe M. Novel electrohydrodynamic preparation of porous chitosan particles for drug delivery. J Mater Sci Mater Med 2008 20:4 2008;20:917–23. doi:https://doi.org/10.1007/S10856-008-3638-4.
28. Kumar Dv, Sethuraman S. Transdermal Drug Delivery System: A Review swapnil deshpande Transdermal Drug Delivery System: A Review. Current Pharma Research n.d.;1.
29. Nikam VK, Kotade KB, Gaware VM, Dolas RT, Dhamak KB, Somwanshi SB, et al. Transungual drug delivery system: a review, vol. 1; 2011.
30. D'souza JI, More HN. Topical anti-inflammatory gels of Fluocinolone Acetonide entrapped in Eudragit based microsponge delivery system. Research J Pharm and Tech. n.d.;1
31. Çomoğlu T, Gönül N, Baykara T. Preparation and in vitro evaluation of modified release ketoprofen microsponges. Il Farmaco. 2003;58:101–6. https://doi.org/10.1016/S0014-827X(02)00007-1.
32. Taghi HS, Abdulbaqi MR, Samein LH, Rahmani MHP. Formulation and in vitro evaluation
33. of a Ramipril entrapped in a microsponge-based drug-delivery system. Int J Pharm Compd. 2023;27(4):340–6.
34. Depypere F, Van Oostveldt P, Pieters JG, Dewettinck K. Quantification of microparticle coating quality by confocal laser scanning microscopy (CLSM). Eur J Pharm Biopharm. 2009;73:179–86. https://doi.org/10.1016/J.EJPB.2009.04.007.
35. Washburn EW. Note on a method of determining the distribution of pore sizes in a porous material. Proc Natl Acad Sci. 1921;7:115–6. https://doi.org/10.1073/PNAS.7.4.115/ASSET/2A938320-7284-4A7E-8413-61B527BA4960/ASSETS/PNAS.7.4.115.FP.PNG.
36. Mahant S, Kumar S, Nanda S, Rao R. Microsponges for dermatological applications: perspectives and challenges. Asian J Pharm Sci. 2020;15:273–91. https://doi.org/10.1016/J.AJPS.2019.05.004.
37. Kawashima Y, Niwa T, Takeuchi H, Hino T, Ito Y. Control of prolonged drug release and compression properties of ibuprofen microsponges with acrylic polymer, Eudragit RS, by changing their Intraparticle porosity. Chem Pharm Bull (Tokyo). 1992;40:196–201. https://doi.org/10.1248/CPB.40.196.
38. Microsponges as Innovative Drug Delivery Systems 2012. doi:10.37285/ijpsn.2012.5.1.2.

39. Singh Y. Martin's physical pharmacy and pharmaceutical sciences physical chemical and biopharmaceutical principles in the pharmaceutical sciences n.d.
40. Lachman L, Lieberman HA. Kanig JL. Theory and practice of industrial pharmacy. 1994;902
41. Jangde R, Ravishankar P. Formulation and characterization of tasteless dispersible tablet containing Ampicilin and Cloxacillin view project "design and development of plant actives based nano lipoidal systems for wound healing" view project microsponges for colon targeted drug delivery system: an overview. Asian J Pharm Tech. 2011;1:87–93.
42. D'Emanuele A, Dinarvand R. Preparation, characterisation, and drug release from thermoresponsive microspheres. Int J Pharm. 1995;118:237–42. https://doi.org/10.1016/0378-5173(94)00384-H.
43. Anderson DL, Chung-Heng C, Nacht S. Flow characteristics of loosely compacted macroporous microsponge® polymeric systems. Powder Technol. 1994;78:15–8. https://doi.org/10.1016/0032-5910(93)02764-2.
44. Kumar S, Tyagi LK. Singh D. Microsponge Delivery System (Mds): A Unique Technology for Delivery of Active Ingredients. IJPSR; 2011. p. 2.
45. Jelvehgari M, Siahi-Shadbad MR, Azarmi S, Martin GP, Nokhodchi A. The microsponge delivery system of benzoyl peroxide: preparation, characterization and release studies. Int J Pharm. 2006;308:124–32. https://doi.org/10.1016/j.ijpharm.2005.11.001.
46. Lee JH, Park TG, Choi HK. Development of oral drug delivery system using floating microspheres. Http://DxDoiOrg/101080/026520499288663. 2008;16:715–29. https://doi.org/10.1080/026520499288663.
47. Feczkó T, Tóth J, Gyenis J. Comparison of the preparation of PLGA–BSA nano- and microparticles by PVA, poloxamer and PVP. Colloids Surf A Physicochem Eng Asp. 2008;319:188–95. https://doi.org/10.1016/J.COLSURFA.2007.07.011.
48. Castellanos IJ, Crespo R, Griebenow K. Poly(ethylene glycol) as stabilizer and emulsifying agent: a novel stabilization approach preventing aggregation and inactivation of proteins upon encapsulation in bioerodible polyester microspheres. J Control Release. 2003;88:135–45. https://doi.org/10.1016/S0168-3659(02)00488-1.
49. Synnott P, Webb M, Chapman R, Ollendorf DA, Pearson SD. Chief Scientific officer institute for clinical and economic review 2017.
50. Sansdrap P, Moës AJ. Influence of manufacturing parameters on the size characteristics and the release profiles of nifedipine from poly(DL-lactide-co-glycolide) microspheres. Int J Pharm. 1993;98:157–64. https://doi.org/10.1016/0378-5173(93)90052-H.
51. Vysloužil J, Doležel P, Kejdušová M, Mašková E, Mašek J, Lukáč R, et al. The influence of different formulations and process parameters during the preparation of drug-loaded PLGA microspheres evaluated by multivariate data analysis. Acta Pharma. 2014;64:403–17. https://doi.org/10.2478/ACPH-2014-0032.
52. Kong T, Wu J, To M, Wai Kwok Yeung K, Cheung Shum H, Wang L. Droplet based microfluidic fabrication of designer microparticles for encapsulation applications. Biomicrofluidics. 2012;6:6. https://doi.org/10.1063/1.4738586.
53. Lombry C, Dujardin N, Préat V. Transdermal delivery of macromolecules using skin electroporation. Pharm Res. 2000;17(1):32–7. https://doi.org/10.1023/A:1007510323344.
54. Kumar R, Philip A, Kumar P&. Modified Transdermal technologies: breaking the barriers of drug permeation via the skin. Trop J Pharm Res. 2007;6:633–44. https://doi.org/10.4314/tjpr.v6i1.14641.
55. Roberts MS. Targeted drug delivery to the skin and deeper tissues: role of physiology, solute structure and disease. Clin Exp Pharmacol Physiol. 1997;24:874–9. https://doi.org/10.1111/J.1440-1681.1997.TB02708.X.
56. Saini V, Kamboj S, Maggon N. Transdermal drug delivery systems: approaches and advancements in drug absorption through. Skin. 2013;20
57. Sharma N, Agarwal G, Rana AC, Ali Bhat Z, Kumar D, Bhat A. A review: Transdermal drug delivery system: a tool for novel drug delivery system. International Journal of Drug Development and Research n.d.;3:0–0.

58. Wokovich AM, Prodduturi S, Doub WH, Hussain AS, Buhse LF. Transdermal drug delivery system (TDDS) adhesion as a critical safety, efficacy and quality attribute. Eur J Pharm Biopharm. 2006;64:1–8. https://doi.org/10.1016/J.EJPB.2006.03.009.
59. Obiedallah MM, Abdel-Mageed AM, Elfaham TH. Ocular administration of acetazolamide microsponges in situ gel formulations. Saudi Pharmaceutical Journal : SPJ. 2018;26:909. https://doi.org/10.1016/J.JSPS.2018.01.005.
60. Kaity S, Maiti S, Ghosh AK, Pal D, Ghosh A, Banerjee S. Microsponges: a novel strategy for drug delivery system. J Adv Pharm Technol Res. 2010;1:283. https://doi.org/10.4103/0110-5558.72416.
61. N NF, S SU, Dharan SS. Microsponge Based In Situ Ocular Gel: A Review n.d.
62. Dhyani A, Kumar G. Ocular delivery of atenolol loaded microsponge in-situ gel: development, characterization and in-vitro evaluation. Indian Journal of Pharmaceutical Education and Research. n.d.;56 https://doi.org/10.5530/ijper.56.1s.45.
63. Tiwari A, Tiwari V, Palaria B, Kumar M, Kaushik D. Microsponges: a breakthrough tool in pharmaceutical research. Future. J Pharm Sci. 2022;8:1–25. https://doi.org/10.1186/S43094-022-00421-9.
64. Muralidharan S. Ramasamy S. A Novel Strategy for Opthalmic Drug Delivery System Containing Ketotifen: Microsponges Enriched Gel (MEGs); 2013.
65. Abd-Elal RMA, Elosaily GH, Gad S, Khafagy ES, Mostafa Y. Full factorial design, optimization, in vitro and ex vivo studies of ocular Timolol-loaded microsponges. J Pharm Innov. 2020;15:651–63. https://doi.org/10.1007/S12247-019-09418-Z/METRICS.
66. Rathnam G, Sangeetha G. Formulation and evaluation of Dorzolamide hydrochloride microsponges loaded in situ gel for ocular administration. Journal of Biology Pharmacy and Health Sciences. 2024;2024:332–42. https://doi.org/10.30574/wjbphs.2024.18.2.0287.
67. Ambikar RB, Bhosale AV. Development and characterization of diclofenac sodium loaded eudragit rs100 polymeric microsponge incorporated into in situ gel for ophthalmic drug delivery system. Int J Pharm Pharm Sci. 2021:63–9. https://doi.org/10.22159/IJPPS.2021V13I9.42405.
68. Zhao CX. Multiphase flow microfluidics for the production of single or multiple emulsions for drug delivery. Adv Drug Deliv Rev. 2013;65:1420–46. https://doi.org/10.1016/J.ADDR.2013.05.009.
69. Optimization and Scale-up Methodology in Preparing Microsponge Loaded with 5-Fluorouracil (5-FU). 2021. doi:10.21203/RS.3.RS-989826/V1.
70. Gusai T, Dhavalkumar M, Soniwala M, Dudhat K, Vasoya J, Chavda J. Formulation and optimization of microsponge-loaded emulgel to improve the transdermal application of acyclovir—a DOE based approach. Drug Deliv Transl Res 2020 11:5 2020;11:2009–29. doi:https://doi.org/10.1007/S13346-020-00862-W.
71. Ladeira BMF, Gomes MC, Custódio CA, Mano JF. High-throughput production of microsponges from platelet lysate for tissue engineering applications. Tissue Eng Part C Methods. 2022;28:325–34. https://doi.org/10.1089/TEN.TEC.2022.0029/ASSET/IMAGES/LARGE/TEN.TEC.2022.0029_FIGURE5.JPEG.
72. Hameed HA, Khan S, Shahid M, Ullah R, Bari A, Ali SS, et al. Engineering of naproxen loaded polymer hybrid enteric microspheres for modified release tablets: development, characterization, in silico modelling and in vivo evaluation. Drug Des Devel Ther. 2020;14:27. https://doi.org/10.2147/DDDT.S232111.